T0299037

Edible and Medicinal Mushrooms of the Himalayas

This book, as part of the "Natural Products Chemistry of Global Plants" series, describes in detail the health-promoting wild edible and medicinal mushrooms specific to the Himalayas region. The focus of the book is to draw on the rich culture, folklore, and environment of the Upper Himalayas, which represents a scientifically significant region. The Himalayas has rich plant resources and a large diversity of plants and mushrooms, which can provide important health benefits as detailed throughout the text. Drawing attention to these mushrooms with detailed scientific descriptions may help in the awareness and in developing sustainable growth of these important resources.

Features

- Provides an opportunity to describe the wild edible and medicinal mushrooms from this scientifically significant region.
- Represents a wider variety of mushrooms than previously published in other books.
- Presents more content related to traditional uses, phytochemistry, pharmacology, distribution, processing, toxicology, conservation, and future prospective of individual mushrooms.
- The plants and mushrooms of the region are valuable resources not only to local populations but to those living outside the region.
- Scientists are monitoring the rich Himalayan plant resources and the consequences of climate change on this precarious ecosystem.

Natural Products Chemistry of Global Plants

Series Editor: Clara Bik-san Lau
Founding Editor: Raymond Cooper

This unique book series focuses on the natural products chemistry of botanical medicines from different countries such as Bangladesh, Borneo, Brazil, Cameroon, China, Ecuador, India, Iran, Laos, Romania, Sri Lanka, Turkey, etc. These fascinating volumes are written by experts from their respective countries. The series will focus on the pharmacognosy, covering recognized areas rich in folklore as well as botanical medicinal uses as a platform to present the natural products and organic chemistry. Where possible, the authors will link these molecules to pharmacological modes of action, reflecting the ethnopharmacological uses. The series intends to trace a route through history from ancient civilizations to the modern day showing the importance to man of natural products in medicines, in foods and a variety of other ways. With special emphasis on plant parts for medicinal uses, phytochemistry and biological activities, this book series will be of useful reference to scientists/pharmacognosists/pharmacists/chemists/graduates/undergraduates/researchers in the fields of natural products, herbal medicines, ethnobotany, pharmacology, chemistry and biology, Furthermore, pharmaceutical companies may also found valuable information on potential herbs and lead compounds for the future development of health supplements and western medicines.

Recent Titles in this Series:

Medicinal Plants of Borneo
Simon Gibbons and Stephen P. Teo

Natural Products and Botanical Medicines of Iran
Reza Eddin Owfi

Natural Products of Silk Road Plants
Raymond Cooper and Jeffrey John Deakin

Brazilian Medicinal Plants
Luzia Modolo and Mary Ann Foglio

Medicinal Plants of Bangladesh and West Bengal: Botany, Natural Products, and Ethnopharmacology
Christophe Wiart

Traditional Herbal Remedies of Sri Lanka
Viduranga Y. Waisundara

Medicinal Plants of Ecuador
Pablo A. Chong Aguirre, Patricia Manzano Santana, Migdalia Miranda Martínez (Eds)

Medicinal Plants of Laos
Djaja Djendoel Soejarto, Bethany Gwen Elkington and Kongmany Sydara

Edible and Medicinal Mushrooms of the Himalayas: Climate Change, Critically Endangered Species and the Call for Sustainable Development
Ajay Sharma, Garima Bhardwaj and Gulzar Ahmad Nayik

Medicinal Plants of Turkey
Ufuk Koca-Caliskan and Esra Akkol

Edible and Medicinal Mushrooms of the Himalayas

Climate Change, Critically Endangered Species, and the Call for Sustainable Development

Edited by

Ajay Sharma
Department of Chemistry, University institute of sciences, Chandigarh University, Gharuan, Mohali, India

Garima Bhardwaj
Department of Chemistry, Sant Longowal Institute of Engineering and Technology, Sangrur, Longowal, Punjab, India

Gulzar Ahmad Nayik
Department of Food Science & Technology, Govt. Degree College Shopian, J&K, India

CRC Press is an imprint of the
Taylor & Francis Group, an **Informa** business

Cover image: Shutterstock

First edition published 2024
by CRC Press
6000 Broken Sound Parkway NW, Suite 300, Boca Raton, FL 33487-2742

and by CRC Press
4 Park Square, Milton Park, Abingdon, Oxon, OX14 4RN

Library of Congress Cataloguing-in-Publication Data
Names: Sharma, Ajay, 1989- editor. | Bhardwaj, Garima, 1988- editor. | Nayik, Gulzar Ahmad, editor.
Title: Edible and medicinal mushrooms of the Himalayas : climate change, critically endangered species and the call for sustainable development /. edited by Ajay Sharma, Garima Bhardwaj, Gulzar Ahmad Nayik.
Description: First edition. | Boca Raton : CRC Press, [2024] | Series: Natural products chemistry of global plants | Includes bibliographical references and index. | Summary: "This book, as part of the "Natural Products Chemistry of Global Plants" series, describes in detail the health promoting wild edible and medicinal mushrooms specific to the Himilayas region. The focus of the book is to draw on the rich culture, folklore, and environment of the Upper Himilayas, which represents a scientifically significant region. The Himilayas has rich plant resources and a large diversity of plants and mushrooms, which can provide important health benefits as detailed throughout the text. Drawing attention to these mushrooms with detailed scientific descriptions may help in the awareness and in developing sustainable growth of these important resources"-- Provided by publisher.
Identifiers: LCCN 2023008122 (print) | LCCN 2023008123 (ebook) | ISBN 9781032195568 (hardback) | ISBN 9781032195520 (paperback) | ISBN 9781003259763 (ebook)
Subjects: LCSH: Mushrooms--Therapeutic use--Himalaya Mountains Region. | Mushrooms--Himalaya Mountains Region. | Mushroom culture--Himalaya Mountains Region.
Classification: LCC RM666.M87 E35 2024 (print) | LCC RM666.M87 (ebook) | DDC 615.3/296--dc23/eng/20230601
LC record available at https://lccn.loc.gov/2023008122
LC ebook record available at https://lccn.loc.gov/2023008123
ISBN: 978-1-032-19556-8 (hbk)
ISBN: 978-1-032-19552-0 (pbk)
ISBN: 978-1-003-25976-3 (ebk)

DOI: 10.1201/9781003259763

Typeset in Times
by MPS Limited, Dehradun

Dedication

This book is dedicated to
My Little Angel Mahalsa Sharma
&
Prof. Raymond Cooper
(Former Chief Editor of The Natural Products Chemistry
of Global Plants Series by Taylor and Francis, CRC Press)

Contents

Preface

This book is a part of the CRC book series on "Natural Products Chemistry of Global Plants." This book epitomizes the different chapters on the health-promoting edible and medicinal mushrooms specific to the upper Himalayas region.

The Himalayan region is not only the reservoir of herbal medicine, rather this region is considered as the prominent hub for wild edible and medicinal mushrooms. Edible and medicinal mushrooms are the gift of mother nature to the human race for fighting against a variety of illnesses. Till date, extensive research on wild edible and medicinal mushrooms is being carried out. The results of these studies have proven them as one of the dietary supplements owing to the presence of many bioactive compounds responsible for boosting one's both growth and immune system. Therefore, the focus of this book is to draw on the rich culture, folklore, ecology, and biodiversity of edible and medicinal mushrooms of the Himalayas, with particular emphasis on the upper Himalayan region. All the mushrooms included in the present book are used as a food by the local tribes and people for their health-promoting properties from primeval times. These traditional health-promoting activities of various mushrooms have been well supported by various scientific articles published in different reputed journals and books.

The current book presents an inclusive guide to edible and medicinal mushrooms of the upper Himalayas, their ecology/distribution, taxonomy/morphology, nutritional composition, bioactive compounds/phytochemistry, medicinal uses/pharmacology, mycology, cultivation, toxicology as well as conclusion and future prospective. This book offers an asset for researchers and graduate students of botany, microbiology, biotechnology, food technology, home science, agricultural sciences, chemistry, and pharmaceutical sciences.

Ajay Sharma, Garima Bhardwaj, Gulzar Ahmad Nayik

Editor Biographies

Dr. Ajay Sharma is a Natural Product Chemist with strong Organic Chemistry, Phytochemistry, Biochemistry, and Ethnobotany background. Presently, he is working as an Assistant Professor at the University Centre for Research and Development, Department of Chemistry, Chandigarh University, Mohali, India. He completed his Graduation (B.Sc. Medical) from Govt. College Hoshiarpur (affiliated to Panjab University, Chandigarh), Master's Degree in Organic Chemistry (Natural Products) from Punjab Agriculture University Ludhiana, and his Ph.D. (Organic Chemistry – Natural product) from SLIET (Sant Longowal Institute of Engineering & Technology, Deemed University) Punjab. He has published over 55 peer-reviewed papers and book chapters in reputed international/national journals and books. Presently, his five edited books are under production in different international publication houses like Springer, Wiley, Taylor & Francis, Elsevier, and Royal Society of Chemistry. He has various honors/Fellowship/Awards to his credit, i.e., Role of honor from PAU Ludhiana (2011), Merit-cum-topper scholarship during M.Sc. (PAU Ludhiana) (2009–10), MHRD fellowship Award for Ph.D. (2012), Award of INSPIRE fellowship (JRF, 2013) from DST, Award of INSPIRE fellowship (SRF, 2015) from DST, 1st Prize in Poster Award at National Conference (NICS-2016), and Best Poster Award in (TSFS-2016). He does research in Organic Chemistry, Natural Products Chemistry, Phytochemistry, Food Chemistry. His research activities have focused on the isolation and characterization of bioactive extracts from traditional medicinal plants and evaluation of their biological potential. Currently, his research group at Chandigarh University is investigating the field of more environmentally friendly synthesis of metal nanoparticles and carbon nanodots using plant extracts/plant-based biomass followed by assessing their biological, catalytic, and sensing potential. Till date, he has successfully guided six M.Sc. dissertation students.

Dr. Garima Bhardwaj is an independent researcher. She earned her Ph.D. in Organic Chemistry from SLIET Longowal, Punjab. The topic of her Ph.D. research was on studies in the utilization of oleo-chemical industry by-products for bio-surfactant synthesis. She has published her PhD work into various reputed international journals. After Ph.D., she joined the Defence Institute of High-Altitude Research (DIHAR), Defence Research and Development Organization (DRDO) as a Research Associate in the Phytochemistry division. During her work period in DIHAR, DRDO, she got an opportunity to work on the nutritional profiling of high-altitude vegetables and medicinal plants where she developed her interest in Phytochemistry. She has published over 22 peer-reviewed papers and book chapters in reputed international/national journals and books. She has received the award of Research fellowships from the Ministry of Human Resource Development (MHRD) and Ministry of Defence, DRDO, India, for her doctoral and post-doctoral research, respectively. She does research in Organic Chemistry, Biosurfactant synthesis, Natural Products Chemistry, Phytochemistry. During Ph.D., her research activities have focused on the isolation, characterization of novel strains to produce biosurfactants and their purification and structural elucidation using modern chromatographic and spectroscopic methods. In her post-doctoral research, she worked on the extraction and purification of bioactive compounds from high-altitude regions and their qualitative and quantitative analysis. She is currently working on the natural product chemistry of the medicinal plants.

Gulzar Ahmad Nayik completed his master's degree in Food Technology from the Islamic University of Science & Technology, Awantipora, Jammu and Kashmir, India, and his Ph.D. from Sant Longowal Institute of Engineering & Technology, Sangrur, Punjab, India. He has published over 80 peer-reviewed research and review papers and 40+ book chapters, and he has edited 13 books with Springer, Elsevier, and Taylor & Francis. Dr. Nayik has also published a textbook on food chemistry and nutrition and has delivered several presentations at various national and international conferences, seminars, workshops, and webinars. Dr. Nayik was shortlisted twice for the prestigious Inspire-Faculty Award in 2017 and 2018 from the Indian National Science Academy, New Delhi, India. He was nominated for India's prestigious National Award (Indian National Science Academy Medal for Young Scientists 2019–2020). Dr. Nayik also fills the roles of editor, associate editor, assistant editor, and reviewer for many food science and technology journals. He has received many awards, appreciations, and recognitions and holds membership in various international societies and organizations. Dr. Nayik is currently editing several book projects with Elsevier, Taylor & Francis, Springer Nature, Royal Society of Chemistry, etc.

Contributors

Muhammad Afzaal
Department of Food Sciences
Government College University
Faisalabad, Pakistan

Gözde Koşarsoy Ağçeli
Hacettepe University
Department of Biology/Biotechnology
Ankara, Turkey

Farak Ali
Girijananda Chowdhury Institute of Pharmaceutical Science - Tezpur
Dekargaon, Tezpur, Assam, India

Swati Allen
Central Procurement Cell, Indian Council of Medical Research
V. Ramalingaswami Bhawan
Ansari Nagar East, New Delhi, India

Shahnaz Alom
Girijananda Chowdhury Institute of Pharmaceutical Science – Tezpur
Dekargaon, Tezpur, Assam, India

Amanjyoti
Chaudhary Devi Lal University
Sirsa, Haryana, India

Poonam Baniwal
Department of Quality Control Food Corporation of India
New Delhi, India

Shahid Bashir
University Institute of Food Science and Technology
The University of Lahore
Pakistan

Nabanita Bhattacharyya
Department of Botany
Gauhati University
Guwahati, Assam, India

Vinod K Bhatt
Ingenious Research and Development Foundation
VIP Enclave, Chandrabani
Dehradun, Uttarakhand, India

Shuvam Bhuyan
Department of Molecular Biology and Biotechnology
Tezpur University
Assam, India

Avnish Chauhan
Department of Environmental Science
Graphic Era Hill University
Dehradun, Uttarakhand, India

Riya Chugh
Department of Chemistry
Chandigarh University
Gharuan, Punjab, India

Barsha Devi
Pandit Deendayal Upadhyaya Adarsha Mahavidyalaya
Tulungia, Assam, India

Kanika Dulta
Department of Food Technology
School of Applied and Life Science
Uttaranchal University
Dehradun, Uttarakhand, India

Ekta
Punjab Agricultural University
Ludhiana, India

Jyoti Gaba
Punjab Agricultural University
Ludhiana, India

R. Gopi
ICAR-Sugarcane Breeding Institute Research Centre
Kannur, Kerala, India

Nilayan Guha
Department of Pharmaceutical Sciences
Faculty of Science and Engineering
Dibrugarh University
Dibrugarh, Assam, India

Arun Kumar Gupta
Department of Food Science and Technology
Graphic Era (Deemed to be University)
Bell Road, Clement Town
Dehradun, Uttarakhand, India

Thiribhuvanamala Gurudevan
Department of Plant Pathology
Centre for Plant Protection Studies
Tamil Nadu Agricultural University
Coimbatore, India

Ajmal Hussen
Wolaita Sodo University
Wolaita, Ethiopia (East Africa)

Ali Ikram
University Institute of Food Science and Technology
The University of Lahore
Pakistan

Swati Jain
Department of Food and Nutrition
Lady Irwin College
University of Delhi
India

Tenzin Jamtsho
Yangchenphug Higher Secondary School
Ministry of Education
Thimphu – District, Bhutan

Chandan Kapoor
ICAR-Indian Agricultural Research Institute
New Delhi, India

Pooja Kapoor
University Institute of Agriculture Science
Chandigarh University
Gharuan, Mohali, India

Gurmeet Kaur
Department of Chemistry
Chandigarh University
Gharuan, Punjab, India

Gurpreet Kaur
Department of Zoology
Mata Gujri College
Fatehgarh Sahib, India

Jagdeep Kaur
Department of Chemistry
Chandigarh University
Gharuan, Punjab, India

Kamalpreet Kaur
Department of Chemistry
Mata Gujri College
Fatehgarh Sahib, India

Mandheer Kaur
Chandigarh College of Technology
Landran, Mohali, India

Pardeep Kaur
Punjab Agricultural University
Ludhiana, India

Simranjot Kaur
Shri Guru Granth Sahib World University
Fatehgarh Sahib, India

Palki Sahib Kaur
Chandigarh College of Technology
Landran, Mohali, India

Rekha Kaushik
MMICTBM (HM)
Maharishi Markandeshwar (Deemed to be University)
Mullana, Ambala, Haryana, India

Abhishek Katoch
University Institute of Agriculture Science
Chandigarh University
Gharuan, Mohali, India

Aditya Kumar
Chandigarh College of Technology
Landran, Mohali, India

Keshav Kumar
Department of Biotechnology
Dr YS Parmar University of Horticulture and Forestry
Nauni, Solan, HP, India

Shiv Kumar
MMICTBM (HM)
Maharishi Markandeshwar (Deemed to be University)
Mullana, Ambala, Haryana, India

Priyanka Kumari
University Institute of Agriculture Science
Chandigarh University
Gharuan, Mohali, India

Rahul Mehra
MMICTBM (HM)
Maharishi Markandeshwar (Deemed to be University)
Mullana, Ambala, Haryana, India

Vikas Menon
Chandigarh College of Technology
Landran, Mohali, India

Poonam Mishra
Department of Food Engineering and Technology
Tezpur University
Assam, India

Bindu Naik
Department of Food Science and Technology
Graphic Era (Deemed to be University)
Bell Road, Clement Town
Dehradun, Uttarakhand, India

Vikash Nain
Chaudhary Devi Lal University
Sirsa, Haryana, India

Manju Nehra
Chaudhary Devi Lal University
Sirsa, Haryana, India

Satyaranjan Padhiary
University Institute of Agriculture Science
Chandigarh University
Gharuan, Mohali, India

Arpita Paul
Department of Pharmaceutical Sciences
Faculty of Science and Engineering
Dibrugarh University
Dibrugarh, Assam, India

Chandramani Raj
ICAR-Indian Institute of Sugarcane Research
Lucknow, Uttar Pradesh, India

Muzamil A. Rather
Department of Molecular Biology and Biotechnology
Tezpur University
Assam, India

Awais Raza
University Institute of Food Science and Technology
The University of Lahore
Pakistan

Farhan Saeed
Department of Food Sciences
Government College University
Faisalabad, Pakistan

Tanvi Sahni
Punjab Agricultural University
Ludhiana, India

Amna Saleem
Department of Food Sciences
Government College University
Faisalabad, Pakistan

Parthasarathy Seethapathy
Department of Plant Pathology
Amrita School of Agricultural Sciences
Amrita Vishwa Vidyapeetham
Coimbatore, India

Kamal Ch. Semwal
Department of Biology
College of Sciences
Eritrea Institute of Technology
Mai Nafhi, Asmara, Eritrea (East Africa)

Muhammad Zia Shahid
University Institute of Food Science and Technology
The University of Lahore
Pakistan

Renuka Sharma
Noida Institute of Engineering & Technology
Noida, India

Sugandha Sharma
MMICTBM (HM)
Maharishi Markandeshwar (Deemed to be University)
Mullana, Ambala, Haryana, India

Susheel Kumar Sharma
ICAR-Indian Agricultural Research Institute
New Delhi, India

Anu Shrivastava
Department of Food and Nutrition
Lady Irwin College
University of Delhi
India

Mukesh S. Sikarwar
College of Pharmacy
Teerthanker Mahaveer University
Moradabad, Uttar Pradesh, India

Divyanshi Singh
Department of Food Technology
Shoolini University
Solan, Himachal Pradesh, India

Harjodh Singh
Chandigarh College of Technology
Landran, Mohali, India

Matber Singh
ICAR-Indian Institute of Soil & Water Conservation
Dehradun, Uttarakhand, India

Shweta Singh
ICAR-Indian Institute of Sugarcane Research
Lucknow, Uttar Pradesh, India

Somvir Singh
University Institute of Biotechnology
Department of Biosciences
Chandigarh University
Mohali, India

Harvinder Kumar Singh
Indira Gandhi Krishi Viswavidyalaya
Raipur, Chhattisgarh, India

Prerna Sood
Punjab Agricultural University
Ludhiana, India

Jeewan Tamang
University Institute of Agriculture Science
Chandigarh University
Gharuan, Mohali, India

Arti Thakur
Department of Botany
Shoolini Institute of Life Sciences and Business Management
Solan, Himachal Pradesh India

Ugyen
Jigme Khesar Strict Nature Reserve (JKSNR)
Department of Forest Park Service
Haa – District, Bhutan

Phurpa Wangchuk
Center of Molecular Therapeutics
Australian Institute of Tropical Health, and Medicine (AITHM)
James Cook University
Cairns – campus
Australia

Yogender Singh Yadav
Department of Dairy Engineering
College of Dairy Science and Technology
Lala Lajpat Rai University of Veterinary and Animal Sciences
Hisar, Haryana, India

Md. Kamaruz Zaman
Girijananda Chowdhury Institute of Pharmaceutical Science – Tezpur
Dekargaon, Tezpur, Assam, India

1 Introduction, Geographical Region, Mushrooms Diversity, Climate, Sustainability

An Overview of Mushrooms

Farak Ali and Shahnaz Alom
Girijananda Chowdhury Institute of Pharmaceutical Science - Tezpur, Dekargaon, Tezpur, Assam, India

Nilayan Guha and Arpita Paul
Department of Pharmaceutical Sciences, Faculty of Science and Engineering, Dibrugarh University, Dibrugarh, Assam, India

Md. Kamaruz Zaman
Girijananda Chowdhury Institute of Pharmaceutical Science - Tezpur, Dekargaon, Tezpur, Assam, India

CONTENTS

INTRODUCTION

The history behind the utilisation of edible and medicinally important mushrooms as food and medicine has been well documented from ancient times in many countries among which the history of India is quite remarkable (Semwal, 2014a). These are macrofungi which are heterotrophic in nature and responsible for maintaining stability and ecological balance (Sheikh et al., 2015). Mushrooms are being considered as an important food material because of their higher nutritive value and minimal side effect (Lillian Barros, Ferreira, et al. 2007; Cheung & Cheung, 2005a). Phytochemically, mushrooms are considered as a reservoir of various phytochemicals which have a greater role in human life, simultaneously plays a significant role in providing nutrition such as minerals, amino acids, vitamins, crude fibre, proteins, low-fat level, etc., which can be easily obtained by consuming edible and medicinally important mushroom in a diet-based manner (Ao et al., 2016; Buswell et al, 1996). Various edible and medicinal mushrooms possess 80–120 g/kg ash value which contains different inorganic elements like phosphorous, magnesium, potassium, calcium, copper, iron, and zinc (Barros et al., 2008; Valverde et al., 2015). Many tribes and their community have used this mushroom and its mycelia as a food item for many centuries. Due to

DOI: 10.1201/9781003259763-1

modernisation of food items, edible mushrooms are being used as a prominent ingredient in food processing such as Pizzas, Burgers, Rolls, Kebabs, etc., whereas other people consumed by making curry of mushroom. As commercialisation of mushrooms is expanding, their rate of production is also increasing. Apart from edible property, it also exhibits various pharmacological properties such as antioxidants, anticancer, radical scavenging, diabetes, ageing, immunomodulating activity etc. (Agrahar-Murugkar & Subbulakshmi, 2005; Lillian Barros, Baptista, et al., 2007; Borchers et al., 2004; Chang, 1996; Cheung & Cheung, 2005b). More than 80 developed and developing countries in the world used to collect wild edible and medicinal mushrooms for earning livelihood and foods, but in the tribal areas, the rate of consumption and production of mushrooms are varied according to indigenous tribe and their locality.

The northwest Himalayan region of India shares a border with Himachal Pradesh and Uttarakhand and both states have huge forest land. It consists of mainly three types of forest such as broadleaf, coniferous, and mixed forest. Among these various important trees are there with whom there is a critical association with mushrooms which is commonly known as ectomycorrhizal association (Semwal, 2014a). Decaying leaves and woods, mosses grow on those trees and provide as a medium for growing mushrooms. In coniferous forests, major trees found are deodar (*Cedrus deodara*), pine (*Pinus roxburghii*), blue pine (*Pinus walliciana*), Himalayan yew (*Taxus baccata*), etc., and the growth of mushrooms is quite higher in woods of these trees. Mycelium of various edible mushrooms is found to be in association with this tree. This symbiotic relationship between the fungi (mushroom) and the ecosystem of forest are important for the maintenance of health of the forest (Semwal, 2014b). Owing to rapid industrialisation, excessive deforestation, and climatic change, there has been a constant and rapid depletion of mushroom diversity in the Himalayan region. In earlier days, edible and medicinal mushrooms have been grown highly in the Pabbar Valley of Shimla district which lies in northwestern Himalaya, but due to the introduction of horticultural crash crops, the place is undergoing a socio-economic transition due to which mushroom growth rate has been depleting day by day (Chauhan, 2021).

A vast number of edible and medicinal mushrooms can be found in the Himalayan region. In the Himalayan region of Jammu and Kashmir, mushroom species like *Agaricus bisporus, Pleurotus ostreatus,* and *Coprinus atramentarius* can be found (Khan et al., 2016). Pabbar Valley of Shimla district has *Alloclavaria purpurea* (Fr.) Dentinger and D. J McLaughlin*, Amanita bisporigera* G.F.Atk, *Auracularia species, Cantheralus lateritius* (Berk.) Singer, *Helvella crispa* (Scop.) Fr., *Helvella compressa* (Synder) N.S. Weber, *Hydnum repandum* (L.) Fr., *Lacterius deliciosus* (Fries) S.F. Grey, *Morchella esculenta* (L.) pers., *Morchella deliciosa* Fries, *Ramaria botrytis* (Pers.) Ricken, *Rhizopogon vulgaris* (Vittad.) M. Lange, and *Sparassis crispa* (Wulf.) Fr. (Chauhan, 2021). Images of these wild and edible mushrooms are shown in Figure 1.1.

GEOGRAPHICAL REGION

The Himalayan geographical region is wide and important due to its abundant mushroom diversity. The region is surrounded by wide swaths of sloping terrain and mountains. It is unique in that it has a wide range of weather conditions, altitudinal variation, topographical features, and soil properties that all work together to support rich mushroom growth. India's Himalayan region is separated into two parts: The North-western and North-eastern Himalayas. Jammu and Kashmir, Uttarakhand, and Himachal Pradesh are the three states that make up India's North-western Himalayan area. These locations‘ distinctive terrain and subtropical to temperate climatic regimes encourage the growth of a wide number of mushrooms. Sikkim, West Bengal, Assam, and Arunachal Pradesh are the four states that make up India's north-eastern Himalayan area. The growth and availability of mushrooms in these states is supported by regional ethnomycological knowledge as well as a diverse biodiversity. These states make up diverse ecological zones in the Himalayas that act as important hotspots for endemism and ectomycorrhizal variety (Chakraborty et al., 2017; Semwal, 2014a).

FIGURE 1.1 Images of wild and edible mushrooms found in the Himalayan region. (a) *Agaricus bisporus,* (b) *Pleurotus ostreatus,* (c) *Coprinus atramentarius,* (d) *Alloclavaria purpurea,* (e) *Amanita bisporigera,* (f) *Auracularia species,* (g) *Cantheralus lateritius*, (h) *Helvella crispa* (Scop.), (i) *Helvella compress*, (j) *Hydnum repandum,* (k) *Lacterius delicious,* (l) *Morchella esculenta,* (m) *Ramaria botrytis,* (n) *Sparassis crispa,* (o) *Rhizopogon vulgaris.*

Source: All the images are collected from Google and they are CC licensed.

MUSHROOM DIVERSITY

Mycotic biological diversity is abundant in the Himalayan ensemble. Uttarakhand is located in the western Himalayas and is one of the world's biodiversity hotspots. The broad weather conditions, plant dispersion, and field features of Uttarakhand's Garhwal region make it ideal for mushroom cultivation. The wide diversity of mushrooms in this area is a source of pride (Vishwakarma et al., 2011). There have been reports of the exploration of various mushroom species in India's North-western Himalayan region. For example, (Sharma & Gautam, 2017) reported six species of Ramaria coral mushrooms (*R. botrytis, R. rubripermanens, R. flava, R. flavescens, R. aurea, and R. stricta*) and six species of Clavaria coral mushrooms (*C. fragilis, C. coralloides, C. purpurea, C. vermicularis, C. amoena, and C. rosea*) during their frequent surveys of North-western Himalayan They have also documented information on the culinary status of mushrooms received from local residents. They have also revealed for the first time through molecular profile and biological activity of mushrooms from this region. Another study found 21 kinds of wild mushrooms in the Garhwal Himalayas, divided into 15 genera and 13 families. *Agaricus augustus, Hericium coralloides, H. erinaceus, Laetiporus sulphureus, Macrolepiota procera, Chlorophyllum rachodes, Pleurotus ostreatus* and *Ramaria sanguinea* were the most important seasonal food species among all the wild edible macrofungi recorded. *Agaricaceae, Russulaceae, Tremellaceae, Gomphaceae,* and *Hericiaceae* were the remaining families (Singh et al., 2017; Singh et al., 2017). In addition, Singh et al. (2017) collected 219 macrofungi specimens from the Garhwal Himalayas and classified them into 15 species, 12 genera, and 08 families. Among the varieties collected and devoured by the locals were *Morchella esculenta, Cantharellus cibarius, Cantharellus minor, Grifola frondosa. Ganoderma lucidum, Agaricus campestris, Hydnum repandum, Coprinus comatus, Morchella esculenta, and Cantharellus cibarius* have all been shown to have therapeutic use in the Garhwal region (Vishwakarma et al., 2011). Uttarakhand has documented 15 species of wild medicinal mushrooms belonging to 15 genera and 14 families. *Cordyceps militaris, Ophiocordyceps sinensis,* and *Morchella esculenta* are *Ascomycota* members, while the remainder are Basidiomycota members (Bhatt et al., 2018). A study was undertaken in two North-western Himalayan states, Uttarakhand and Himachal Pradesh, from 2000 to 2013. The information was acquired on edible mushrooms that were consumed by the locals as well as for trading purposes. As a result, it was discovered that *Cordyceps sinensis* and numerous species of *Morchella* were harvested specifically for trade reasons in the Himalayan highlands during the spring season. *Amanita, Agaricus, Astraeus, Hericium, Macrolepiota, Morchella, Pleurotus,* and *Termitomyces* species have been reported to be very commonly consumed by the local people, whereas *Auricularia, Cantharellus, Sparassis, Lactarius, Ramaria,* and *Russula* genera have been reported to be less commonly collected and consumed (Semwal, 2014a). *Morchella* is one of the most expensive edible mushrooms known on the planet (Nitha et al., 2007). It costs between Rs. 2000 (29.22 $) and Rs. 2500 (36.53 $) per kilogram in the market (Negi, 2006). *Morchella esculenta*, also known as Guchhi, morels, common morel, real morel, yellow morel, and sponge morel, is one of Morchellaceae's most expensive fungi (Ajmal et al., 2015). It is regarded as a delicacy among vegetarians due to its distinct flavour, taste, and texture, and is used in a variety of cuisines all around the world. Morels can be found in temperate woods throughout India, particularly in the Northwestern states of Himachal Pradesh, Punjab, Jammu and Kashmir, and Uttarakhand. Locals in these locations think that the emergence of morel mushrooms is linked to thunder and lightning (Negi, 2006). Apart from that, Morels have long been used to treat illnesses like pneumonia, fever, cough and stomach ache, as well as in pregnant and breastfeeding moms. It is also said to be a cure-all for all respiratory problems. Furthermore, morels have antibacterial, anti-inflammatory, immunostimulatory, and other pharmacological properties (Bala et al., 2017). The numerous species of Morels have a lot of variation. Although morphological research and traditional taxonomy are useful in identifying morels, molecular approaches such as DNA sequencing may be of considerable assistance in resolving the taxonomic difficulties surrounding several species of *Morchella*. Kanwal et al. (2011) have gathered *Morchella* from the Western Himalayas of India in this regard. Yellow and black morels, as well as *Verpa spp.*, were determined to be abundant based on sequencing analysis. Furthermore, phylogenetic analysis using maximum parsimony, maximum likelihood, and Bayesian inference indicated a strong differentiation between yellow and black morels. Morels have also been studied in Uttaranchal's Kumaun Himalaya, in the Darma valley of Pithoragarh district. The goal of the study was to throw light on the need for more research into these macrofungi. This study (Negi, 2006) also documented the characteristics of common edible Morchella species such as *M. semilibera, M. angusticeps, M. deliciosa, M. crassipes, M. esculenta,* and *M. conica*. The Indian state of Jammu and Kashmir, which is also located in the Northwest Himalayas, is bounded by the main Himalayan mountains on the north and east, and the Punjab plains on the south. This state provides a conducive environment for mushroom growth due to its diverse meteorological and geographic characteristics (Kumar & Sharma, 2011). The abundance of macrofungal species in the state is strongly tied to the state's different forest communities and weather patterns. The majority of this state's regions haven't been well investigated for mushroom diversity (Pala et al., 2012). As a result, a systematic survey of various parts of the state has been conducted on a regular basis by various personnel. As a result, new mushroom species have been discovered. Some of the mushroom species have been discovered for the first time in Jammu and Kashmir, and some have even been reported for the first time in India. 150 wild mushroom samples were gathered and analysed for macro–micro morphological and ethnomycological traits in a study (Kumar & Sharma, 2011). As a result, 66 taxa of wild mushrooms belonging to 33 genera were classified into 22 families, 10 orders, and three classes. The identified mushrooms belonged to the following genera, viz. *Agaricus, Astraeus, Amanita, Auricularia, Boletus, Bovista, Cantharellus, Chalciporus, Clavaria, Clavulina, Coprinus, Flammulina, Geopora, Gyromitra, Helvella, Lactarius, Lentinus, Leucopaxillus, Lycoperdon, Macrolepiota, Macrolepiota, Mac* Ethnomycological data was acquired from tribal men, women, village heads, and Ayurvedic hakims in order to have a better knowledge of these fungi and their economic significance. It was found that some of the varieties of mushrooms are consumed as fresh vegetables including *Agaricus arvensis, Boletus spp., Coprinus comatus, Peziza badia, Clavaria vermicularis, Clavulina spp., Geopora arenicola, Gyromitra spp., Helvella spp., Macrolepiota procera, Morchella spp., Pleurotus spp., Pleurotus spp.* While other types, such as *Morchella spp., Geopora arenicola, Sepultaria sumneriana, Sparassis spp., Pleurotus spp.,* and *Verpa conica,* are specifically consumed throughout the winter by residents of hilly areas when vegetables are few and travel is restricted owing to harsh weather conditions. Aside from the types of mushrooms described above, Kumar and Sharma (2011) have looked on the variety of boletoid macrofungi in Jammu and Kashmir. *Austroboletus malaccensis, Boletus edulis, Boletus formosus, Boletus granulatus, Boletus luridus,* and *Suillus cavipes* were among the six boletoid taxa studied, which included *Austroboletus malaccensis, Boletus edulis, Boletus formosus, Boletus formosus, Boletus granulatus, Boletust* luridus, and *Suillus cavipes.* Kumar and Sharma (2009) have presented the habit, habitat, macro- and microscopic features of mushrooms belonging to the order *Pezizales* with nine taxa distributed over four families using images as part of their effort on exploring numerous species of mushrooms in Jammu and Kashmir state. These included *Helvella atra, Helvella crispa, Helvella elastica, Geopora arenicola, Geopyxis catinus, Morchella esculenta, Otidea leporina, Peziza badia, Sepultaria arenosa, Geopora arenicola, Geopyxis catinus, Morchella esculenta, Otidea leporina, Peziza badia,* and *Sepultaria arenosa.*

Sikkim, a small Himalayan state in India, is known for its meaty wild mushrooms. A wide range of microclimatic conditions, as well as altitudinal variance, favour and nourish the growth of a great number of mushrooms in the state. The ethnomycological knowledge and mushroom variety in this state has been accumulated by a number of scholars. Das (2010) recorded 126 wild mushrooms from the Barsey Rhododendron Sanctuary in the state of Sikkim, along with their distribution, growing time, and edibility status. 19 of the 126 species found belong to the *Ascomycota* class, while 107 belong to the Basidiomycota class. It has also been classified roughly 46 medicinally significant mushrooms, including *Cordyceps, Xylaria, Coprinus, Ganoderma, Hygrocybe, Lycoperdon, Russula, Trametes,* and *Xylaria* as the most common genera. In addition, during his macrofungal study of Sikkim from 2014 to 2015, he rediscovered a *Strobilomyces* species, *Strobilomyces polypyramis*, from the state's North district after a 164-year absence (Das et al., 2014). It has also been described two *Cortinariaceae* species, *Cortinarius varicolor* (subg. *Phlegmacium*), and *Cortinarius salor* (subg. *Myxacium*), for the first time from India, along with full macro- and micromorphological characterisation (Kanad Das & Chakraborty, 2015). According to Panda and Swain (2011), *Cordyceps sinensis*, also known as Yarsa gumba/Keera Jhar, is a purported longevity-promoting herb. Its anticancer activity is aided by its pharmacological potential in modifying hepatic, renal, cardiovascular, and endocrine systems, as well as erythropoiesis and immunomodulation. Apart from the aforementioned mushroom species, (Chakraborty et al., 2017) it has been described for the first time two boletoid mushroom species, *Aureoboletus nephrosporus* and *Strobilomyces mirandus,* with full morphological descriptions. During an ethnomycological examination of the forest region of East district in Sikkim, Dutta et al. (2015), discovered a marasmioid fungus, *Marasmius indopurpureostriatus*, which they named *Marasmius indopurpureostriatus*. These are some of the reports on the mushroom diversity compilation in this state, which reflect the current state of mushroom richness and productivity. Assam is a state in northeast India with a diversified landscape of hills and valleys. It has a wide range of geoclimatic conditions and altitudinal differences, which support a diverse range of natural habitats such as woods, meadows, and wetlands, all of which promote the luxuriant growth of wild mushrooms. Furthermore, reserve forests such as the Eastern Himalayan upper Bhabar Sal forest, Eastern Himalayan lower Bhabar Sal forest, Eastern Terai Sal forest, Eastern heavy alluvial Plain Sal forest, Eastern Hill Sal forest, Northern Secondary moist deciduous forest, Evergreen forest, Lower alluvial Savanah, Woodland, Eastern west alluvial grass land, Riparian Fringe Forest, and Khoir Sissoo forest can be found there. The majority of these forests are home to a diverse range of ethnic groups, each with unique knowledge of mushroom diversity and domestication. As a result, an attempt has been made to benchmark macrofungi diversity within the state in terms of morphological distribution, habitat, and edibility. The diversity and evaluation of a few edible mushrooms used by several ethnic tribes of Western Assam were described in a study undertaken (Sarma et al., 2010) and has been discovered 26 different mushroom species from 14 genera and 13 families. *Ganoderma lucidum* had the highest distribution frequency (83.33 percent), followed by *Cantharellus tubaeformis* and *Agaricus bisporus*, and species of *Agaricus, Boletus, Lenzites, Lycoperdon,* and *Termitomyces* had the lowest (16.66 percent). It has also been summarised the edibility of 25 mushrooms from the genera *Auricularia, Agaricus, Boletus, Calvatia, Cantherallus, Ganoderma, Lentinus, Laetiporus, Lycoperdon, Morchella, Schizophyllum, Termitomyces,* and *Tricholoma*. The importance of mushroom taxonomy and diversity was also investigated (Parveen et al., 2017). 44 samples were obtained from various sites during macrofungal excursion and tested for edibility, pharmacological characteristics, and industrial applications. *Lentinus, Ganoderma, Pleurotus, Agaricus, Lycoperdon,* and *Macrolepiota* were among the most common genera found. In addition, a study conducted (Nath & Sarma, 2018) revealed 14 kinds of edible mushrooms as well as their preferred substrata, offering light on the link between ethnobotanical knowledge and traditional folk practises. Among the species employed by the tribes to overcome dietary and nutraceutical deficits are *Auricularia auricula, Ganoderma lucidum,* and *Trametes pubescens*. According to several employees, the use of wild edible mushrooms by numerous ethnic communities has improved the socioeconomic situation of the state's tribal people. West Bengal, a part of the Eastern Himalayas, is physically located in one of the world's top 25 biodiversity hotspots. It functions as an 'oriental region,' with a diverse range of phytogeographical circumstances. It stretches from the Himalayas in the north to the Bay of Bengal in the south in terms of topography. It has a wide range of climatic patterns and hydrological regimes that are conducive to the growth of mushrooms in a tropical forest ecosystem. Acharya and his colleagues have made a significant contribution to assessing mushroom diversity in West Bengal. (Arun Kumar Dutta & Acharya, 2014) described an extensive study to investigate knowledge about mushroom use for medical and culinary uses. A five-year macrofungal foray was done in eight West Bengal districts. The study looked at 34 macro fungi, with *Amanita, Astraeus, Russula, Termitomyces, Armillaria, Auricularia, Fistulina, Grifola, Hericeum, Coprinus, Pholiota, Meripilus, Pleurotus, Calocybe, Lentinus, Tricholoma,* and *Volvariella* being consumed as food and Cordyceps, Ganoderma (Giri et al., 2013) evaluation of the nutraceutical potential of three different wild edible mushrooms: *Lentinus squarrosulus, Russula albonigra,* and *Tricholoma giganteum* (Khatua et al., 2015). Another investigation was done on *Russula senecis*, which not only revealed the species' medicinal potential but also taxonomically categorised it based on macro- and micromorphological characteristics (Bhatt et al., 2017; Singh et al., 2017). Conductance of a thorough study of wild mushrooms in West Bengal's Gurguripal Eco forest, resulting in a detailed investigation of nine mushroom species (*Termitomyces heimii, Astraeus hygrometricus, Leucopaxilus sp., Amanita vaginata, Agaricus campestris, Russula delica, Schizophyllum commune, Pleurotus ostreatus, and Cantharellus sp.*). I have been explained why they're important and how they might be used to improve human health, nutrition, and disease prevention. It has also been researched on mushroom eco-diversity, production, and distribution frequency; a total of 71 species were identified in the state's Paschim Medinipur region. The species recorded came from 41 genera and 24 families, with 32 edible species, 39 inedible species, and 19 species with

TABLE 1.1
Diversity of Mushrooms in the North-Western and North-Eastern Himalayan Region of India

North-western Himalaya			
Mushrooms found	**Zone of occurrence**	**State**	**References**
Lactarius, Laetiporus, Laccaria, Stropharia, Marasmius, Cortinarius, Ramaria, Russula, and *Strobilomyces*	Kangra, Kullu, Shimla, Solan, and Lahaul Spiti	Himachal Pradesh	(Chaudhary et al., 2016)
Cantharellus cibarius, Coprinus comatus, Geopora arenicola, Ramaria Formosa, Ramaria flavo-brunnescens, Sparassis crispa and *Termitomycetes striatus Phallus macrosporus, Phallus rubicundus* and *Phallus hadriani*	Jammu	Jammu and Kashmir	(Koul et al., 2019; Kumar & Sharma, 2009)
Amanita ceciliae, Amanita flavoconia, Amanita muscaria, Amanita pantherina, Amanita phalloides, Amanita vaginata, Amanita virosa, Russula aeruginea, Russula atropurpurea, Russula aurea, Russula cyanoxantha, Russula delica, Russula emetica and *Russula nobilis Gyromitra sphaerospora (Peck) Sacc. Mutinus caninus Cortinarius flexipes, Cortinarius fulvoconicus,* and *Cortinarius infractus*	Kashmir Himalayas		(Itoo et al., 2015; Pala et al., 2012)
Collybia chrysoropha and *Russula albida*	North Kashmir		(Kaur & Rather, 2016)
Verpa bohemica	Rajouri		(Anand et al., 2013)
Morchella esculenta, Coprinus comatus, Fomes fomentarius, Ganoderma lucidum, Neolentinus sp., *Suillus sibiricus, Suillus granulates, Lactarius deliciosus, Russula atropurpurea, Russula aurea, Calvatia* sp., *Lycoperdon* sp., *Agaricus bisporus,* and *Cantharellus cibarius*	Yusmarg, Gulmarg, Mammer, Kellar and Pahalgam		(Farooq et al., 2017)
Cyathus olla, Laetiporus sulphurous, Peziza ammophila, Peziza ampliata, Peziza badia, Peziza succosa, Peziza vesiculosa,Inocybe curvipes, Inocybe sororia Geopora arenicola, G. sepulta, Pulvinula convexella, P. Miltina, Anthracobia macrocystis, and *Geopyxis majalis*	Ladakh		(Dorjey et al., 2016; R Yangdol et al., 2016; Yangdol et al., 2014)
Strobilomyces echinocephalus, Strobilomyces mollis, Clitocybe dilatata, Clitocybe hydrogramma, and *Clitocybe nebularis*	Poonch		(Kour et al., 2013; Kour et al., 2015)
Ganoderma lucidum, Agaricus campestris, Hydnum repandum, Coprinus comatus, Morchella esculenta, and *Cantharellus cibarius*	Garhwal	Uttarakhand	(Vishwakarma et al., 2011)
Agaricus augustus, Hericium coralloides, H. erinaceus, Laetiporus sulphureus, Macrolepiota procera, Chlorophyllum	Garhwal Himalaya		(Bhatt et al., 2017; Singh et al., 2017)

(Continued)

TABLE 1.1 ***(Continued)***
Diversity of Mushrooms in the North-Western and North-Eastern Himalayan Region of India

rachodes, Pleurotus ostreatus, Ramaria sanguinea, Coprinus comatus, Macrolepiota procera, Cantharellus lateritius, Ramaria botrytis, Ramaria sanguinea, Helvella crispa, Laetiporus sulphureus, Psathyrella candolleana, Aleuria aurantia, Lactifluus volemus, Lactifluus corrugis, Russula cyanoxantha, Stropharia rugosoannulata, Tremella foliacea, T. mesenterica, and *T. fuciformis*			
	North-eastern Himalaya		
Agaricus sp., Cortinarius sp., Clitocybe sp., Collybia sp., Gilera sp., Hygrophorus sp., Hypholoma sp., Flammula sp., Lactarius sp., Lentinus sp., Mycena sp., Pleurotus sp., Pluteus sp., and *Russula sp.*	Himalayan range	Sikkim	(Acharya, et al., 2010a; Acharya et al., 2010b)
Lysurus sp.	Midnapur district	West Bengal	
Tulostoma chudaei	Howrah and South 24 Parganas		(Chakraborty et al., 2013)
Amanita hemibapha, A. vaginata, A. vaginata var. *alba, Astraeus hygrometricus, Russula albonigra, R. brevipes, R. cyanoxantha, Russula* sp. *R. senecis, R. lepida, Termitomyces clypeatus, T. heimii,* and *T. microcarpus*	Lateritic		(Dutta & Acharya, 2014)
Armillaria mellea, Auricularia auricula, Fistulina hepatica, Grifola frondosa, Hericeum sp., *Coprinus comatus, Pholiota squarrosa, Meripilus giganteus,* and *Pleurotus* sp.	Himalayan		
Calocybe indica, Lentinus squarrosulus, Pleurotus ostreatus, M. gigantea, Macrocybe lobayensis, and Volvariella volvacea	Coastal		
Russula sp.	Darjeeling hills		(Paloi et al., 2015)

pharmacological active compounds. Other reports on mushroom diversity from the northwest and northeast have also been compiled in Table 1.1.

CLIMATE CONDITION

Wild edible Himalayan mushrooms are widely distributed all over the Himalayan region in different climate conditions. They are mainly grown on soil covered with humus, decayed organic matter, and dead stumps of some plant species. They are saprophytic on soil and litter and found under broadleaf and conifers trees in association with different other mushroom species. Some of the common edible Himalayan mushrooms along with their climate condition are given in Table 1.2 (Chauhan, 2021; Semwal, 2014b; Singh et al., 2017).

TABLE 1.2
Description on Climate Conditions of Various Edible and Medicinal Mushrooms Found in the Himalayan Region

Mushroom	Family	Common name	Climate conditions	References
Agaricus campestris	Agaricaceae	Meadow mushroom	They are terrestrial, grow among grasses and forest edges also, mainly found during June to October i.e., during rainy season, they form fairy rings and found commonly at the lower elevation of Himalaya.	(Chauhan, 2021; Semwal, 2014b)
Alloclavaria purpurea (Fr.) Dentinger & D.J. McLaughlin	Clavariaceae	Shuntoo	They grow on soil. The season of collection of this mushroom is in between July and September.	
Amanita bisporigera G.F. Atk.	Amanitaceae	Chhatar or Mundothra	They mainly grow on dead stumps of *Pinus*. The season of collection of this mushroom is in between June and August.	
Amanita chepangiana	Amanitaceae	Chepang slender Caesar	They are terrestrial, restricted in the lower and middle hills of North-western Himalaya, mainly found in the 600m to 2200m elevation in the *Tectona grandis* and *Shorea robusta* dominated broadleaf forests and also in the areas of Himalaya dominated by oak trees. Till now they are not recorded in regions of forest with coniferous trees.	
Amanita hemibapha (Berk. & Br.)	Amanitaceae	Half-dyed slender Caesar, Caesar's mushroom	They are mainly solitary to scattered or sometimes gregarious, terrestrial, found in mixed forests mainly form symbiotic associated with *Rhododendron arboretum*, Shorea robusta, *Tectona grandis, Myrica esculenta* and *Quercus leucotrichophora, Pinus roxburghii, Lyonia ovalifolia,* etc. In Northwestern Himalayan region they are widely distributed from 900 m to 2800 m elevation.	(Singh et al., 2017)
Amanita vaginata	Amanitaceae	Grisette	They are terrestrial, found widely in broadleaf, coniferous to mixed forests. They are distributed in a wide range from lower, middle to high hills of Himalayas. They live in association with *Abies pindrow, Betula utilis, Cedrus deodar, Cupressus torulosa, Cinnamomum tamala, Myrica esculenta, Lyonia ovalifolia, Pinus roxburghii, Pinus wallichiana, Rhododendron arboretum, Quercus leucotrichophora,* and *Quercus semecarpipholia.*	
Astraeus hygrometricus	Diplocystaceae	Earthstar	They are terrestrial, mainly grow on soil in the broadleaf forests of lower Shiwalik hills, Garhwal Himalaya. They grow mainly in association with *Tectona grandis* and *Shorea robusta.*	(Chauhan, 2021; Khan et al., 2016)

(Continued)

TABLE 1.2 *(Continued)*

Description on Climate Conditions of Various Edible and Medicinal Mushrooms Found in the Himalayan Region

Auricularia auricula-judae	Auriculariaceae	Tree ear, Jew's ear, wood ear, jelly ear	They are solitary to gregarious, occasionally caespitose, mainly grow on dead and living wood. Mainly found in association with *Bauhinia malabarica*, *Delonix regia*, *Grievillea robusta* and *Quercus leucotrichophora*. The season of collection of this mushroom is in between June and October during monsoon season.	
Auricularia polytricha	Auriculariaceae	Cloud ear fungus	They are solitary to gregarious, ligniccolous, occasionally caespitose, mainly grow on dead and living wood. Mainly found in association with *Bauhinia malabarica*, *Delonix regia*, *Grievillea robusta* and *Quercus leucotrichophora*.	
Auricularia Spp.	Auriculariaceae	Kan	They grow on soil. The season of collection of this mushroom is between July and August.	
Boletus edulis	Boletaceae	King mushroom	They are solitary and scattered to gregarious, found under *Quercus semecarpifolia*. They are mainly found under a variety of deciduous and conifers trees.	
Cantharellus cibarius	Cantharellaceae	Golden Chanterelle	They are terrestrial, solitary to highly distributed, sometimes gregarious, found in broadleaf, coniferous and mixed forests, mainly found in association with *Cedrus deodara*, *Cupressus torulosa*, *Myrica esculenta*, *Rhododendron arboretum*, and *Quercus leucotrichophora*. Mainly found in the higher elevation from 1500 m to 2400 m in the Garhwal Himalaya.	
Cantharellus lateritius (Berk.) Singer	Cantherellaceae	Maran kee Chhatri	They grow on the rotten logs of *Ulmus* trees. The season of collection of this mushroom is between July and September.	(Chandrawati & Narendra Kumar, 2014; Semwal, 2014b; Singh et al., 2017)
Cantharellus minor	Cantharellaceae	Small chanterelle	They are scattered, sometimes found in clusters, terrestrial. They are grown on soil covered with moss, decaying organic matter, decomposing leaf litter. They are found in oak, coniferous and mixed forests at higher elevations in the Himalayas.	
Cordyceps sinensis	Ophiocordycipitaceae	Caterpillar fungus, keeda ghaas, keeda jari, yarshagumba	They are solitary, terrestrial. Grow on soil during early monsoon i.e., from May to July, in the Rudraprayag alpine region. They also grow on the larvae produced by *Hepialus oblifurcus* insect and also grow in relation with dwarf shrubs of *Rhododendron anthopogon*.	
Craterellus cornucopioides	Cantharellaceae	The Horn of Plenty and Black Trumpet	They are mainly scattered also found in clusters. They are mainly grown on the grounds under oak trees.	

Grifola frondosa	Meripilaceae	Hen of the Woods, Maitake	They are solitary, commonly found at the base of conifers and broadleaf trees and are found in association with *Cedrus deodara* and *Quercus semecarpifolia*.	
Helvella crispa (Scop.) Fr.	Helvellaceae	Bakra	They grow on soil. The season of collection of this mushroom is between July and August.	
Helvella compressa (Synder) N.S. Weber	Helvellaceae	Baktu	They grow on soil. The season of collection of this mushroom is between July and August.	
Hericium erinaceus	Hericiaceae	Lion's mane, Hedgehog mushroom	Mainly grown in the crack of living but decaying broadleaf trees, most commonly found in *Quercus leucotrichophora* in the Garhwal Himalaya.	(Chauhan, 2021; Arun Kumar Dutta & Acharya, 2014; Semwal, 2014b)
Hydnum repandum	Hydnaceae	Chhatri, Hedgehog mushroom, The sweet tooth.	They grow on soil, solitary and scattered to gregarious, they are mycorrhizal associated with conifers or broadleaf trees. The season of collection of this mushroom is between July and September.	
Lactarius azonites	Russulaceae	Milky mushroom	They are solitary to scattered, terrestrial, grow with broadleaf trees such as *Myrica esculenta, Rhododendron arboretum* and *Quercus leucotrichophora*.	
Lactarius camphoratus	Russulaceae	Milky mushroom	They are solitary to scattered, terrestrial, grow in humid places or on soggy banks under *Cedrus deodara, Pinus wallichiana, Rhododendron arboretum* and *Quercus leucotrichophora*.	
Lactarius delicious (Fries) S.F. Grey	Russulaceae	Chhatri, Delicious milkcap, Saffron milkcap	They are saprophytic on the soil. They are gregarious to highly scattered, found under *Abides pindrow and Quercus semecarpifolia*. The season of collection of this mushroom is in monsoon season.	
Lactarius subindigo	Russulaceae		They are solitary to highly scattered, found mainly on humus-rich soil, they live by forming ectomycorrhizae with *Quercus leucotrichophora* in mixed coniferous and mild deciduous forests of Garhwal Himalaya.	
Lactifluus hygrophoroides	Russulaceae	Hygrophorus Milky	They are solitary to highly scattered, gregarious, found mainly on humus-rich soil, they live in broadleaf forests in association with *Lyonia ovalifolia, Myrica esculenta, Rhododendron arboretum,* and *Quercus leucotrichophora*.	
Macrolepiota procera	Agaricaceae	Parasol mushroom	They are terrestrial, commonly grow in grasslands and at forests edges from lower to middle hills of Himalaya commonly in *Pinus roxburghii* forest.	
Marasmius oreades	Marasmiaceae	Fairy ring mushroom	They are gregarious, found in groups or in fairy rings on grasslands, lawns and pastures.	
Morchella esculenta (L.) Pers	Morchellaceae		They are terrestrial, saprophytic on litter and soil. Found in high elevations of Garhwal region, Kullu and Chamoli district of the Himalayan region. They	

(Continued)

TABLE 1.2 *(Continued)*
Description on Climate Conditions of Various Edible and Medicinal Mushrooms Found in the Himalayan Region

		Cheyauun, Gucchi, Chunchuroo, Sponge mushroom, yellow morel	are solitary to widely distributed, found either in cluster or gregarious form mainly under conifers. They grow on damp and humid soil with the leaf litter of *Abies pindrow*, *Cedrus deodara*, *Cupressus torulosa*, and *Picea smithiana*. The season of collection of this mushroom is between March and August.
Morchella deliciosa Fries	Morchellaceae	Cheyauun	They are saprophytic on litter and soil. The season of collection of this mushroom is in between April and November.
Pleurotus cornucopiae	Pleurotaceae	Oyster mushroom	They are lignicolous, mainly grow on the species of Eurporbia and dead wood of some unidentified tree in the lower hills of Garhwal Himalayan region.
Ramaria botrytis (Pers.) Ricken	Gomphaeceae	Shoontu, Coral fungus	They are terrestrial, saprophytic on the soil. Grow in the mixed forests of Garhwal Himalaya mainly in association with oak, *Myrica* and *Pinus roxburghii*. The season of collection of this mushroom is in monsoon season.
Rhizopogon vulgaris (Vittad.) M. Lange	Rhizopogonaceae	Zanda	They are saprophytic on the soil. The season of collection of this mushroom is between September and November.
Russula brevipes	Russulaceae	The stubby brittlegill	They grow on grounds in broadleaf and coniferous forests. They are solitary to highly scattered. Present in association with *Cedrus deodar*, *Cupressus torulosa*, *Rhododendron arboretum*, *Myrica esculenta*, *Pinus waliichiana*, *Pinus roxburghii*, and *Quercus leucotrichophora*.
Russula lepida	Russulaceae	Rosy russula	They are solitary to scattered, terrestrial. Grow in the lower Shiwalik hills of Himalaya in humus-rich soil of *Shorea robusta* forests.
Russula virescens	Russulaceae	Quilted green russula, green cracking russula, green brittlegill	They are present in solitary to highly scattered, mainly on humus-rich soil, found in association with *Cedrus deodar*, *Cupressus torulosa*, *Rhododendron arboretum*, *Myrica esculenta*, *Lyonia ovalifolia* and *Quercus leucotrichophora* under broadleaf and conifers trees.
Sparassis crispa (Wulf.) Fr.	Sparassidaceae	Chinchadoo, Cauliflower fungus	They grow on conifer roots mainly at the base of *Cedrus deodara*. They are rare species. The season of collection of this mushroom is between July and August. Found in the coniferous forests of Garhwal Himalaya.
Strobilomyces floccopus	Boletaceae	Old Man of the Woods	They are solitary to abundantly distributed in broadleaf or mixed forests, found under *Rhododendron arboreum* and *Quercus leucotrichophora*.

Termitomyces eurrhizus	Lyophyllaceae	They are solitary, terrestrial, caespitose, mainly grow on the mounds made up by ants in the broadleaf forests. They grow in association with *Pinus roxburghii* and *Shorea robusta* in the broadleaf forests of the Shiwalik region, Himalaya.
Termitomyces heimii	Lyophyllaceae	They are solitary, mainly grow on the mounds made up by ants in the forests. They grow underneath the canopy of *Pinus roxburghii* and *Shorea robusta* in the Shiwalik region, Garhwal Himalaya.
Termitomyces microcarpus	Lyophyllaceae	They are terrestrial, gregarious to scattered near termite hills, grow in the troops of fruit bodies, in deciduous forests of the lower Shiwalik Hills, Garhwal Himalaya in association with *Tectona grandis* and *Shorea robusta*.
Termitomyces species	Lyophyllaceae	They are solitary to scattered, terrestrial, grow on high hills of Himalaya, in the soil of mound prepared by ants, in the oak-mixed forest.

ETHNOBOTANICAL USE

Science ancient times, local people of the Himalayan region use mushrooms as a food supplement, ornamentals and also provide a source of income to many local rural communities. Mushrooms are a rich source of antioxidant, protein, fibres, minerals, and are low in fats, therefore local people use mushrooms as an alternative source of meat (Gregori et al., 2007). Also, in some region of Himalayas traditional healers uses mushroom in different illness, but till now there is no written evidence available about these traditional uses. In the Pauri region of Uttarakhand, some traditional healer uses wild mushrooms especially those grown on cow dung, for illness associated with children. Some traditional healers of the Pauri Garhwal region, Uttarakhand believe that some of the mushrooms they use in healing therapy have some psychoactive activities. Most of the traditional use of mushrooms in Northwestern Himalaya vanished with the death of the traditional healers, since they were not recorded or documented at that time (Semwal, 2014b). In the Himalayan Alpine region, mainly in Uttarakhand and Arunachal Pradesh region of India, Tibet and Nepal, a very important traditional medicine is found known as Keeda jadi, Yarsa gumba, Yarsa gunbu, Keeda ghaas, etc. This Keeda jadi is an entomo-fungal combination of a parasite fungus *Cordyceps sinesis* and a larva of *Hepialus armoricanus* (caterpillar host). Traditionally, *Cordyceps sinesis* is used to improve the function of lungs, liver and kidneys and treat different disease related to heart. *Cordyceps sinesis* is used by local people to improve their livelihood by sealing them and also, they use it to improve sexual activities and physical stamina. In traditional folklore of Jammu district, different species of mushrooms such as *Agaricus californicus* Peck, *Morchella esculenta* (L.) Pers, *Termitomyces clypeatus* R. Heim, etc., are used as medicine to treat gastrointestinal disorders, heart disease, skin disease, improve immunity and strength and various illness and also as culinary. Local people of Himalaya collect *Morchella esculenta* (L.) Pers and then dried and stored it for consumption during the scarcity of protein-rich food, especially during extreme cold seasons (Rai et al., 2005). Traditionally *Termitomyces heimii* is used as blood tonic and in fever, common cold and various fungal infections (Chandrawati & Narendra Kumar, 2014). *Termitomyces eurrhizus* is used in diarrhoea, hypertension and pain related with rheumatic arthritis. Also, traditionally *Podoxis pistillaria* is used in different skin diseases, sunburn and inflammation. Traditionally in Western Himalayas mainly in Kashmir region, *Auricularia auricula-judae* (Bull.) Quel is consumed after childbirth to decrease weakness and also it is used in hypertension and wound healing, *Morchella tridentina* Bres, *Morchella esculenta* Pers is used to treat cold and cough and *Sparassis spathulate* (Schwein.) Fr. is used by local people to treat stomach-related problems (Ullah et al., 2022). *Morchella esculenta* is the most expensive mushroom of morchellaceae family. Because of its delicious taste and flavour, people used it in different recipes all over the world. Traditionally, *Morchella esculenta* is used as hallucinogenic, purgative, laxative, body tonic, emollient, immune booster and also in weakness, arthritis, gastrointestinal disorders, wound healing and different stomach-related problems (Ajmal et al., 2015). Also, it is used in thatching and basket making by the local people of Swat (Ali et al., 2011).

CONCLUSION AND FUTURE ASPECTS

Mushrooms are a prominent source of bioactive compounds and also gaining popularity as fancy food materials. Wild edible and medicinal mushrooms are widely distributed in the Himalayan regions such as Uttarakhand, Shimla, and Jammu & Kashmir and in some parts of North-eastern Himalayan region. Apart from food materials, mushrooms are being used traditionally by many tribes for treatment of various ailments such as heart disease, infection in liver, kidney, etc. Mushrooms have got the potential to make a significant contribution in the field of health and nutrition. Additionally, they also contribute significantly to a society's economic growth. From 2010 to 2017, mushroom industries in India grew at a pace of 4.3% per year. India's unique climate condition makes it more suitable for breeding place for wide variety of mushrooms. While the majority of these mushroom varieties are employed for ethnomedicinal purposes by tribal communities and several of these remain undiscovered. There is an urgent need for the conservation of the rich diversity of mushrooms. Modern strategies should be implemented to improve the conservation and long-term production of edible and medicinal mushrooms. For instance, to boost mushroom output, cold storage facilities with well-equipped processing units should be established. Additionally, research and development to extent the shelf-life and the quality of mushrooms should also be consider. Further, public knowledge about the health benefits of mushrooms should be raised to promote their cultivation and use. Moreover, a relationship between regional agrochemical industries and universities should be formed to ensure the effective commercialisation of mushroom farming. The identification of previously unidentified mushroom species will provide a new avenue for research in the field of nutrition and medicine. In depth, studies into the physiology, biochemistry, and molecular biology of mushrooms should be done in order to increase our understanding and techniques for screening medicinal mushrooms at the preclinical and clinical levels, hence enabling their therapeutic applications.

REFERENCES

Acharya, K., Pradhan, P., Chakraborty, N., Dutta, A. K., Saha, S., Sarkar, S., & Giri, S. (2010a). Two species of Lysurus Fr.: Additions to the macrofungi of West Bengal. *Journal of the Botanical Society of Bengal*, *64*(2), 175–178.

Acharya, K., Rai, M., & Pradhan, P. (2010b). Agaricales of Sikkim Himalaya: A review. *Researcher*, *2*(5), 29–38. 10.5530/pj.2017.1.6

Agrahar-Murugkar, D., & Subbulakshmi, G. (2005). Nutritional value of edible wild mushrooms collected from the Khasi hills of Meghalaya. *Food Chemistry*, *89*(4), 599–603. 10.1016/j.foodchem.2004.03.042

Ajmal, M., Akram, A., Ara, A., Akhund, S., & Gagosh Nayyar, B. (2015). *Morchella esculenta*: An edible and health beneficial mushroom. *Pakistan Journal of Food Sciences*, *25*(2), 71–78.

Ali, H., Sannai, J., Sher, H., & Rashid, A. (2011). Ethnobotanical profile of some plant resources in Malam Jabba valley of Swat, Pakistan. *Journal of Medicinal Plant Research*, *5*(18), 4676–4687.

Anand, N., & Chowdhry, P. N. (2013). Taxonomic and molecular identification of Verpa bohemica: A newly explored fungi from Rajouri (J&K), India. *Recent Research in Science and Technology*, *5*(1), 09–12.

Ao, T., Seb, J., Ajungla, T., & Deb, C. R. (2016). Diversity of wild mushrooms in Nagaland, India. *Open Journal of Forestry*, *06*(05), 404–419. 10.4236/ojf.2016.65032

Bala, P., Gupta, D., Sharma, Y. P. (2017). Mycotoxin research and mycoflora in some dried edible morels marketed in Jammu and Kashmir, India. *Journal of Plant Development Sciences*, *9*(8), 771–778.

Barros, L., Ferreira, I. C. F. R., & Baptista, P. (2008). Phenolics and antioxidant activity of mushroom *Leucopaxillus giganteus* mycelium at different carbon sources. *Food Science and Technology International*, *14*(1), 47–55. 10.1177/1082013208090094

Barros, L., Baptista, P., Estevinho, L. M., & Ferreira, I. C. F. R. (2007). Bioactive properties of the medicinal mushroom *Leucopaxillus giganteus* mycelium obtained in the presence of different nitrogen sources. *Food Chemistry*, *105*(1), 179–186. 10.1016/j.foodchem.2007.03.063

Barros, L., Ferreira, M. J., Queirós, B., Ferreira, I. C. F. R., & Baptista, P. (2007). Total phenols, ascorbic acid, β-carotene and lycopene in Portuguese wild edible mushrooms and their antioxidant activities. *Food Chemistry*, *103*(2), 413–419. 10.1016/j.foodchem.2006.07.038

Bhatt, R. P., Singh, U., & Uniyal, P. (2018). Healing mushrooms of Uttarakhand Himalaya, India. *Current Research in Environmental and Applied Mycology*, *8*(1), 1–23. 10.5943/cream/8/1/1

Bhatt, R., Singh, R. P., Stephenson, S. L., Uniyal, P., & Mehmood, T. (2017). Wild edible mushrooms from high elevations in the Garhwal Himalaya-II. *Current Research in Environmental & Applied Mycology*, *7*(3), 208–226. 10.5943/cream/6/2/6

Borchers, A. T., Keen, C. L., & Gershwin, M. E. (2004). Mushrooms, tumors, and immunity: an update. *Experimental Biology and Medicine*, *229*(5), 393–406. 10.1177/153537020422900507

Buswell, J. A., Cai, Y. J., Chang, S. T., Peberdy, J. F., Fu, S. Y., & Yu, H. S. (1996). Lignocellulolytic enzyme profiles of edible mushroom fungi. *World Journal of Microbiology and Biotechnology*, *12*(5), 537–542. 10.1007/BF00419469

Chakraborty, N., Dutta, A. K., Pradhan, P., & Acharya, K. (2013). Tulostoma chudaei Pat an addition to macrofungal flora of India. *Journal of Mycopathological Research*, *51*(1), 185–187.

Chakraborty, D., Semwal, K. C., Adhikari, S., Mukherjee, S. K., & Das, K. (2017). Morphology and phylogeny reveal two new records of boletoid mushrooms for the Indian mycobiota. *Tropical Plant Research*, *4*(1), 62–70. 10.22271/TPR.2017.V4.I1.009

Chandrawati, P. S., & Narendra Kumar, N. N. T. (2014). Macrofungal wealth of Kusumhi forest of Gorakhpur UP, India. *American International Journal of Research in Formal Applied and Natural Sciences*, *5*(1), 71–75.

Chang, R. (1996). Functional properties of edible mushrooms. *Nutrition Reviews*, *54*(11), S91–S93. 10.1111/j.1753-4887.1996.tb03825.x

Chaudhary, R, & Tripathy, A (2016). Diversity of wild mushroom in Himachal Pradesh (India). *International Journal of Innovative Science, Engineering and Technology*, *5*(6), 10859–10886.

Chauhan, P. P. (2021). An ethnobotanical survey of wild edible mushrooms – A potential resource of food and income generation in Pabbar Valley, Himachal Pradesh, India. *Plant Archives*, *21*(2). 10.51470/plantarchives.2021.v21.no2.041

Cheung, L. M., & Cheung, P. C. K. (2005a). Mushroom extracts with antioxidant activity against lipid peroxidation. *Food Chemistry*, *89*(3), 403–409. 10.1016/j.foodchem.2004.02.049

Cheung, L. M., & Cheung, P. C. K. (2005b). Mushroom extracts with antioxidant activity against lipid peroxidation. *Food Chemistry*, *89*(3), 403–409. 10.1016/j.foodchem.2004.02.049

Das, K., Hembrom, M., & Parihar, A. (2014). *Strobilomyces polypyramis*–Rediscovery of a wild mushroom from Sikkim, India. *Indian Journal of Plant Sciences*, *3*(2), 13–18.

Das, K. (2010). Diversity and conservation of wild mushrooms in Sikkim with special reference to Barsey Rhododendron Sanctuary. *NeBIO*, *1*(2), 1–13. 10.4081/dr.2010.e1

Das, K., & Chakraborty, D. (2015). Two new records of Cortinarius from Sikkim (India). *New Biology Report*, *4*(1), 1–6.

Dorjey, K., Kumar, S., & Sharma, Y. P. (2016). Studies on genus Peziza from Ladakh (Jammu & Kashmir), India. *Kavaka*, *46*, 18–22.

Dutta, A. K., Das, K., & Acharya, K. (2015). A new species of Marasmius sect. Globulares from Indian Himalaya with tall basidiomata. *Mycosphere*, *6*(5), 560–567. 10.5943/mycosphere/6/5/5

Dutta, A. K., & Acharya, K. (2014). Traditional and ethno-medicinal knowledge of mushrooms in West Bengal, India. *Asian Journal of Pharmaceutical and Clinical Research*, *7*(4), 36–41.

Farooq, R., Shah, M. A., & Reshi, Z. A. (2017). Morphological and molecular characterization of some mushrooms in Kashmir Himalayan forests. International Conference on Recent Innovations in Science, Agriculture, Engineering and Management. ISBN 978-93-86171-80-1335-342.

Giri, S., Mandal, S. C., & Acharya, K. (2013). Proximate analysis of three wild edible mushrooms of West Bengal, India. *International Journal of PharmTech Research*, *5*(2), 365–369.

Gregori, A., Švagelf, M., & Pohleven, J. (2007). Cultivation techniques and medicinal properties of Pleurotus spp. *Food Technology and Biotechnology*, *45*(3), 238–249.

Itoo, Z. A., Reshi, Z. A., Basharat, Q., Majeed, S. T., & Andrabi, K. I. (2015). Identification and characterization of ectomycorrhizal *Cortinarius* species (Agaricales, Basidiomycetes) from temperate Kashmir Himalaya, India, by ITS barcoding. *Advances in Molecular Biology*, 2015, 1–9. 10.1155/2015/507684

Kanwal, H. K., Acharya, K., Ramesh, G., & Reddy, M. S. (2011). Molecular characterization of Morchella species from the Western Himalayan Region of India. *Current Microbiology*, *62*(4), 1245–1252. 10.1007/S00284-010-9849-1

Kaul, S., Choudhary, M., Gupta, S., Agrawal, D. C., & Dhar, M. K. (2019). Diversity and medicinal value of mushrooms from the Himalayan region, India. *Medicinal Mushrooms: Recent Progress in Research and Development*, 371–389.

Kaur, M, & Rather, HA (2016). Records of agarics: New to Jammu and Kashmir. *Journal on New Biological Reports*, *5*(3), 165–169.

Khan, A. A., Gani, A., Ahmad, M., Masoodi, F. A., Amin, F., & Kousar, S. (2016). Mushroom varieties found in the Himalayan regions of India: Antioxidant, antimicrobial, and antiproliferative activities. *Food Science and Biotechnology*, *25*(4), 1095–1100. 10.1007/s10068-016-0176-6

Khatua, S., Dutta, A. K., & Acharya, K. (2015). Prospecting *Russula senecis*: A delicacy among the tribes of West Bengal. *PeerJ*, *3*(3), e810. 10.7717/peerj.810

Kour, H. (2013). Two species of Strobilomyces from Jammu and Kashmir, India. *Mycosphere*, *4*, 1006–1013. 10.5943/mycosphere/4/5/14

Kour, S., Kour, H., Kumar, S., & Sharma, Y. P. (2015). New records of Clitocybe species from Jammu and Kashmir, India. *Indian Journal of Forestry*, *38*(1), 43–46.

Kumar, S., & Sharma, Y. P. (2011). Additions to boletes from Jammu and Kashmir. *Journal of Mycology and Plant Pathology*, *41*(4), 579–583.

Kumar, S., & Sharma, Y. P. (2009). Some potential wild edible macrofungi of Jammu province (Jammu and Kashmir), India. *Indian Journal of Forestry*, *32*(1), 113–118.

Nath, R. K., & Sarma, T. C. (2018). Edible macrofungi of Kaliabar sub-division of Nagaon district, Assam, India. *Annals of Plant Sciences*, *7*(3), 2161. 10.21746/aps.2018.7.3.12

Negi, C. (2006). Morels (*Morchella* spp.) in Kumaun Himalayas. *Natural Product Radiance*, *5*(4), 306–310.

Nitha, B., Meera, C. R., & Janardhanan, K. K. (2007). Anti-inflammatory and antitumour activities of cultured mycelium of morel mushroom, *Morchella esculenta*. *Current Science*, *92*(2), 235–239.

Pala, S. A., Wani, A. H., Boda, R. H., & Mir, R. A. (2012). Three hitherto unreported macro-fungi from Kashmir Himalaya. *Pakistan Journal of Botany*, *44*(6), 2111–2115.

Paloi, S., Dutta, A. K., & Acharya, K. (2015). A new species of Russula (Russulales) from Eastern Himalaya, India. *Phytotaxa*, *234*(3), 255–262. 10.11646/phytotaxa.234.3.6

Panda, A. K., & Swain, K. C. (2011). Traditional uses and medicinal potential of *Cordyceps sinensis* of Sikkim. *Journal of Ayurveda and Integrative Medicine*, *2*(1), 9. 10.4103/0975-9476.78183

Parveen, A., Khataniar, L., Goswami, G., Jyoti Hazarika, D., Das, P., Gautom, T., Barooah, M., & Chandra Boro, R. (2017). A study on the diversity and habitat specificity of macrofungi of Assam, India. *International Journal of Current Microbiology and Applied Sciences*, *6*(12), 275–297. 10.20546/ijcmas.2017.612.034

Rai, M., Tidke, G., & Wasser, S. P. (2005). Therapeutic potential of mushrooms. *Natural Product Radiance*, *4*(4), 246–257.

Sarma, T. C., Sarma, I., & Patiri, B. N. (2010). Edible mushroom used by some ethnic tribes of western Assam. *The Bioscan*, *3*, 613–625.

Semwal, K. (2014a). Edible mushrooms of the Northwestern Himalaya, India: A study of indigenous knowledge, distribution and diversity. *Mycosphere*, *5*(3), 440–461. 10.5943/mycosphere/5/3/7

Semwal, K. (2014b). Edible mushrooms of the Northwestern Himalaya, India: A study of indigenous knowledge, distribution and diversity. *Mycosphere*, *5*(3), 440–461. 10.5943/mycosphere/5/3/7

Sharma, S. K., & Gautam, N. (2017). Chemical and bioactive profiling, and biological activities of coral fungi from Northwestern Himalayas. *Scientific Reports*, *7*(1), 1–13. 10.1038/srep46570

Sheikh, P. A., Dar, G. H., Dar, W. A., Shah, S., Bhat, K. A., & Kousar, S. (2015). Chemical composition and anti-oxidant activities of some edible mushrooms of Western Himalayas of India. *Vegetos - An International Journal of Plant Research*, *28*(2), 124. 10.5958/2229-4473.2015.00046.4

Singh, U., Bhatt, R. P., Stephenson, S. L., Uniyal, P., & Mehmood, T. (2017). Wild edible mushrooms from high elevations in the Garhwal Himalaya-II. *Current Research in Environmental and Applied Mycology*, *7*(3), 208–226. 10.5943/cream/7/3/8

Singh, V., Bedi, G. K., & Shri, R. (2017). In vitro and in vivo antidiabetic evaluation of selected culinary –Medicinal mushrooms (Agaricomycetes). *International Journal of Medicinal Mushrooms*, *19*(1), 17–25. 10.1615/IntJMedMushrooms.v19.i1.20

Ullah, T. S., Firdous, S. S., Shier, W. T., Hussain, J., Shaheen, H., Usman, M., Akram, M., & Khalid, A. N. (2022). Diversity and ethnomycological importance of mushrooms from Western Himalayas, Kashmir. *Journal of Ethnobiology and Ethnomedicine*, *18*(1), 32. 10.1186/s13002-022-00527-7

Valverde, M. E., Hernández-Pérez, T., & Paredes-López, O. (2015). Edible mushrooms: Improving human health and promoting quality life. *International Journal of Microbiology*, *2015*, 1–14. 10.1155/2015/376387

Vishwakarma, M. P., Bhatt, R. P., & Gairola, S. (2011). Some medicinal mushrooms of Garhwal Himalaya, Uttarakhand, India. *International Journal of Medicinal and Aromatic Plants*, *1*(1), 33–40.

Yangdol, R., Kumar, S., & Sharma, Y. P. (2014). A new edible variety of Laetiporus sulphureus from the cold desert of Ladakh. *Journal of Mycology and Plant Pathology*, *44*(4), 463–465.

Yangdol, R., Kumar, S., Lalotra, P., & Sharma, Y. P. (2016). Two species of Inocybe from Trans– Himalayan Ladakh (J&K), India. *Current Research in Environmental and Applied Mycology*, *6*(4), 305–311.

2 *Auricularia auricula-judae (Bull.)* *Major Bioactive Compounds, Health Benefits, and Pharmacological Activities*

Tenzin Jamtsho
Yangchenphug Higher Secondary School, Ministry of Education, Thimphu – District, Bhutan

Ugyen
Jigme Khesar Strict Nature Reserve (JKSNR), Department of Forest Park Service, Haa – District, Bhutan

Phurpa Wangchuk
Center of Molecular Therapeutics, Australian Institute of Tropical Health, and Medicine (AITHM), James Cook University, Cairns – Campus, Australia

CONTENTS

DOI: 10.1201/9781003259763-2

INTRODUCTION

Mushrooms (also known as macrofungi) are considered nutritional, anti-oxidative, and valued as culinary ingredients. Studies have proven them as one of the dietary supplements owing to the presence of many bioactive compounds responsible for boosting one's both growth and immune system (Kadnikova et al., 2015). *Auricularia auricula-judae* is one such mushroom, a non-toxic and medicinal fungus that belongs to the heterobasidiae of the basidiomycete family. It is also known by a few common names such as Jew's ear, free ear, black fungus, wood ear, and tree jellyfish (Kadnikova et al., 2015; Jo et al., 2014). *A. auricular-judae* typically grows on decomposed logs found in the moist area of the mountainside. They have waxy and cartilaginous fruiting bodies with the color becoming black when dried from fresh purplish-brown. Further, fresh fruiting bodies are rubbery and gelatinous, resembling a human ear (Oli et al., 2020). *A. auricular-judae* is either collected from the wild or cultivated for local delicacies, contributing to livelihood. It is nutritious as it contains carbohydrates, protein, amino acids, trace elements, and vitamins within the fruiting bodies (Zhang et al., 2010; Packialakshmi et al., 2017). Therefore, this mushroom has proven as an integral part of both traditional food and therapeutic purposes in Central, East, and West Africa, the Americas, Europe, Southeast Asia, and China (Wu et al., 2015; Oli et al., 2020; Li et al., 2021; Odamtten et al., 2021)

Besides its nutritional value, *A. auricular-judae* has shown numerous pharmacological activities, including antioxidants, anticancer, anti-radiation, and blood lipid-lowering potential in human and animal in-vitro cell assays (Zhuan-Yun et al., 2015) Thus, it is used as raw material for producing medicinal and therapeutic herbal products (Kadnikova et al., 2015). In Asia, it is famous for its flavor and medicinal properties, for instance, in the Chinese traditional medicine formulations (Cai et al., 2015). Similarly, in European folk medicine, *A. auricula-judae* is well-known as an astringent to treat throat and eye infections (Sękara, 2015). This chapter presents an overview of *A. auricula-judae* mushroom, including its distribution, taxonomy, ethnobotany, phytochemistry, and role in functional food, pharmaceuticals, and cosmetics industries.

ECOLOGY AND DISTRIBUTION

A. auricula-judae is versatile and grows in various ecological conditions and is geographically diverse in distribution, spreading from North America, Europe, and Asia to other tropical and temperate areas (Li et al., 2021). They commonly grow as solitary or densely in clusters on dead trees and fallen logs of elm, banyan, poplar, oak, acacia, sawdust, and other cellulosic-rich materials (Jo et al., 2014; Qiu et al., 2019). Their habitat preference may differ slightly based on their geographical distribution, but not much difference in the preferable environmental variables.

A. auricula-judae is found growing on the fallen trunk of *Carpinus spp* and dead branch of *Sambucus spp* in the Czech Republic, on the growing tree of *Cornus* and *Platanus* species in Denmark and France, on the fallen log of *Quercus spp* in China, on decayed wood of Sambucus, *Humbie* and fallen twigs of Acer plants in the United Kingdom, on the fallen logs of *Ulmus* in Germany, and the trunk of Euonymus, and fallen branch of Fraxius plants in Russia (Wu et al., 2015). In the Pacific and the Gulf of Mexico, this fungus grows on the branches and stumps of various plants species (namely *Coffea arabica, Prunus dulcis, Hevea brasiliensis, Mangifera indica, Tabebuia donnell-smithii, Ficus benjamina* L. and *Yucca* spp) from May to February *A. auricula-judae* is gathered from wild or cultivated in Asia, like in China, Taiwan, Thailand, Indonesia, Korea, Vietnam, Japan, and Bhutan (Xu et al., 2018; Pak et al., 2021). In Europe and Australia, it grows throughout the year; however, they are common in late summer and autumn (Li et al., 2021). The studies have also reported the occurrence of *A. auricula-judae* in the African countries namely,

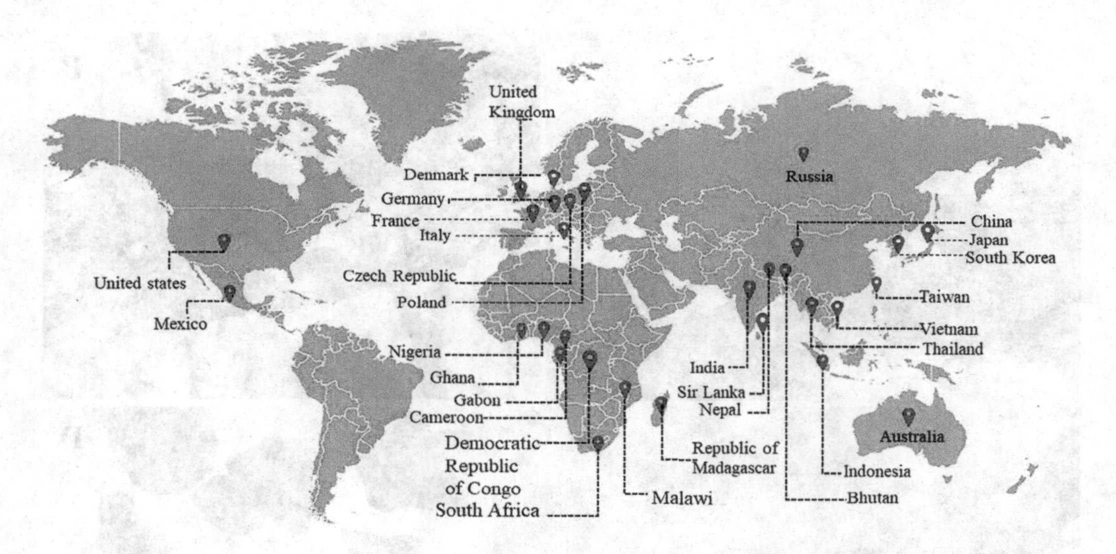

FIGURE 2.1 Distribution of *Auricularia auricula-judae* (Bull.) as a native and cultivated species.

Gabon, Congo, Malawi, South Africa, Nigeria, and Ghana) (Odamtten et al., 2021). It is also reported in Italy, Colombia, and Switzerland (Bandara, 2019). The *A. auricula-judae* is adaptable to a wide range of hosts. Besides common English names, *A. auricula-judae* has other vernacular names by the region where they grow. For instance, it is known as *Kikurage* or tree jellyfish in Japan, *Jeli Namchu* in Bhutan, *Thalthaley chyau* in Nepal, *Black fungus* in Russia, *Mokibeoseot* in Korea, and "Mu Er or Wood ear" in China (Jo et al., 2014; Kadnikova et al., 2015; Packialakshmi et al., 2017). *A. Auricula-judae* is considered the most prevalent species of *Auricula* genus for cultivation in East Asia of which China has topped as the fourth-largest cultivator in the world (Kadnikova et al., 2015). Overall, this species of fungus has demonstrated resilience to both human and climatic treatments as they can withstand the cold temperature and still produce growth (Figure 2.1).

TAXONOMY AND MORPHOLOGICAL FEATURES

A. auricula-judae is commonly called as Juda's or Jew's ear and was formally considered a single species. Recent studies have identified *A. auricula-judae* as an ambiguous species due to the fruiting body's morphological resemblance to other species and its vulnerability to environmental influences (Li et al., 2014). This fungus species known as Heimuer from China was reported as unique from *A. auricula-judae* in Europe and subsequently identified as *A. heimuer*, a new species (Montoya-Alvarez et al., 2011; Wu et al., 2014). Similarly, Looney (2013) readdressed the previously reported *A. auricula-judae* as *A. americana* from southeastern USA. Further, *A. auricula-judae* reported from the European part of Russia was morphologically similar to *A. americana* and *A. villosula* in the Russian Far East (Malysheva and Bulakh, 2014). Thus, the taxonomist recommends considering the proportions of basidiospores and the occurrence and nonappearance of the medulla to distinguish the species from other similar species (Wu et al., 2015).

Another attribute to help confirm its identity identification of *A. auricula-judae* is the morphology, which is ear shape with yellowish-brown or reddish-brown and darker appearance with age. The colorless and transparent mycelium gives rise to the fruiting body, which is irregularly disc-shaped (Liu et al., 2021). Typically, the fungus is 3 to 8 cm in height and approximately 12 cm wide on maturation (Nadir et al., 2020). It is gelatinous and shallow when fresh but shrinks considerably and becomes very tough on desiccation (Montoya-Alvarez et al., 2011). The hymenium on the fertile layer is often smooth or pilose and appears golden brown to reddish-brown when fresh and

FIGURE 2.2 *Auricularia auricula-judae* (Bull.) growing in natural habitat; (A) Fresh young fruiting body on live branch; (B) Matured fruiting body on the branch; (C) Fruiting body on dry branch; (D) Different shape and size of fruiting body.

yellow-green on drying. On the other hand, the lower sterile surface is slightly folded and appears pale brown to dark brown. Basidia are cylindrical-shaped with three septa measuring 50–62 x 3–6 μm wide to barrel-shaped with about three transverse septa, 50–62 x 3–5.5 μm wide with three septa. The basidiospores are often oblong, allantoid, thin-walled, and measure 9–18 μm long x 4.6–7.5 μm wide (Malysheva and Bulakh, 2014) (Figure 2.2).

METABOLOMICS AND DNA SEQUENCING – SOLUTION FOR TAXONOMICAL DISCREPANCIES

Recent studies have identified *A. auricula-judae* as a species complex due to their ambiguous classification and diversity due to the fruiting body's greater morphological resemblance and its vulnerability to environmental influences (Li et al., 2014). Misidentification of any species will contribute to adulteration in food and herbal products. Molecular tools and techniques have proven more precise in identifying fungal species of both wild and cultivated varieties (Zhao et al., 2019). Li et al. (2014) analyzed 32 Chinese *A. auricula-judae* (commercial cultivars) using IGS (nuclear ribosomal DNA intergenic spacer) technique. The analysis showed that nucleotide sequence variation in the IGS supports the phylogenetic studies that correlate to their geographical origin, an excellent molecular marker for identifying genetic variation (Li et al., 2014).

A. auricula-judae domesticated in China exists in three varieties, namely Quanjin, Banjin, and Wujin. Comparative analysis of transcriptome was conducted using fruiting bodies to trace their origin and genetic composition (Zhao et al., 2019). Quanjin showed significant differences in gene expression from Wujin and Banjin. The fungi have either cluster or chrysanthemum types of fruiting patterns that pose taxonomic problems (Yao et al., 2018). For precise species identification, analyzing molecular biomarkers for a fruiting pattern is recommended. For instance, a sequence characterized by amplified region (SCAR) molecular marker "SCL-18" is associated with *A. auricula-judae's* cluster-type fruiting body pattern (Yao et al., 2018).

MEDICINAL USES

The *A. auricula-judae* is a nutritional and functional food source. This fungus has been prevalent in Chinese and European folk medicines until the 19th century (Sun et al., 2016). In Chinese and West African folk medicines, the fungus is reported to be used as an astringent for treating sore throats, jaundice, and sore eyes (Sun et al., 2016). The topical application of fresh *Auricularia auricula-judae* is also indicated for treating tonsillitis, ophthalmia, laryngocele, and staphylococcus infections (Kadnikova et al., 2015). Similarly, in European folk medicine, the fruiting body resembles the throat folds and is used for throat ailments after boiling with milk, beer, or vinegar. Furthermore, the gelatinous consistency of the fungus has been favored in treating eye ailments (Sękara, 2015; Xu et al., 2018). Their use as a blood tonic is also reported in Ghanaian folklore medicine (Apetorgbor et al., 2008) (Table 2.1).

The consumption of this fungus is also known to have benefited people who have high blood pressure, diabetes, cancer, cardiac problems, renal failure, and constipation (Zhao et al., 2019). In China and other parts of Asia, *A. auricula-judae* has exhibited significant value in treating various health conditions, including hemorrhoids, hemoptysis, angina, diarrhea, and gastrointestinal ailments (Sánchez Vázquez et al., 2018). A study by Acharya K (2004) has shown that polysaccharides of this fungus have stimulatory effects on our immune system. In some cases, it causes the production of anti-proliferative interferon and interleukins. Mapoung et al. (2021) also describe its potential as a therapeutic agent in promoting the skin wound-healing effects (Mapoung et al., 2021) (Table 2.1).

CULINARY USES

The *A. auricula-judae* has a soft, jelly-like texture. Due to its high fiber content, essential amino acids, polysaccharides, vitamins, and trace elements, it is popular in Asian (Vietnam, Nepal, India, and Bhutan) food shops (Liu et al., 2021; Chhoeda and Yangchen, 2017). They are eaten as a salad,

TABLE 2.1
Distribution and Ethnopharmacological Uses of *Auricularia auricula-judae*

Ailment/Use	Plant Part(s) Used	Country	Mode of Administration/Treatment	References
Cold and fever	Fruiting body	China, Japan, and Korea,	Fruiting body together with chicken, pak choi, and ginger	(Sękara, 2015; Perera and Li, 2011)
Constipation	Whole	India	It can be eaten as fibre supplement as a raw or as an aqueous soup.	
Eye	Fruiting body		Used as an astringent.	(Sękara, 2015)
Ear puss	Fruiting body	India	Fried and grounded fruiting bodies are taken with any liquid.	(Sękara, 2015)
Menstrual problem	Fruiting body	India	It is eaten as chatni (in Indian Hindu language – like salad).	(Sękara, 2015; Perera and Li, 2011)
Throat	Whole	European folk medicine	Used as an astringent after being boiled in beer, milk, or vinegar.	(Sękara, 2015)

pork belly with bell peppers, and egg meatloaf in Vietnam. Chinese prepare sour and hot soup out of this mushroom, similar to Ghanaian cuisine (Apetorgbor et al., 2008). The gelatinous nature of this mushroom has additional commercial value as it is easier to dry and transport (Liu et al., 2021), and its freshness can be restored by wetting it with water.

CULTIVATION – SUBSTRATE AND NUTRITIONAL REQUIREMENTS

A. auricula-judae has progressively evolved into domestic cultivation from the wild habitat and is documented as the first species of edible fungus cultivated by humankind (Li et al., 2021; Sánchez Vázquez et al., 2018). *A. auricula-judae* has been recorded as the oldest edible mushroom being cultivated by humans, and its cultivation in China using logs as substrate dates back to 300 BC (Sánchez Vázquez et al., 2018). It is now largely cultivated in several countries, including China, Indonesia, Thailand, Taiwan, Japan, Korea, and Russia, and is considered as four major cultivated edible mushrooms in the world (Jo et al., 2014; Pak et al., 2021). According to Zhao et al., (2019), *A. auricula-judae* is cultivated on a large scale in northern China, making it the fourth large producer globally. Initially, wooden logs have preferred as the substrate for cultivation in China. In recent years with the introduction of spawn production, which incorporates fruit body production from mycelia, using logs or polybag method has been instituted (Priya et al., 2016). Various raw materials are used as a substrate, such as woodchips, sawdust, wheat brans, straws, plaster stones, rice brans, and lime as a supplementary source of minerals (Verma P, 2017). However, different materials exhibited differential influences on the nutrient content. For example, maize cobs and wheat bran displayed a better impact on nutrient content than rice bran and sawdust (Onyango et al., 2013). In another report, corn stalk as a substitute for sawdust has significantly increased the total ash, iron, protein, and copper contents, while fats content was reduced in the black fungus. At the same time, corn stalks were suggested as more promising substitute materials as they are available at a low cost, and also the use of sawdust or woodchip demonstrates an adverse effect on wood forests (Yao et al., 2019).

Besides growth substrate, ideal temperature, pH, and supplementation of minerals and nutrients also determine the yield (Carrasco et al., 2018). Jo et al. (2014) reported the use of culture media such as PDA (potato dextrose agar) and MCM (mushroom complete medium) at the temperature ranging 25–30°C with the pH value within 6–9 as optimal growth conditions for better biomass production. Further, Yu and Jo et al. (2013 and 2014) asserts that carbon and nitrogen ratio ranging from 10 to 1, along with the addition of organic acid (succinic acid and lactic acid) and minerals (hydrous magnesium sulfate and Monopotassium phosphate) as the ideal condition for obtaining optimal growth (Yu et al., 2013; Jo et al., 2014). Among many sources, mannose and fructose are suggested as the best carbon source, while peptone for nitrogen. While, thiamine-HCl and biotin are the best supplement source of vitamins to enhance mycelial growth (Kim et al., 2014).

BIOACTIVE COMPOUNDS AND THEIR PHYTOCHEMICAL PROPERTIES

Several studies have evaluated the principal chemical and biochemical components of *A. auricula-judae* and suggested their potential as an alternative source of polysaccharides, minerals, and protein (Bach et al., 2017).

POLYSACCHARIDES

A. auricula-judae polysaccharides are considered the most potent bioactive compound (Bao et al., 2016; Yao et al., 2019). Through various methods of extraction, such as gel chromatography (e.g., Sephadex G, Sephacryl S, and Sepharose CL methods), different types of polysaccharides from dried, fresh, mycelium of the solid culture and submerged culture extract, and pickled fruiting body (Ma et al., 2010; Zeng et al., 2012; Khaskheli et al., 2015; Zhang et al., 2018). Mainly,

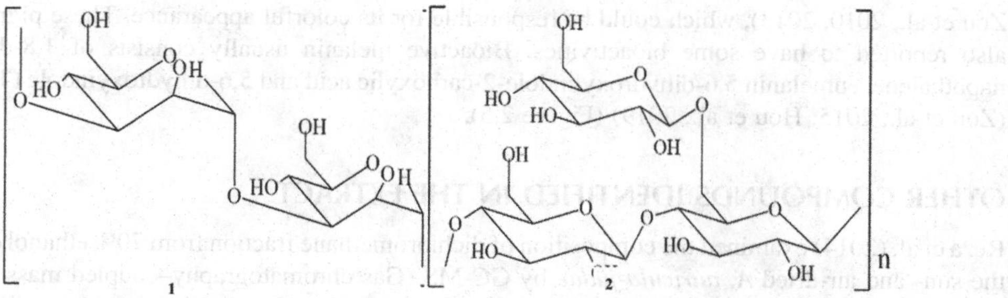

FIGURE 2.3 Backbone chemical structures of heteropolysaccharides. (1) α-(1→3)-linked D-mannopyranosyl: (2) Water soluble β-D glucan (AAG).

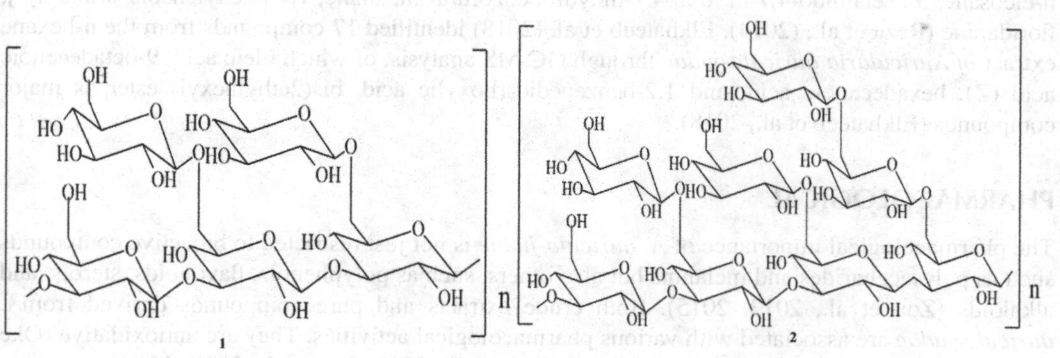

FIGURE 2.4 Chemical structures of homopolysaccharides from *A. auricula-judae:* (1) Glucan I and (2) Glucan II.

homopolysaccharides such as Glucan I and II were reported from the fruiting body of *A. auricula-judae* with the chain structure of β-(1→3)-linked D-glucopyranosyl and replaced by (1→6)-linked β-Dglucopyranosyl groups (Misaki et al., 1981; Sone et al., 1978). A study by Bandara et al. (2015) reported that various chemical, physical, and bioactive properties of *Auricularia* polysaccharides correspond to the variation in numbers of (l→6)-D-glycosidic linkages attached to the β-(1→3)-linked D-glucopyranosyl (Table 2.1) (Bandara et al., 2015).

Most of the reported polysaccharides have indicated the varying composition of monosaccharides corresponding to the extraction method and types of solvent used. For example, the polysaccharides obtained from *A. auricula-judae* are mainly composed of galactose, mannose and glucose, mannose, when extracted using hot water with α (1 → 4)-D-glucopyranosyl as the structural backbone of the group, whereas glucuronic acid structure with O6 glucopyranose side group is obtained with 70% ethanol extraction (Du et al., 2015; Wang et al., 2013). According to Sone et al. (1978), the fruiting body is mainly comprised of β-D-configuration of water-soluble Glucan I and β-D-glucosidic linkages of alkali-insoluble Glucan II (Figure 2.3 and 2.4), and acidic heteropolysaccharides, which contains mannose, glucose, xylose and glucuronic acid in the molar ratio of 4.1:1.3:1:1.3 (Sone et al., 1978) (Figure 2.3 and 2.4). Overall, arabinose, xylose, mannose, glucose, rhamnose, and galactose are the chief components of the *A. auricula-judae*-derived polysaccharides (Cai et al., 2015; Zhou et al., 2017; Xia et al., 2019).

MELANIN

Several studies have reported the presence of melanin pigments such as eumelanin and pheomelanin in the fruiting body of *Auricularia auricula-judae* (Sun et al., 2016; Prados-Rosales et al., 2015;

Zou et al., 2010, 2013), which could be responsible for its colorful appearance. These pigments are also reported to have some bioactivities. Bioactive melanin usually consists of 1,8-dihydroxy naphthalene, eumelanin 5,6-dihydroxylindole-2-carboxylic acid and 5,6-dihydroxyindole (Table 2.2) (Zou et al., 2015; Hou et al., 2019) (Figure 2.5).

OTHER COMPOUNDS IDENTIFIED IN THE EXTRACT

Reza et al. (2014) examined the composition of dichloromethane fraction from 70% ethanol extract of the sun- and air-dried *A. auricula-judae* by GC-MS (Gas chromatography–coupled mass spectroscopy). They identified daizane and 5-, 11-, 17-, 23-tetrakis (1,1-dimethyl)-28-methoxypentacyclo as major compounds. Other bioactive compounds are 2-mercaptobenzoic acid, 2,4-di-tert-butylphenol, n-hentriacontane, n-hexacosane, n-ethoxycarbonyl-3-hydroxyindoline, n-heptacosane, gibberellin A3, n-eicosane, 5-acetamido-4,7-di¬oxo-4,7-dihydrobenzofurazan, silane, 1,4-phenylenebis [trimethy-], floridanine (Reza et al., (2014). Elkhateeb et al. (2018) identified 17 compounds from the n-hexane extract *of Auricularia auricula-judae* through GC-MS analysis, of which oleic acid, 9-octadecenoic acid (Z), hexadecanoic acid, and 1,2-benzenedicarboxylic acid, bis(2ethylhexyl) ester as major compounds (Elkhateeb et al., 2018).

PHARMACOLOGICAL

The pharmacological importance of *A. auricula-judae* is not just restricted to bioactive compounds such as polysaccharides and melanins but also others, such as polyphenols, flavonoids, sterols, and alkaloids (Zou et al., 2013, 2015). Both crude extracts and pure compounds derived from*A. auricula-judae* are associated with various pharmacological activities. They are antioxidative (Oke and Aslim, 2011; Cai et al., 2015; Packialakshmi et al., 2017; Xu et al., 2018; Mapoung et al., 2021), anti-cancer (cytotoxic) (Sone et al., 1978; Misaki et al., 1981; Ukai et al., 1983; Misaki and Kakuta, 1995; Reza et al., 2011; Reza et al., 2012), antimicrobial (Cai et al., 2015; Deka et al., 2017; Oli et al., 2020), anti-inflammatory (Ukai et al., 1983; Damte et al., 2011), antidiabetic, anticoagulant, and hypocholesterolemic (Luo et al., 2009; Looney, 2013; Elkhateeb et al., 2018), and hepatoprotective (Zhang et al., 2018; Hou et al., 2019) (Table 2.3).

ANTIOXIDANT ACTIVITY

Antioxidant activities of *A. auricula-judae* polysaccharides were evaluated by free radical-scavenging ABTS assay, where dose-dependent activity was observed with an EC50 value of 226.67 ± 10.41 μg/mL as the most effective concentration (Mapoung et al., 2021). Xu et al., (2018) investigated the effects of heteropolysaccharides using catalase (CAT) and superoxide dismutases (SOD) activity assays in Caenorhabditis elegans, and found to inhibit reactive oxygen species (ROS) better than the control (Xu et al., 2018).

The study by Oke and Aslim (2011) reported catechin, caffeic acid, gallic acid, and *p*-hydroxybenzoic acid, as principal phenolic components in crude methanolic (AME) and aqueous extract (AAE) of dried *A. auricula-judae* via HPLC analysis. The extract was evaluated for scavenging activity using a 2,2-diphenylpicrylhydrazyl (DPPH) assay on BHK 21 (baby hamster kidney fibroblast) cell line. *A. auricula-judae* aqueous extract inhibited 50% of DPPH free radical at an IC50 value of 0.309 ± 0.021 mg/ ml, while methanol extract at IC50 value of 0.855 ± 0.062 mg/mL. Overall, aqueous extracts displayed potent scavenging activity compared to the methanol extracts on baby hamster fibroblast cells Oke and Aslim (2011). Similarly, Elkhateeb et al. (2018) examined the various extracts of *A. auricula-judae* for the antioxidant of DPPH radical with ascorbic acid as a positive control. The ethyl acetate extract showed activity with an IC50 value of 333.1μg/mL followed by the methanol extract with an IC50 value of 371.7 μg/mL (Elkhateeb et al., 2018).

TABLE 2.2

Structural Elucidation, Extraction Method, and Monomeric Composition of Polysaccharides and Melanin from *Auricularia auricula-judae*

A. auricula polysaccharides fractions/melanin	Source	Collected from	Extraction Purification	chain structure	MW (kDa)	Monosaccharides/elements composition	References
A. auricula polysaccharides fresh (AAPF) and *A. auricula* polysaccharides pickle (AAPP)	Fresh and pickled fruiting body	Hubei, China.	• Extraction by hot water • Purified by dialysis.	β-glycosidic linkage		• Galactose, mannose, arabinose, rhamnose and glucose, (Molar ratio= 1: 1:1.18:1:16.74) • Xylose, mannose, galactose and atabinose (Molar ratio = 1:1.52:4.76:15.59)	(Khaskhei et al., 2015)
A. auricular polysaccharide (AAP-I)	Mycelium was grown under solid-state fermentation	Fuzhou, China	• Extracted by hot water, • Purified by DEAE Sephadex A-50	α-glycoside linkage	–	–	(Zeng et al., 2013)
Acid hydrolyzed AAP fraction (AAPHs-F)	Fruiting body	Heilongjiang, China	Extraction by 1% NaOH	(1 → 3)-linked-α-D-glucopyranose	885.37 Da	Mannose, glucose, and galactose (Molar ratio = 12.7: 3.25: 1)	(Fang et al., 2019)
Auricularia auricula polysaccharides (AAP)	Fruiting body	Heilong Jiang province, China	• Extracted by boiling water • Purified by dialysis	Sulfated glycoprotein	–	Xylose, rhamnose, glucose, mannose, galactose, and arabinose (molar ratio = 15.2 12.4 12.5 1.3 0.8 0.3)	(Chen et al., 2008)
Auricularia auricula-judae acidic polysaccharides (aAAP-1)	Grounded fruiting body	Heilongjiang, China	• Extracted with alkali-soluble alcohol precipitation • Purified by ion-exchange chromatography	–	1538 Da	Glucuronic acid, Mannose, xylose, glucose and galactose (Molar ratio = 9.88:80.63:1.31:2.25:1.31)	(Bian et al., 2020)
Auricularia auricula-judae polysaccharide (AAPs)	Fruiting body	Harbin, China	• Extraction with hot water • Purified by gel filtration chromatography	–	3.30×10^5 Da	D-galactose, D-glucose, D-mannose and D-xylose (Molar ratio = 0.21:9.91:4.67:1.91)	(Bai et al., 2014)

(*Continued*)

TABLE 2.2 ***(Continued)***

Structural Elucidation, Extraction Method, and Monomeric Composition of Polysaccharides and Melanin from *Auricularia auricula-judae*

A. auricula polysaccharides fractions/melanin	Source	Collected from	Extraction Purification	chain structure	MW (kDa)	Monosaccharides/elements composition	References
Carboxy methylated *Auricularia auricula* polysaccharides (CMAAP22)	Fruiting body	Qingyuan, China	• Extraction by boiling water • Purified by Ion-exchange chromatography	B-(1 → 3)-D-mannose	3.4 ×10^6 Da	Mannose and glucose (Molar ratio = 1.06:1)	(Yang et al., 2011)
Crude polysaccharide (S2)	Dried fruiting body	Seoul, Korea	• Extracted by using distilled water • Purified by Sephacryl S-400 HR column	–	160 kDa	Mannose, glucose, xylose, and hexuronic acid (Molar ratio = 0.35:0.26:0.25:0.14)	(Yoon et al., 2003)
Eumelanin	Fruiting body	China	• Ultrasound-assisted 1.25 M NaOH extraction • Purified by washing with chloroform, ethyl acetate, ethanol	–	48.99 kDa	–	(Hou et al., 2019)
Exopolysaccharides (CEPSN-1 and CEPSN-2)	Submerged culture	Helongjiang, China	• Precipitation with four-fold ice-cold ethanol • Purified by DEAE 650 M and Sephacryl S-300 HR columns	Back bone chain: (1→ 4)-α-glucan Branched residues: (1→4, 6)-α-glucan	4.6 6.7	• Glucuronic acid, galactose, mannose and glucose (Molar ratio = 0.61:0.38:0.11:98.9) • Galacturonic acid, glucuronic acid, fucose, arabinose, galactose, mannose, and glucose (molar ratio = 0.20:0.50:0.45:0.11:1.04:97.56)	(Zhang et al., 2018)
Heteropolysaccharide (AAPs-F)	Dried fruiting body	Helongjiang, China	• Extracted with 1% NaOH, • Purified by Sephadex G-10 column.	β-glucan	143.15	Glucose, galactose, and fucose with the molar ratio of 51: 2: 1	(Xu et al., 2018)

Immunologically active glucuronoxylomannan (AAPs)	Fruiting body	China	• Extraction with water • Purified by anion exchange column	(→3)-D-Man*p*-(α1→3)-[D-Xyl*p*-(β1→6)]-D-Man*p*-(α1→3)-D-Man*p*-(α1→3)-[D-GlcA*p*-(b1→2)]-D-Man*p*-(α1→3)-D-Man*p*– (α1→)	70	Mannose, glucuronic acid xylose with molar ratio of 65:15:10	(Perera et al., 2018)
Melanin from *Auricularia auricula*							
Pheomelanin	Fruiting body	Heilongjiang, China	• Ultrasound-assisted 1 M NaOH extraction • Purified by Sephadex G-100 column	–	–	• Elemental composition: C, N, H, O, and S • Metal composition: Ca, Fe, Cu and Zn (Concentration = 2.12, 1.43, 0.19, and 0.18 mg/g)	(Zou et al., 2013)
Polysaccharides and their artificial gastrointestinal fluid hydrolysates (AAPHs – AAPHs1, AAPHs2, and AAPHs3).	Fruiting body	China	• Hydrolysis of AAPs in simulated gastric medium, • Purified by dialysis.	–	320, 169, and 62 kDa	Arabinose, xylose, mannose, 2-deoxy-glucose, glucose and glucosamine (molar ratio = 2.85:0.78:4.31:1.49:1.00:0.34)	(Lu et al., 2018)
Sulfated neutral *Auricularia auricular* polysaccharides (SNAAP)	Fruiting bodies	China	• Extracted using water and ethanol precipitation	–	4.95×10^5 Da	Mannose and glucose (Molar ratio of 1.7:1)	(Chen et al., 2019)
The eumelanin	Powered	Maypro, New York	• Enzyme-assisted extraction • washed with phosphate-buffered saline	–		C, H, N and O with percentage of 41.18%, 5.56% 1.66%, and 51.60%, respectively	(Prados-Rosales et al., 2015)
Ultrasonic-degraded *Auricularia auricula* polysaccharides (AAP)	Fruiting body	Hubei Province, China	• Extracted using a 0.1 mol/L NaOH solution • Purified by DEAE-Sephadex A-25 column	β-(1 →3)-glucose	1.4×10^3	Glucose	(Qiu et al., 2019)
		Hubei, China			201–215		

(*Continued*)

TABLE 2.2 *(Continued)*

Structural Elucidation, Extraction Method, and Monomeric Composition of Polysaccharides and Melanin from *Auricularia auricula-judae*

A. auricula polysaccharides fractions/melanin	Source	Collected from	Extraction Purification	chain structure	MW (kDa)	Monosaccharides/elements composition	References
Water soluble neutral polysaccharides (AF1)	Crushed dried fruiting bodies		• Extraction by hot 0.15 M NaCl • Purified by dialysis	β-(1→3)-d-glucan with two β-(1→6)-d-glucosyl residues		Terminal glucose, 1,3,6-linked glucose, and 1,3-linked glucose (Molar ratios = 1.9:1:1.9)	(Xu et al., 2012)
Water soluble β-D glucan (AAG)	Dry fruiting body	(Hubei, China	• Extraction by 70% ethanol/water solution • Purified by dialysis	Backbone structure: β-(1→4)-linked-D-glucopyranosyl Substituted by: β-(1→6)-linked glucopyranosyl Side chain: 0–6	288	Glucuronic acid and glucose (Molar ratio = 1:6)	(Ma et al., 2008)
water-soluble acidic AAP (WAF)	Fruiting body	(Hubei, China	• Extraction by 0.9% NaCl, • Purified by dialysis	β-(1→ 4)-D-glucan with b-(1 → 6)-D-glucose	$2.8x\ 10^5$	Mannose, xylose, and glucose with a molar ratio of 1.00:0.23:0.21	(Ma et al., 2010)
Water-soluble polysaccharide (AAP-Ia)	Fruiting body	Heilongjiang, China	• Ultrasound-assisted extraction • Purified by anion-exchange and gel-permeation chromatography	-	3500 Da	D-galactose, D-glucose, D-mannose, D-xylose, L-arabinose, L-rhamnose (Molar ratio = 0.4:1:3.6:0.4:2.6:0.2)	(Zhang et al., 2011)

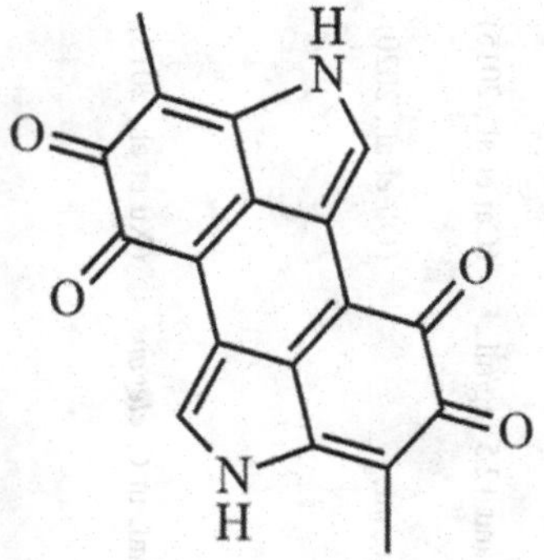

FIGURE 2.5 Chemical structures of natural melanin extracted from *A. auricula.*

Packialakshmi et al. (2017) reported the flavonoids, total phenolic contents, and the total antioxidant capacity of the aqueous extract using DPPH free-radical scavenging activity, hydroxyl radical, and ferric ion-reducing antioxidant power (FRAP) assays. The crude extract showed free radical reducing activities with EC50 values of 3.65 mg/mL, 6.11 mg/mL, and 7.35 mg/mL, respectively. Similar work by Acharya et al. (2004) stated that crude, boiled, and ethanolic extracts of *A. auricula-judae* can impede lipid peroxidation with IC50 values of 310, 572, and 398 μg/mL, respectively.

Eumelanin is hepatoprotective in mice with ethanol-induced liver injury (Hou et al., 2019). Oral administration of eumelanin in mice (i.e., prophylactically once a day for three weeks) significantly reduced liver injury index in mice, such as serum alanine aminotransferase, aspartate aminotransferase, and liver malondialdehyde levels (Hou et al., 2019). In contrast, the levels of ADH (alcohol dehydrogenase), CAT, and SOD levels markedly decreased in the melanin-treated mice (Hou et al., 2019). Another type of melanin called pheomelanin from the fruiting bodies showed significant free-radical scavenging activities with IC50 values of 0.18 (for DPPH radical), 0.34 (for hydroxyl radical), and 0.59 mg/mL (superoxide radical) suggesting its potential as a natural antioxidant (Zou et al., 2015).

Anti-inflammatory

The anti-inflammatory activity of dichloromethane extract of *A. auricula-judae* was evaluated in bacterial derived-lipopolysaccharide (LPS)-induced murine RAW 264.7 macrophages (Damte et al., 2011). The dichloromethane extract decreases the expressions of inflammatory cytokines (IL-6, TNF-α, and IL-1β) mRNA in a dose-dependent manner (Damte et al., 2011). Zhuan-Yun et al. (2015) evaluated LPS-induced acute lung injury in adult Sprague Dawley rats treated with A. auricular-judae polysaccharide. The results showed significant improvement in LPS induced lungs condition, decreased protein content in the bronchoalveolar lavage fluid and malondialdehyde level in the lungs, and inhibited the release of inflammatory cytokines (TNF α and IL 6) in blood, suggesting potent anti-inflammatory potential.

Antimicrobial and Antifungal Activity

Several studies have investigated the antimicrobial activity of crude *A. auricular-judae* polysaccharide on some bacteria such as *Escherichia coli, Bacillus subtilis, Micrococcus luteus, Staphylococcus aureus*, and fungi *Aspergillus niger* and *Saccharomyces cerevisiae* (Cai et al., 2015). The crude polysaccharide was more effective against *S. aureus* than *E. coli*, with the inhibition zones of 9.84 ± 0.076 mm and 5.55 ± 0.182 and MICs values of 6.25 mg/mL and 12.5 mg/mL, respectively (Cai et al., 2015). A similar study conducted by Oli et al. (2020) showed antimicrobial and antifungal activities of the tris and warm aqueous protein extracts from dry

TABLE 2.3
Reported Pharmacological Activities of Polysaccharides, Melanin and Extracts of A. auricula-judae

Reported pharmacological activities	Tested materials/ Isolated compound/ fraction	Model organism(s)/Cell line(s) used	Effective dosage/ Result	References
Anticoagulant activity (ex vivo)	Crude polysaccharides	Sprague–Dawley (SD) male rats	Orally fed with crude polysaccharide in 50% DMSO at 300 mg/kg body weight per day, and showed thrombin inhibition by antithrombin and inhibits platelet aggregation and blood clotting. Aspirin was a positive control.	(Yoon et al., 2003)
Antidiabetic activity (In vitro)	Ethyl acetate extract	α-amylase activity	IC50 value of 14.05±3.2 ppm.	(Elkhateeb et al., 2018)
Anti-inflammatory activity	Dichloromethane extract	Murine macrophage (RAW 264.7 cells)	Significant inhibition of LPS-induced NO production in a dose-dependent fashion.	(Damte et al., 2011)
Anti-inflammatory activity	Dichloromethane extract	LPS-stimulated murine RAW 264.7 macrophages (RAW264.7 Cells)	Extract brought down the expressions of inflammatory cytokines mRNA.	(Damte et al., 2011)
Anti-inflammatory activity	Glucuronoxylomannan (MEA and MHA)	Sprague Dawley mice: Carrageenin edema, Scald hyperalgesia	Intraperitoneal administration at a dosage of 50 mg/kg for 2 times and showed: • Hind paw thickness 0% and pain threshold 6% by MEA; • Hind paw thickness 33%, Pain threshold 19% by MHA.	(Ukai et al., 1983)
Anti-inflammatory activity	Polysaccharides	Acute lung injury (ALI) caused by LPS in Adult Sprague-Dawley rats	Attenuated protein concentration in the bronchoalveolar lavage fluid (BALF), inhibited myeloperoxidase (MPO) activity, and reduced the malondialdehyde (MDA) level and lung W/D weight ratio, and inhibited the blood TNF-α and IL-6 levels.	(Zhuan-Yun et al., 2015)
Antimicrobial activity	Crude polysaccharides (AAP)	*Escherichia coli, Staphylococcus aureus*	MICs of AAP on *S. aureus* and *E. coli* were 6.25 mg/mL and 12.5 mg/mL, respectively.	(Cai et al., 2015)
Antimicrobial activity	Tris buffer and warm water protein extracts fruiting bodies	*Staphylococcus aureus, Bacillus subtilis Pseudomonas aeruginosa, Klebsiella pneumoniae Escherichia coli,*	MIC of 5 μg/mL except for *E. coli*, whose, MIC was 2.5 μg/mL	(Oli et al., 2020)
Antioxidant activity	Polysaccharide hydrolysate (AAPs-F)	*Caenorhabditis elegans* strains N2 (wild-type)	Protected the injury induced by H_2O_2 at a dose of 0.2 mg/mL in *C. elegans*.	(Xu et al., 2018)

Antioxidant activity	Crude polysaccharides (AAP)	ABTS Radical Scavenging Assay	Scavenging % increased from 9.73 to 78.01% when test concentration was increased from 0.3125 to 5 mg/mL.	(Cai et al., 2015)
		Superoxide Radical Assay	Can scavenge superoxide radicals in a concentration-dependent manner.	
Antioxidant activity (in vitro)	Water-soluble polysaccharide	Human fibroblasts ABTS assay method	AAP exhibited antioxidant activity in a dose-dependent (0–400 μg/mL) manner with an effective concentration (EC50) of 226.67 ±10.41 μg/mL.	(Mapoung et al., 2021)
Antioxidant activity	Hot water extract of *Auricularia auricula-judae*	DPPH (1,1-diphenyl-2-picryl-hydrazyl) radical scavenging assay	scavenging activity of *Auricularia auricula* the EC50 value of 3.65 mg/mL.	(Packialakshmi et al., 2017)
		OH radical scavenging assay	with EC50 value of 6.11 mg/mL.	
		FRAP (ferric reducing antioxidant power) assay	with EC50 values 7.35 mg/mL	
Antioxidant activity	Ethyl acetate and methanol extracts	DPPH radical scavenging assay	IC50m value of 333.1 μg/mL and 371.7 μg/mL, respectively.	(Elkhateeb et al., 2018)
Antioxidant activity (The DPPH radical scavenging effect)	Methanolic (AME) Aqueous (AAE) extract	BHK 21	*A. auricula-judae* aqueous extract (AAE) at the IC50 value of 0.309 ± 0.021 mg/ mL) and Methanol extracts exhibited weak antioxidant activity with IC50 value of 0.855 ± 0.062 mg/mL.	(Oke and Aslim, 2011)
Antioxidant and nitric oxide synthase activation properties	Crude, boiled, and ethanolic extract of *A. Auricula*	RBC-lipid peroxidation	IC50 value of 403, 510, and 373 μg/mL, respectively.	(Acharya et al., 2004)
Anti-tumor activity	Dichloromethane fraction	NSCLC (Human non-small cell lung cancer) NCI H358 (bronchioalveolar) and SNU1 (gastric carcinoma) cells	IC50 value at 57.2 mg/mL. IC50 vale at 73.2 mg/mL.	(Reza et al., 2014) (Reza et al., 2014)
Anti-tumor activity (in vivo)	A water-soluble b-D-glucan (AAG)	Xenograft Sarcoma 180 tumor cells	Inhibited proliferation by 23.5% and 34.1% at concentrations of 0.005 and 0.05 mg/L, respectively.	(Ma et al., 2010)
		Acinar cell carcinoma tumor cells (in vitro)	AAG prevents tumor growth in dose-dependent fashion.	(Ma et al., 2010)
Anti-tumor activity	Glucan I with backbone chain β-(1→3)-linked D-glucopyranosyl	Sarcoma 180 cells	Intraperitoneal administration at a dosage of 8 mg per kg of ICR-JCL mouse per day, for 10 days; Average tumor weight of 0.20 g and inhibition ratio of 96.6%.	(Sone et al., 1978; Misaki et al., 1981; Misaki and Kakuta, 1995)

(Continued)

TABLE 2.3 *(Continued)*

Reported Pharmacological Activities of Polysaccharides, Melanin and Extracts of A. auricula-judae

Reported pharmacological activities	Tested materials/ Isolated compound/ fraction	Model organism(s)/Cell line(s) used	Effective dosage/ Result	References
Anti-tumor Activity	Glucan II with backbone chain β-(1→3)-linked D-glucopyranosyl	Sarcoma 180 cells	Intraperitoneal administration at a dosage of 8 mg per kg of ICR-JCL mouse per day, for 10 days. Average tumor weight of 0.20 g and inhibition ratio of 96.6%.	(Sone et al., 1978; Misaki et al., 1981; Misaki and Kakuta, 1995)
Anti-tumor activity	Heteropolysaccharide: α-(1→3)-linked D mannopyranosyl	Sarcoma 180 cells	Intraperitoneal administration at a dosage of 10 mg per kg of ICR-JCL mouse per day, for 10 days, Average tumor weight 4.75 g and inhibition ratio of 18.5%.	(Misaki et al., 1981)
Anti-tumor activity	Acidic heteropolysaccharide (MEA and MHA)	Sarcoma 180 cells (S-180) induced in ddY male mice	Intraperitoneal administration at a dosage of 25 mg per kg per day, for 10 days, mean tumor weight 5.93±2.63 g, 7.23±4.34 g, and inhibition ratio of 42% and 29%, respectively.	(Ukai et al., 1983)
Antitumor activity (In vitro)	Dichloromethane	MTT and SRB assay method: P388D1 macrophage Sarcoma 180 cells in BALB/c mice	IC50 values of 38.32 μg/mL and 94.20 μg/mL significantly ($p < 0.05$) reduced the tumor size than control group, with inhibition % of 76.13%.	(Reza et al., 2011)
Antitumor activity (In vitro)	70% ethanol extracts	Sarcoma 180 cells	IC50 at 0.72 mg/mL	(Reza et al., 2012)
Antitumor activity (In-vitro)	Ethyl acetate	P388D1 macrophage Sarcoma 180 cells	IC50 value of 43.80 μg/mL IC50 value of 108.90 μg/mL	(Reza et al., 2011)
Antitumor activity (In-vitro)	Butanol	P388D1 macrophage Sarcoma 180 cells	IC50 value of 94.62 μg/mL IC50 value of 34.10 μg/mL	(Reza et al., 2011)
Antitumor activity (In-vitro)	Water fractions from 70% ethyl acetate extract	P388D1 macrophage Sarcoma 180 cells	IC50 value of 28.19 μg/mL	(Reza et al., 2011)

			IC50 value of 102.30 μg/mL	
Cytotoxic activity (In-vitro)	Nonpolar (n-hexane) extract has Oleic acid;	HCT116 (Human colon cancer) cell line	IC50 value of 43.5 μg/mL.	(Elkhateeb et al., 2018)
Hypocholesterolemic effect (in vivo)	Polysaccharides	Hyperlipidemic ICR mice, induced by cholesterol-enriched diet (CED) ()	300mg/kg AAP was administered through gastric infusion for 12 weeks. Lowered serum TC level by 10.5% (9.86±1.25 mmol/L) than that of CED (11.02±1.51 mmol/L) treatment group.	(Luo et al., 2009)
Hypocholesterolemic effects	Ethanol extract	ICR mice	AAE orally administered 150 mg/kg/d b.w, for 8-week showed hypocholesterolemic effect in cholesterol-enriched diet (CED) fed mice, improving antioxidant status, decreasing the TCL and atherosclerosis index, increasing HDL-C level and fecal bile acids.	(Chen et al., 2011)
Hypoglycemic activity	Water-soluble polysaccharides from fruiting bodies	Genetically diabetic mice (KK-A^y)	Oral administration of 30 g of FA/kg Decrease plasma glucose, insulin, urinary glucose.	(Yuan et al., 1998)
Hypolipidemic activity	Ethanol extract of Dried fruit body	male C57BL/6 mice *in vivo*	A normal diet (ND) of 0.1%, high fat diet of 0.3%, with 1% (w/w) of AAE were given for 8 weeks, mice indicate reduced body weight and adipose tissue mass.	(Reza et al., 2015)
Inhibition of lipid peroxidation activity	Crude, boiled, and ethanolic extract of *A. Auricula*	human red blood cells (RBC)	310, 572, and 398 μg/mL, respectively	(Acharya et al., 2004)
inhibitory activity	70% ethanol extracts	NSCLC (Human non-small cell lung cancer) NCI H358 (bronchioalveolar) and SNU1 (gastric carcinoma) cells	Inhibiton of 69.76%, and 68.01%, respectively by the crude extract.	(Reza et al., 2012)
Super oxide scavenging activity	Crude, boiled, and ethanolic extract of *A. Auricula*		47 μg/mL, 80 and (50 μg/mL), respectively	(Acharya et al., 2004)

fruiting bodies of *A. auricular-judae* against infectious bacterial and fungal isolates. The Tris protein extract was potent against *E. coli*, *P. aeruginosa*, and *S. aureus* with an inhibition zone diameter of 6.00 ± 0.00 mm, 4.00 ± 0.00 mm, and 3.66 ± 0.53 mm, respectively. While warm aqueous extract revealed the most potent effect against *E. coli* and *S. aureus* with an inhibition zone of 66 ± 0.53 mm each. Both the extract also indicated effective antifungal activity against *Candida albicans* and *Trichophyton schoenleinii* with a MIC value of 5 μg/m (Oli et al., 2020).

Immunomodulatory Activity

The cold-water-soluble glucuronoxylomannan (MAE) extract from the fruiting bodies of *A. auricula-judae* was evaluated for immunostimulatory activity (Perera et al., 2018). The investigation treated the Raw 264.7 macrophages with cold-water-soluble glucuronoxylomannan extract (which contains polysaccharides) at a 1, 3, and 10 μg/mL concentrations. It significantly ($p < 0.01$) augmented the production of NO (nitric oxide), TNF-alpha, and IL-6, which are important immune mediators (Perera et al., 2018), responsible for triggering inflammations.

Three sulfated *A. auricula-judae* polysaccharides coded as sAAPt, sAAP1, and sAAP2 (prepared by the chlorosulfonic acid-pyridine method), were evaluated for in-vitro immuno-enhancing activity in 14 days old white roman chickens vaccinated with ND (Newcastle disease) vaccine (test treatment administered once a day for three consecutive days at concentrations ranging from 1.953–0.244 μg/mL). Lymphocyte proliferation was highest for sAAP1, followed by sAAPt than the control group, indicating that sAAP1 and sAAPt can potentially enhance the protective immune system (Nguyen et al., 2012).

Anticoagulant Activity

Sprague–Dawley male rats were orally administered *A. auricula-judae*-derived crude polysaccharide (coded as aAAP-1) in 50% of DMSO (dimethyl sulfoxide) with dose of 300 mg/kg body weight/day for the duration of four weeks. Blood samples collected an hour after the last feeding showed inhibition of platelet aggregation and thus preventing the clotting (Yoon et al., 2003).

Bian et al. (2020) examined the in vitro anticoagulant activity of *A. auricula-judae*-derived acidic polysaccharides on normal human plasma through APTT (activated partial thromboplastin time) PT (prothrombin time), and TT (thrombin time) test experiments. In the APTT test, 50.0 μg/mL of aAAP-1 significantly prolonged the clotting time (47.3 s). Similarly, the concentration of aAAP-1 at 50.0 μg/mL also significantly ($p<0.01$) prolonged PT (73.7s) and TT (28.2) compared to saline control (13.3s) confirming the anticoagulant property of aAAP-1 (Bian et al., 2020).

Anti-diabetic Activity

A. auricula-judae polysaccharide (AAP-I) extracted by hot water and purified by chromatographic technique (DEAE Sephadex A-50 and Sephadex G200) was examined for in vivo hypolipidemic properties (Zeng et al., 2013). In the study, male Kunming mice with hyperlipidemia induced by a high-fat diet were orally administered with AAP-I at 50.0, 100.0, and 200.0 mg/kg concentrations. After 14 and 28 days, the AAP-I significantly lowered triglyceride, low-density lipoprotein (LDL) cholesterol, and total cholesterol in the mice, indicating hypolipidemic properties of AAP-I (Zeng et al., 2013). In another study, male Sprague-Dawley rats administered with pulverized *A. auricula-judae* of 5g/100g for four weeks showed a compelling effect in lowering serum low-density lipoprotein (LDL) cholesterol and total serum cholesterol levels. It also significantly ($p < 0.05$) lowered total cholesterol and lipid level in the liver compared to the control, thus indicating its anti-hyperlipidemic activity (Cheung, 1996). Luo et al., (2009), reported similar findings when hyperlipidemic ICR mice induced by a cholesterol-enriched diet (CED) were treated with 300mg/kg

AAP via gastric infusion for 12 weeks. The study recorded a significant decrease in total cholesterol (TC) level, exhibiting hypolipidemic activity (Luo et al., 2009).

For eight weeks, hypercholesterolemia ICR mice (induced by CED), when orally administered crude ethanol extract of *A. auricula-judae* (AAE-150 mg/kg/d b.w.), showed a decrease in the total cholesterol level, while high-density lipoprotein cholesterol level increased, signifying a notable hypocholesterolemic effect (Chen et al., 2011). The diabetic mice strain (KK-Ay), when supplemented with the water-soluble polysaccharides (FA) derived from fruiting bodies, significantly reduced insulin and urinary glucose level and, on the contrary, increased the hepatic glycogen content, suggesting the hypocholesterolemic effect of FA (Yuan et al., 1998).

Elkhateeb et al. (2018) screened n-hexane, ethyl acetate, methanol, and chloroform extracts of *A. auricula-judae* for their inhibitory effect on α-amylase enzymatic activity. The ethyl acetate extract exhibited the maximum inhibitory activity with IC50 value of 14.05±3.2 ppm, followed by the chloroform extract with IC50 value of 14.3±4.49 ppm (Elkhateeb et al., 2018), thus suggesting a potential source of anti-diabetic pure compounds.

Radio-protective Properties

Bai et al. (2014) studied the radio-protective property of the *A. auricular-judae* polysaccharides (AAP IV) together with grape seed procyanidins (GSP) in splenocytes of Wistar rats exposed to radiation. The study showed that AAP IV+GSP together reduced spleen glutathione level and spleen index ($p < 0.005$) and also showed significant inhibitions of malondialdehyde upsurge ($p < 0.005$) levels in the radiation-induced damaged rat splenocytes (Bai et al., 2014; Zhuan-Yun et al., 2015).

The 60Co-γ-radiated mice usually shows high blood glucose level, low insulin and hepatic glycogen, improved blood glucose tolerance, and dysfunctional hepatic and pancreatic (Kaneto et al., 2004; Burns et al., 2011). However, when treated with sulfated neutral polysaccharides (SNAAP) at the concentration of 50, 100, and 200 mg/kg·bw, 60Co-γ-radiated mice could restore distorted glucose metabolism-for example, increasing the phosphorylation of JNK (c-Jun N-terminal kinase) and Fox01 (Forkhead box 01) while decreasing the phosphorylation of Akt (Serine/Threonine Kinase 1) and GSK-3β (Glycogen synthase kinase-3β). Similarly, expression of PEPCK (Phosphoenolpyruvate carboxykinase), G6Pase (Glucose 6-phosphatase), and GYS2 (Glycogen phosphorylase 2) were elevated in the liver while expression of PDX1 (Pancreatic and Duodenal Homeobox 1), GLUT2 (Glucose transporter 2) and IRS1 (Insulin Receptor Substrate 1) in the pancreas decreased in the radiated mice upon treatment with SNAAP. Thus, this evidence of improved gluconeogenesis indicates the antiradiation property of SNAAP (Chen et al., 2019).

Anti-cancer and Cytotoxic Activity

Water-soluble glucan I with branched (1→3)-β-D-glucan, when examine demonstrated significant antitumor effects against sarcoma in mice (Table 4) (Sone et al., 1978; Misaki et al., 1981). The cytotoxic and antitumor effects of fractionated crude extracts of ethyl acetate (EtOAc), dichloromethane (DCM), ethanol (EtOH), butanol (BuOH), and water from *A. auricula-judae* were examined on sarcoma 180 and P388D1 cell line by MTT (3-(4,5-dimethylthiazol-2yl)-2,5-diphenyltetrazolium bromide) and SRB assays (Reza et al., 2011) with doxorubicin as a positive control. The DCM fraction demonstrated the highest cytotoxicity, followed by BuOH, EtOH, EtOAc, and water fractions at the dose of 1 mg/mL in P388D1 cells. With regard to antitumor activity, EtOAc, BuOH, DCM, EtOH and water fractions exhibited significant ($P < 0.05$) activity against P388D1 cells with IC50 value of 143.80 μg/mL, 094.62 μg/mL, 38.32 μg/mL, 044.03 μg/mL and 028.19 μg/mL, respectively. Similarly, IC50 values for antitumor activity against Sarcoma 180 cell line were 108.90 μg/mL (DCM fraction), 134.10 μg/mL (BuOH fraction), 94.20 μg/mL (EtOH fraction), 133.00 μg/mL (EtOAc fraction), and 102.30 μg/mL (water fraction)

(Reza et al., 2011). Reza et al. (2012) reported the antiproliferative activity of *A. auricula-judae* extract as 65.71%, 69.76%, and 68.01% against Sarcoma 180, NCI H358, and SNU 1 cell lines, respectively. Water-dissoluble polysaccharides composed of b-(1 - 4)-D-glucan with b-(1 - 6)-D-glucose side groups (AGG) were isolated from 70% ethanol extract of *A. auricula-judae*, and evaluated for in vivo antitumor activities. For this test, sarcoma 180 tumor cells were introduced into BALB/c mice (eight weeks old) subcutaneously, and after 24 hours, treatment (both sample-AGG and positive control-5-fluorouracil) was given intraperitoneally (concentrations at 5, 20, and 40 mg/kg, once a day for eight days). AAG showed antitumor activities in a dose-dependent fashion, of which the 20 mg/kg dose was the best with the highest inhibitory effect (Ma et al., 2010). According to Novakovic et al. (2016), water and crude ethanol extracts of wild *A. auricula-judae* have a significant cytotoxicity effect at IC50 values of 285.7 and 333.3 μg/mL, respectively, against MCF-7 (human breast cancer) cell line. Different extracts of *A. auricula-judae*, including methanol, hexane, chloroform, and ethyl acetate extracts, showed in vitro cytotoxicity effect against colon carcinoma human tumor cell lines (HCT116) (Elkhateeb et al., 2018). Of many tested, n-hexane was the most potent one, with the IC50 value of 43.5 μg/mL, and it is attributed to the presence of oleic acid as the major constituent of n-hexene extract (Elkhateeb et al., 2018).

APPLICATION IN THE FOOD PRODUCTION AND ITS TOXICITY

A. auricula-judae is also used in food production due to its antioxidant and probiotic properties. The use of its aqueous extract in yogurt production enhances the *Bifidobacterium bifidum* Bb-12 and *Lactobacillus acidophilus* La-5 to improve the gut microbiome. Likewise, supplementation of fungus aqueous extract (0.1%) significantly improved the antioxidant properties of yogurt (Faraki et al., 2020). In the study by Fan et al. (2007), the finding reveals that using *A. auricula-judae* polysaccharide flour as an ingredient in bread preparation improves DPPH radical scavenging activity with minimal effect on the sensory index when used up to 9%. Several studies have confirmed *A. auricula-judae*, as a constituent in functional wheat bread products, enhances water retention and adsorption ability of wheat flour (Yuan et al., 2017). *A. auricula-judae* is also used to make wood-ear sausage with carrageenan as a meat substitute (Hermawan et al., 2020).

The permissible dose concentration of toxic elements such as cadmium (Cd), lead (Pb), and arsenic (As) are reported as 0.007, 0.025, and 0.003 mg/kg of human body weight (FAO/WHO, 1991). However, levels of Pb, Cd, and As in fresh *A. auricula-judae* is not more than 0.01, 0.1, and 0.2 mg/kg, which is significantly low compared to the standard parameters specified by Russian Federation for edible mushrooms, illustrating a safe fungus for human consumption (Kadnikova et al., 2015).

CONCLUSION AND FUTURE PROSPECTIVE

Auricularia auricula-judea is a non-toxic, edible mushroom popular as a dietary supplement and in traditional folk medicines. This mushroom has high nutritional contents of carbohydrates and protein, amino acids, trace elements, and vitamins and is valued for local delicacies. This mushroom is cosmopolitan, with a higher success rate of adaptation to varying ecological conditions from tropical to temperate regions.

Several researchers have recorded their micro and macrohabitat and substrate preferences, such as dead trees, fallen logs, stumps, live branches, dried branches, and partially or fully decomposed logs. Their substrate choices differ based on the ecological region; for example, *A. auricula-judae* is found growing on the fallen trunk of *Carpinus spp* and dead branch of *Sambucus spp* in the Czech Republic, on the growing tree of *Cornus* and *Platanus* species in Denmark and France, on the fallen log of *Quercus spp* in China, on decayed wood of *Sambucus*, Humbie and fallen twigs of Acer plants in the United Kingdom, on the trunk of *Euonymus*, and fallen branch of *Fraxinus* plants in Russia. However, the relationship between the growth substrate and bioactive compounds in *A. auricula-judae* is less studied; hence it would be interesting to find out if their growth substrate influences phytochemical constituents, particularly bioactive compounds.

Polysaccharides and melanins are widely studied bioactive compounds from *A. auricula-judae*. Different, crude *A. auricula-judae* polysaccharides showed significant biological activity, including antioxidant, antimicrobial, anticancer, antidiabetic, anti-inflammatory, immunomodulatory, anti-hypercholesterolemic, and radio-protective properties. Furthermore, *A. auricula-judae* is also a source of proteins, which could be a potential ingredient in developing novel drugs, but it requires more extensive work in the future. Phytochemical analysis of a different extract of *A. auricula-judae* has shown the presence of hydrocarbon, and methyl ester compound, diazane as a principal compound of ethanol extract, oleic acid as a major compound of n-hexane extract, and also contains flavonoids, steroids, terpenoids, tannins, and saponins. However, the study concerning their potentiality in developing novel drugs and pharmacological activity is less known, which is another opportunity to research more.

REFERENCES

Acharya, K., Samui, K., Rai, M., Dutta, B.B., Acharya, R. 2004. Antioxidant and nitric oxide synthase activation properties of*Auricularia auricula*. *Indian J. Exp. Biol.* 42: 538–540.

Apetorgbor, M., Apetorgbor, A., Obodai, M. 2008. Indigenous knowledge and utilization of edible mushrooms in parts of Southern Ghana. *Ghana J. Forestry* 19(1): 20–34. 10.4314/gjf.v19i1.36908

Bach, F., Helm, C.V., Bellettini, M.B., Maciel, G.M., Haminiuk, C.W.I. 2017. Edible mushrooms: A potential source of essential amino acids, glucans, and minerals. *Int. J. Food Sci. 52*(11): 2382–2392. 10.1111/ijfs.13522

Bai, H., Wang, Z., Cui, J., Yun, K., Zhang, H., Liu, R.H., Fan, Z., Cheng, C. 2014. Synergistic radiation protective effect of purified *Auricularia auricular-judae* polysaccharide (AAP IV) with grape seed procyanidins. *Molecules* 19(12): 20675–20694. 10.3390/molecules191220675

Bandara, A. 2019. A review of the polysaccharide, protein and selected nutrient content of *Auricularia,* and their potential pharmacological value. *Mycosphere* 10(1): 579–607. 10.5943/mycosphere/10/1/10

Bandara, A.R., Chen, J., Karunarathna, S., Hyde, K.D., Kakumyan, P. 2015. *Auricularia thailandica* sp. Nov. (*Auriculariaceae, Auriculariales*): A widely distributed species from Southeastern Asia. *Phytotaxa* 208(2): 579–607. 10.11646/phytotaxa.208.2.3

Bao, H., You, S., Cao, L., Zhou, R., Wang, Q., Cui, S.W. 2016. Chemical and rheological properties of polysaccharides from fruit body of *Auricularia auricular-judae*. *Food Hydrocolloids* 57: 30–37. 10.1016/j.foodhyd.2015.12.031

Bian, C., Wang, Z., Shi, J. 2020. Extraction optimization, structural characterization, and anticoagulant activity of acidic polysaccharides from *Auricularia auricula-judae*. *Molecules* 25(3): 710. 10.3390/molecules25030710

Burns, S.F., Bacha, F., Lee, S.J., Tfayli, H., Gungor, N., Arslanian, S.A. 2011. Declining β-Cell function relative to insulin sensitivity with escalating OGTT 2-h glucose concentrations in the nondiabetic through the diabetic range in overweight youth. *Diabetes Care* 34(9): 2033–2040. 10.2337/dc11-0423

Cai, M., Lin, Y., Luo, Y., Liang, H., Sun, P. 2015. Extraction, antimicrobial, and antioxidant activities of crude polysaccharides from the wood ear medicinal mushroom *Auricularia auricula-judae* (Higher Basidiomycetes). *Int. J. Med. Mushrooms* 17(6): 591–600. 10.1615/IntJMedMushrooms.v17.i6.90

Carrasco, J., Zied, D.C., Pardo, J.E., Preston, G.M., Pardo-Giménez, A. 2018. Supplementation in mushroom crops and its impact on yield and quality. *AMB Express* 8(1): 146. 10.1186/s13568-018-0678-0

Chen, G., Luo, Y.C., Ji, B.P., Li, B., Guo, Y., Li, Y., Su, W., Xiao, Z.L. 2008. Effect of polysaccharide from *Auricularia auricula* on blood lipid metabolism and lipoprotein lipase activity of ICR Mice fed a cholesterol-enriched diet. *J. Food Sci.* 73(6): 103–108. 10.1111/j.1750-3841.2008.00821.x

Chen, G., Luo, Y.C., Ji, B.-P., Li, B., Su, W., Xiao, Z.L., Zhang, G.Z. 2011. Hypocholesterolemic effects of *Auricularia auricula* ethanol extract in ICR mice fed a cholesterol-enriched diet. *J. Food Sci. Technol.* 48(6): 692–698. 10.1007/s13197-010-0196-9

Chen, Z., Wang, J., Fan, Z., Qiu, J., Rumbani, M., Yang, X., Zhang, H., Wang, Z. 2019. Effects of polysaccharide from the fruiting bodies of *Auricularia auricul*ar on glucose metabolism in 60Co-γ-radiated mice. *Int. J. Biol. Macromol.* 135: 887–897. 10.1016/j.ijbiomac.2019.05.136

Cheung, P.C.K. (1996). The hypocholesterolemic effect of two edible mushrooms: *Auricularia auricula* (tree-ear) and *Tremella fuciformis* (white jelly-leaf) in hypercholesterolemic rats. *Nutr. Res.* 16(10): 1721–1725.

Chhoeda Yangchen, U. (2017). Wild vegetable diversity and their contribution to household income at Patshaling Gewog, Tsirang. *Bhutan J. Res. Dev., 4*(1), 30–38. 10.17102/cnr.2017.04.

Damte, D., Reza, Md. A., Lee, S.J., Jo, W.S., Park, S.C. 2011. Anti-inflammatory activity of dichloromethane extract of *Auricularia auricula-judae* in RAW264.7 cells. *Toxicol. Res.* 27(1): 11–14. 10.5487/TR.2011.27.1.011

Deka, A.C., Indrani, S., Sneha, D., Sarma, T. 2017. Antimicrobial properties and phytochemical screening of some wild macrofungi of rani—Garbhanga reserve forest area of Assam, India. *Adv. Appl. Sci. Res.* 8(3): 17–22.

Du, X., Zhang, Y., Mu, H., Lv, Z., Yang, Y., Zhang, J. 2015. Structural elucidation and antioxidant activity of a novel polysaccharide (TAPB1) from *Tremella aurantialba. Food Hydrocoll.* 43:459–464. 10.1016/j.foodhyd.2014.07.004

Elkhateeb, W.A., El-Hagrassi, A.M., Fayad, W., Daba, G.M., Ahmed, E.F. 2018. Cytotoxicity and hypoglycemic effect of the Japanese jelly mushroom *Auricularia auricula-judae. Chem. Res. J.* 3(4): 123–133.

Fan, L., Zhang, S., Yu, L., Ma, L. 2007. Evaluation of antioxidant property and quality of breads containing *Auricularia auricula* polysaccharide flour. *Food Chem.* 101(3): 1158–1163. 10.1016/j.foodchem.2006.03.017

Fang, Z., Chen, Y., Wang, G., Feng, T., Shen, M., Xiao, B., Gu, J., Wang, W., Li, J., Zhang, Y. 2019. Evaluation of the antioxidant effects of acid hydrolysates from *Auricularia auricular* polysaccharides using a *Caenorhabditis elegans* model. *Food & Function* 10(9): 5531–5543. 10.1039/C8FO02589D

FAO/WHO. 1991. *Protein quality evaluation (Report of a jo ined FAO/WHO export conclusion)*, FAO Rome.

Faraki, A., Noori, N., Gandomi, H., Banuree, S.A.H., Rahmani, F. 2020. Effect of *Auricularia auricula* aqueous extract on survival of *Lactobacillus acidophilus La-5* and *Bifidobacterium bifidum Bb-12* and on sensorial and functional properties of synbiotic yogurt. *Food Sci. Nutr.* 8(2): 1254–1263. 10.1002/fsn3.1414

Hermawan, N., Romulo, A., Wardana, A.A. 2020. Development and texture profile of wood-ear mushroom (*Auricularia auricula*) sausage formulated with carrageenan. IOP Conference Series: *Earth and Environmental Science* 426(1): 012182. 10.1088/1755-1315/426/1/012182

Hou, R., Liu, X., Yan, J., Xiang, K., Wu, X., Lin, W., Chen, G., Zheng, M., Fu, J. 2019. Characterization of natural melanin from *Auricularia auricula* and its hepatoprotective effect on acute alcohol liver injury in mice. *Food Funct.* 10(2): 1017–1027. 10.1039/C8FO01624K

Jo, W.S., Kim, D., Seok, S.J., Jung, H.Y., Park, S.C. 2014. The culture conditions for the mycelial growth of *Auricularia auricula-judae. J. Mushrooms* 12(2): 88–95. 10.14480/JM.2014.12.2.88

Kadnikova, I.A., Costa, R., Kalenik, T.K., Guruleva, O.N., Yanguo, S. 2015. Chemical composition and nutritional value of the mushroom *Auricularia auricula-judae. J. Food Nutr. Res.* 3(8): 478–482. 10.12691/jfnr-3-8-1

Kaneto, H., Nakatani, Y., Kawamori, D., Miyatsuka, T., Matsuoka, T. 2004. Involvement of oxidative stress and the JNK pathway in glucose toxicity. *Rev. Diabet. Stud.* 1(4): 165–165. 10.1900/RDS.2004.1.165

Khaskheli, S.G., Zheng, W., Sheikh, S.A., Khaskheli, A.A., Liu, Y., Soomro, A.H., Feng, X., Sauer, M.B., Wang, Y.F., Huang, W. 2015. Characterization of *Auricularia auricula* polysaccharides and its antioxidant properties in fresh and pickled product. *Int. J. Biol. Macromol.* 81: 387–395. 10.1016/j.ijbiomac.2015.08.020

Kim, J.H., Lee, Y.H., Jang, M.J., Won, S.Y., Joo, Y. 2014. Selection of superior strains from collected ear mushrooms for artificial cultivation and their optimal condition of mycelial growth. *Kor. J. Mycol.* 42(1): 57–63. 10.4489/KJM.2014.42.1.57

Li, J., Li, Z., Zhao, T., Yan, X., Pang, Q. 2021. Proteomic analysis of *Auricularia auricula-judae* under freezing treatment revealed proteins and pathways associated with melanin reduction. *Front. Microbiol.* 11: 610173. 10.3389/fmicb.2020.610173

Li, L., Zhong, C., Bian, Y. 2014. The molecular diversity analysis of *Auricularia auricula-judae* in China by nuclear ribosomal DNA intergenic spacer. *Electron. J. Biotechnol.* 17(1): 27–33. 10.1016/j.ejbt.2013.12.005

Liu, E., Ji, Y., Zhang, F., Liu, B., Meng, X. 2021. Review on *Auricularia auricula-judae* as a functional food: Growth, chemical composition, and biological activities. *J. Agric. Food Chem.* 69(6): 1739–1750. 10.1021/acs.jafc.0c05934

Looney, B. 2013. Systematics of the genus *Auricularia* with an emphasis on species from the southeastern United States. *North American Fungi* 8(6): 1–25. 10.2509/naf2013.008.006

Lu, A., Yu, M., Shen, M., Fang, Z., Xu, Y., Wang, S., Zhang, Y., Wang, W. 2018. Antioxidant and anti-diabetic effects of *Auricularia auricular* polysaccharides and their degradation by artificial gastrointestinal digestion—Bioactivity of *Auricularia auricular* polysaccharides and their hydrolysates. *Acta Sci. Pol. Technol. Aliment.* 17(3): 277–288. 10.17306/J.AFS.0557

Luo, Y., Chen, G., Li, B., Ji, B., Guo, Y., Tian, F. 2009. Evaluation of antioxidative and hypolipidemic properties of a novel functional diet formulation of *Auricularia auricula* and Hawthorn. *Innov. Food Sci. Emerg. Technol.* 10(2): 215–221. 10.1016/j.ifset.2008.06.004

Ma, Z., Wang, J., Zhang, L. 2008. Structure and chain conformation of β-glucan isolated from *Auricularia auricula-judae*. *Biopolymers* 89(7):614–622. 10.1002/bip.20971

Ma, Z., Wang, J., Zhang, L., Zhang, Y., Ding, K. 2010. Evaluation of water soluble β-d-glucan from *Auricularia auricular-judae* as potential anti-tumor agent. *Carbohydr. Polym.* 80(3): 977–983. 10.1016/j.carbpol.2010.01.015

Malysheva, V.F., Bulakh, E.M. 2014. Contribution to the study of the genus *Auricularia* (*Auriculariales*, Basidiomycota) in Russia. *Novosti Sist. Nizsh. Rast* 48: 164–180.

Mapoung, S., Umsumarng, S., Semmarath, W., Arjsri, P., Thippraphan, P., Yodkeeree, S., Limtrakul Dejkriengkraiku, L.P. 2021. Skin wound-healing potential of polysaccharides from medicinal mushroom *Auricularia auricula-judae* (Bull.). *J. Fungi* 7(4): 247. 10.3390/jof7040247

Misaki, A., Kakuta, M. 1995. Kikurage (Tree-ear) and Shirokikurage (white Jelly-leaf): *Auricularia auricula* and *Tremella fuciformis*. *Food Rev. Int.* 11(1): 211–218. 10.1080/87559129509541035

Misaki, A., Kakuta, M., Sasaki, T., Tanka, M., Miyaji, I. 1981. Studies on interrelation of structure and antitumor effects of polysaccharide: Antitumor action of periodate modified, branched (1 -3)-p-d-glucan of *Auricularin auricrrla-judae*, and other polysaccharides containing (l-3)-glycosidic linkages. *Carbohydr. Res.* 92: 115–129.

Montoya-Alvarez, A.F., Hayakawa, H., Minamya, Y., Fukuda, T., López-Quintero, C.A., Franco-Molano, A.E. 2011. Phylogenetic relationships and review of the species of *Auricularia* (fungi: basidiomycetes). *Colombia* 33(1): 55–66.

Nadir, H.A., Ali, A.J., Salih, S.A. 2020. *Auricularia nigricans* (Auriculariaceae, Basidiomycota) is first introduced from Halabja Province, Iraq. *J. Fungus* 11(1): 68–74. 10.30708.mantar.588958

Nguyen, T.L., Wang, D., Hu, Y., Fan, Y., Wang, J., Abula, S., Guo, L., Zhang, J., Khakame, S.K., Dang, B.K. 2012. Immuno-enhancing activity of sulfated *Auricularia auricula* polysaccharides. *Carbohydr. Polym.* 89(4): 1117–1122. 10.1016/j.carbpol.2012.03.082

Novakovic, A., Karaman, M., Kaisarevic, S., Radusin, T., Beribaka, M., Ilic, N. 2016. *Auricularia auricula—Judae* (Bull.:Fr.) Wettst. 1885 cytotoxicity on breast cancer cell line (MCF 7). III International Congress. *'Food Qual. Saf'* 112–115.

Odamtten, G.T., Addo, J., Wiafe-Kwagyan, M. 2021. Record of medicinal Jew's (*Auricularia auricula-judae* (Bull.) Quél) ear mushroom growing in the Greater Accra Region, Ghana and its possible health values. *Ghana J. Sci.* 62(2), 25–35. 10.4314/gjs.v62i2.3

Oke, F., Aslim, B. 2011. Protective effect of two edible mushrooms against oxidative cell damage and their phenolic composition. *Food Chem.* 128(3): 613–619. 10.1016/j.foodchem.2011.03.036

Oli, A.N., Edeh, P.A., Al-Mosawi, R.M., Mbachu, N.A., Al-Dahmoshi, H.O.M., Al-Khafaji, N.S.K., Ekuma, U.O., Okezie, U.M., Saki, M. 2020. Evaluation of the phytoconstituents of *Auricularia auricula-judae* mushroom and antimicrobial activity of its protein extract. *Eur. J. Integr. Med.* 38: 101176. 10.1016/j.eujim.2020.101176

Onyango, B.O., Otieno, C.A., Palapala, V.A. 2013. Effect of wheat bran supplementation with fresh and composted agricultural wastes on the growth of Kenyan native wood ear mushrooms [*Auricularia auricula* (L. ex Hook.) Underw.]. *Afr. J. Biotechnol* 12(19): 2692–2698. 10.5897/AJB2011.4163

Packialakshmi, B., Sudha, G., Charumathy, M. 2017. Studies on phytochemical compounds and antioxidant potential of *Auricularia auricula-judae*. *Int. J. Pharm. Sci* 8(8): 508–3515. 10.13040/IJPSR.0975-8232.8(8).3508-15

Pak, S., Chen, F., Ma, L., Hu, X., Ji, J. 2021. Functional perspective of black fungi (*Auricularia auricula*): Major bioactive components, health benefits and potential mechanisms. *Trends Food Sci. Technol.* 114: 245–261. 10.1016/j.tifs.2021.05.013

Perera, N., Yang, F.L., Chern, J., Chiu, H.W., Hsieh, C.Y., Li, L.H., Zhang, Y.L., Hua, K.F., Wu, S.H. 2018. Carboxylic and *O*-acetyl moieties are essential for the immunostimulatory activity of glucuronoxylomannan: A novel TLR4 specific immunostimulator from *Auricularia auricula-judae*. *ChemComm* 54(51): 6995–6998. 10.1039/C7CC09927D

Perera, P.K., Li, Y. 2011. Mushrooms as a functional food mediator in Preventing and ameliorating diabetes. *J. funct. food health dis* 1(4): 161. 10.31989/ffhd.v1i4.133

Prados-Rosales, R., Toriola, S., Nakouzi, A., Chatterjee, S., Stark, R., Gerfen, G., Tumpowsky, P., Dadachova, E., Casadevall, A. 2015. Structural characterization of melanin pigments from commercial preparations of the edible mushroom *Auricularia auricula*. *J. Agric. Food Chem.* 63(33): 7326–7332. 10.1021/acs.jafc.5b02713

Priya, R.U., Geetha, D., Darshan, S. 2016. Biology and cultivation of black ear mushroom. *Auricularia spp* 5(22): 10252–10254.

Qiu, J., Zhang, H., Wang, Z. 2019. Ultrasonic degradation of polysaccharides from *Auricularia auricula* and the antioxidant activity of their degradation products. *Lebensm Wiss Technol.* 113: 108266. 10.1016/j.lwt.2019.108266

Reza, A., Choi, M.J., Damte, D., Jo, W.S., Lee, S.J., Lee, J.S., Park, S.C. 2011. Comparative antitumor activity of different solvent fractions from an *Auricularia auricula-judae* ethanol extract in P388D1 and Sarcoma 180 Cells. *Toxicol. Res.* 27(2): 77–83. 10.5487/TR.2011.27.2.077

Reza, M.A., Hossain, Md. A., Damte, D., Jo, W., Hsu, W.H., Park, S.C. 2015. Hypolipidemic and hepatic steatosis preventing activities of the wood ear medicinal mushroom *Auricularia auricula-judea* (Higher basidiomycetes) ethanol extract in vivo and in vitro. *Int. J. Med. Mushrooms* 17(8): 723–734.

Reza, Md. A., Hossain, Md. A., Lee, S.J., Yohannes, S.B., Damte, D., Rhee, M., Jo, W.S., Suh, J.W., Park, S.C. 2014. Dichlormethane extract of the jelly ear mushroom *Auricularia auricula-judae* (Higher Basidiomycetes) inhibits tumor cell growth in vitro. *Int. J. Med. Mushrooms* 16(1): 37–47. 10.1615/IntJMedMushr.v16.i1.40

Reza, Md. A., Jo, W.S., Park, S.C. 2012. Comparative antitumor activity of jelly ear culinary-medicinal mushroom, *Auricularia auricula-judae* (Bull.) J. Schrot. (Higher Basidiomycetes) extracts against tumor cells in vitro. *Int. J. Med. Mushrooms* 14(4): 403–409. 10.1615/IntJMedMushr.v14.i4.80

Sánchez Vázquez, J.E., Mata, G., Royse, D.J. ECOSUR (Institution: Mexico) (Eds.). 2018. Updates on tropical mushrooms: *Basic Appl. Res.* (1st. edition). ECOSUR.

Sękara, A. 2015. *Auricularia* spp. – Mushrooms as novel food and therapeutic agents – A review. *Sydowia* 67: 1–10. 10.12905/0380.sydowia67-2015-0001.

Sone, Y., Kakuta, M., Misaki, A. 1978. Isolation and characterization of polysaccharides of "Kikurage," Fruit body of *Auricularia auricula-judae*. *Agric. Biol. Chem.* 42(2): 417–425. 10.1080/00021369.1978.10862990

Sun, S., Zhang, X., Chen, W., Zhang, L., Zhu, H. 2016. Production of natural edible melanin by *Auricularia auricula* and its physicochemical properties. *Food Chem.* 196: 486–492. 10.1016/j.foodchem.2015.09.069

Ukai, S., Kiho, T., Hara, C., Kuruma, I., Tanaka, Y. 1983. Polysaccharides in fungi. XIV. Anti-inflammatory effects of the polysaccharides from fruit bodies of several fungi. *J. Pharm. Dyn.* 6: 983–990.

Verma P. R.K. 2017. Diversity of macro-fungi in central India-IV: *Auricularia auricula-judae*, a neutraceutical jelly mushroom. *Van Sangyan* 4: 23–31.

Wang, X.F., Li, Q.Z., Bao, T.W., Cong, W.R., Song, W.X., Zhou, X.W. 2013. In vitro rapid evolution of fungal immunomodulatory proteins by DNA family shuffling. *Appl. Microbiol. Biotechnol.* 97(6): 2455–2465. 10.1007/s00253-012-4131-z

Wu, F., Yuan, Y., He, S.H., Bandara, A.R., Hyde, K.D., Malysheva, V.F., Li, D.W., Dai, Y.C. 2015. Global diversity and taxonomy of the *Auricularia auricula-judae* complex (Auriculariales, Basidiomycota). *Mycol. Prog.* 14(10): 95. 10.1007/s11557-015-1113-4

Wu, F., Yuan, Y., Malysheva, V.F., Du, P., Dai, Y.C. 2014. Species clarification of the most important and cultivated *Auricularia mushroom* "Heimuer": Evidence from morphological and molecular data. *Phytotaxa* 186(5): 241. 10.11646/phytotaxa.186.5.1

Xia, Y.G., Yu, L.S., Liang, J., Yang, B.Y., Kuang, H.X. 2019. Chromatography and mass spectrometry-based approaches for perception of polysaccharides in wild and cultured fruit bodies of *Auricularia auricular-judae*. *Int. J. Biol. Macromol.* 37: 1232–1244. 10.1016/j.ijbiomac.2019.06.176

Xu, S., Xu, X., Zhang, L. 2012. Branching structure and chain conformation of water-soluble glucan extracted from *Auricularia auricula-judae*. *J. Agric. Food Chem.* 60(13): 3498–3506. 10.1021/jf300423z

Xu, Y., Shen, M., Chen, Y., Lou, Y., Luo, R., Chen, J., Zhang, Y., Li, J., Wang, W. 2018. Optimization of the polysaccharide hydrolysate from *Auricularia auricula* with antioxidant activity by response surface methodology. *Int. J. Biol. Macromol.* 113: 543–549. 10.1016/j.ijbiomac.2018.02.059

Yang, L., Zhao, T., Wei, H., Zhang, M., Zou, Y., Mao, G., Wu, X. 2011. Carboxymethylation of polysaccharides from *Auricularia auricula* and their antioxidant activities in vitro. *Int. J. Biol. Macromol.* 49(5): 1124–1130. 10.1016/j.ijbiomac.2011.09.011

Yao, F.-J., Lu, L.-X., Wang, P., Fang, M., Zhang, Y.-M., Chen, Y., Zhang, W.-T., Kong, X.-H., Lu, J., Honda, Y. 2018. Development of a molecular marker for fruiting body pattern in *Auricularia auricula-judae*. *Mycobiology* 46(1): 72–78. 10.1080/12298093.2018.1454004

Yao, H., Liu, Y., Ma, Z.F., Zhang, H., Fu, T., Li, Z., Li, Y., Hu, W., Han, S., Zhao, F., Wu, H., Zhang, X. 2019. Analysis of nutritional quality of black fungus cultivated with corn stalks. *J. Food Qual.* 2019: 1–5. 10.1155/2019/9590251

Yoon, S.J., Yu, M.A., Pyun, Y.R., Hwang, J.K., Chu, D.C., Juneja, L.R., Mourao, P.A.S. 2003. The nontoxic mushroom *Auricularia auricula* contains a polysaccharide with anticoagulant activity mediated by antithrombin. *Thromb. Res.* 112(3): 151–158. doi:10.1016/j.thromres.2003.10.022

Yu, Y.J., Choi, K.H., Jeong, J.S., Lee, K.K., Choi, S.R. 2013. Study on characteristic of mycelial culture in ear mushroom. *J. Mushrooms* 11(1): 15–20. 10.14480/JM.2013.11.1.015

Yuan, B., Zhao, L., Yang, W., McClements, D.J., Hu, Q. 2017. Enrichment of bread with nutraceutical-rich mushrooms: Impact of *Auricularia auricula* (Mushroom) flour upon quality attributes of wheat dough and bread: enrichment of bread with mushrooms. *J. Food Sci.* 82(9): 2041–2050. 10.1111/1750-3841.13812

Yuan, Z., He, P., Cui, J., Takeuchi, H. 1998. Hypoglycemic effects of water-soluble polysaccharides from *Auricularia auricula-judae* Quel. On geneically diabetic KK.A y Mice. *Biosci. Biotechnol. Biochem.* 62(10): 1898–1903.

Zeng, F., Zhao, C., Pang, J., Lin, Z., Huang, Y., Liu, B. 2013. Chemical properties of a polysaccharide purified from solid-state fermentation of *Auricularia Auricular* and its biological activity as a hypolipidemic agent: Polysaccharide of *Auricularia auricular*. *J. Food Sci.* 78(9): 1470–1475. 10.1111/1750-3841.12226

Zeng, W.C., Zhang, Z., Gao, H., Jia, L.R., Chen, W.Y. 2012. Characterization of antioxidant polysaccharides from *Auricularia auricular* using microwave-assisted extraction. *Carbohydr. Polym.* 89(2): 694–700. 10.1016/j.carbpol.2012.03.078

Zhang, H., Wang, Z.Y., Zhang, Z., Wang, X. 2010. Purified *Auricularia auricular-judae* polysaccharide (AAP I-a) prevents oxidative stress in an ageing mouse model. *Carbohydr. Polym.* 84(2011): 638–648. 10.1016/j.carbpol.2010.12.044

Zhang, H., Wang, Z.Y., Zhang, Z., Wang, X. 2011. Purified *Auricularia auricular-judae* polysaccharide (AAP I-a) prevents oxidative stress in an ageing mouse model. *Carbohydr. Polym.* 84(1): 638–648. 10.1016/j.carbpol.2010.12.044

Zhang, Y., Zeng, Y., Men, Y., Zhang, J., Liu, H., Sun, Y. 2018. Structural characterization and immunomodulatory activity of exopolysaccharides from submerged culture of *Auricularia auricula-judae*. *Int. J. Biol. Macromol.* 115: 978–984. 10.1016/j.ijbiomac.2018.04.145

Zhao, Y., Wang, L., Zhang, D., Li, R., Cheng, T., Zhang, Y., Liu, X., Wong, G., Tang, Y., Wang, H., Gao, S. 2019. Comparative transcriptome analysis reveals relationship of three major domesticated varieties of *Auricularia auricula-judae*. *Sci. Rep.* 9(1): 78. 10.1038/s41598-018-36984-y

Zhou, R. Bao, H., Kang, Y.H. 2017. Synergistic rheological behavior and morphology of yam starch and *Auricularia auricula-judae* polysaccharide-composite gels under processing conditions. *Food Sci. Biotechnol.* 26(4): 883–891. 10.1007/s10068-017-0122-2

Zhuan-Yun, L., Xue-Ping, Y., Bin, L., Reheman, H.N., Yang, G., Zhan, S., Qi, M. 2015. *Auricularia auricular-judae* polysaccharide attenuates lipopolysaccharide-induced acute lung injury by inhibiting oxidative stress and inflammation. *Biomed. Rep.* 3(4): 478–482. 10.3892/br.2015.470

Zou, Y., Xie, C., Fan, G., Gu, Z., Han, Y. 2010. Optimization of ultrasound-assisted extraction of melanin from *Auricularia auricula* fruit bodies. *Innov. Food Sci. Emerg. T.* 11(4): 611–615. 10.1016/j.ifset.2010.07.002

Zou, Y., Yang, Y., Zeng, B., Gu, Z., Han, Y. 2013. Comparison of physicochemical properties and antioxidant activities of melanins from fruit-bodies and fermentation broths of *Auricularia auricula*. *Int. J. Food Prop.* 16(4): 803–813. 10.1080/10942912.2011.567433

Zou, Y., Zhao, Y., Hu, W. 2015. Chemical composition and radical scavenging activity of melanin from *Auricularia auricula* fruiting bodies. *Food Sci. Technol.* 35(2): 253–258. 10.1590/1678-457X.6482

3 *Boletus edulis Bull., Boletus bicolor*

Gurpreet Kaur
Department of Zoology, Mata Gujri College, Fatehgarh Sahib, India

Kamalpreet Kaur
Department of Chemistry, Mata Gujri College, Fatehgarh Sahib, India

CONTENTS

INTRODUCTION

Edible mushrooms have long been a source of wealth for ethnic tribes from many communities. Mushrooms are popular throughout the world due to their delicacy, particularly their distinct scent, texture, and low-fat content. Mushrooms are regarded to be extremely nutritious and therapeutic. Mushrooms are considered invaluable resources for the production of several key biomolecules with commercial uses in the food and pharmaceutical industry (Kalac, 2016; Borthakur and Joshi, 2019). Additionally, mushrooms are an excellent source of food for combating malnutrition in impoverished countries. This low-priced edible mushroom is not only high in nutrients such as vitamin D, however, it also exerts enormous health-promoting effects such as anti-neoplastic, anti-HIV, and anti-diabetic. It is an ecologically friendly, cost-effective crop that can be cultivated anywhere in the world and at any time of year using low-priced starting ingredients (Mahajan, 2022). The purpose of the chapter is to explore the bioactive

DOI: 10.1201/9781003259763-3

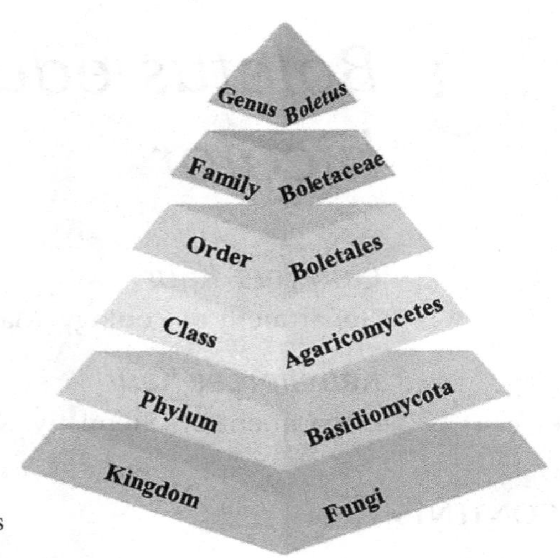

FIGURE 3.1 Taxonomic classification of the genus *Boletus*.

constituents, nutritional, and therapeutic potential of *Boletus edulis* and *Boletus bicolor*, as well as their uses as functional foods or a source of medicine. The mushroom belonging to the genus *Boletus* (Figure 3.1) is a popular edible wild mushroom for human consumption all over the World. Phytochemical studies demonstrated that species belonging to *the Boletus* genus contained polysaccharides, phenolic compounds, flavonoids, anthocyanins, and lectins. *B. edulis* is commonly known as porcini and king bolete and contained a low amount of fat and more quantity of carbohydrates, proteins, fibers, amino acids, and minerals (Zhang *et al.*, 2011). *B. edulis* is a rich source of zinc, copper, manganese, iron, and several vitamins like ascorbic acid, vitamin B, and tocopherols (Balta *et al.*, 2019). *B. edulis* can be used as an additive or flavoring agent due to its odor or taste and is also utilized in the meat industry for extending the shelf life of frankfurter products (Novakovic, 2021). Numerous bioactivities from *B. edulis* mushrooms have been described such as antioxidant, anti-cancer, anti-microbial, anti-hyperglycaemic, anti-inflammatory, and antihypertensive (Zhang *et al.*, 2018). *Boletus bicolor* is known by the name colored bolete and is extensively utilized traditionally for the cure of numerous human ailments like malnutrition, gastric ulcers, and hypercholesterolemia. Various studies demonstrated that this mushroom species exhibited antioxidant, anticancer, and anti-microbial activity as well as revealed negligible cytotoxicity against the normal cells (Tibuhwa, 2017). Taxonomic classification of *Boletus* is presented in figure 3.1.

TRADITIONAL USES

B. edulis is utilized in traditional medicine for the cure of low back and leg pain, paresthesia in limbs, and tendon soreness. *B. bicolor* has robust antioxidant activity, which could be linked with its traditional uses. It is traditionally utilized for the cure of several ailments such as malnutrition, intestinal ulcers, and hypercholesterolemia, and enhances the immunity in long-ill people and breast-feeding mothers. Mostly *Boletus* is cooked or fried and acquired as a stew in the belief that it boosts the immune system. Among the phytoconstituents, phenolic compounds, beta carotenes, and vitamins were observed to be very high in the *Boletus*. These active components caused the induction of the immune system against cancer and gap junctional communication. The bioactivities are responsible for the traditional medicinal uses by ethnic communities (Grabmann, 2005; Tibuhwa, 2017).

DISTRIBUTION AND MORPHOLOGY

BOLETUS EDULIS

B. edulis is distributed worldwide and is mainly found in cold temperate and subtropical areas. Usually, it finds its existence in Europe extending from Northern Scandinavia, Greece, Italy, Morocco, North America, Mexico, extending from Northeastern areas of Heilongjiang to Yunnan Province in China, Sagarmatha National Park in Nepal, Arunachal Pradesh in India and Northwest Iran (Wang *et al.*, 1995; Hall *et al.*, 1998; Adhikary *et al.*, 1999; Cui *et al.*, 2016). Morphologically, *Boletus edulis* has a sticky, yellow-brown/red-brown cap having a diameter of 7–22 cm, which has a convex or flat shape and is lighter than the remaining part at the edges. The stipe is fleshy, white/ light brown in colour, has a length of 7 to 18 cm and is thick but swollen in the middle. The tubes are white in the beginning but become green-yellow on maturity and have 1–2 cm of depth. Spores of *B. edulis* are elliptical/spindle-shaped and have dimensions of 6×16 micrometres (Kalač, 2009; Argyropoulos *et al.*, 2011; Tapwal *et al.*, 2021).

BOLETUS BICOLOR

B. bicolor is mainly distributed in the Great Lakes area of Southeast Canada, Florida peninsular region, Wisconsin state of Midwestern United States and some parts of Nepal and China. It finds its presence in deciduous woods and exists either in isolation or in clusters. Mostly it grows beneath or in proximity to trees bearing broad leaves like oak. Its growing season is from June to October. Cap of *B. bicolor* is sticky, convex or flat in shape, has a diameter of 5 to 6 cm and is reddish brown/brick red in colour. Red/ purplish red stipe of *B. bicolor* is fleshy, 3 to 4 cm in length and it tapers near the top. Spores have sub-fusiform and smooth surfaces, have a size of 10×6 micrometres and the cytoplasm is yellowish brown with a germ pore which is open and convex in shape. Body size of the fruit is 8 to 15 cm and is hemispherical, campanulate and glabrous (Christensen *et al.*, 2008; Nguyen *et al.*, 2022).

PHYTOCHEMISTRY OF *BOLETUS EDULIS*

Mushrooms contain plentiful primary metabolites and secondary metabolites including polysaccharides, fibres, proteins and minerals (Manzi *et al.*, 1999; Kalač, 2009). The most important bioactive constituents of mushrooms are polysaccharides and their protein complexes which have excellent immune-regulatory properties and are used as anti-inflammatory, anti-tumour and antioxidant agents (Ruthes *et al.*, 2015). A number of primary metabolites and secondary metabolites which are biologically active phytochemicals are present in *B. edulis* like organic acids, polysaccharides, phenolic compounds, vitamin D, vitamin C, minerals, various tocopherols and proteins (Dentinger *et al.*, 2010).

PRIMARY METABOLITES PRESENT IN *B. EDULIS*

Carbohydrates and proteins are the most predominant primary metabolites of *B. edulis*. Mushrooms have two fruiting bodies viz. cap and stipe (Table 3.1). The nutritional significance of mushrooms is governed by the dry matter/moisture content ratio (Mattila *et al.*, 2002). The range of the moisture content on fresh weight (FW) basis was 85.56- 89.15% and that on dry weight (DW) basis fell in the range of 6.56–28.99%. The ash contents ranged from 1.15–22.06% on DW basis. The ashes are made up of minerals such as P, K, Mg and S and the variation in ash content could be because of the type and mineral content of the soil (Kalač, 2009). The protein and carbohydrate contents investigated by various researchers were between 7.39–36.24% and 9.23–74.70% on a dry weight (DW) basis. *B. edulis* was investigated to be extremely low in fats

TABLE 3.1
Primary Metabolites Present in *B. edulis*

S. No.	*B. edulis* cap/stipe	Moisture	Ash (% DW)	Proteins (% DW)	Fat (% DW)	Carbohydrates (% DW)	References
1	Cap	ND	5.3	27	2.8	65	Ouzouni and Riganakos, 2007
2	Cap	89% FW	5.5	21.1	2.5	70	Heleno *et al.*, 2011
3	Cap	ND	7.1	17.2	4.6	71	Barros *et al.*, 2008
4	Stipe	07.8% DW	6.2	17.9	1.1	74	García *et al.*, 2021
5	Cap	6.6% DW	7.8	31.5	1.6	59	García *et al.*, 2021
6	Cap	85% FW	1.2	7.4	1.7	9	Çaglarlrmak et al., 2002
7	Cap	7% DW	8.4	36.2	1.9	46	Fogarasi *et al.*, 2018
8	Cap	11.9% DW	5.8	18.5	5.7	56	S.-Y. Tsai *et al.*, 2008
9	Cap	ND	6.4	31	ND	ND	Lalotra *et al.*, 2018
10	Cap	29% DW	22.1	5.2	7.4	25	Gyar *et al.*, 2011

and the fat content of the cap of the mushroom was in the range of 1.64–7.44% DW basis. The fibre content of the mushroom had been investigated by very few researchers and it was found to be in the range of 4.3–13.70%. *B. edulis* acts as a low-energy food and its energetic value per 100 g DW ranges from 347.5–399 kcal. García *et al.*, 2021 studied the influence of various culinary actions such as grill, roast, marinate and vacuum confit on the nutritional profile of cap and stipe of *B. edulis*. The protein content of raw cap and raw stipe differed significantly. 75% more protein content was found in the raw cap in contrast to the raw stipe (17.98 g of lyophilized *B. edulis* stipe sample in contrast to 31.46 g of lyophilized cap sample per 100 g DW). The cap was found to have a considerably lower carbohydrate amount than that contained in the stipe (59.10 and 74.70 g per 100 g DW of lyophilized samples respectively). Similarly, the moisture content of the stipe (7.87g/100 g DW) was slightly higher than that of the cap (6.56 g/100 g DW). But the lipid and ash content of the cap (1.64 and 7.80 g per 100 g DW, respectively) was more than that present in the stipe (1.13 and 6.18 g per 100 g DW, respectively, of the lyophilized samples).

The impact of different culinary treatments on primary metabolites was also studied (Figure 3.2). Irrespective of the cooking treatments, metabolites were reduced; marination caused the smallest difference whereas confit produced the largest reduction in the primary metabolites probably due to the absorption of oils during the treatment. Fernandes *et al.* (2015) investigated the effect of irradiation of electron beams on fibrous composition of dried samples of *B. edulis*. Non-irradiated samples of *B. edulis* contained soluble and insoluble fibres present in 7.4 g/100 g DW and 21.0 g/100 g DW amounts. Significant decrease in the number of insoluble fibres (7.7 g per 100 g DW) and total fibre content was detected when 10 kGy strength of electron beam was used for irradiation. Moreover, no remarkable change was observed in the content of soluble fibre.

Secondary Metabolites

Polysaccharides

Polysaccharides are the predominant component of mushrooms and are known for their brilliant anti-oxidant nature (Siu *et al.*, 2016; Zhang *et al.*, 2018). But the bioactive potential of polysaccharides depends upon the composition of monosaccharides, degrees of sulfation and

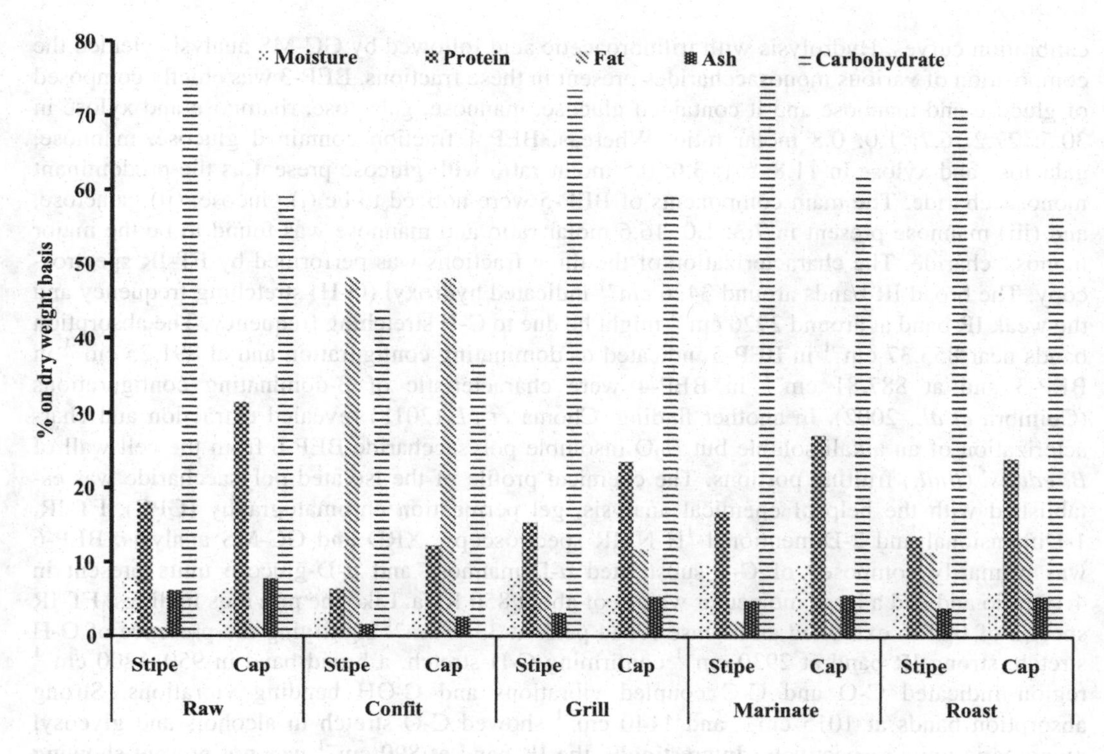

FIGURE 3.2 Effect of culinary techniques on primary metabolites.

branching and distribution of molecular weight (Sun *et al.*, 2009). The bioactive potential depends upon the ratio of different monosaccharides present in the polysaccharide (Lo *et al.*, 2011). Furthermore, it has been noticed that polysaccharides with smaller molecular masses showed outstanding immunomodulatory effects (Sun *et al.*, 2012). Meng *et al.*, 2021 isolated a water-soluble polysaccharide (BEP-1) having a large molecular weight of about 6.0×10^6 Da from *Boletus edulis* by employing cold water method. Most of the researchers used hot water for the extraction of neutral polysaccharides from *B. edulis*. The predominant components of the extracted polysaccharide (BEP) were sugar and uronic acid present in 92.20 and 12.62%, respectively. Additionally, traces of nucleic acids and proteins were also detected. The acid polysaccharide nature of BEP was confirmed by Fourier Transform-Infrared (FT-IR) spectral studies. The spectra showed strong IR peaks at 3397.31 cm^{-1} and 2926.10 cm^{-1} due to O-H bond and C-H stretching vibrations. Whereas, IR peaks at 1643.69 and 1412.56 cm^{-1} displayed carbonyl stretching of the carboxyl group indicating presence of uronic acid moiety in BEP-1. The occurrence of pyranose rings was confirmed by IR peaks at 1135.91 and 1053 cm^{-1} (Zhang *et al.*, 2018). Moreover, α- and β-glycosidic bonds present in BEP-1 were confirmed by IR bands at 917.25 and 826.96 cm^{-1} (Kong *et al.*, 2015). The monosaccharide composition as elucidated by ion chromatography (IC) indicated the presence of (i) galactose, (ii) glucose, (iii) xylose, (iv) mannose, (v) glucuronic acid, and (vi) galacturonic acid in BEP-1 existing in 0.34:0.28: 0.28:2.57:1.00:0.44 ratio. Luo *et al.* (2012) attempted extraction followed by purification and then characterization of a water-soluble polysaccharide (BEP-2) from *B. edulis Bull.* Hot H_2O was utilised for the extraction process and ethanol was used for precipitation. BEP-2 was subsequently purified by employing Ion exchange chromatography. Three fractions namely BEP-3, BEP-4 and BEP-5 were obtained from diethylaminoethyl (DEAE)-Cellulose column eluted with H_2O, 0.1 M and 0.3 M sodium chloride solutions respectively. The molecular weights of the fractions BEP-3 (25.0 KDa), BEP-4 (9.6 KDa) and BEP-5 (7.3 KDa) were determined from the calibration curves. Hydrolysis with trifluoroacetic acid followed by GC-MS analysis yielded the composition of various monosaccharides present in these fractions. BEP-3 was chiefly composed of glucose and mannose and it contained glucose, mannose, galactose, rhamnose and xylose in 30.5: 27.2: 6.7: 1.0: 0.8 molar ratio. Whereas, BEP-4 fraction contained glucose, mannose, galactose and xylose in 11.8: 5.1: 3.6: 0.5 molar ratio with glucose present as the predominant monosaccharide. The main components of BEP-5 were noticed to be (i) glucose, (ii) galactose, and (iii) mannose present in 7.3: 1.0: 16.6 molar ratio and mannose was found to be the major monosaccharide. The characterization of the three fractions was performed by FT-IR spectroscopy. The broad IR bands around 3426 cm^{-1} indicated hydroxyl (O-H) stretching frequency and the weak IR band at around 2926 cm^{-1} might be due to C-H stretching frequency. The absorption bands near 853.37 cm^{-1} in BEP-5 indicated α- dominating configuration and at 891.25 cm^{-1} in BEP-3 and at 887.81 cm^{-1} in BEP-4 were characteristic of β-dominating configurations (Coimbra *et al.*, 2002). In another finding, Choma *et al.* (2018) revealed extraction and characterization of an alkali soluble but H_2O insoluble polysaccharide BEP-6 from the cell wall of *B. edulis*. (*Bull.*) fruiting portions. The chemical profile of the isolated polysaccharide was established with the help of chemical analysis, gel permeation chromatography (GPC), FT-IR, 1-Dimensional and 2-Dimentional ^{1}H NMR spectroscopy, XRD and GC-MS analysis. BEP-6 was primarily composed of C-3 substituted α-D-mannose and α-D-glucose units present in 4:10 ratio and had a large molecular weight of about 850 KDa. Like the previous findings, FT-IR spectra of BEP-6 exhibited an intense broad peak at 3350 cm^{-1} signifying the presence of O-H stretch, strong IR band at 2930 cm^{-1} confirming C-H stretch, a broad band in 950–1200 cm^{-1} region indicated C-O and C-C coupled vibrations and C-OH bending vibrations. Strong absorption bands at 1015 cm^{-1} and 1140 cm^{-1} showed C-O stretch in alcohols and glycosyl monomeric units respectively. Interestingly, the IR band at 890 cm^{-1} was not present showing the absence of β-glucan units in BEP-6. And the presence of α-glycosidic linkages in the polymeric chain was confirmed by the presence of IR bands at 820, 840 and 930 cm^{-1}. In order to establish the anomeric configuration of sugars (glucose, mannose) present in BEP-6, various spectroscopic experiments comprising of ^{I}H NMR (ID and 2D) like COSY, NOSEY and TOCSY and ^{1}H/^{13}C HMBC and HSQC were done. Two spin systems belonging to α-configuration of D-glucopyranose and D-mannopyranose were identified. Furthermore, a polymer with high molecular weight composed of glucose and mannose chains connected via (1→3) glycosidic bonds was established. The polymer contained an insignificant number of xylans in it. The XRD analysis of BEP-6 indicated well-defined diffraction peaks at 18.3° and 21.5° 2Θ values and two low intensity broad bands at 10° and 40° 2Θ values. This demonstrates that the structure of α-(1→3)-D-glucomannans isolated from the mushroom resembled that of cellulose I and was composed of crystalline and some partly amorphous regions. Moreover, the polysaccharide exhibited remarkable affinity for biosorption of some heavy metal ions especially lead (up to 96%) and cadmium (up to 79%) from water solution by making weak interactions with polymeric dipoles (O-H or C-O) or through metal ion-induced dipolar interactions. Zhang *et al.*, 2018 also noticed that the composition of the monosaccharides varied significantly in the two fruiting bodies (caps and stipes) of *B. edulis* as well as the samples of mushrooms obtained from diverse geographies and climates. The polysaccharides isolated from *B. edulis* contained (i) arabinose, (ii) mannose, (iii) xylose, (iv) galactose, (v) glucose, and (vi) rhamnose monosaccharides; and glucose was found to be the predominant monosaccharide present in the caps of *B. edulis* mushrooms collected from Qingchuan, Guangyuan, Sichuan region; Nanhua, Chuxiong, Yunnan region; and Pingwu, Mianyang, Sichuan region of China containing 40.1%, 50.28% and 39.00% of glucose, respectively. Moreover, mannose was the most abundant monosaccharide (53.10 %) present in the cap of *B. edulis* collected from Luquan, Kunming, and Yunnan regions of China. The stipes of *B. edulis* obtained from Nanhua, Chuxiong, Yunnan region and Luquan, Kunming, Yunnan areas of China had the highest glucose content (61.04–63.34%).

Organic Acids

Organic acids perform a significant role in deciding the organoleptic characteristics of vegetables and fruits (Vaugham and Geissler, 1997). Acids have antioxidant potential and act as protecting agents against many diseases and do not alter their properties during storage and processing unlike pigments and other flavouring agents (Silva *et al.*, 2004; Cámara *et al.*, 1994). Valentão *et al.* (2005) identified and quantified various organic acids present in various wild mushrooms belonging to *Boletus edulis*, *Xerocomus chrysenteron*, *Amanita caesarea*, *Lactarius deliciosus*, *Gyroporus castaneus* and *Suillus collinitus* species by employing HPLC-UV detection. All the mushroom species contained a minimum of five organic acids including (i) malic acid, (ii) citric acid, (iii) succinic acid, (iv) ketoglutaric acid, and (v) fumaric acid. Malic acid along with citric-ketoglutaric acid pair was detected as the predominant fraction except in *A. Caesarea* which contained ascorbic acid along with malic acid. *B. edulis* contained malic acid-quinic acid pairs in 66% to 91% amounts, along with oxalic, fumaric and citric-ketoglutaric acid pairs whereas shikimic and ascorbic acid were not detected. Likewise, Ribeiro *et al.* (2006) attempted quantification of various organic acids and phenolic compounds present in wide varieties of wild but edible mushrooms including *Boletus edulis*, *Russula cyanoxantha*, *Suillus bellini*, *Amanita rubescens*, *Hygrophorus agathosmus*, *Tricholoma equestre*, *Suillus granulatus*, *Tricholomopsis rutilans*, and *Suillus luteus* with the help of HPLC-DAD and HPLC-UV detection. Each species displayed a distinctive profile but all the species contained oxalic, citric and fumaric acids along with malic acid and quinic acid present as the predominant fractions making up to 35% to 84% of the analysed non-aromatic acids and citric acid representing approximately 30% of the non-aromatics. The main acids existing in *B. edulis* included (i) oxalic acid, (ii) malic acid, (iii) citric acid, (iv) fumaric acid, (v) quinic acid, and (vi) aconitic acid a newly reported compound present in 48–58 g/kg amount. The malic acid-quinic acid pair represented 71% of the aromatic acid and was the main component whereas oxalic acid was present in minimum amounts representing only 3 % of the non-aromatic acids. In comparison to the previous reports by Valentão *et al.*, 2005, ketoglutaric and succinic acids were not present whereas aconitic acid was reported for the first time. However, no phenolic compounds were reported in *B. edulis* species. Ribeiro *et al.* (2008) carried out comparative analysis of the phytochemicals (organic acids, total phenolic compounds and alkaloid content) existing in the caps and stipes of wild edible mushrooms belonging to *Russula cyanoxantha*, *Suillus granulatus*, *Amanita rubescens* and *Boletus edulis* species. Organic acid profile of *B. edulis* consisted of citric, succinic, malic, oxalic and fumaric acids. Moreover, the highest content of these acids prevailed in the caps while the stipe and the whole mushroom contained similar organic acid content. Quantitatively, malic acid was the main compound whereas fumaric acid and succinic acids were present in minimum amounts. Furthermore, higher concentrations of citric and malic acids were present in the stipe and cap of mushroom, respectively. *B. edulis* mushroom contained significant amounts of total alkaloids accumulated in the cap and/or stipe in comparison to other species which had more or less similar amounts of total alkaloids concentrated in their caps. No phenolic compounds were detected in *B. edulis* whereas *S. granulatus* and *A. rubescens* contained significant amounts of p-hydroxybenzoic acid accumulated in the cap and entire mushroom. Moreover, excellent antioxidant activity was displayed by *B. edulis* for DPPH-scavenging.

Phenolic Compounds

Another very important class of secondary metabolites is phenolic compounds which includes phenolic acids like hydroxybenzoic acid, hydroxycinnamic acid, hydroxyphenylacetic acid; flavonoids such as anthocyanins, flavones, Isoflavones, and non-flavonoid compounds. The differences in the type and amounts of various phenolic compounds obtained from different samples of *B. edulis* is because of the variable climatic and geographical conditions; methods of extraction; techniques used for quantification and analytical approach. The phenolic compounds exhibit remarkable antioxidant activity which depends upon the total phenolic content (TPC) present in the mushrooms. Palacios *et al.*, 2011 reported nine phenolic compounds including (i) caffeic acid, (ii) chlorogenic acid, (iii) *p*-coumaric acid, (iv) gallic acid, (v) gentisic acid, (vi) *p*-hydroxybenzoic acid, (vii) homogentisic acid, (viii) myricetin, and (ix) protocatechuic acid present in 15.09 mg, 62.79 mg, 0.87 mg, 212.96 mg, 60.85 mg, 24.07 mg, 2290.97 mg, 17.98 mg, and 168.46 mg, respectively per 1g of dried *B. edulis* mushroom. Özyürek *et al.* (2014) employed microwave-assisted extraction (MAE) method for extracting polyphenolics and other secondary metabolites from B. *edulis* mushrooms. Conditions of MAE were optimised as 80°C as the microwave temperature, 5 minutes as the extraction time from 80% methanolic solution to get desired composition and maximum extraction of the phenolic compounds. Total phenolic content (TPC) was observed to be 357.7 μmol Trolox equivalents (TR) per g of the methanolic extract of *B. edulis*. The polyphenols as analysed by ultra-performance liquid chromatography (UPLC) and ESI-MS/MS indicated presence of ten polyphenolic compounds including (i) gallic acid (0.19 μg/g of dried mushroom), (ii) vanillic acid (0.41 μg/g of dried mushroom), (iii) protocatechuic acid (0.21 μg/g of dried mushroom), (iv) hesperidin (0.77 μg/g of dried mushroom), (v) rutin (0.17 μg/g of dried mushroom), (vi) rosmarinic acid (1.71 μg/g of dried mushroom), (vii) naringenin (0.29 μg/g of dried mushroom), (viii) quercetin (0.07 μg/g of dried mushroom), (ix) kaempferol (0.28 μg/g of dried mushroom), and (x) apigenin (0.48 μg/g of dried mushroom) in the extract of *B. edulis*. Fogarasi *et al.* (2018) through HPLC-MS analysis found that *B. edulis* was enriched with phenolic compounds such as flavonoids and phenolic acids. The major phenolic compounds present in *B. edulis* per 100 g fresh weight included (i) 4-hydroxybenzoic acid (209.87 mg), (ii) 2,4-dihydroxybenzoic acid (69.13 mg), (iii) 4-hydroxyphenylacetic acid (25.30 mg), (iv) protocatechuic acid (43.58 mg), (v) *p*-coumaric acid (23.11 mg), (vi) sinapic acid (27.38 mg), (vii) *o*-coumaric acid (11.42 mg), (viii) cinnamic acid (168.61 mg), (ix) 3,5 dicaffeoylquinic acid (31.55 mg), (x) catechin (145.57 mg), and (xi) gallocatechin (26.63 mg). Sarikurkcu *et al.* (2008) analysed four edible mushrooms of Turkey region including *Boletus edulis*, *Lactarius deterrimus*, *Xerocomus chrysenteron* and *Suillus collitinus* and reported that *B. edulis* possessed the highest phenolic content of 31.64 μg gallic acid equivalents/mg of the extract whereas the total flavonoid content of *B. edulis* (0.458 μg quercetin equivalents/mg of extract) was smaller than that possessed by *Suillus collitinus* (0.510 μg quercetin equivalents/mg of extract). Li *et al.* (2020) estimated polyphenolic content present in various morphological parts viz. caps and stipes of *B. edulis* by using FT-MIR spectroscopy. The polyphenolic content in stipes (17.33 mg per g DW) was more than that found in the caps (14.76 mg per g DW) of *Boletus edulis* indicating that phenolic compounds accumulated more easily in stipes than in the caps. Jaworska *et al.*, 2014 studied the impact of air and freeze drying on the total polyphenolic and total flavonoid content of blanched and unblanched *B. edulis* mushrooms. Flavonoid content was significantly reduced on drying (2.8 mg/g DW unblanched *B. edulis* (2.9 mg/g DW)whereas the total polyphenolic content was increased on air drying from 17.6 mg/g DW found in unblanched mushroom to 20.6 mg/g of DW in air dried unblanched *B. edulis*. Whereas freeze drying caused a decrease in polyphenols in *B. edulis* (16.7 mg/g DW of unblanched mushroom). Moreover, storage further relegated the quality of *B. edulis*.

Vitamins and Their Precursors

Jaworska *et al.* (2014) reported that fresh *B. edulis* contained plenty of vitamin C (1.9 mg/g DW), β-carotene (1.23 mg/g DW) and total tocopherols (49.51 μg/g DW). Tocopherols were further composed of (i) α-tocopherol (65%), (ii) β-tocopherol (32%), (iii)Υ-tocopherol (2%), and (iv) δ-tocopherol (3%). Vitamin C content was significantly decreased in unblanched freeze dried (1.82 mg/g DW) and air dried (1.31 mg/g DW) *Boletus edulis* mushrooms. Air drying reduces vitamin C content more drastically than freeze drying. Blanching before air drying and freeze drying further reduced the contents of vitamins. Storage of dried *B. edulis* for a period of 12 months resulted in significant loss of phyto-nutrients including tocopherols and vitamin C and no traces of these nutrients could be detected when stored for 24 months irrespective of the storage conditions. Teichmann *et al.* (2007) investigated the existence of ergosterol as the major sterol present in *B. edulis* which acted as a precursor of Vitamin D_2 and on irradiation with ultraviolet light is converted into Vitamin D_2. The king bolete was found to be enriched with vitamin D_2 and its precursor ergosterol containing 10 µg and 192.2 mg respectively present in 100 g FW of these two. Heleno *et al.* (2011) quantified tocopherols and ascorbic acid present in *B. edulis*. Total tocopherol content present in *B. edulis* was estimated to be 5.80 µg/g DW and α-, Υ- and δ-tocopherols were present in 0.12, 5.17 and 0.51 µg/g DW amount, respectively. Interestingly, β-tocopherol was not detected. Ascorbic acid present in *B. edulis* was estimated to be 5.32 mg/g DW. Tsai *et al.* (2007) concluded that the antioxidant ability of *B. edulis* obtained from Taiwan was related to its tocopherol content. Both the hot water and the ethanolic extracts of *B. edulis* were plentiful of tocopherols. The ethanolic extract was composed of α, Υ and Δ tocopherols in 4.65, 0.75 and 0.78 mg/g of the extract whereas the hot water extract contained α and Υ tocopherols in 2.28 and 0.90 mg/g of the extract. Δ-tocopherol was not at all detected. The analysis of various nutraceuticals in *B. edulis* as performed by Barros *et al.*, 2008 with the help of HPLC fluorescence and other spectrophotometric methods revealed the presence of total tocopherols along with α, β and Υ-tocopherols in 10.65, 0.32, 8.90 and 1.42 µg/g of the mushroom respectively. In addition to this, β-carotene and lycopene were also present in 2.73 and 1.14 µg/g of *Boletus edulis* mushroom respectively.

Indole Compounds

Quantification of indole compounds by HPLC analysis of the methanolic extract of processed *B. edulis* was performed by Muszyńska *et al.*, 2012. The processed *B. edulis* contained three indole compounds namely L-tryptophan which acts as a precursor to indole structured neurotransmitters in 8.35 mg amount, indole acetonitrile in 0.73 mg amount and indole in 1.60 mg per 100 g DW of the mushroom. Thermal processing significantly influenced the composition as well as the content of the indole compounds in *B. edulis*. The comparative analysis of the chemical profile of thermally processed and unprocessed *B. edulis* indicated the absence of serotonin on thermal processing whereas a significant amount of serotonin in unprocessed *B. edulis* was observed to be 10.14 mg/100 g DW. 100 g DW of the unprocessed sample of *B. edulis* contained 0.18 mg of 5-hydroxytryptophan, 1.17 mg of tryptamine and 0.68 mg of melatonin which were not detected in processed sample. Interestingly, the amount of L-tryptophan was larger in the processed sample (8.35 mg/100 g DW) than that present in the unprocessed one (0.39 mg/100 g DW). The breakdown of the compounds at elevated temperatures was the major cause of the variation in the amounts and profile of the indole compounds in processed and unprocessed samples. Exactly similar results were reported by the same researchers in 2011 confirming the presence of (i) 5-hydroxytryptophan (0.18 mg), (ii) serotonin (10.14 mg), (iii) L-tryptophan (0.39 mg), (iv) melatonin (0.68 mg), and (v) tryptamine (1.17 mg) in the methanolic extract of 100 g DW of *B. edulis* (Muszyńska *et al.*, 2011).

Lectins

Lectins are non-immune proteins capable of binding selectively with carbohydrates without their enzymatic modification (Sharon, 2007). They are universal in nature and are present in all the living organisms. Sugar selectivity of lectins has been extensively used in sciences (Bovi *et al.*, 2013). Mushrooms are now recognized for the presence of lectins in them and their potential applications in glycobiology. Bovi *et al.*, 2013 purified a novel protein named *Boletus edulis* lectin (BEL) β-trefoil from the fruiting body of *B. edulis* mushroom which was a homodimer and each protomeric subunit of the protein had β-trefoil folding. The X-ray structure of BEL β-trefoil was established prior to the determination of sequence of amino acids in it. The protein exhibited strong anti-proliferative influence against human cancer cells. The apo-protein existed in different crystal forms but 1.12 Å resolution provided the crystal structure belonging to the $P2_1$ space group composed of asymmetric dimer. The lectin had three carbohydrate binding sites namely α, β and Υ whose binding with various carbohydrates like lactose, N-acetyl-galactosamine, galactose and T-antigen disaccharide was studied with the help of X-ray diffraction studies. All the binding sites were capable of binding with sugars. Zheng *et al.* (2007) reported isolation of a homodimer lectin specific for xylose and melibiose from *B. edulis*. The lectin dimer was composed of two subunits each having a molecular weight of 16.3 kDa and exhibited hemagglutinating activity and could not be agglutinated by acids and bases up to a limit of 25 mM concentration and was stable for agglutination up to a temperature of 40°C.

Lovastatin, ϒ-aminobutyric Acid (GABA) and Ergothioneine

The lovastatin, ergothioneine and ϒ-aminobutyric acid (GABA) plays a vital role in lowering the cholesterol levels and thus decreasing the chances of coronary heart diseases (Aarons *et al.*, 2007). GABA regulates neuron excitability and is a key neurotransmitter inhibitor and antihypertensive agent (Shizuka *et al.*, 2004; Tanaka *et al.*, 2009). Ergothioneine acts as an antioxidant and provides protection against oxidative stress (Dubost *et al.*, 2007; Hartman, 1990; Aruoma *et al.*, 1999). Chen *et al.* (2012) reported presence of lovastatin (327.3 mg/kg DW), GABA (202.1 mg/kg DW) and ergothioneine (494.4 mg/kg DW) in the fruiting bodies of *Boletus edulis* mushroom. Furthermore, Lo *et al.* investigated the mycelia of *B. edulis* and found sufficient amounts of lovastatin (131.4 mg/kg DW), GABA (1274.3 mg/kg DW) and ergothioneine (258.0 mg/kg DW) in it. Tsai *et al.* (2008) also quantified *ϒ*-amino butyric acid (110 mg/kg DW) in the fruiting body of *B. edulis* mushroom in a relatively lesser amount.

MINERAL CONTENT OF *B. EDULIS*

Mushrooms act as a potential source of minerals for humans as they possess mineral accumulation ability. Reports are available in literature confirming higher concentration of potassium in mushrooms than that present in fruits/vegetables (Liu *et al.*, 2016; Vetter, 2005; Liu *et al.*, 2012). Moreover, mushrooms can be a health threat due to their heavy metal absorption and tolerance ability (Hall *et al.*, 2012). However, the content of heavy metals is seldom above the acceptable range (Lau *et al.*, 2013; Liu *et al.*, 2015; Liu *et al.*, 2012). Su *et al.* (2018) investigated the mineral content and mineral accumulation ability of the cap and stipe of *B. edulis* belonging to nine different areas of Yunnan Province of China. Fourteen mineral elements including Na, Mg, Ca, Ba, V, Cr, Mn, Fe, Ni, Co, Cu, Zn, Cd and Sr were noticed. The mineral content of the two morphological areas viz. cap and stipe and the site of collection of *B. edulis* varied significantly. The main elements present in *B. edulis* were found to be Na, Mg, Fe and Ca. Additionally, Ba, Mn, Cr, Zn and Cu were present in moderate amounts whereas Ni, Cd, Co, V and Sr were present in relatively smaller amounts. The two morphological parts accumulated minerals to different extents. The caps were enriched with Na, Mg, Cu, Cd and Zn; whereas Ba, Sr, Mn, Co and Ni accumulated preferably in stipes. The calcium content of different geographical areas ranged between 118–468 and 131–572 mg/kg DW respectively in the caps and stipes of *B. edulis* and the quotient of different elements present in the cap and the stipe ($Q_{c/s}$) value ranged between 0.45–1.26. The Mg and Na mean contents were found to be larger in the caps (421–655 and 48–2720 mg/kg DW respectively) than that found in the stipes (246–748 and 19–631 mg/kg DW respectively) and $Q_{c/s}$ values were more than 1 for all the explored regions except Yimen, Yuxi and Weixi, Diqing regions of Yunnan for Mg and Na respectively. The mushrooms act as a rich source of iron. The iron content of *B. edulis* was extended from 395–1550 mg/kg DW for the caps and 332–3568 mg/kg DW for the stipes. Moreover, the content was different for different geographical areas of Yunnan, China. The mean values of the trace mineral elements Ba, Co, Cr, Cu, Mn, Ni, Sr, V and Zn for the caps of *B. edulis* ranged between 20–159 mg/kg DW, 1.1–7.2 mg/kg DW, 18–161 mg/kg DW, 24–53 mg/kg DW, 15–41 mg/kg DW, 14–42 mg/kg DW, 5.5– 143.6 mg/kg DW, 4.3–32.2 mg/kg DW and 86–120 mg/kg DW and for stipes it was found to be present between 18–182 mg/kg DW, 3.2–13.2 mg/kg DW, 10–203 mg/kg DW, 16–84 mg/kg DW, 19–71 mg/kg DW, 20–76 mg/kg DW, 5.4–176.9 mg/kg DW, 5.7–60.3 mg/kg DW and 40–135 mg/kg DW, respectively. Cadmium acts as a toxic element and is added in the food chain and therefore affects human health. Cd content in the cap of *B. edulis* was found to be in the range of 7.7–85.4 mg/kg DW and in the stipes it was ranged between 3.7 and 91.5 mg/kg DW. The authors claimed that Zn and Cd were the only elements which exhibited bioconcentration (BCF greater than one) whereas all other detected elements showed bio-exclusion (BCF less than one).

Similarly, Wang *et al.* (2015) analysed mineral content of cap and stalk of *Boletus* mushrooms of Yunnan Province including *B. edulis*. Ca, Mg, Na, K, P, Fe, Cu, Zn and Mn were the major elements present in *Boletus edulis*. Cap contained larger amounts of K (cap= 9413 mg/kg DW; stipe = 4865 mg/kg DW), Na (cap= 191 mg/kg DW; stipe= 122 mg/kg DW), Ca (cap=320 mg/kg DW; stipe=317 mg/kg DW), Mg (cap=429 mg/kg DW; stipe=297 mg/kg DW), Zn (cap=88 mg/kg DW; stipe=41 mg/kg DW), P (cap=6615 mg/kg DW; stipe=2429 mg/kg DW) and Cu (cap=51 mg/kg DW; stipe=32 mg/kg DW) than the stipe. Whereas Mn (cap=26 mg/kg DW; stipe= 70 mg/kg DW) and Fe (cap= 931 mg/kg DW; stipe= 4200 mg/kg DW) were more concentrated in the stalk rather than the cap. Lalotra *et al.*, 2018 investigated the mineral content of *B. edulis* collected from Jammu and Kashmir regions of India. The mushroom was enriched with four minerals namely Mn, Fe, Cu and Zn. The Cu content present in *B. edulis* was found to be 34.4 mg/kg DW and was comparatively smaller than that detected by Su *et al.* (2018). Of these, Fe (812 mg/kg DW) was the most predominant mineral followed by Zn (96.3 mg/kg DW) and Mn (54.4 mg/kg DW).

In a similar manner, Liu *et al.* (2016) assessed the mineral content of *B. edulis* collected from four different regions of Sichuan and Yunnan Provinces of China. The most important minerals present in *B. edulis* from four different regions were ranged between K (15744–27944 μg/g DW), Na (482–1184 μg/g DW), Ca (384–863 μg/g DW) and Mg (574–1083 μg/g DW). The K/Na ratio rested between 13.3 and 58.0 whereas most fruits and vegetables have a K/Na ratio greater than 2. The high K/Na ratio indicated the nutritional significance of *B. edulis*. Zn, Cu, Mn and Fe act as components of many enzyme complexes in the human body. *B. edulis* also contained Fe (221–524 μg/g DW), Zn (76–159 μg/g DW), Mn (33–69 μg/g DW) and Cu (19–73 μg/g DW). Furthermore, the toxic metals including Pb, Cd and As were also detected. The content of Pb was more in the two samples collected from Yunnan (2.5 and 5.5 μg/g DW) than from the Sichuan region of China (1.3 and 1.9 μg/g DW). The amount of Cd ranged from 1.7 to 2.8 μg/g DW.

PHARMACOLOGICAL PROPERTIES

Boletus showed a variety of biological activities such as antioxidant, antiviral, antineoplastic, hepato-protective, anti-inflammatory, and antibacterial. Moreover, silver nanoparticles derived from *B. edulis* showed an average size of $<$100 nm and displayed robust anti-fungal, anti-neoplastic, antibacterial, and Wound healing properties (Kaplan *et al.*, 2021). *B. edulis* has gained the most attention as compared to other species of the genus due to their commercial value (Table 3.2).

ANTIOXIDANT ACTIVITY

Boletus edulis is a very popular mushroom and contains a marked amount of antioxidant constituents. *B. edulis* has been revealed to exhibit *in vivo* antioxidant efficacy in mice or rats with declined oxidative stress, which is mostly connected with enhanced health improvement. ROS are produced internally in the organism due to reactions of metabolic pathways and from outside sources such as UV light, ionizing radiation, and environmental pollutants (Kozarski *et al.*, 2015). These ROS can harm biological molecules like DNA and proteins and cause apoptosis or dysfunction which contributes to the severity of fatal diseases like cancer, cardiovascular and renal problems (Kozarski *et al.*, 2015). Thus, antioxidants comprising foods have the potential to neutralize the free radicals. It was observed that *B. edulis* revealed maximum inhibition against free radical production at low IC_{50} value. Moreover, the antimicrobial potential of *B. edulis* was noticed against *Staphylococcus aureus*, *Klebsiella*, and *Escherichia coli* (Rosa *et al.*, 2020). The fruiting bodies of *B. edulis* have been screened for the extraction of its polysaccharides (BEPF30, BEPF60,

TABLE 3.2
Biological Activities of the Genus *Boletus*

S. No.	Biological activities	Species of *Boletus*	Mode of experiment	Mechanism of action	References
1.	Antioxidant	*B. edulis* *B. edulis* *B. bicolor*	*In vitro* *In vivo* *In vitro*	EC_{50}=0.43–1.80 mg/mL The antioxidant potential was associated with bioactive components such as polysaccharides, phenolics, and lutein. Declining oxidative Stress and MDA Concentration, IC_{50} = 44.23 μg/mL	Huang *et al.*, 2021 Guo et al., 2020a Zhu & Jia, 2018 Fernandes *et al.*, 2015 Zheng *et al.*, 2019 Cui *et al.*, 2014 Luo *et al.*, 2012 Ahmed *et al.*, 2015 Tibuhwa, 2017
2.	Anticancer	*B. edulis* *B. bicolor*	*In vivo* (Sarcoma tumor) *In vivo* (Ranca cells) *In vitro* *In vitro* *In vitro* *In vitro* *In vitro* *In vitro &* *In vivo* *In vitro*	Enhanced spleen index and lymphocytes, More levels of SOD Restoration of the thymus and spleen indices, Amplified the concentration of cytokines and activation of T and B lymphocytes The IC_{50} against MDA-MB-231 cells= 152 μg/mL IC_{50} against Ca761 cells=134 μg/mL Inhibition against cancer cells =77–92% Hampered the proliferation of HepG-2, MCF-7, HeLa, HT-29, CaCo-2, and A549. Arrested 60% LS-180 cancer cell in the G_1 phase Caused 24% arrest of HT-29 tumor cells in the S phase Hindered the growth of A549 cells, triggering apoptosis and repressed the tumor cells about 64.72%. Upregulation in the levels of cytochrome c, P21, Bax, caspase-3, and P16, while the levels of CDK4, p-Rb, Bcl-2, and cyclin D was markedly downregulated.	Tang & Lu (1999) Wang *et al.* (2014) Meng *et al.*, 2021 Bovi *et al.*, 2011 Bovi *et al.*, 2013 Lemieszek *et al.*, 2013 Lemieszek *et al.*, 2013 Zhang *et al.*, 2021a Zhang *et al.*, 2021b

3	Antiinflammatory	*B. edulis*	*In vivo*	Decreased airway resistance, lesions and proinflammatory Responses.	Wu *et al.*, 2016
4	Hepatoprotective	*B. edulis*	*In vivo*	Lessened the hepatic damage by reinstating the hepatic function.	Xiao *et al.*, 2018
		B. edulis	*In vivo*	Diminished the SGOT & SGPT levels, Amplified the concentration of GSH and SOD in the liver, thus dropping the amount of MDA and hepatocyte lesions.	Zheng *et al.*, 2019
5.	Antimicrobial	*B. edulis*	*In vitro*	Robust inhibition against *E.coli, B. subtilis* and *Proteus*	Guo et al., 2020a
		B. edulis	*In vitro*	Inhibition zone against gram +ve bacteria= 10.12–19.03 mm.	Salihović *et al.*, 2019
		B. bicolor	*In vitro*	Inhibition zone against *Bacillus subtilis* 2.3 mm	Tibuhwa, 2017
6	Antihypertensive	*B. edulis*	*In vitro*	Protein hydrolysate exerted 21.1% ACE inhibitory activity	Khongdetch *et al.*, 2022 Kaprasob *et al.*, 2022
7	Antiviral	*B. edulis*	*In vitro*	IC_{50} against HIV= 14 μM.	Zheng *et al.*, 2007
		B. edulis	*In vitro*	Selective index against HSV is 13.96.	Santoyo *et al.*, 2012
8	Immunostimulatory	*B. edulis*	*In vitro*	Increased the microbicidal activity of macrophages and Nitric Oxide secretion	Yu *et al.*, 2022

and BEPF80). Further *in vitro* antioxidant activity of polysaccharides was examined using hydroxyl assays, superoxide radical assays, reducing power, and chelating activity. BEPF60 exerted the maximum reducing power and chelating activity, as well as superoxide and hydroxyl radical inhibition (Zhang *et al.*, 2011). In another investigation, it was observed that *Boletus edulis* contain a significant number of B-group vitamins and this quantity of vitamins fulfilled the recommended daily intake for healthy adults. In addition to it, mushrooms were found to be rich in polyphenols, flavonoids, ascorbic acid, β-carotene, lycopene, and tocopherols and displayed remarkable antioxidant activity in DPPH, ABTS, and FRAP assay (Jaworska *et al.*, 2015). The antioxidant property of this mushroom is primarily due to the presence of polysaccharides and phenolic components. Various assays demonstrated that maximum antioxidant action were observed in alcoholic, water, petroleum ether and butanolic extracts of *B. edulis* (Guo *et al.*, 2020a; Vamanu and Nita, 2013). In another study, the scavenging action of cold and hot aqueous extracts of *B. edulis* on DPPH was observed to be more in comparison to the acetone and ethanolic extract (Wang and Xu, 2014). Additionally, Guo *et al.* (2020a) described that butanol, petroleum ether, and ethyl acetate *B. edulis* extracts showed robust scavenging activity on DPPH radicals in contrast to the water extracts. The aqueous extract of *B. edulis* unveiled more FRAP and TEAC concentration as compared to the lipid-soluble fraction (Guo *et al.*, 2012). The antioxidant action of *B. edulis* is strongly associated with the presence of saccharide content in it. Moreover, polyphenols and flavonoids isolated from *B. edulis* have also shown antioxidant activity (Wang *et al.*, 2020; Guo *et al.*, 2020b). Additionally, enzymatic hydrolysate from *B. edulis* possessed scavenging action against DPPH radicals and the total antioxidant activity was observed to be 18 U/mL (Huang *et al.*, 2021). It was observed that maximum antioxidant activity was shown by the cap of *B. edulis* followed by the stem and whole mushroom (Ribeiro *et al.*, 2008). It depicted that some phytoconstituents of the cap and stem caused the variation in the antioxidant ability of the entire mushroom. Polysaccharides purified from *B. edulis* enhanced the antioxidant markers of red blood cells and boosted the reduction of superoxide radicals (Tang & Lu, 1999). Luo *et al.*, 2012 demonstrated that the efficiency of SOD and its potential to decrease the malondialdehyde (MDA) were remarkably heightened in mice after intraperitoneal treatment with the *B. edulis* polysaccharide. Likewise in another investigation, *B. edulis* polysaccharide efficiently repressed hepatic lipid peroxidation in mice through the improvement in the levels of SOD, and glutathione, thus declining the quantity of MDA and circumventing hepatic necrosis (Zheng *et al.*, 2019). In another study, treatment of *B. edulis* blocked oxidative stress by the enhancement in the antioxidant capacity and reduction of MDA concentration (Cui *et al.*, 2014).

Antineoplastic Activity

B. edulis contains anticancer bioactive components such as polysaccharides, lectin, and glycoproteins, making it a good anticancer drug. A novel anti-cancer BEAP protein was purified from the fruiting portion of the *B. edulis*. *In vitro* and *in vivo* studies revealed that BEAP protein displayed a robust anti-cancer impact on A549 cells. This protein caused the induction of apoptosis through the increment in the levels of caspases, and Bax/Bcl-2 ratio (Zhang *et al.*, 2021). Moreover, in another study, a polysaccharide isolated from *Boletus edulis* enhanced the thymus and spleen index, increased the proliferation of splenocytes and boosted the activity of Cytotoxic killer cells in Spleen. Treatment with polysaccharide (BEP) isolated from *Boletus edulis* caused the increment in the concentration of IL-2, and IL-6 in tumor-bearing mice. Moreover, BEP treatment (100 and 400 mg/kg) recovered all the abnormal levels of haematological and biochemical enzymes of tumor-bearing mice. It indicated the immunomodulatory potential of *Boletus* polysaccharide in the prevention of renal cancer (Wang *et al.*, 2014). Moreover, a polysaccharide of *Boletus edulis* (BEP) has the marked potential to hamper the proliferation of breast cancer cells. Morphological studies showed that treatment of BEP caused apoptotic and necrotic changes in breast cancer cells. Flow cytometric studies demonstrated the cell cycle block in S phase, change in mitochondrial membrane potential, and apoptosis. Moreover, Western blotting revealed the increment in Bax/Bcl-2 ratios in the BEP treated group which might enhance the secretion of cytochrome C, and caspases in breast cancer cells (Meng *et al.*, 2021). In another investigation, lectins purified from *B. edulis* enhanced marked antiproliferative activity against several cancer cell lines (Bovi *et al.*, 2013). Lemieszek *et al.* (2013) demonstrated that glycoproteins isolated from *B. edulis* caused the reduction in the multiplication of colon adenocarcinoma cells in the G_1 phase. However, these biopolymers exhibited no adverse reaction on the normal colon cells.

Anti-inflammatory Activity

Polysaccharides derived from *Boletus* (BEP) showed anti-inflammatory activity against ovalbumin-stimulated asthma in BALB/c mice. Treatment of Polysaccharide from *B. edulis* augmented the IL-4 and IFN-gamma concentration in the alveolar fluid and improved the alveolar wall thickening. Moreover, polysaccharides induced a remarkable increment in the percentage of anti-inflammatory Treg cells. However, BEP treatment caused a marked decline in the expression of cyclophilin A protein. It was observed that the efficacy of polysaccharides was found to be similar to the standard drug dexamethasone (Wu *et al.*, 2016).

Hepatoprotective Activity

The hepatoprotective effects of *B. edulis* polysaccharides (BEP) were monitored in a streptozocin-induced diabetic rat. The polysaccharides caused the marked decline in blood glucose, serum transaminases (SGPT and SGOT) and lipids in serum. Moreover, treatment of *B. edulis* polysaccharides enhanced the levels of catalase and glutathione in the liver while lowering MDA level, indicating its hepatoprotective effect (Xiao *et al.*, 2018). Similarly, Zheng *et al.* (2019) showed improvement in the abnormal levels of hepatic enzymes in mice with CCl_4-induced hepatic injury after the treatment of BEP. Moreover, BEP treatment normalized the histological architecture of the hepatic portion which depicts the hepatoprotective impact of polysaccharides.

Antibacterial and Antiviral Activities

Numerous *B. edulis* preparations have been shown to exhibit antimicrobial and antiviral properties. The antibacterial action of *B. edulis* was demonstrated against *Klebsiella*, *Escherichia*, *Acinetobacter, Staphylococcus* and *Pseudomonas* (Naimushina *et al.*, 2020). Guo *et al.* (2020a) investigated the protective efficacy of different *B. edulis* solvent extracts against the different bacterial strains. The *B. edulis* extracts remarkably repressed the growth to varying degrees against *E. coli*, *Proteus vulgaris*, *Bacillus subtilis*, *P. aeruginosa*, *S. aureus*, and *Porphyromonas vulgaris*. Among all these extracts, ethanol acetate extract showed the maximum bacteriostatic potential against the tested strains and displayed the inhibitory concentration in the range of 25–50 mg/mL. In another study, A lectin obtained from *B. edulis* inhibited human immunodeficiency virus 1 reverse transcriptase by 50% at 14 μM (Zheng *et al.*, 2007). In addition to it, the water extract from *B. edulis* can decrease the number of plaques in cells at 35 mg/mL, hence impeding the multiplication of Herpes simplex virus. However, porcini mushroom was observed to be less potent in comparison to the water extract against replication of Herpes simplex virus (Santoyo *et al.*, 2012).

Other Therapeutic Effects

B. edulis contains a huge number of polysaccharides, chiefly dietary fibers, which may be useful in avoiding and treating constipation (Xin *et al.*, 2018; Zhu and Jia, 2018). Peptides isolated from *B. edulis* showed maximum antioxidant potential and modest angiotensin I-converting enzyme (ACE) inhibitory activity. Similarly, protein hydrolysate of *B. edulis* exerted 21.1% against the angiotensin I-converting enzyme (Khongdetch *et al.*, 2022; Kaprasob *et al.*, 2022). The *B. edulis* polysaccharide (BEP) revealed immunostimulatory activity by increasing the phagocytosis of macrophages and secretion of microbicidal molecules. It was observed that treatment of BEP caused the upregulation in the concentration of IL-6 and TNF-α by 191% and 196% in contrast to the control group, respectively (Yu *et al.*, 2022).

Boletus bicolor

There has been little bioactivity research on other *Boletus* species, *Boletus bicolor*. A novel D1 protein (40 kDa) purified from *Boletus bicolor* remarkably hindered the multiplication of mammalian lung adenocarcinoma cell lines and showed a negligible impact on the normal renal cells. Flow cytometric studies showed that D1 protein tempted cell death and blocked the lung adenocarcinoma cells in the G1 phase. Moreover, Western blot studies exhibited the upregulation in the levels of cytochrome c, P21, Bax, caspase-3, and P16, while the levels of CDK4, p-Rb, Bcl-2, and cyclin D were markedly downregulated in D1 protein treated cells (Zhang *et al.*, 2021). In addition to it, methanol extracts of *B. bicolor* revealed protective efficacy against *Bacillus subtilis* with an inhibition zone of 2.3 mm (Tibuhwa, 2017). In another study, *B. bicolor* showed antioxidant activity with the IC_{50} value of 44.23 μg/ml (Ahmed *et al.*, 2015).

SAFE PROFILE OF *BOLETUS EDULIS*

The safety of *B. edulis* mushroom ingestion is linked to the identification of the correct mushroom species, noxiousness, and quality. It was observed that this mushroom species is commonly misinterpreted as poisonous mushroom species (Li *et al.*, 2021). *Boletus edulis* is sometimes confused with *B. huronensis* (fake king bolete), which can cause serious gastrointestinal issues (Bakaitis, 2019). *B. edulis*, on the other hand, is considered the safest wild mushroom to eat since deadly species that seem similar may be simply identified by careful study. As a result, it is critical to accurately recognize *B. edulis* for the sake of the customers' health. Moreover, it has been noticed that *B. edulis* has the potential to absorb heavy metals or harmful compounds if these are present in the surrounding environment. Širić *et al.*, 2017 reported that *B. edulis* collected contain the maximum concentration of mercury in the cap and stem region and followed by cadmium (Cd) and lead (Pb). Likewise, Su *et al.*, 2018 noticed the significant amount of Cd in caps and stems of porcini mushroom from China. Similarly, Falandysz (2022) displayed the presence of Hg and Pb in the cap and stem region of *B. edulis* from Poland. However, it has been noticed that the amount of heavy metals was found to be more in the *B. edulis* collected from Croatia in contrast to the *B. edulis* from Poland. If an adult of 60 kg weight consumes three hundred gram of *B. edulis* in a week, the concentration of Cd was observed to be smaller than the recommended weekly intake and exhibited no adverse effects on health risk. As heavy metals contamination in the soil varies widely between the different regions, more intake of *B. edulis* can harm the consumers (Širić *et al.*, 2017; Su *et al.*, 2018). Moreover, *B. edulis* has been stored for the preparation of dry products which may result in microbial development and variations in physical and chemical characteristics. *B. edulis* mushroom is highly utilized in cooking due to its unique flavor. However, incidents of food sickness from eating this mushroom have been recorded recently. Screening of *B. edulis* showed the ribotoxin-like proteins Edulitin 1 and Edulitin 2. It was found that Edulitin 2 can only be degraded at higher temperatures. Thus, it is recommended to eat mushrooms after proper cooking to decrease food poisoning from mushrooms (Landi *et al.*, 2021).

ROLE IN FOOD INDUSTRY

B. edulis can be used as an additive or flavouring agent due to its odor, flavour, and taste like earthy, fantastic-nutty, and meaty all at once (Tsai *et al.*, 2008). It showed efficient anti-oxidant and anti-bacterial activity *in vitro*, thus they were utilized in the manufacture of frankfurters for the enhancement of the shelf life. Sausages with *B. edulis* were found to be firm throughout two months of cold storage when compared to the shelf life of frankfurters without *B. edulis*. It might be an exciting finding for the meat industry as it can lower expenditures and enhance profitability. Moreover, it has been seen that the *B. edulis* caused a marked improvement in the texture of sausages (Novakovic, 2021).

CONCLUSION

Mushrooms are regarded as invaluable resources to produce several key biomolecules with commercial uses in the various industries. The numerous pharmacotherapeutic applications of *Boletus* mushroom has allured it as a functional food and a reservoir of bioactive constituents. *Boletus* mushroom has good nutraceutical potential due to the presence of a greater number of proteins and low-fat content. Several studies have shown the numerous biological effects of *Boletus* mushroom like anticancer, antioxidant, anti-microbial, antiviral, and hepatoprotective. These bioactivities lay a theoretical foundation for further isolation of bioactive constituents from *B. edulis*, as well as a novel research area for the growth of pharmaceutical medications. Further mechanistic studies are still required for the evaluation of the mechanism of the known isolated bioactive constituents of *Boletus* mushrooms.

REFERENCES

Aarons, C. B., Cohen, P. A., Gower, A., Reed, K. L., Leeman, S. E., Stucchi, A. F., & Becker, J. M. (2007). Statins (HMG-CoA reductase inhibitors) decrease postoperative adhesions by increasing peritoneal fibrinolytic activity. *Annals of Surgery*, 245(2), 176.

Adhikary, R. K., Baruah, P., Kalita, P., & Bordoloi, D. (1999). Edible mushrooms growing in the forests of Arunachal Pradesh. *Advances in Horticulture and Forestry*, 6, 119–123.

Ahmed, A., Kamran Taj, M., Yonglıang, Z., Taj, I., Hassani, M. T., Sajıd, M., … & Liping, S. (2015). Chemical composition and antioxidant properties of eight wild edible Boletaceae mushrooms growing in Yunnan Province of China. *Jökull Journal*, 65(12), 146–167.

Argyropoulos, D., Khan, M. T., & Müller, J. (2011). Effect of air temperature and pre-treatment on color changes and texture of dried *Boletus edulis* mushroom. *Drying Technology*, 29(16), 1890–1900.

Aruoma, O. I., Spencer, J. P. E., & Mahmood, N. (1999). Protection against oxidative damage and cell death by the natural antioxidant ergothioneine. *Food and Chemical Toxicology*, 37(11), 1043–1053.

Bakaitis, B. (2019). *Boletus huronensis*: Comments on its toxicity with diagnostic images of its field characteristics and staining reactions. *North American Mycological Association*.

Balta, I., Miresan, V., Raducu, C., Longodor, A. L., Eugenia, B., Marchis, Z., … & Aurelia, C. (2019). The Physico-chemical Composition and the Level of Metals from (*Boletus edulis*) and (*Cantharellus cibarius*) from the Vatra Dornei Area. *ProEnvironment Promediu*, 11(36).

Barros, L., Cruz, T., Baptista, P., Estevinho, L. M., & Ferreira, I. C. (2008). Wild and commercial mushrooms as source of nutrients and nutraceuticals. *Food and Chemical Toxicology*, 46(8), 2742–2747.

Borthakur, M., & Joshi, S. R. (2019). Wild mushrooms as functional foods: The significance of inherent perilous metabolites. In *New and Future Developments in Microbial Biotechnology and Bioengineering* (pp. 1–12). Elsevier.

Bovi, M., Carrizo, M. E., Capaldi, S., Perduca, M., Chiarelli, L. R., Galliano, M., & Monaco, H. L. (2011). Structure of a lectin with antitumoral properties in king bolete (*Boletus edulis*) mushrooms. *Glycobiology*, 21(8), 1000–1009.

Bovi, M., Cenci, L., Perduca, M., Capaldi, S., Carrizo, M. E., Civiero, L., … & Monaco, H. L. (2013). BEL β-trefoil: A novel lectin with antineoplastic properties in king bolete (*Boletus edulis*) mushrooms. *Glycobiology*, 23(5), 578–592.

Bovi, M., Cenci, L., Perduca, M., Capaldi, S., Carrizo, M. E., Civiero, L., Chiarelli, L. R., Galliano, M., & Monaco, H. L. (2013). BEL β-trefoil: A novel lectin with antineoplastic properties in king bolete (*Boletus edulis*) mushrooms. *Glycobiology*, 23(5), 578–592.

Çaglarlrmak, N., Ünal, K., & Ötles, S. (2002). Nutritional value of edible wild mushrooms collected from the Black Sea region of Turkey. *Micologia Aplicada International*, 14, 1–5.

Cámara, M., Díez, C., Torija, M. E., & Cano, M. P. (1994). HPLC determination of organic acids in pineapple juices and nectars. *Zeitschrift für Lebensmittel-Untersuchung und Forschung*, 198(1), 52–56.

Chen, S. Y., Ho, K. J., Hsieh, Y. J., Wang, L. T., & Mau, J. L. (2012). Contents of lovastatin, γ-aminobutyric acid and ergothioneine in mushroom fruiting bodies and mycelia. *Lwt*, 47(2), 274–278.

Choma, A., Nowak, K., Komaniecka, I., Waśko, A., Pleszczyńska, M., Siwulski, M., & Wiater, A. (2018). Chemical characterization of alkali-soluble polysaccharides isolated from a *Boletus edulis* (*Bull.*) fruiting body and their potential for heavy metal biosorption. *Food Chemistry*, 266, 329–334.

Christensen, M., Bhattarai, S., Devkota, S., & Larsen, H. O. (2008). Collection and use of wild edible fungi in Nepal. *Economic Botany*, 62(1), 12–23.

Coimbra, M. A., Gonçalves, F., Barros, A. S., & Delgadillo, I. (2002). Fourier transform infrared spectroscopy and chemometric analysis of white wine polysaccharide extracts. *Journal of Agricultural and Food Chemistry*, 50(12), 3405–3411.

Cui, F. S., Jin, C. X., & Cui, C. B. (2013). Study on scavenging free radial activity of extracts from *Boletus edulis* of Changbai Mountain. *Food Ind*, 34(5), 133–136.

Cui, F. S., Zhang, H., Li, G. H., & Li, Z. H. (2014). The antioxidant activities *in vivo* of the total flavonoids from *Boletus edulis* of Changbai mountain. *Journal of Food Science and Technology*, 39(8), 201–205.

Cui, Y. Y., Feng, B., Wu, G., Xu, J., & Yang, Z. L. (2016). Porcini mushrooms (*Boletus* sect. *Boletus*) from China. *Fungal Diversity*, 81(1), 189–212.

Dentinger, B. T., Ammirati, J. F., Both, E. E., Desjardin, D. E., Halling, R. E., Henkel, T. W., Moreau, P. A., Nagasawa, E., Soytong, K., Taylor, A. F., & Watling, R. (2010). Molecular phylogenetics of porcini mushrooms (*Boletus* section *Boletus*). *Molecular Phylogenetics and Evolution*, 57(3), 1276–1292.

Dubost, N. J., Ou, B., & Beelman, R. B. (2007). Quantification of polyphenols and ergothioneine in cultivated mushrooms and correlation to total antioxidant capacity. *Food Chemistry*, 105(2), 727–735.

Falandysz, J., Meloni, D., Fernandes, A. R., & Saniewski, M. (2022). Effect of drying, blanching, pickling and maceration on the fate of 40K, total K and 137Cs in bolete mushrooms and dietary intake. *Environmental Science and Pollution Research*, 29(1), 742–754.

Fernandes, Â., Barreira, J. C., Antonio, A. L., Morales, P., Férnandez-Ruiz, V., Martins, A., Oliveira, M. B. P., & Ferreira, I. C. (2015). Exquisite wild mushrooms as a source of dietary fiber: Analysis in electron-beam irradiated samples. *LWT-Food Science and Technology*, 60(2), 855–859.

Fogarasi, M., Socaci, S. A., Dulf, F. V., Diaconeasa, Z. M., Fărcaş, A. C., Tofană, M., & Semeniuc, C. A. (2018). Bioactive compounds and volatile profiles of five Transylvanian wild edible mushrooms. *Molecules*, 23(12), 3272.

García, M. M., Paula, V. B., Olloqui, N. D., García, D. F., Combarros-Fuertes, P., Estevinho, L. M., Árias, L. G., Bañuelos, E. R., & Baro, J. M. F. (2021). Effect of different cooking methods on the total phenolic content, antioxidant activity and sensory properties of wild *Boletus edulis* mushroom. *International Journal of Gastronomy and Food Science*, 26, 100416.

Grabmann, J. (2005). Terpenoids as plant antioxidants. *Vitamins & Hormones*, 72, 505–535.

Guo, L., Qian, X., Dai, H., Xu, H., Kong, D., Liu, H.u., …Hua, Y. (2020a). Antioxidant and antibacterial properties of extracts from different polar solvents from *Boletus edulis Bull. Food Science and Technology*, 45(04), 175–181.

Guo, L., Jaydar, N., Qun, J., Yin, X., & Kan, H. (2020b). Study on purification process of total flavonoids from *Boletus edulis* with macroporous resin and its antioxidant activity. *China Food Additives*, 31(01), 85–91.

Guo, Y. J., Deng, G. F., Xu, X. R., Wu, S., Li, S., Xia, E. Q., … & Li, H. B. (2012). Antioxidant capacities, phenolic compounds and polysaccharide contents of 49 edible macro-fungi. *Food & Function*, 3(11), 1195–1205.

Gyar, S. D., & Owaku, G. (2011). Estimation of some metal elements and proximate properties of Boletus edulis (Fr), a wild mushroom species in the Nigerian savannah. *Trakia Journal of Sciences*, 9(2), 22–25.

Hall, I. R., & Zambonelli, A. (2012). *Edible Ectomycorrhizal Mushrooms: Current Knowledge and 417 Future Prospects*. New York: Springer-Verlag (Part I: Chapter 1).

Hall, I. R., Lyon, A. J. E., Wang, Y., and Sinclair, L. (1998) Ectomycorrhizal fungi with edible fruiting bodies 2. *Boletus edulis*. *Economic Botany*, 52(1), 44–56.

Hartman, P. E. (1990). Ergothioneine as antioxidant. In *Methods in Enzymology* (Vol. 186, pp. 310–318). Academic Press.

Heleno, S. A., Barros, L., Sousa, M. J., Martins, A., Santos-Buelga, C., & Ferreira, I. C. (2011). Targeted metabolites analysis in wild *Boletus* species. *LWT-Food Science and Technology*, 44(6), 1343–1348.

Huang, D., Gao, Y., Liu, L., Zhang, H., Zhang, Y., Chen, H., …Zeng, Y. (2021). Optimization of preparation technology and antioxidant activity of enzymatic hydrolysate from *Boletus edulis* hydrolyzed by protease. *Science and Technology of Food Industry*, 42(12), 209–217.

Jaworska, G., Pogoń, K., Bernaś, E., & Skrzypczak, A. (2014). Effect of different drying methods and 24-month storage on water activity, rehydration capacity, and antioxidants in *Boletus edulis* mushrooms. *Drying Technology*, 32(3), 291–300.

Jaworska, G., Pogoń, K., Skrzypczak, A., & Bernaś, E. (2015). Composition and antioxidant properties of wild mushrooms *Boletus edulis* and *Xerocomus badius* prepared for consumption. *Journal of Food Science and Technology*, 52(12), 7944–7953.

Kalac, P. (2016). *Edible mushrooms: chemical composition and nutritional value*. Academic Press.

Kalač, P. (2009). Chemical composition and nutritional value of European species of wild growing mushrooms: A review. *Food Chemistry*, 113(1), 9–16.

Kaplan, Ö., Tosun, N. G., Özgür, A., Tayhan, S. E., Bilgin, S., Türkekul, İ., & Gökce, İ. (2021). Microwave-assisted green synthesis of silver nanoparticles using crude extracts of *Boletus edulis* and *Coriolus versicolor*: Characterization, anticancer, antimicrobial and wound healing activities. *Journal of Drug Delivery Science and Technology*, 64, 102641.

Kaprasob, R., Khongdetch, J., Laohakunjit, N., Selamassakul, O., & Kaisangsri, N. (2022). Isolation and characterization, antioxidant, and antihypertensive activity of novel bioactive peptides derived from hydrolysis of King *Boletus* mushroom. *LWT*, 160, 113287.

Khongdetch, J., Laohakunjit, N., & Kaprasob, R. (2022). King *Boletus* mushroom-derived bioactive protein hydrolysate: Characterisation, antioxidant, ACE inhibitory and cytotoxic activities. *International Journal of Food Science & Technology*, 57(3), 1399–1410.

Kong, L., Yu, L., Feng, T., Yin, X., Liu, T., & Dong, L. (2015). Physicochemical characterization of the polysaccharide from*Bletilla striata*: Effect of drying method. *Carbohydrate Polymers*, 125, 1–8.

Kozarski, M., Klaus, A., Jakovljevic, D., Todorovic, N., Vunduk, J., Petrović, P., … & Van Griensven, L. (2015). Antioxidants of edible mushrooms. *Molecules*, 20(10), 19489–19525.

Lalotra, P., Bala, P., Kumar, S., & Sharma, Y. P. (2018). Biochemical characterization of some wild edible mushrooms from Jammu and Kashmir. *Proceedings of the National Academy of Sciences, India Section B: Biological Sciences*, 88(2), 539–545.

Landi, N., Ragucci, S., Culurciello, R., Russo, R., Valletta, M., Pedone, P. V., … & Di Maro, A. (2021). Ribotoxin-like proteins from *Boletus edulis*: Structural properties, cytotoxicity and *in vitro* digestibility. *Food Chemistry*, 359, 129931.

Lau, B. F., Abdullah, N., & Aminudin, N. (2013). Chemical composition of the tiger's milk mushroom, *Lignosus rhinocerotis* (Cooke) Ryvarden, from different developmental stages. *Journal of Agricultural and Food Chemistry*, 61(20), 4890–4897.

Lemieszek, M. K., Cardoso, C., Nunes, F. H. F. M., Marques, G., Pożarowski, P., & Rzeski, W. (2013). *Boletus edulis* biologically active biopolymers induce cell cycle arrest in human colon adenocarcinoma cells. *Food & Function*, 4(4), 575–585.

Li, W., Pires, S. M., Liu, Z., Liang, J., Wang, Y., Chen, W., … & Guo, Y. (2021). Mushroom poisoning outbreaks—China, 2010–2020. *China CDC Weekly*, 3(24), 518.

Li, X. P., Li, J., Li, T., Liu, H., & Wang, Y. (2020). Species discrimination and total polyphenol prediction of porcini mushrooms by Fourier transform mid-infrared (FT-MIR) spectrometry combined with multivariate statistical analysis. *Food Science & Nutrition*, 8(2), 754–766.

Liu, B., Huang, Q., Cai, H., Guo, X., Wang, T., & Gui, M. (2015). Study of heavy metal concentrations in wild edible mushrooms in Yunnan Province, China. *Food Chemistry*, 188, 294–300.

Liu, H., Zhang, J., Li, T., Shi, Y., & Wang, Y. (2012). Mineral element levels in wild edible mushrooms from Yunnan, China. *Biological Trace Element Research*, 147(1), 341–345.

Liu, Y., Chen, D., You, Y., Zeng, S., Li, Y., Tang, Q., Han, G., Liu, A., Feng, C., & Chen, D. (2016). Nutritional composition of *Boletus* mushrooms from Southwest China and their antihyperglycemic and antioxidant activities. *Food Chemistry*, 211, 83–91.

Liu, Y., Chen, D., You, Y., Zeng, S., Li, Y., Tang, Q., … & Chen, D. (2016). Nutritional composition of *Boletus* mushrooms from Southwest China and their antihyperglycemic and antioxidant activities. *Food Chemistry*, 211, 83–91.

Lo, T. C. T., Chang, C. A., Chiu, K. H., Tsay, P. K., & Jen, J. F. (2011). Correlation evaluation of antioxidant properties on the monosaccharide components and glycosyl linkages of polysaccharide with different measuring methods. *Carbohydrate Polymers*, 86(1), 320–327.

Luo, A., Luo, A., Huang, J., & Fan, Y. (2012). Purification, characterization and antioxidant activities in vitro and in vivo of the polysaccharides from *Boletus edulis Bull*. *Molecules*, 17(7), 8079–8090.

Mahajan, G. (2022). A comprehensive review on potential and prospects of mushroom cultivation. 10.2139/ssrn.4105173

Manzi, P., Gambelli, L., Marconi, S., Vivanti, V., & Pizzoferrato, L. (1999). Nutrients in edible mushrooms: An inter-species comparative study. *Food Chemistry*, 65(4), 477–482.

Mattila, P., Salo-Väänänen, P., Könkö, K., Aro, H., & Jalava, T. (2002). Basic composition and amino acid contents of mushrooms cultivated in Finland. *Journal of Agricultural and Food Chemistry*, 50(22), 6419–6422.

Meng, T., Yu, S. S., Ji, H. Y., Xu, X. M., & Liu, A. J. (2021). A novel acid polysaccharide from *Boletus edulis*: Extraction, characteristics and antitumor activities in vitro. *Glycoconjugate Journal*, 38(1), 13–24.

Muszyńska, B., & Sułkowska-Ziaja, K. (2012). Analysis of indole compounds in edible Basidiomycota species after thermal processing. *Food Chemistry*, 132(1), 455–459.

Muszyńska, B., Sułkowska-Ziaja, K., & Ekiert, H. (2011). Indole compounds in fruiting bodies of some edible Basidiomycota species. *Food Chemistry*, 125(4), 1306–1308.

Naimushina, L. V., Zykova, I. D., Gubanenko, G. A., Rechkina, E. A., & Kondratyuk, T. A. (2020). Comparative analysis of antiradical and antibacterial activity of *Boletus edulis* basidiomycetes growing in different climatic zones. In *IOP Conference Series: Earth and Environmental Science* (Vol. 421, No. 7, p. 072004). IOP Publishing.

Nguyen, N. P. D., Kien, N. H., Hien, T. T. T., Huong, D. T. T., Sinh, N. V., Nguyen, N. T., … & Hien, P. T. T. (2022). Species Diversity of *Boletus* Dill. ex Fr in Chu Yang Sin, National Park, Dak Lak, Vietnam. *Advanced Studies in Biology*, 14(1), 41–50.

Novakovic, S. (2021). The potential of the application of *Boletus edulis*, *Cantharellus cibarius* and *Craterellus cornucopioides* in frankfurters: A review. In *IOP Conference Series: Earth and Environmental Science* (Vol. 854, No. 1, p. 012068). IOP Publishing.

Ouzouni, P., & Riganakos, K. (2007). Nutritional value and metal content profile of Greek wild edible fungi. *Acta Alimentaria*, 36(1), 99–110.

Özyürek, M., Bener, M., Güçlü, K., & Apak, R. (2014). Antioxidant/antiradical properties of microwave-assisted extracts of three wild edible mushrooms. *Food Chemistry*, 157, 323–331.

Palacios, I., Lozano, M., Moro, C., D'arrigo, M., Rostagno, M. A., Martínez, J. A., … & Villares, A. (2011). Antioxidant properties of phenolic compounds occurring in edible mushrooms. *Food Chemistry*, 128(3), 674–678.

Palacios, I., Lozano, M., Moro, C., D'arrigo, M., Rostagno, M. A., Martínez, J. A., García-Lafuente, A., Guillamón, E., & Villares, A. (2011). Antioxidant properties of phenolic compounds occurring in edible mushrooms. *Food Chemistry*, 128(3), 674–678.

Ribeiro, B., Lopes, R., Andrade, P. B., Seabra, R. M., Gonçalves, R. F., Baptista, P., & Quelhas, I. (2008). Comparative study of phytochemicals and antioxidant potential of wild edible mushroom caps and stipes. *Food Chemistry*, 110(1), 47–56.

Ribeiro, B., Rangel, J., Valentao, P., Baptista, P., Seabra, R. M., & Andrade, P. B. (2006). Contents of carboxylic acids and two phenolics and antioxidant activity of dried Portuguese wild edible mushrooms. *Journal of Agricultural and Food Chemistry*, 54(22), 8530–8537.

Rosa, G. B., Sganzerla, W. G., Ferreira, A. L. A., Xavier, L. O., Veloso, N. C., da Silva, J., … & Ferrareze, J. P. (2020). Investigation of nutritional composition, antioxidant compounds, and antimicrobial activity of wild culinary-medicinal mushrooms *Boletus edulis* and *Lactarius deliciosus* (Agaricomycetes) from Brazil. *International Journal of Medicinal Mushrooms*, 22(10).

Ruthes, A. C., Smiderle, F. R., & Iacomini, M. (2015). D-Glucans from edible mushrooms: A review on the extraction, purification and chemical characterization approaches. *Carbohydrate Polymers*, 117, 753–761.

Salihović, M., Šapčanin, A., Špirtović-Halilović, S., Mahmutović-Dizdarević, I., Jerković-Mujkić, A., Veljović, E., … & Zećiri, S. (2019). Antimicrobial activity of selected wild mushrooms from different areas of Bosnia and Herzegovina. In *International Conference on Medical and Biological Engineering* (pp. 539–542). Springer, Cham.

Santoyo, S., Ramírez-Anguiano, A. C., Aldars-García, L., Reglero, G., & Soler-Rivas, C. (2012). Antiviral activities of *Boletus edulis, Pleurotus ostreatus* and *Lentinus edodes* extracts and polysaccharide fractions against Herpes simplex virus type 1. *Journal of Food and Nutrition Research*, 51, 225–235.

Sarikurkcu, C., Tepe, B., & Yamac, M. (2008). Evaluation of the antioxidant activity of four edible mushrooms from the Central Anatolia, Eskisehir – Turkey: *Lactarius deterrimus, Suillus collitinus, Boletus edulis. Xerocomus chrysenteron. Bioresource Technology*, 99(14), 6651–6655.

Sharon, N. (2007). Lectins: Carbohydrate-specific reagents and biological recognition molecules. *Journal of Biological Chemistry*, 282(5), 2753–2764.

Shizuka, F., Kido, Y., Nakazawa, T., Kitajima, H., Aizawa, C., Kayamura, H., & Ichijo, N. (2004). Antihypertensive effect of γ-amino butyric acid enriched soy products in spontaneously hypertensive rats. *Biofactors*, 22(1–4), 165–167.

Silva, B. M., Andrade, P. B., Valentão, P., Ferreres, F., Seabra, R. M., & Ferreira, M. A. (2004). Quince (*Cydonia oblonga* Miller) fruit (pulp, peel, and seed) and jam: Antioxidant activity. *Journal of Agricultural and Food Chemistry*, 52(15), 4705–4712.

Širić, I., Kasap, A., Bedeković, D., & Falandysz, J. (2017). Lead, cadmium and mercury contents and bioaccumulation potential of wild edible saprophytic and ectomycorrhizal mushrooms, Croatia. *Journal of Environmental Science and Health, Part B*, 52(3), 156–165.

Siu, K. C., Xu, L. J., Chen, X., & Wu J. Y. (2016). Molecular properties and antioxidant activities of polysaccharides isolated from alkaline extract of wild*Armillaria ostoyae* mushrooms. *Carbohydrate Polymers*, 137, 739–746.

Su, J., Zhang, J., Li, J., Li, T., Liu, H., & Wang, Y. (2018). Determination of mineral contents of wild *Boletus edulis* mushroom and its edible safety assessment. *Journal of Environmental Science and Health, Part B*, 53(7), 454–463.

Sun, L., Wang, C., Shi, Q., & Ma, C. (2009). Preparation of different molecular weight polysaccharides from *Porphyridium cruentum* and their antioxidant activities. *International Journal of Biological Macromolecules*, 45(1), 42–47.

Sun, L., Wang, L., & Zhou, Y. (2012). Immunomodulation and antitumor activities of different-molecular-weight polysaccharides from *Porphyridium cruentum*. *Carbohydrate Polymers*, 87(2), 1206–1210.

Tanaka, H., Watanabe, K., Ma, M., Hirayama, M., Kobayashi, T., Oyama, H., Sakaguchi, Y., Kanda, M., Kodama, M., & Aizawa, Y. (2009). The effects of γ-aminobutyric acid, vinegar, and dried bonito on blood pressure in normotensive and mildly or moderately hypertensive volunteers. *Journal of Clinical Biochemistry and Nutrition*, 45(1), 93–100.

Tang, W., & Lu, X. (1999). Study of the biological activity and the antitumor activity of *Boletus edulis* against Sarcoma-180. *Journal of Southwest China Normal University (Natural Science)*, 24(04), 478–481.

Tapwal, A., Kapoor, K. S., Thakur, Y., & Kumar, A. (2021). Ectomycorrhizal fungi associated with *Pinus gerardiana* in Kinnaur district of Himachal Pradesh, India. *Studies in Fungi*, 6(1), 425–436.

Teichmann, A., Dutta, P. C., Staffas, A., & Jägerstad, M. (2007). Sterol and vitamin D2 concentrations in cultivated and wild grown mushrooms: Effects of UV irradiation. *LWT-Food Science and Technology*, 40(5), 815–822.

Tibuhwa, D. (2017). Cytotoxicity, antimicrobial and antioxidant activities of *Boletus bicolor*, a basidiomycetes mushroom indigenous to Tanzania. *Tanzania Journal of Science*, 43(1), 151–163.

Tsai, S. Y., Tsai, H. L., & Mau, J. L. (2007). Antioxidant properties of *Agaricus blazei*, *Agrocybe cylindracea*, and *Boletus edulis*. *LWT-Food Science and Technology*, 40(8), 1392–1402.

Tsai, S. Y., Tsai, H. L., & Mau, J. L. (2008). Non-volatile taste components of *Agaricus blazei*, *Agrocybe cylindracea* and *Boletus edulis*. *Food Chemistry*, 107(3), 977–983.

Valentão, P., Lopes, G., Valente, M., Barbosa, P., Andrade, P. B., Silva, B. M., Baptista, P., & Seabra, R. M. (2005). Quantitation of nine organic acids in wild mushrooms. *Journal of Agricultural and Food Chemistry*, 53(9), 3626–3630.

Vamanu, E., & Nita, S. (2013). Antioxidant capacity and the correlation with major phenolic compounds, anthocyanin, and tocopherol content in various extracts from the wild edible *Boletus edulis* mushroom. *BioMed Research International*, 2013, 313905. 10.1155/2013/313905

Vaugham, J. G., & Geissler, C. A. (1997). *The New Oxford Book of Food Plants*. Oxford University Press.

Vetter, J. (2005). Mineral composition of basidiomes of Amanita species. *Mycological Research*, 109(6), 746–750.

Wang, J., Liu, T., Ren, S., Yang, Z., Qin, W., Wang, L., Zhuo, Q., Gong, Z., & Shen, S. (2020). Analysis and evaluation of nutritional composition and antioxidant activities in five varieties of *Boletus spp.* from Yunnan. *Edible Fungi of China*, 39(10), 87–91.

Wang, X. M., Zhang, J., Li, T., Wang, Y. Z., & Liu, H. G. (2015). Content and bioaccumulation of nine mineral elements in ten mushroom species of the genus *Boletus*. *Journal of Analytical Methods in Chemistry*, 2015.

Wang, Y., & Xu, B. (2014). Distribution of antioxidant activities and total phenolic contents in acetone, ethanol, water and hot water extracts from 20 edible mushrooms via sequential extraction. *Austin Journal of Nutrition and Food Sciences*, 2(1), 5.

Wang, Y., Sinclair, L., Hall, I. R., & Cole, A. L. J. (1995). *Boletus edulis* sensu lato: A new record for New Zealand. *New Zealand Journal of Crop and Horticultural Science*, 23(2), 227–231.

Wu, S., Wang, G., Yang, R., & Cui, Y. (2016). Anti-inflammatory effects of *Boletus edulis* polysaccharide on asthma pathology. *American Journal of Translational Research*, 8(10), 4478.

Xiao, Y., Xu, Q., Zhou, X., Liu, N., Gou, Z., & Li, S. (2018). The improvement effect of *Boletus edulis* polysaccharides on the liver injury in type 2 diabetic rats. *Science and Technology of Food Industry*, 39(11), 297–300.

Xin, X., Zheng, K., Niu, Y., Song, M., & Kang, W. (2018). Effect of *Flammulina velutipes* (golden needle mushroom, eno-kitake) polysaccharides on constipation. *Open Chemistry*, 16(1), 155–162.

Yu, S., Ma, R., Dong, X., Ji, H., & Liu, A. (2022). A novel polysaccharide from *Boletus edulis*: Extraction, purification, characterization and immunologic activity. *Industrial Crops and Products*, 186, 115206.

Zhang, A., Xiao, N., He, P., & Sun, P. (2011). Chemical analysis and antioxidant activity *in vitro* of polysaccharides extracted from *Boletus edulis*. *International Journal of Biological Macromolecules*, 49(5), 1092–1095.

Zhang, L., Hu, Y., Duan, X., Tang, T., Shen, Y., Hu, B., … & Liu, Y. (2018). Characterization and antioxidant activities of polysaccharides from thirteen *Boletus* mushrooms. *International Journal of Biological Macromolecules*, 113, 1–7.

Zhang, L., Hu, Y., Duan, X., Tang, T., Shen, Y., Hu, B., Liu, A., Chen, H., Li, C., & Liu, Y. (2018). Characterization and antioxidant activities of polysaccharides from thirteen *Boletus* mushrooms. *International Journal of Biological Macromolecules*, 113, 1–7.

Zhang, M. H., Zhou, R., Liu, F., & Ng, T. B. (2021). Purification of a novel protein with cytotoxicity against non-small-cell lung cancer cells from *Boletus bicolor*. *Archiv der Pharmazie*, 354(9), 2100135.

Zheng, Q., Zhang, H., Li, W., Tao, W., Zhou, F., & Gao, X. (2019). Hepatoprotective effect of *Boletus edulis* polysaccharides on mice with acute hepatic damage. *Food and Machinery*, 35(12), 141–145.

Zheng, S., Li, C., Ng, T. B., & Wang, H. X. (2007). A lectin with mitogenic activity from the edible wild mushroom *Boletus edulis*. *Process Biochemistry*, 42(12), 1620–1624.

Zhu, J., & Jia, Y. (2018). Research on extraction technology and intervention of constipation of dietary fiber from *Boletus edulis*. *Strait Pharmaceutical Journal*, 30(06), 31–33.

4 *Cordyceps sinensis* (Caterpillar Fungus) – Yarsagumba and *Cordyceps militaris*

Jyoti Gaba, Prerna Sood, Tanvi Sahni, and Pardeep Kaur
Punjab Agricultural University, Ludhiana, India

CONTENTS

INTRODUCTION

Throughout mankind history, natural bioresources have been crucial for feeding a substantial section of the world's population. According to estimates, around 1.6 billion world's population, including 60 million autochthonous ethnic groups, is directly dependent on natural products for their livings. Autochthonous forest dwellers rely on natural resources not just for their survival, but also for a significant portion of their revenue. According to Angelsen et al. (2014), forests

DOI: 10.1201/9781003259763-4

provide more than 20 percent of family income in emergent nations and more than 90 percent of those who are extremely poor, and rely on woodland for all or part of their livelihood (Pouliot and Treue, 2013). In alpine regions, particularly the Himalayas, gathering non-timber forest produce and herbs from their natural habitats is an important occupation (Negi et al., 2018). The Himalaya, which is a Global Species Hotspot, is widely known for its variety, unique biodiversity, and a wide range of commodities and services. Bioresources have long been used for food, housing, and medicine by local and indigenous ethnic groups in this region. Indigenous ethnic tribes in the Himalaya have deep knowledge about the traditional uses of medicinal plant and health management (Negi et al., 2018). People in various sections of the Himalayas use several folk medical systems like Unani, Amchi, Ayurveda, Siddha, and others. In India, more than 1100 medicinal plant varieties are traded, mostly for use in various medicinal systems with many of these varieties having their native habitats in the Himalayan highlands. Himalayan mushrooms have found a special place in traditional medicinal systems with a plethora of nutritional and health-promising applications.

Genus *Cordyceps* ('cordy' means club and 'ceps' means head) is ascomycete fungi, parasitic on arthropods at different development stages insects. More than 300 *Cordyceps* fungi species have been described in the literature including *Cordyceps militaris* and *Ophiocordyceps sinensis* (*aka Cordyceps sinensis*) as the most important ones. When this fungus attacks an insect, the fungal mycelium completely damages the host, and an elongated stem, *aka* ascocarp, emerges from the head of the dead insect in the form of a branch (Gu et al., 2006). The ascocarp or 'the caterpillar fungus' is collected by the local population, which has been documented vastly for being used in traditional Chinese and Indian medicinal systems. Its local names are Yartsa gunbu, Yarcha Gumba Yarsa Gumba, Sanjeevani Bhooti, Jeeva buti, Keera jhar, keeda ghass, Chyou Kvia, etc. In Chinese medicinal system, *Cordyceps* is known for treatment of chronic cough, lower back pain, constipation, asthma, bacterial infections, high blood pressure, kidney, and heart problems (Zhou et al., 2009). It is consumed as an additive to hen/duck soup with the purpose of enhancing immunity and body strengthening of physically weak patients. *C. militaris* and *O. sinensis* have been reported to exhibit antitumor (Yuan et al., 2005), immunomodulating (Koh et al., 2002), antihypertensive (Xu et al., 2000) with protective influence on the liver (Liu and Shen, 2003), kidney and heart (Xu et al., 2000). Main constituents of *Cordyceps* having antitumor activity include polysaccharides, adenosine, and sterols. Owing to such a wide range of medicinal properties, the rise in demand of these Himalayan fungal species is obvious. In current times, *O. sinensis* and *C. militaris* are among the main constituents of immunity boosting herbal tablets and capsules. *O. sinensis* is among the world's costliest fungi with high nutritional, economic and pharmacological potential. When such a single species emerges as high value, as a result of rising demand, its natural numbers begin to experience severe strain.

In recent decades, a variety of *O. sinensis* from high-peak environments of the Himalaya, has been documented. In the last three decades, collecting high-value, caterpillar fungus has become a significant source of income and livelihood in Bhutan (Wangchuk et al., 2012), Nepal (Kuniyal and Sundriyal, 2013; Shrestha and Bawa, 2014), India (Kuniyal and Sundriyal, 2013), and China (Winkler, 2008, 2009). Contemporary concerns about ending hunger and guaranteeing food and health security have prompted new methods of gathering, consuming, and trading natural resources (Shrestha et al., 2019). Conservation and management of economically and biologically important natural species should be a pressing priority for mankind.

In this chapter, we have discussed in detail the importance, distribution, phytochemistry, pharmacology, cultivation, and conservation of two *Cordyceps* species *viz. O. sinensis* and *C. militaris*.

TAXONOMY

	O. sinensis	*C. militaris*
Kingdom	fungi	fungi
Division	Ascomycota	Ascomycota
Class	Sordariomycetes	Sordariomycetes
Order	Hypocreales	Hypocreales
Family	Ophiocordycipitaceae	Cordycipitaceae
Genesis	Ophiocordycipitaceae	*Cordyceps*
Species	*O. sinensis*	*C. militaris*

COMMON NAMES

Cordyceps is known by different local names in different languages and countries.

Tibet: Caterpillar Fungus, Yartsa gunbu
Local Name: Yarcha Gumba Yarsa Gumba,
Nepali Name: Keera jhar, Jeeva buti, keeda ghass, Chyou Kvia, Sanjeevani Bhooti
Chinese Name: Dong Chong Xi Cao
Japanese: Tocheikasa
English Name: cordyceps mushroom, Caterpillar fungus

MORPHOLOGY AND FRUITING BODIES

Caterpillar fungus that develops on the larvae of host insects and spends the entirety of its three- to four-year or longer larval period underground. When infected with the fungus in the spring, these larvae often die after consuming the roots and caudexes of alpine plants. In the early spring, when the outside temperature rises, the endosclerotium begins to germinate, extrudes through the larva's head portion, and eventually protrudes through the soil, termed stroma (fruiting bodies). The adult perithecia on its head are packed with ascospores that resemble threads. When the fungus reaches maturity, its fruiting body releases thousands of ascospores into the environment (mainly soil), where newly born caterpillars come into touch with them and get infected. Since their hatching time overlaps with the discharges and dispersion of the ascorpores, the infection most likely happens at the first instar stage. The fungus infects the larvae's hemocoel before multiplying and fragmenting into fusiform hyphae. All of the larva's internal organs, with the exception of its exoskeleton, are consumed by the fungi as they propagate via the circulatory system throughout the body. The infected larva goes to a depth of 2 to 5 cm beneath the soil's surface where it passes away, its head facing upward. A yeast-like stage then emerges after a period of hibernation, spreading through the hemocoel and condensing within the insect's lipid reserve. Therefore, as the fungus grows, the host's food supply is depleted, thereby starving the caterpillar to death. Before the earth freezes, a little stroma bud often breaks through the sclerotium's (host larvae) head. It is a stage of the life cycle known as dormancy that may withstand unfavorable snow-cold circumstances. *O. sinensis* hyphae can grow at temperatures as low as 2 degrees, with 15 to 18 degrees being the optimal range.

DISTRIBUTION

Cordyceps genus is cosmopolitan with high occurrence in east and south-east Asia. However, *O. sinensis* is found only in very limited areas of the Tibetan Plateau at an altitude of 3500 to 5000m above sea level (asl). *C. militaris* is widely distributed in sub-tropical and temperate regions of

TABLE 4.1
Areas Where the Caterpillar Fungus has Been Found in Four Different Nations

Country	Altitudinal range (m asl)	Reported regions of occurrence
Bhutan	4200–5200	Bumthang Valley (North Central Bhutan), Namna (North Western Bhutan), and Bumdeling Wildlife Sanctuary
China	2260–5000	Xinjiang, Yunan, Shanxi, Jilin, Shaanxi, Hubei, Taiwan, Guangxi, Zhejiang, Jiangxi, Guizhou, Guangdong, Sichuan and Hainan Province, and Lhasa and Shannan in Tibet
India	3200–4800	Uttarakhand (Darma valley, Ralamdhura, Choudans valley, Panchachuli base, Moist alpine areas of Dharchula and Munsyari Blocks especially, Pindari catchment in Bageshwar district, Nanda Devi Biosphere Reserve, Niti valley, Kanol in Chamoli district, Sutol, Arunachal Pradesh and Sikkim (North and East Sikkim i.e., Luchung, Khangchendzanga National Park and Wildlife Sanctuary, etc.
Nepal	3540–5050	Darchula, Bajura, Dolpa, Kalikot, Mugu, Humla, Jumla, Rukum, Bajhang, Manang, Mustang, Rasuwa, Gorkha, Lamjung, Dhading, Dolakha, Solukhumbu, Sindhupalchowk, Sankhuwasabha, and Taplejung districts

American, European, and Asian continents. Table 4.1 provides the information about altitude range and occurrence of *O. sinensis* and *C. militaris* in specific areas (state/district) of a country.

MEDICINAL USES

Cordyceps is considered both animal and vegetable in ancient medicinal texts and documents. It has been used extensively in conventional medicinal systems to treat different ailments and diseases. It can be stored in the form of dried powder as well as in the form of vinegar and ethanol extract. *O. sinensis* is known for boosting energy *via* enhanced production of ATP and hence improving exercise performance (Manabe et al., 1996; Kumar et al., 2011; Chen et al., 2014). It has been used as dietary supplement to reduce fatigue and increase sex drive since generations (Huang et al., 2001; Hsu et al., 2011). It shows anti-ageing properties and anti-tumor effects *via* suppressing the growth of lung, colon, skin, and liver cancer cells. Due to its insulin-mimicking action, it has been utilised to manage type-2 diabetes (Guo and Friedman, 2010). *C. militaris* is counted among the oldest medicinal fungal species for sourcing of important secondary metabolites. It has a long history of being used for conditions like asthma, chronic cough, low immunity, night time urination, high cholesterol, ear ringing, sexual problems, opium addiction, weakness and dizziness, anemia and excess weight loss, etc. (Olatunji et al., 2018; Panda and Swain, 2011; Zhu et al., 1998). It has shown high biological potential with anti-inflammatory, anti-tumor, anti-oxidant, anti-proliferative, pro-sexual, antiviral, antibacterial, insecticidal, lavicidal, anti-diabetic, liver protective, and neuroprotective activities.

CULTIVATION AND HARVESTING

The insect species prone to *O. sinensis* infection live underground at the dept of nearly 6 inches at an altitude of 3000–5000 m asl in the Himalayas and Tibetan plateau (Liu et al., 2002). Caterpillar belonging to genera *Thitarodes*, *Bipectilus*, *Endoclita*, *Gazoryctra,* and *Pharacis* are potential hosts of *O. sinensis* (Lo et al., 2013). A branch like dark brown fungal elongate grows from the caterpillar's head and emerges out of the ground before spring season (Stone, 2008). This fungal elongate disperses spores at the end of the summer season, which release the mycelia and infect the caterpillar in late autumn *via* interaction with skin chemicals.

Harvesting of *Cordyceps* starts in the mid of May and continues until the beginning of August. The search for *Cordyceps* starts in the morning, collectors scan the area very carefully by laying on the ground and keep on searching till evening. As the required species is located, grasses and surrounding dust are removed and fungal elongation is taken out carefully from the soil to avoid any damage to it. After gently brushing away any dirt particles that had adhered to the *Cordyceps*, the collected materials are placed air dried until evening. The high economic importance of *Cordyceps* fungus can be understood from the fact that local people stop sending their children to schools during the harvesting season and engage them in fungus-collecting tasks in order to get more money. The collected materials are given to a sub-local dealer on a piece-by-piece basis whenever they arrived in the community. The dealer stretched them out on a newspaper, wash and dry them properly at room temperature. After that, the dried goods are folded in tissue paper and kept in a sealed container. In Gnathang, it was customary to wrap the cleaned items in muslin fabric and hang them over the bhukari (room heater) to dry and store them.

Very high demand and rising prices have got the attention of researchers towards the synthetic production of *O. sinensis*, although there has been no success in the field of rearing of fungi on infected and cultivated caterpillars (Jiang and Yao, 2003).

PHYTOCHEMISTRY

C. sinensis is a well-known and most expensive species of genus *Cordyceps* that possess unbeatable pharmacological properties like healing dysfunction of kidney and lungs, fatigue, anticancer, aphrodisiac effect, glucose homeostasis, immune-regulatory, nephroprotective effects, etc. Hence there is a need to explore the constituents of *C. sinensis* responsible for these marvelous health benefits. Till Now, chemical constituents of *C. sinensis* were broadly categorized into nucleosides, polysaccharides, sterols, proteins, amino acids, polypeptides, and many others. A detailed description of the structure, corresponding subtypes, and their biological activity are summarized in Table 4.2. Cordycepin is one of the major bioactive nucleosides whose structure is analogous to adenosine with OH absent at the 3'position of the ribose sugar. It possesses anti-proliferative, proapoptotic, anti-inflammatory activity, antibacterial, anti-viral, and insecticidal activity. It prevents focal cerebral ischemic/reperfusion (IR) injury and is also used for the treatment of heart diseases like myocardial infarction; anti-inflammatory and analgesic medicine (Wong et al., 2010). Other nucleosides are also present with nitrogenous base pairs like adenosine, cytosine, uracil, thymine, guanine, and hypoxanthine. These nucleosides enhance immune response; increase iron absorption in the gut; repair gastrointestinal injury; control urethral inflammation; increase blood circulation and improve brain functioning (Fan et al., 2007). Other components responsible for the anti-inflammatory property of *Cordyceps* are cordysinin A-E. Cordysinin A is cyclo (L-leucine-L-hydroxyproline). Cordysinin B is structurally different from adenosine nucleoside by a methoxy group at 2' position. Cordysinin C and D are enantiomeric mixtures of 1-(9H-β-carbolin-1-yl)ethanol. Polysaccharides constitute 3–8% weight of the fungal body. It is well known that acid phosphatase activity and phagocytosis are both increased by cordysinocan (Cheung et al., 2009; Zheng et al., 2011). The exopolysaccharide fraction of *C. sinensis* is found responsible for antioxidative property; immunomodulatory; anti-tumor effects and scavenging of free radicals (Sheng et al., 2011). APS (acid polysaccharide) reduces H_2O_2-induced cell death; stimulate macrophages and imparts immunomodulatory effect (Shen et al., 2011). Bioactive steroids are also isolated from *C. sinensis* namely, 5α,8α-epidoxy-24(R)-methylcholesta-6,22-dien-3β-D-glucopyranoside and 5,6-epoxy-24(R)-methylcholesta-7,22-dien-3β-ol with anti-tumor activity (Matsuda et al., 2009). Apoptosis inducing sterols are sitosterol, 5α-8α-epidioxy-22E-ergosta-6,22-dien-3-β-ol, 5α-8α-epidioxy-22E-ergosta-6,9,22-trien-3β-ol, 5α,6α-epoxy-5α-ergosta-7,22-dien-3β-ol and ergosterol. Ergosterol is an active ingredient in steroid

TABLE 4.2
Bioactive Constituents of *C. sinensis*

S.No.	Constituents	Structure	Biological Activity	References
1.	**Nucleosides**			
1.1	Cordycepin	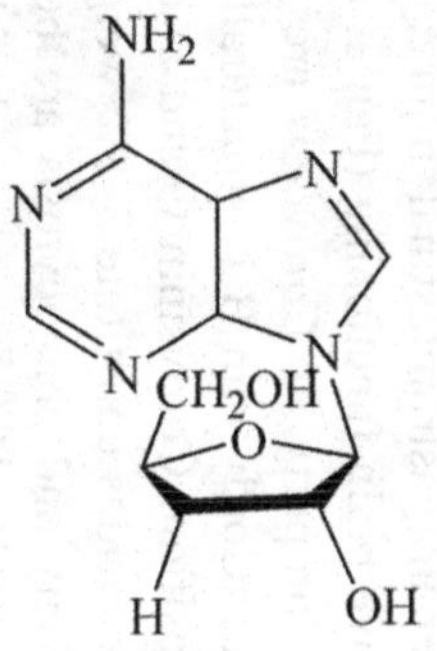	Antiproliferative, Proapoptotic, anti-inflammatory; prevent focal cerebral ischemic/reperfusion (IR) injury (13,14); For treating heart diseases like myocardial infraction; anti-inflammatory and analgesic medicine; antibacterial, anti-viral, insecticidal activity	Ng and Wang, 2005; Wong et al., 2010; Wang et al., 2012; Qian et al., 2012
1.2	Adenosine	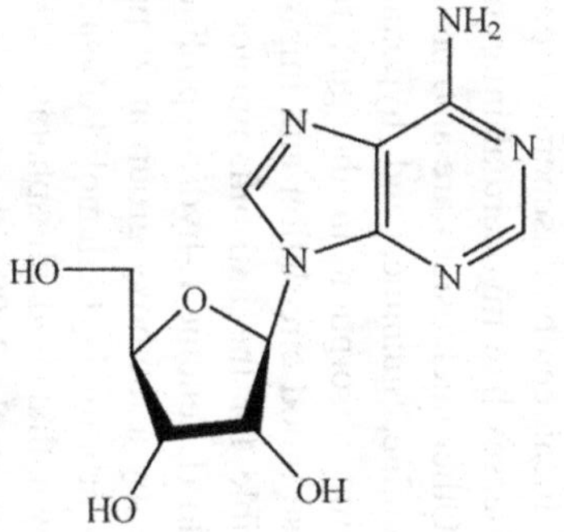	Higher in *C. sinensis* than *C. militaris*; helps in energy transfer in cells; prevents cytoprotection and tissue damage in chronic heart failures; anti-inflammatory and anticonvulsant activity	Kitakaze and Hori, 2000; Nakav et al., 2008; Manfredi and Sparks, 1982; Ontyd and Schrader, 1984

1.3	Other base pairs Cytosine Uracil Thymine Guanine hypoxanthine	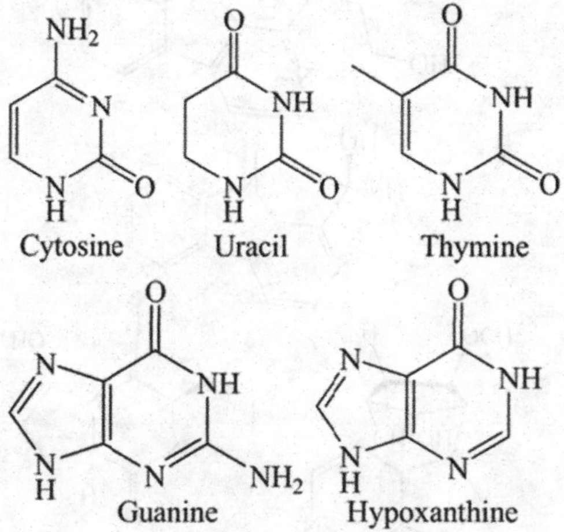 Cytosine, Uracil, Thymine, Guanine, Hypoxanthine	1. nucleosides were found in higher quantity in cultured *C. sinensis* as compared to naturally occurring species. 2. Nucleosides are used as markers for evaluating nucleoside-containing material. 3. Nucleotides like AMP, GMP, and UMP enhance immune response; increase iron absorption in the gut; repair gastrointestinal injury; controls urethral inflammation; increases blood circulation; improve brain functioning.	Yang et al., 2009; Fan et al., 2007; Struck-Lewicka et al., 2014; Yang et al., 2010; Shaoping et al., 2001; Zhou et al., 2009

(*Continued*)

TABLE 4.2 *(Continued)*
Bioactive Constituents of *C. sinensis*

S.No.	Constituents	Structure	Biological Activity	References
1.4	Cordysinins A-E (*C. sinensis*)	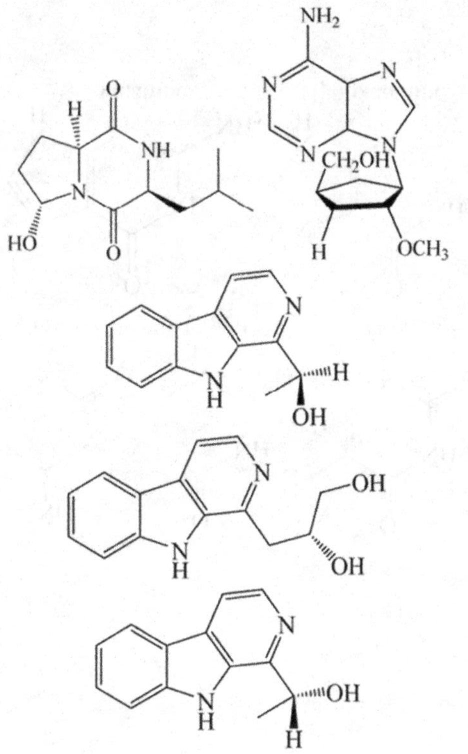	Anti-inflammatory	Yang et al., 2011
2	**Polysaccharides**	3–8% of total weight	Antitumor; Hypoglycemic, Hypolipidemic, anticancer; anti-influenza virus; immunopotentiation; hypoglycemic; hypocholesterolemic; anti-oxidant effects	Chen et al., 1997; Li et al., 2002; Ohta et al., 2007; Nakamura et al., 1999; Kiho, 1999; Koh et al., 2003
2.1	Cordysinocan	Molecular weight 82kDa Glucose, Mannose, Glactose (2.4:2:1)	Induce Cell proliferation	Cheung et al., 2009

2.2	EPSF (Exo polysaccharide fraction)	Molecular weight -1.04 X 10^5 Da Mannose, Glucose, Glactose (23:1:2.6)	Antioxidative property; immunomodulatory; anti-tumor effects.	Sheng et al., 2011; Movassagh et al., 2004; Song et al., 2011
2.3	Cyclopeptide-cordyhepta peptide A		Antimalarial, cyctotoxic activities	Rukachaisirikul et al., 2006
2.4	APS (Acid polysaccharide)	Mannose: glucose: glactose (3.3:2.3:1)	Reduced H_2O_2 induced cell death; stimulate macrophages; immunomodulatory effect	Shen et al., 2011; Chen et al., 2010
3	**Steroids**			
3.1	5α,8α-epidoxy-24(R)-methylcholesta-6,22-dien-3β-D-glucopyranoside	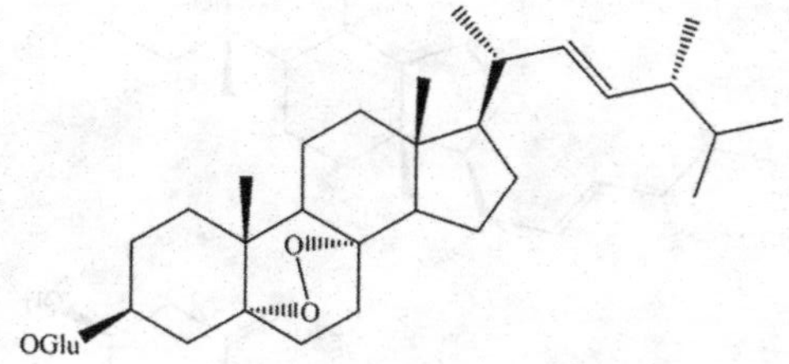	Antitumor activity	Matsuda et al., 2009; Bok et al., 1999
3.2	5,6-epoxy-24(R)-methylcholesta-7,22-dien-3β-ol	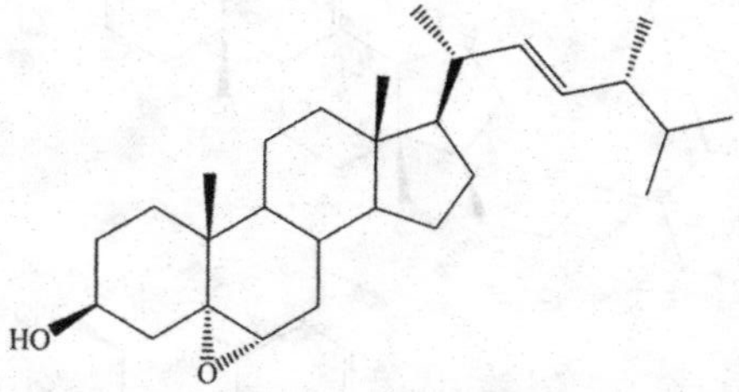		Bok et al., 1999
3.3	Ergosterol-3-*O*-β-D-glucopyranoside	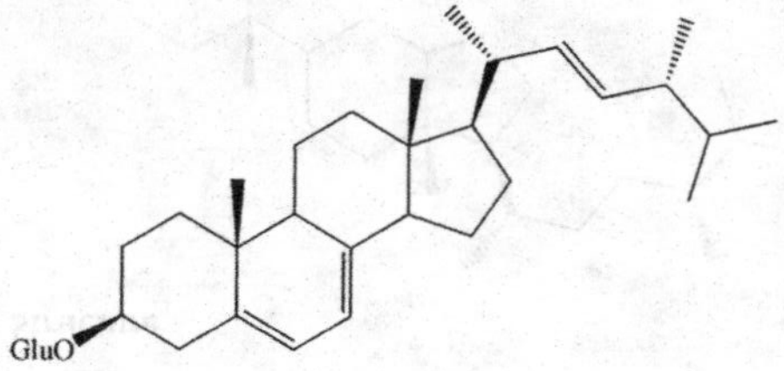		Bok et al., 1999

(*Continued*)

TABLE 4.2 *(Continued)*
Bioactive Constituents of *C. sinensis*

S.No.	Constituents	Structure	Biological Activity	References
3.4	22-dihydroergosteryl-3-O-β-D-glucopyranoside	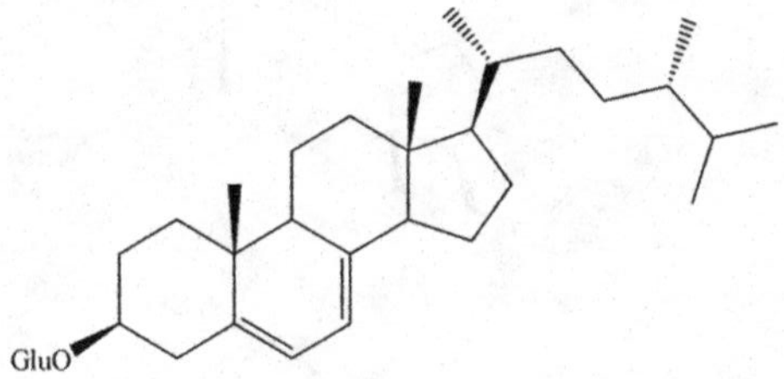 GluO		Bok et al., 1999
3.5	Sitosterol	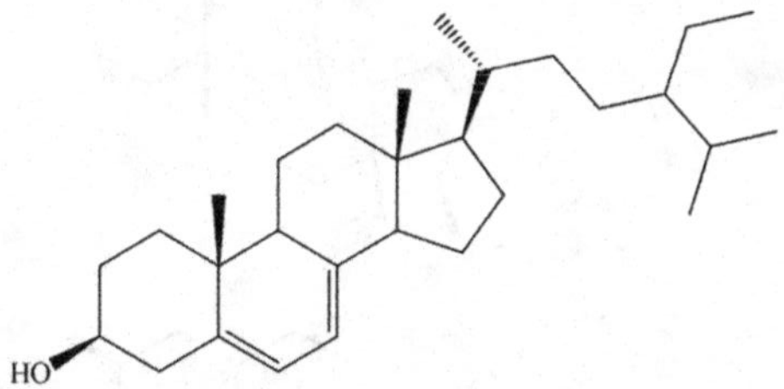 HO		Matsuda et al., 2009
3.6	5α-8α-epidioxy-22E-ergosta-6,22-dien-3-β-ol	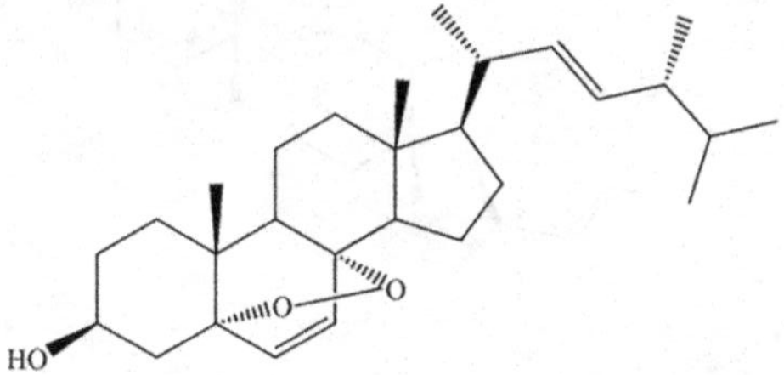 HO, O, O		Matsuda et al., 2009
3.7	5α-8α-epidioxy-22E-ergosta-6,9,22-trien-3β-ol	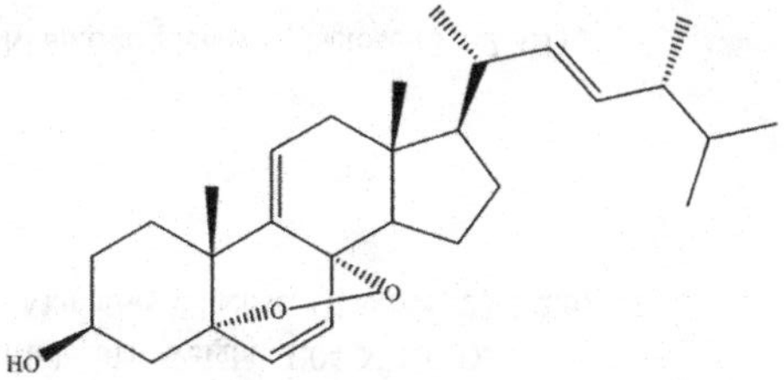 HO, O, O		Matsuda et al., 2009

3.8	5α,6α-epoxy-5α-ergosta-7,22-dien-3β-ol		Anticancer activity	Matsuda et al., 2009
3.9	Ergosterol		Active ingredients in steroid hormone drug: possessed antibacterial activity against *P. aeruginosa*; *E. aerogenes*; *C. albicans*	Kitchawalit et al., 2014; Zheng et al., 2013; Matsuda et al., 2009
3.10	H1A		Manage autoimmune disorders; Suppress activated HMC and alleviated IgAN (Bergers' disease)	Yang et al., 2003; Lin et al., 1999; Yang et al., 1999
4	**Proteins**	Mostly intracellular and extracellular protease enzymes		
5	**Amino acids and Polypeptides**		Reduce arterial pressure of rats; Vasorelaxant effect; Treatment of Hypertension	Chiou et al., 2000
5.1	Cordymin		Effect on diabetic osteopenia by decreasing ALP and TRAP activity; mediated recovery of β-cells	Vestergaard et al., 2009; Ahmed et al., 2006
5.2	Cordycedipeptide A (3-acetamino-6-isobutyl-2,5-dioxopiperazine)		Cytotoxic to L-929, A375 and Hela cell lines	Jia et al., 2005

(Continued)

TABLE 4.2 ***(Continued)***
Bioactive Constituents of *C. sinensis*

S.No.	Constituents	Structure	Biological Activity	References
5.3	Cordyceamides A and B	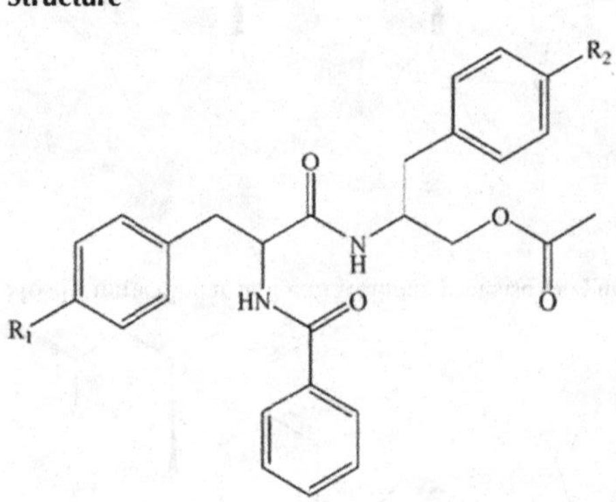 HN, O, N H, O, O, R_1, R_2 (A) R_1 = OH, R_2=H (B) R_1= OH, R_2= OH	Cytotoxic to L929, A375 and Hella cell lines	Jia et al., 2009
5.4	Tryptophan (AA)		Precursor of serotonin which induce insomnia; sedative-hypnotic effect; Immune inhibition	Zhang et al., 1991
5.5	GABA	Non-protein aminoacids	Inhibits neurotransmitter in CNS (Cerebellum, hippocampus, hypothalamus, striatum, spinal cord); regulate sleep, memory, learning process, emotional process like anxiety and stress; mypconvulsant and anti-relaxant activity	Chan et al., 2015; Boonstra et al., 2015; Hepsomali et al., 2020
5.6	Ergothioneine	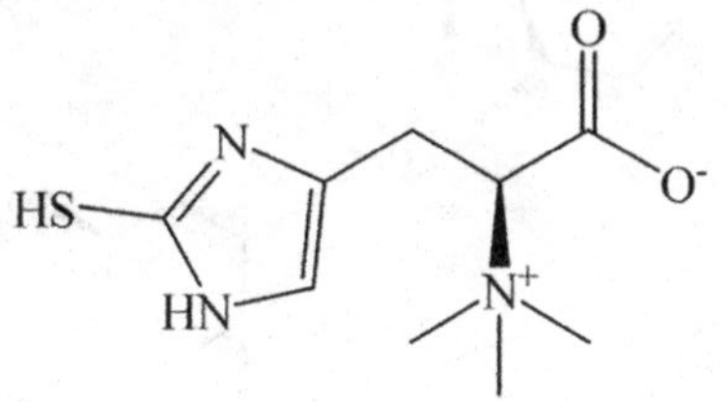 HS, N, HN, O, O^-, N^+	Antioxidant, Cyto-protective, Radioprotective	Borodina et al., 2020

6	Cordycepic acid		Treat liver fibrosis, diuretic effect; improves plasma osmotic pressure	Chatterjee et al., 1957; Guo and Friedman, 2010; Ouyang et al., 2013
7	Carotenoids	β-carotenes; Lycopene; Lutein; Zeaxanthine; Cordyxanthine (I-IV)	Lycopene boots endothelium of cardiovascular disease patients; used in chemotherapy of prostate cancer; lutein and zeaxanthine improves cognitive functions; level of cortisols, stress; also act as antioxidants	Dong et al., 2013; Bovier and Hammond, 2015; Stringham et al., 2018; Renzi-Hammond et al., 2017
8	Lavostatin		Drugs with pleiotropic effect; protect vascular endothelium; treat Hypercholestrolemia in patients; reduces coronary artherosclerosis in coronary artery patients	Cohen et al., 2014; Chen et al., 2012; Oesterle et al., 2017; Tobert, 1988
	Other compounds	Phenolic compounds like vanillic, caffeic acid, p-hydroybenzoic acid, gallic acid, protocatechuic acid	Antibacterial, antifungal, antiviral, anti-inflammatory	

hormone drugs, possessing antibacterial activity against *Pseudomonas aeruginosa*; *Enterobacter aerogenes* and *Candida albicans*. Fatty acids are also identified in GC-MS (Gas chromatography-mass spectrometry) studies. Unsaturated fatty acids constitute 57.84% namely, myristic, lauric, pentadecanoic, linoleic, stearic, oleic, lignoceric, docosanoic, palmitic and palmitoleic acid. Linoleic acid is present in maximum proportion (38.44%) (Zhou et al., 2009). Secondary metabolites like amino acids and polypeptides are also present which reduce arterial pressure of rats, having vasorelaxant and anti-hypertension effects. Cordymin effects on diabetic osteopenia by decreasing ALP (Alkaline phosphatase) and TRAP (Tartrate resistant acid phosphatase) activity-mediated recovery of β-cells. *In vitro* cultivated *C. militaris* acts as an alternative of *C. sinensis* due to the similar chemical components in their fruiting bodies (Jędrejko et al., 2021). *C. militaris* is reported to constitute cordycepin and adenosine in higher amounts as compared to *C. sinensis*. Presence of other bioactive compounds like lectins, cordycepic acid, carotenoids- lutein, zea-xanthin, ergosterol, lovastatin, GABA (γ-amino butyric acid), phenolics, flavonoids have also been described in the literature (Chan et al., 2015, Chen et al., 2012, Ohen et al., 2014).

PHARMACOLOGY

C. sinensis is an entomopathogenic fungus with high medicinal value and hence is used widely in health sector and pharmaceutical industries (Russell and Paterson, 2008; Ng and Wang, 2005). It is used for the treatment of reproductive disorders, kidney diseases, cardiac diseases, and various tumorous growths (Panda and Swain, 2011). It is also used as a source of energy by many athletes and elderly peoples (Zhu and Rippe et al., 2004). Several reports confirm that this fungus increases energy via producing more ATP, which is used by cell for various activities (Siu et al., 2004; Dai et al., 2001). Cordycepin, the major chemical constituent of *Cordyceps,* and other derivatives of cordycepin like ergosterols, polysaccharides, and glycoproteins have shown immunostimulatory, antioxidant, anti-fungal, anti-tumorous, and anti-metastasis effects (Yang et al., 2006), summarized in Table 4.3.

Anti-fatigue Effects

Fatigue is a feeling of tiredness or difficulty in doing any activity and it can be physical and mental fatigue (Chaudhuri and Behan, 2004; Mizuno et al., 2008). *Cordyceps* is used to restore the exhausted energy because of its potential to increase the strength and endurance (Bucci, 2002).

Polysaccharides from *Cordyceps* have shown its anti-fatigue, anti-inflammatory, anti-allergic, and hyperglycemic effects (Kuo et al., 2007). When mice were treated with the 200 mg/kg polysaccharide part of *Cordyceps* for 21 days, it enhanced the swimming fortitude time of mice, deferred the increase of lactic acid in the blood, increased liver and muscle glycogen (Li and Li, 2009). Ma et al. (2008) reported that the glycogen of liver and muscle are good parameters of fatigue and exhaustion. Dohm et al. (1983) also suggested that the reduction of muscle glycogen during strenuous exercise was an important factor for fatigue. *Cordyceps* can significantly increase the muscle and liver glycogen level after heavy exercise. Song et al. (2015) observed the effect of *C. militaris* extract (fruiting body extract) on antifatigue in mouse for two weeks which resulted in delayed fatigue phenomenon confirmed by forced running and forced swimming test. Besides this *C. militaris* also increased energy level, antioxidant enzyme level, and reduced reactive oxygen species and lactic acid.

Anti-angiogenic Effects

Angiogenesis is the process of the emergence of new blood vessels from existing vessels (Folk man, 1971). Nowadays anti-angiogenetic technologies are being used for the treatment of cancer. Different species of *Cordyceps* and their extracts have been reported to show anti-angiogenic and tumor growth effects. Among the herbal ingredients for this treatment most common is *C. militaris*

TABLE 4.3
Biological Activities of *C. sinensis/C.militaris*

Pharmacological activity	Extract or Part Used	Model Animals/ Tissues	Dose and Duration	References
Antifatigue	Polysaccharide *C. sinensis*	Mice	200 mg/kg, 21 days	Trigg et al., 1971; Li and Li, 2009
Anti-angiogenic	Extract of *C. militaris*	HUVECs	100 mg/L, 3–6 hours	Park et al., 2009a; Park et al., 2009b; Park et al., 2009c; Yoo et al., 2004
Immunomodulatory	Polysaccharide *C. sinensis*	Human blood	0.025–0.1 mg	Kuo et al., 2007
Spermatogenic	Mycelium powder of *C. militaris*	Boars	10g/boar, 2 months	Lin et al., 2007; Yu et al., 2006
Anti-aging	*C. sinensis* extract	Mice	2 and 4 g/kg, 6 weeks	Ji et al., 2009
Antioxidant	Water and ethanol extract of *C. sinensis*	Rats	2–4 ml, 28 days	Ra et al., 2008; Dong and Yao, 2008; Wang et al., 2005; Won and Park, 2005
Anti-metastasis	Water extract (*C. sinensis*)	Lewis lung carcinoma (LLC) and B16 melanoma cells	100 mg/kg, 26 days	Nakamura et al., 1999
Anti-diabetic or Hypoglycemic	Crude extract and polysaccharide of *C. militaris*	HUVECs	25 microgram/ ml, 12–36 hrs	Chu et al., 2011
Hepatoprotective and Antifibrotic	*C. militaris* Extract	Mice	200 mg/kg/ daily, 4 weeks	Manabe et al., 2000
Renal Protective	Powder of *C. militaris*	Lupus Nephritis (LN) Patients	2–4 g/day, 3 years	Lu, 2002
Cardio vascular	CS-4	Isolated aorta	50ug/ml, 1–15 min	Zhu et al., 1998; Liu et al., 2014
Hypolipidemic	Polysaccharide *C. sinensis*	Rats	50–100 mg/kg, 2 weeks	Yang et al., 1999
Anti-inflammatory	Water Extract of *C. militaris*	Macrophages of mice	1250ug/ml, 24 hrs	Jo et al., 2010
Anticancer	*Cordycepin of C. militaris*	Mice	15 mg/kg/day (2 weeks	Yoshikawa et al., 2004

(Stevan, 1999, Carmeliet and Jain, 2000). Yoo et al. (2004) reported that when 100–200 mg/L extract of *C. militaris* was induced to Human tumor and Umbilical Vein Endothelical cells (HUVEC) (HT 1080) *in vitro*, it inhibited the growth of HUVEC's after 3 and 6 hours of treatment. This was also evidenced by reduced expression of MMP-2 gene which is directly related with tumor growth in various types of cancer (Nomura et al., 1995).

Immunomodulatory Effects

The immune system provides protection from any type of infection or pathogen infecting the human body. Immunomodulating drugs are used to reduce the infections and restore the immune system to its normal function. Immunosuppressive drugs are used to suppress or control autoimmune system when excessive tissue damage occurs (Taylor et al., 2005). Polysaccharides from *Cordyceps* have shown antioxidant, antitumor, anti-inflammatory and immunomodulatory effects (Zhang et al., 2005; Aman et al., 2000; Yang et al., 2006). Several studies found *C. sinensis* as immunomodulator with potentiating or modulatory effect on immune system (Feng et al., 2008; Taylor et al., 2005). Zhu et al. (2012) reported the immunostimulatory effects of polysaccharides extracted from *C. gunni.* Similar results of enhanced immunomodulatory and immunostimulatory effects were shown by Kuo et al. (2007) by testing the polysaccharides from *C. sinensis* on human peripheral blood at a dose of 0.025–0.1 mg. Jung et al. (2019) found the increased immunoregulatory functions after a clinical trial of *C. sinensis* on healthy Korean subjects for 8 weeks.

Sexual Enhancement Activity or Spermatogenic Effects

Testosterone is the main hormone involved in spermatogenesis. As it initiates the development of sperm by activating the Sertoli or nurse cells. Various species of *Cordyceps* have been used as nutritious food to enhance sexual activity. Evidences showed that *C. militaris* and *C. sinensis* increase the sexual activity, libido activity and improve the reproductive health in humans (Zhu et al., 1998).

In one study mycelial culture of *C. sinensis* was given to male mice at a dose of 0.2mg/g for 7 days, which resulted in increased production of plasma testosterone (Huang et al., 2004). Similar study was conducted on castrated rats and were given cultured *C. sinensis* (0.5–2 g/kg for 21 days), which resulted in decreased penis erection latency (Ji et al., 2009). *C. militaris* has large amount of cordycepin as compare to *C. sinensis* (Yu et al., 2006). From ancient times Cordycepin obtained from *Cordyceps* has been used to cure male feebleness and other related disorders. Some studies have also explained the steroidogenic effects of *Cordyceps* species (Wan et al., 1988). *C. sinensis* has many important medicinal properties in relation to reproductive function in both males and females. It stimulated spermatogenesis (Huang et al., 2000). Lin et al. (2007) found the spermatogenic effect of *C. militaris* when given as diet supplement for two months to subfertile boars. Spermatogenesis was found to be enhanced and the quality and quantity of sperms was improved after two months.

Anti-aging Effects

Aging is defined as the decline in memory and brain functioning (Zhan et al., 1990). During the process of aging in mammals, there is an accretion of free radicals, which causes the oxidative damage to cells and organelles. *Cordyceps* showed anti-aging activity through its activity of scavenging of oxygen free radicals (Zhang et al., 1995). Ji et al. (2009) found that *C. sinensis* extract increased the memory in mice treated with D-galactose, and in castrated rats it increased the sexual function. It also improved the activity of anti-oxidative enzymes related to aging. It was found that *C. sinensis* had effect on hippocampus of brain and it can delay the decline of brain function in aging mice. This study also showed that *C. sinensis* extract enhanced the brain and antioxidative enzyme activity as aging is related with variance between oxidative damage and antioxidative defense function.

Antioxidant Activity

Reactive oxygen species (ROS) are known for their role in biological systems as these are useful as well as harmful to living systems (Valko et al., 2004). ROS can be increased or decreased while exposure to environmental contaminants and can cause damage to proteins, lipids, and membranes of cells. And this damage can lead to serious diseases like aging, tumors, cancer, respiratory diseases, and brain disorders (Valko et al., 2007; Zhong, 2006; Rahman, 2003). Ethanol and water extract the fruiting body of *C. sinensis* showed its antioxidant activity *in vitro* in various studies (Li et al., 2006; Li et al. 2002; Ra et al., 2008; Dong and Yao, 2008; Wang et al., 2005; Won and Park, 2005). Nowadays cultured *C. sinensis* is preferred more for its antioxidant activity over the natural one (Dong and Yao, 2008). *C. sinensis* and *C. militaris* helped in reducing the oxidative damage of biomolecules because of the free radical scavenging activity of both the species of *Cordyceps.*

Anti-metastasis and Antitumor Effects

Tumor or cancer is the most critical disease-causing mortalities worldwide (Xiao and Zhong, 2007). Cordycepin is one of the major components of *C. sinensis* (Hsu et al., 2017) which is commonly used for its anti-tumor, anti-metastasis, anti-fatigue, and anti-inflammatory effects (Wang et al., 2017a, 2017b; Hwang et al., 2017; Zeng et al., 2017). Studies have shown the antitumor activity of Cordycepin on tumor cells through different pathways. Natural and fermented *Cordyceps* have also shown antitumor effects (Zhou et al., 2009; Feng et al., 2008). Liu et al. (2008) have also suggested that *Cordyceps* can also be used for the treatment of cancer along with other chemotherapy methods.

Anti-diabetic or Hypoglycemic Effects

Cordyceps has shown its hypoglycemic activity in animals. In one study rats were given *Cordyceps* carbohydrate extract for 25 days, which resulted in increased insulin sensitivity, decreased insulin secretion, and decreased insulin in the plasma of rats (Balon et al., 2002). High glucose content induced oxidative stress, which may lead to vascular diabetic impediments (Susztak et al., 2006). Chu et al. (2011) found the ameliorative effect of *C. sinensis* and *C. militaris* against the glucose-induced oxidative stress in HUVEC.

Hepatoprotective and Antifibrotic Effects

Cordyceps extracts can be used to treat acute and chronic hepatitis (Zhao, 2000). Nguyen et al. (2021) showed the hepatoprotective effect of exopolysaccharide (EPS) of *C. sinensis* against the hepatotoxicity caused by CCl_4 in rats. Histopathological changes in liver tissue and oxidative effect in rats caused by CCl_4 was recovered by EPS extract of *C. sinensis*. Wang et al. (2015) also showed hepatoprotective ability of polysaccharide residue of *C. militaris* on mice. Results of histopathology of liver tissue showed that the residue of polysaccharide had the ability to attenuate liver cell damage. Polysaccharides extracted from mycelial culture of *Cordyceps* has shown its anti-liver injury effects in liver fibrosis in rats after CCl_4 administration (Peng et al., 2013). Nan et al. (2001) carried out a study on rats to induce liver fibrosis through bile duct ligation and scission. The lyophilized mycelial culture of *C. militaris* was given to rats through oral intubation (30 mg/kg/day for 28 days). After 28 days rats were anesthetized and liver was removed, weighed and processed for histochemical, immunohistochemical and biochemical (liver marker enzymes) examination. Results showed that *C. militaris-treated* rats showed reduced damage of liver architecture, bile duct proliferation and fibrosis as compared to control rats. Similar results of anti-fibrotic potential of *C. sinensis* were also shown by Boker et al. (1991).

Renal Protective Effects

Renal disease is the condition which shows the problems in urine/blood tests with decreased filtration efficiency of nephron of kidney (Nugent et al., 2011). *C. sinensis* is widely used for the treatment of renal dysfunction, chronic and acute nephritis or kidney failure (Feng et al., 2008). Many clinical studies have investigated the potential of *C. sinensis* for treating people with chronic kidney diseases (Deng et al., 2001; Jin and Chen, 2004; Yu and Tan, 2003). Zhang et al. (2011)

observed the effect of *C. sinensis* on kidney functions of patients with chronic allograft nephropathy. In this study, patients were treated with immunosuppressive drugs along with and *C. sinensis* and after six months of treatment, renal function was observed to be improved in most of the patients.

Cardiovascular Effects

Now days fermented *C. sinensis* is used as a substitute for natural *C. sinensis*. Wu et al. (2018) reported the protective effect of *C. sinensis* against doxorubicin-induced cardiotoxicity in male Sprague Wistar rats. In this study male rats were treated with doxorubicin, captopril and fermented *C. sinensis* and results showed decreased the heart weight index, serum phosphodiesterase, and increased myocardial content, increased mitochondrial and anti-oxidant enzymes activities. *C. sinensis* was found to inhibit the myocardial hypertrophy and myocardial damage and it was concluded that it can be used to prevent cardiotoxicity induced by doxorubicin. Yan et al. (2012) showed the cardioprotective effect of *C. sinensis* through coronary perfusion pressure and ventricular function and adenosine receptor initiation. Liu et al. (2014) also reported the protective effect of *C. sinensis*, when it was given through oral intubation, it attenuated the cardiac and liver injury in rats.

Hypolipidemic Effects

Hyperlipidemia is the major risk factor for the patients suffering from cardiovascular diseases. Nowadays people are favoring for diets having high fat, cholesterol, high energy, and reduced fiber content which ultimately leads to coronary diseases (Bush et al., 1988; Kang and Song, 1997). Yang et al., 1999 found the hypolipidemic effect of exopolysaccharide of *C. sinensis* when given (50–100mg/kg) to rats for two weeks. Exopolysaccharide of *C. sinensis* effectively reduced the plasma triglyceride and total cholesterol, thereby showing the potential of *C. sinensis* in combating hyperlipidemia in rats.

TOXICOLOGY

C. sinensis and *C. militaris* has medicinal and pharmaceutical properties. Only few studies have reported its toxicity as this fungus is safe for medicinal use and safe for long-term use. In a clinical trial, rabbits were fed with *Cordyceps* (10 g/kg/day for 3 months), and no mortality or other toxic effects on liver, kidney, or other organs functioning were observed (Huang et al., 1987). Patients with some types of diseases like diabetes, hypoglycemia, liver, and kidney diseases are suggested to avoid its use. Besides some published data on toxicity of *Cordyceps* but still *Cordyceps* can be considered as non-toxic mushrooms (Tuli et al., 2014). Suwannasaroj et al. (2021) carried out a study to test the acute and subchronic toxicity of *C. militaris* in Wistar rats. For acute toxicity testing, doses of *C. militaris* were 300 and 2,000 mg/kg body weight of rats. The results of acute toxicity showed no severe toxic symptoms and no mortality in rats. For subacute toxicity testing, doses were 5, 20, and 80 mg/kg body weight of rats and the results revealed that, No Observed Adverse Effect Level (NOAEL) of *C. militaris* was 80 mg/kg body weight /day for rats.

Some studies found the adverse gastrointestinal behavior and other symptoms like nausea, diarrhea, and mouth ulcers due to use of *Cordyceps* fungi (Zhou et al., 1998). In another study allergic responses were also found in patients treated with *C. sinensis*. *Cordyceps* have been found to have effect on plasma testosterone levels in males. In females its effect was lacking (Wong et al., 2007; Huang et al., 2004;). Meena et al., 2013 determined the toxicity of mycelia culture of *C. sinensis* in rats. *C. sinensis* mycelium was cultured in the laboratory and ground to form powder. This powder of laboratory culture of mycelia (LCM) was given to rats through oral intubation by cannula for 28 days. After 1, 2.5 and 4 hours of administration of LCM toxic effects and other behavioural changes were observed in rats. Body weight and feed intake was significantly increased in test rats. This increase in feed intake and body weight was due to increased growth of animals (Zhu and Rippe, 2004). No death was recorded in rats treated with LCM. Hematological parameters like RBC (red blood cells), WBC (white blood cells), PCV (packet cell volume) and Hb (hemoglobin) were found higher in LCM-treated rats than in control rats. WBC count was also found to be increased while neutrophils were significantly decreased after LCM treatment. Adeyemi et al. (2008) suggested that increased WBC count is indicative of boosted immune system.

Aramwit et al. (2015) reported *in vivo* toxicity and *in vitro* mutagenecity of cordycein of *C. sinensis*. The toxicity testing was done on rats and rats were given cordycepin for 30 days. Although cordycepin has anticancer effect but showed non-toxic effects to non-cancer cells. Haematological parameters like RBC, platelets and WBCs were observed to be increased in cordycepin treated rats than control rats (Meena et al., 2013). Histopathological changes were observed in liver tissue of rats after cordycepin treatment as compared to other tissues. Haemoglobin values were also differed significantly in cordycepin fed rats as compared to control rats. Zhou and Yao (2013) reported hepatotoxicity and nephrotoxicity of powder of *C.militaris*. For this repeated toxicity test of 28 days was done on male and female rats and were fed with *C. militaris* powder orally at 0, 1,2 and 3 g/kg/day. Increased concentration of liver marker enzymes at 3 g/kg/day showed hepatoxicity in both the sexes. The kidney toxicity was also shown by genetic expression of markers (kidney injury molecule-1 (KIM-1)) of kidney toxicity (Joy and Nair, 2008). Significant decrease was observed in kidney SOD (superoxide dismutase) and CAT (catalase) activities in male and female rats fed with *C. militaris* powder (Romeu et al., 2002). Nephrotoxicity is characterised by necrotic tubules, destruction of mitochondria and antioxidant defence system (Wojcikowski et al., 2007). Table 4.4 summarizes the toxicology of *C. sinensis* and *C. militaris*.

ECONOMIC IMPORTANCE

Owing to the presence of large number of important phytoconstituents and numerous biological activities, *Cordyceps* demand is increasing at high speed in American, Asian, European, and African countries. As per an estimate, the global *C. sinensis* market is expected to have a worth of more than thousand million dollars by the year 2027. Increase in number of patients with chronic diseases like diabetes, heart, and respiratory problems have also contributed to growing demand of this species. Both *C. sinensis* and *C. militaris* are available in liquid, powder and tablet/capsule form, which can be used as food additive, dietary supplement, and pharma/health care products. Some of the companies operating in the global *C. sinensis* and *C. militaris* market are- The Lubrizol Corporation, Naturalin bio-resources co., Ltd, Xi'an Saina Biological Technology Co., Ltd. (Herbsino), Shanghai Kangzhou Fungi Extract Co., Ltd., Quyuan sunnycare Inc., Nutra Green Biotechnology Co., Ltd., Health Choice Limited, Dalong Biotechnology Co., Ltd, Nutrastar International Inc., etc. Table 4.5 enlists some of the commercially available products containing *Cordyceps*.

CONCLUSION

Cordyceps is one of the most important and expensive species growing at high altitudes of the Himalayas. With a range of bioactive constituents, these fungal species have been considered valuable in traditional as well as modern medicinal systems. Collection, processing and trading, etc., of *Cordyceps* provides high economic benefits to the local people of the Himalayan region. However, uncontrolled harvesting of this highly valued natural medicine is exerting pressure on the sustainability of the ecosystem. Therefore, there is a need for maintaining a balance between its habitat conservation and economic benefits for a sustainable development. For this, a rational

TABLE 4.4
Toxicology of *C. sinensis/militaris*

Toxicity	Part of *Cordyceps* Used	Model Animal Used	Route, Dose and Duration	Toxic Effects	References
General Toxicity	mycelium of *C. sinensis*	Rats	Oral, 30 days	Body weight and feed intake significantly increased	Zhu and Rippe, 2004
Acute toxicity	Cordycepin of *C. militaris*	Wistar rats	Oral, 300 and 2,000 mg/kg body weight, 15 days	No severe toxicity symptoms and no mortality reported	Suwannasaroj et al., 2021
Subchronic toxicity	Cordycepin of *C. militaris*	Wistar rats	Oral, 5, 20 and 80 mg/kg body weight, 90 days	Toxicity reported at these doses	Suwannasaroj et al., 2021
Haematotoxicity	Laboratory culture of mycelium of *C. sinensis*	Rats	Oral, 28 days	RBC, WBC count significantly increased	Meena et al., 2013
Hepatotoxicity	Codycepin	Rats	Oral, 30 days	Liver marker enzymes (ALT, AST) significantly increased	Aramwit et al., 2015
Nephrotoxicity	Powder of *C. militaris.*	Rats	Oral, 28 days	Degeneration in cellular organelles and necrotic tubules of kidney	Zhou and Yao 2013

TABLE 4.5
Commercially Available Products Containing *C. sinensis* and *C. militaris*

Sr. No.	Name of Product	Brand name	Formulation	Health benefits
1	*Cordyceps* extract	Bulk supplements.com	Powder	Immune support
2	*Cordyceps*	VH Herbals	Capsules	Improve anxiety and cognitive support, relieve stress, boosts athletic performance, improves heart health
3	Organic *Cordyceps*	Nutricost	Powder	Promotes immune support
4	Genius mushrooms	The Genius Brand	Capsules	Liver support, boost energy, brain health
5	*Cordyceps* mushroom	Fresh Nutrition	Capsules	Supports healthy natural immune system
6	Cordychi	Host Defence	Capsules	Fatigue reduction
7	*Cordyceps* mushroom	Horbaach	Capsules	Promote health and wellness

harvesting approach is suggested, where different areas having*Cordyceps* species will be harvested in a particular sequence/cycle and with a minimum specified time gap. Also, research in the field in artificial cultivation and *in vitro* cultivation should be encouraged to decrease the burden on the natural habitat of *Cordyceps*.

REFERENCES

O.O. Adeyemi, A.J. Akindele, K.I. Nwumeh. 2008. Acute and subchronic toxicological assessment of *Byrsocarpus coccineus* Schum and Thonn (Connaraceaae) aqueous leaf extract. *Planta Med* 74: A-26.

L.A. Ahmed, R.M. Joakimsen, G.K. Berntsen, V. Fønnebø, H. Schirmer. 2006. Diabetes mellitus and the risk of non-vertebral fractures: the Tromsø study. *Osteoporosis International* 17(4): 495–500.

S. Aman, D.J. Anderson, T.J. Connolly, A.J. Crittall, J. Guijun. 2000. From adenosine to 3′-deoxyadenosine: development and scale-up. *Org Process Res Dev* 4: 601–605.

A. Angelsen, P. Jagger, R. Babigumira, B. Belcher, N.J. Hogarth, S. Bauch, J. Börner, C. Smith-Hall, S. Wunder. 2014. Environmental income and rural livelihoods: A Global-comparative analysis. *World Dev* 64: S12–S28.

P. Aramwit, S. Porasuphatana, T. Srichana, T. Nakpheng. 2015. Toxicity evaluation of cordycepin and its delivery system for sustained *in vitro* anti-lung cancer activity. *Nanoscale Res Lett* 10: 152.

T.W. Balon, A.P. Jasman, J.S. Zhu. 2002. A fermentation product of *Cordyceps sinensis* increases whole-body insulin sensitivity in rats. *J Altern Complement Med* 8: 315–323.

J.W. Bok, L. Lermer, J. Chilton, H.G. Klingeman, G.N. Towers. 1999. Antitumor sterols from the mycelia of *Cordyceps sinensis*. *Phytochemistry* 51(7): 891–898.

K. Boker, G. Schwarting, G. Kaule, V. Gunzler, E. Schmidt. 1991. Fibrosis of the liver in rats induced by bile duct ligation. Effects of inhibition by prolyl 4-hydroxylase. *J Hepatol* 13: 35–40.

E. Boonstra, R. De Kleijn, L.S. Colzato, A. Alkemade, B.U. Forstmann, S. Nieuwenhuis. 2015. Neurotransmitters as food supplements: the effects of GABA on brain and behavior. *Front Psychol* 6: 1–6.

I. Borodina, L.C. Kenny, C.M. McCarthy, K. Paramasivan, E. Pretorius, T.J. Roberts, D.B. Kell. 2020. The biology of ergothioneine, an antioxidant nutraceutical. *Nutr Res Rev* 33(2): 190–217.

E.R. Bovier, B.R. Hammond. 2015. A randomized placebo-controlled study on the effects of lutein and zeaxanthin on visual processing speed in young healthy subjects. *Arch Biochem Biophys* 572: 54–57.

L.R. Bucci 2002. Selected herbals and human exercise performance. *Am J Clin Nutr* 72: 624S–636S.

T.L. Bush, L.P. Fried, E. Barett-Connor. 1988. Cholesterol, lipoproteins, and coronary heart disease in women. *Clin Chem* 34: B60–B70.

P. Carmeliet, R.K. Jain. 2000. Angiogennesis in cancer and other diseases. *Nature* 407: 249–257.

J.S.L. Chan, G.S. Barseghyan, M.D. Asatiani, S.P. Wasser. 2015. Chemical composition and medicinal value of fruiting bodies and submerged cultured mycelia of caterpillar medicinal fungus *Cordyceps militaris* CBS-132098 (Ascomycetes) *Int J Med Mushrooms* 17: 649–659.

R. Chatterjee, K.S. Srinivasan, P.C. Maiti. 1957. *Cordyceps sinensis* (Berkeley) saccardo: structure of cordycepic acid. *J Am Pharm Assoc* 46(2): 114–118.

A. Chaudhuri, P.O. Behan. 2004. Fatigue in neurological disorders. *Lancet* 363: 978–988.

S.Y. Chen, K.J. Ho, Y.J. Hsieh, L.T. Wang, J.L. Mau. 2012. Contents of lovastatin, γ-aminobutyric acid and ergothioneine in mushroom fruiting bodies and mycelia. *LWT* 47: 274–278.

C.Y. Chen, C.W. Hou, J.R. Bernard, C.C. Chen, T.C. Hung, L.L. Cheng, C.H. Kuo. 2014. Rhodiolacrenulata-and *Cordyceps sinensis*-based supplement boosts aerobic exercise performance after short-term high altitude training. *High Alt Med Biol* 15(3): 371–379.

W. Chen, F. Yuan, K. Wang, D. Song, W. Zhang. 2012. Modulatory effects of the acid polysaccharide fraction from one of anamorph of *Cordyceps sinensis* on Ana-1 cells. *J Ethnopharmacol* 142(3): 739–745.

W. Chen, W. Zhang, W. Shen, K. Wang. 2010. Effects of the acid polysaccharide fraction isolated from a cultivated *Cordyceps sinensis* on macrophages in vitro. *Cell Immunol* 262(1): 69–74.

Y.J. Chen, M.S. Shiao, S.S. Lee, S.Y. Wang. 1997. Effect of *Cordyceps sinensis* on the proliferation and differentiation of human leukemic U937 cells. *Life Sci* 60(25): 2349–2359.

J.K.H. Cheung, J. Li, A.W.H. Cheung, Y. Zhu, K.Y.Z. Zheng, C.W.C. Bi, R. Duan, R.C.Y. Choi, D.T.W. Lau, T.T.X. Dong, B.W.C. Lau, K.W.K. Tsim. 2009. Cordysinocan, a polysaccharide isolated from cultured *Cordyceps*, activates immune responses in cultured T-lymphocytes and macrophages: signaling cascade and induction of cytokines. *J Ethnopharmacol* 124(1): 61–68.

W.F. Chiou, P.C. Chang, C.J. Chou, C.F. Chen. 2000. Protein constituent contributes to the hypotensive and vasorelaxant acttvties of *Cordyceps sinensis*. *Life Sci* 66(14): 1369–1376.

H.L. Chu, J.C. Chien, P.D. Duh. 2011. Protective effect of *Cordyceps militaris* against high glucose-induced oxidative stress in human umbilical vein endothelial cells. *Food Chem* 129: 871–876.

N. Cohen, J. Cohen, M.D. Asatiani, V.K. Varshney, H.T. Yu, Y.C. Yang, Y.H. Li, J.L. Mau, S.P. Wasser. 2014. Chemical composition and nutritional and medicinal value of fruit bodies and submerged cultured mycelia of culinary-medicinal higher Basidiomycetes mushrooms. *Int J Med Mushrooms* 16(3): 273–291.

G. Dai, T. Bao, C. Xu, R. Cooper, J.X. Zhu. 2001. CordyMaxTM Cs-4 improves steady-state bioenergy status in mouse liver. *J Altern Complem Med* 7: 231–240.

Y.Y. Deng, Y.P. Chen, X.L. He, L. Li. 2001. Study of *Cordyceps* on mechanism in delaying chronic renal failure. *Chin J Integrated Traditional West Nephrol* 2(7): 381–383.

G.L. Dohm, E.B. Tapscott, H.A. Barakat, G.J. Kasperek. 1983. Influence of fasting on glycogen depletion in rats during exercise. *J Appl Physiol* 55: 830–833.

C.H. Dong, Y.J. Yao. 2008. *In vitro* evaluation of antioxidant activities of aqueous extracts from natural and cultured mycelia of *Cordyceps sinensis*. *LWT-Food Sci Technol* 41: 669–677.

J.Z. Dong, S.H. Wang, X.R. Ai, L. Yao, Z.W. Sun, C. Lei, Y., Wang, Q. Wang. 2013. Composition and characterization of cordyxanthins from *Cordyceps militaris* fruit bodies. *J Funct Foods* 5(3): 1450–1455.

H. Fan, F.Q. Yang, S.P. Li. (2007). Determination of purine and pyrimidine bases in natural and cultured *Cordyceps* using optimum acid hydrolysis followed by high performance liquid chromatography. *J Pharm Biomed Anal* 45(1): 141–144.

K. Feng, Y.Q. Yang, S.P. Li. 2008. Renggongchongcao. In: Li S.P., Wang Y.T., editors. Pharmacological Activity-Based Quality Control of Chinese Herbs. New York: Nova Science Publisher, Inc, 2008. pp. 155–178.

J. Folk man 1971. Tumor angiogenesis: therapeutic implications. *N Engl J Med* 285: 1182–1186.

D.X. Gu, G. Zhang, J. Wang, L. Xin. 2006. A review and prospect on the studies of *Cordyceps sinensis* (Berk) Sacc. *J Chin Inst Food Sci Technol* 6: 137–141.

J. Guo, S.L. Friedman 2010. Toll-like receptor 4 signaling in liver injury and hepatic fibrogenesis. *Fibrogenesis Tissue Repair* 3(1): 1–19.

P. Hepsomali, J.A. Groeger, J. Nishihira, A. Scholey 2020. Effects of oral gamma-aminobutyric acid (GABA) administration on stress and sleep in humans: A systematic review. *Front Neurosci* 14: 923–935.

C.C. Hsu, Y.A. Lin, B. Su, J.H. Li, H.Y. Huang, M.C. Hsu. 2011. No effect of *Cordyceps sinensis* supplementation on testosterone level and muscle strength in healthy young adults for resistance training. *Biol Sport* 28(2): 107–110.

P.Y. Hsu, Y.H. Lin, E.L. Yeh, H.C. Lo, T.H. Hsu, C.C. Su. 2017. Cordycepin and a preparation from *Cordyceps militaris* inhibit malignant transformation and proliferation by decreasing EGFR and IL-17RA signaling in a murine oral cancer model. *Oncotarget* 8: 93712–93728.

Y. Huang, J. Lu, B. Zhu, Q. Wen, F. Jia, S. Zeng, T. Chen, Y. Li, G. Cheng, Z. Yi. 1987. Toxicity study of fermentation *Cordyceps mycelia* B414. *Chin Tradit Pat Med* 10: 24–25.

Y.L. Huang, S.F. Leu, B.C. Liu, C.C. Sheu, B.M. Huang. 2004. *In vivo* stimulatory effect of *Cordyceps sinensis* mycelium and its fractions on reproductive functions in male mouse. *Life Sci* 75: 1051–1062.

B.M. Huang, C.C. Hsu, S.J. Tsai, C.C. Sheu, S.F. Leu. 2001. Effects of *Cordyceps sinensis* on testosterone production in normal mouse Leydig cells. *Life Sciences* 69(22): 2593–2602.

B.M. Huang, Y.M. Chuang, C.F. Chen, S.F. Leu. 2000. Effects of extracts from mycelium of *Cordyceps sinensis* on steroidogenesis in MA-10 mouse Leydig tumor cells. *Bio Phram Bull* 23(12): 1532–1535.

I.H. Hwang, S.Y. Oh, H.J. Jang, E. Jo, J.C. Joo, K.B. Lee, H.S. Yoo, M.Y. Lee, S.J. Park, I.S. Jang. 2017. Cordycepin promotes apoptosis in renal carcinoma cells by activating the MKK7-JNK signaling pathway through inhibition of c-FLIPL expression. *PLoS One* 12: e0186489.

K.J. Jędrejko, J. Lazur, B. Muszyńska. 2021. *Cordyceps militaris*: An overview of its chemical constituents in relation to biological activity. *Foods* 10(11): 2634.

D.B. Ji, J. Ye, C.L. Li, Y.H..Wang, J. Zhao, S.Q. Cai. 2009. Anti-aging effect of *Cordyceps sinensis* extract. *Phytother Res* 23: 116–122.

J.M. Jia, X.C. Ma, C.F. Wu, L.J. Wu, G.S. Hu. 2005. Cordycedipeptide A, a new cyclodipeptide from the culture liquid of *Cordyceps sinensis* (B ERK.) S ACC. *Chem Pharm Bull* 53(5): 582–583.

J.M. Jia, H.H. Tao, B.M. Feng. 2009. Cordyceamides A and B from the Culture Liquid of *Cordyceps sinensis* (B ERK.) S ACC. *Chem Pharm Bull* 57(1): 99–101.

Y. Jiang, Y.J. Yao. 2003. Anamorphic fungi related to *Cordyceps sinensis*. *Mycosystema*. 22: 161–176.

Z.H. Jin, Y.P. Chen. 2004. Clinical study on the effect of *Cordyceps sinensis* mycelium on delaying progression of chronic kidney failure. *J Tradit Chin Med* 20(3): 155–157.

W.S. Jo, Y.J. Choi, H.J. Kim, J.Y. Lee, B.H. Nam, J.D. Lee, S.W. Lee, S.Y. Seo, M.H. Jeong. 2010. The anti-inflammatory effects of water extract from *Cordyceps militaris* in murine macrophage. *Mycobiol* 38: 46–51.

J. Joy, C.K.K. Nair. 2008. Amelioration of cisplatin induced nephrotoxicity in Swiss albino mice by Rubia cordifolia extract. *J Cancer Res Ther* 4(3): 111–115.

S.J. Jung, E.S. Jung, E.K. Choi, H.S. Sin, K.C. Ha, S.W. Chae. 2019. Immunomodulatory effects of a mycelium extract of *Cordyceps* (Paecilomyceshepiali; CBG-CS-2): A randomized and double-blind clinical trial. *BMC Complement Altern Med* 19(1): 77–84.

H.J. Kang, Y.S. Song. 1997. Dietary fiber and cholesterol metabolism. *J Korean Soc Food Sci Nutr* 26: 358–369.

T. Kiho, K. Ookubo, S. Usui, S. Ukai, K. Hirano. 1999. Structural features and hypoglycemic activity of a polysaccharide (CS-F10) from the cultured mycelium of *Cordyceps sinensis*. *Biol Pharm Bull* 22(9): 966–970.

M. Kitakaze, M. Hori. 2000. Adenosine therapy: A new approach to chronic heart failure. *Expert opinion on investigational drugs* 9(11): 2519–2535.

S. Kitchawalit, K. Kanokmedhakul, S. Kanokmedhakul, K. Soytong. 2014. A new benzyl ester and ergosterol derivatives from the fungus Gymnoascusreessii. *Nat Prod Res* 28(14): 1045–1051.

J.H. Koh, K.W. Yu, H.J. Suh, Y.M. Choi, T.S. Ahn. 2002. Activation of macrophages and the intestinal immune system by an orally administered decoction from cultured medium of *Corydyceps sinensis*. *Biosci Biotechnol Biochem* 66: 407–411.

J.H. Koh, J.M. Kim, U.J. Chang, H.J. Suh. 2003. Hypocholesterolemic effect of hot-water extract from mycelia of *Cordyceps sinensis*. *Biol Pharm Bull* 26(1): 84–87.

R. Kumar, P.S. Negi, B. Singh, G. Ilavazhagan, K. Bhargava, N.K. Sethy. 2011. *Cordyceps sinensis* promotes exercise endurance capacity of rats by activating skeletal muscle metabolic regulators. *J Ethnopharm* 136(1): 260–266.

C.P. Kuniyal, R.C. Sundriyal. 2013. Conservation salvage of *Cordyceps sinensis* collection in the Himalayan mountains is neglected. *Ecosys Serv* 3: E40–E43.

M.C. Kuo, C.Y. Chang, T.L. Cheng, M.J. Wu. 2007. Immunomodulatory effect of exo-polysaccharides from submerged cultured *Cordyceps sinensis*: Enhancement of cytokine synthesis, CD11b expression, and phagocytosis. *Appl Microbiol Biotechnol* 75: 769–775.

S.P. Li, Z.R. Su, T.T. Dong, K.W. Tsim. 2002. The fruiting body and its caterpillar host of *Cordyceps sinensis* show close resemblance in main constituents and anti-oxidation activity. *Phytomed* 9: 319–324.

T. Li, W. Li. 2009. Impact of polysaccharides from *Cordyceps* on antifatigue in mice. *Sci Res Essays* 4: 705–709.

S.P. Li, F.Q. Yang, K.W.K. Tsim. 2006. Quality control of *Cordyceps sinensis*, a valued traditional Chinese medicine. *J Pharm Biomed Anal* 41: 1571–1584.

W.H. Lin, M.T. Tsai, R.C.W. Hou, H.F. Hung, C.H. Li, H.K. Wang, M.N. Lai, K.C.G. Jeng. 2007. Improvement of sperm production in subfertile boars by *Cordyceps militaris* supplement. *Am J Chin Med* 35(4): 631–641.

C.Y. Lin, F.M. Ku, Y.C. Kuo, C.F. Chen, W.P. Chen, A. Chen, M.S. Shiao. 1999. Inhibition of activated human mesangial cell proliferation by the natural product of *Cordyceps sinensis* (H1-A): An implication for treatment of IgA mesangial nephropathy. *J Lab Clin Med* 133(1): 55–63.

W.C. Liu, W.L. Chuang, M.L. Tsai, J.H. Hong, W.H. McBride, C.S. Chiang. 2008. *Cordyceps sinensis* health supplement enhances recovery from taxol-induced leucopenia. *Exp Biol Med* 233: 447–455.

X. Liu, F. Zhong, X.L. Tang, F.L. Lian, Q. Zhou, S.M. Guo, J.F. Liu, P. Sun, X. Hao, Y. Lu. 2014. *Cordyceps sinensis* protects against liver and heart injuries in a rat model of chronic kidney disease: A metabolomic analysis. *Acta Pharmacol Sin* 35: 697–706.

Y.K. Liu, W. Shen. 2003. Inhibitive effect of *Cordyceps sinensis* on experimental hepatic fibrosis and its possible mechanism. *World J Gastroenterol* 9: 529–533.

Z. Liu, Z. Liang, A. Liu, Y. Yao, K.D. Hyde, Z. Yu. 2002. Molecular evidence for teleomorph-anamorph connections in Cordyceps based on ITS-5.8S rDNA sequences. *Mycol Res* 106(9): 1100–1108.

H.C. Lo, C. Hsieh, F.Y. Lin, T.H. Hsu. 2013. A systematic review of the mysterious caterpillar fungus *Ophiocordyceps sinensis* in Dong-Chong XiaCao and related bioactive ingredients. *J Tradit Complement Med* 3: 16–32.

L. Lu. 2002. Study on effect of *Cordyceps sinensis* and artemisinin in preventing recurrence of lupus nephritis. *Chin J Integr Med* 22: 169–171.

K.J. Ma, F.G. Li, J. Wang. 2008. Study on chronic fatigue syndrome with traditional Chinese medicine. *China Mod Dr* 10: 131–133.

N. Manabe, Y. Azuma, M. Sugimoto, K. Uchio, M. Miyamoto, N. Taketomo, H. Tsuchita, H. Miyamoto. 2000. Effects of the mycelial extract of cultured *Cordyceps sinensis* on *in vivo* hepatic energy metabolism and blood flow in dietary hypoferric anaemic mice. *Br J Nutr* 83: 197–204.

N. Manabe, M. Sugimoto, Y. Azuma, N. Taketomo, A. Yamashita, H. Tsuboi, H. Miyamoto. 1996. Effects of the mycelial extract of cultured *Cordyceps sinensis* on *in vivo* hepatic energy metabolism in the mouse. *Jap J Pharm* 70(1): 85–88.

J.P. Manfredi, H.V. Sparks Jr. 1982. Adenosine's role in coronary vasodilation induced by atrial pacing and norepinephrine. *Am J Physiol Heart Circ Physiol* 243(4): H536–H545.

H. Matsuda, J. Akaki, S. Nakamura, Y. Okazaki, H. Kojima, M. Tamesada, M. Yoshikawa. 2009. Apoptosis-inducing effects of sterols from the dried powder of cultured mycelium of *Cordyceps sinensis*. *Chem Pharm Bull* 57(4): 411–414.

H. Meena, K.P. Singh, P.S. Negi, Z. Ahmed. 2013. Sub-acute toxicity of cultured mycelia of Himalayan entomogenous funfus *Cordyceps sinensis* (Berk.) SACC in rats. *Indian J Exp Biol* 51: 381–387.

K. Mizuno, M. Tanaka, S. Nozaki, H. Mizuma, S. Ataka, T. Tahara, T. Sugino, T. Shirai, Y. Kajimoto, H. Kuratsune, O. Kajimoto. 2008. Antifatigue effects of coenzyme Q10 during physical fatigue. *Nutrition* 24(4): 293–299.

M. Movassagh, A. Spatz, J. Davoust, S. Lebecque, P. Romero, M. Pittet, D. Rimoldi, D. Liénard, O. Gugerli, L. Ferradini, C. Robert. 2004. Selective accumulation of mature DC-Lamp+ dendritic cells in tumor sites is associated with efficient T-cell-mediated antitumor response and control of metastatic dissemination in melanoma. *Cancer Res* 64(6): 2192–2198.

K. Nakamura, Y. Yamaguchi, S. Kagota, Y.M. Kwon, K. Shinozuka, M. Kunitomo. 1999. Inhibitory effect of *Cordyceps sinensis* on spontaneous liver metastasis of lewis lung carcinoma and B16 melanoma cells in syngeneic mice. *Jpn J Pharmacol* 79: 335–341.

S. Nakav, C. Chaimovitz, Y. Sufaro, E.C. Lewis, G. Shaked, D. Czeiger, M. Zlotnik, A. Douvdevani. 2008. Anti-inflammatory preconditioning by agonists of adenosine A1 receptor. *PLoS One 3*(5): e2107.

J.X. Nan, E.J. Park, B.K. Yang, C.H. Song, G. Ko, D.H. Sohn. 2001. Antifibroticeffect of extracellular biopolymer from submerged mycelial cultures of *Cordyceps militaris* on liver fibrosis induced by bile duct ligation and scission in rats. *Arch Pharm Res* 24: 327–332.

V.S. Negi, B.C. Joshi, R. Pathak, R.S. Rawal, K.C. Sekar. 2018. Assessment of fuelwood diversity and consumption patterns in cold desert part of Indian Himalaya: Implication for conservation and quality of life. *Journal Clean Prod* 196: 23–31.

T.B. Ng, H.X. Wang. 2005. Pharmacological actions of *Cordyceps*, a prized folk medicine. *J Pharm Pharmacol* 57: 1509–1519.

Q.V. Nguyen, T.T. Vu, M.T. Tran, P.T.H. Thi, H. Thu, T.H.L. Thi, H.V. Chuyen, M.H. Dinh. (2021). Antioxidant activity and hepatoprotective effect of exopolysaccharides from cultivated *Ophiocordyceps sinensis* against CCl4-induced liver damages. *Nat Prod Commun* 16(2): 1–9.

H. Nomura, H. Sato, M. Seiki, M. Mai, Y. Okada. 1995. Expression of membrane type matrix metalloproteinase in human gastric carcinomas. *Cancer Res* 55: 3263–3266.

R.A. Nugent, S.F. Fathima, A.B. Feigl, D. Chyung. 2011. The burden of chronic kidney disease on developing nations: a 21st century challenge in global health. *Nephron* 118(3): c269–c277.

A. Oesterle, U. Laufs, J.K. Liao. 2017. Pleiotropic effects of statins on the cardiovascular system. *Circ res* 120(1): 229–243.

N. Ohen, J. Cohen, M.D. Asatiani, V.K..Varshney, H.T. Yu, Y.C. Yang, Y.H. Li, J.L. Mau, S.P. Wasser. 2014. Chemical composition and nutritional and medicinal value of fruit bodies and submerged cultured mycelia of culinary-medicinal higher basidiomycetes mushrooms. *Int J Med Mushrooms* 16: 273–291.

Y. Ohta, J.B. Lee, K. Hayashi, A. Fujita, D.K. Park, T. Hayashi. 2007. In vivo anti-influenza virus activity of an immunomodulatory acidic polysaccharide isolated from *Cordyceps militaris* grown on germinated soybeans. *J Agric Food Chem* 55(25): 10194–10199.

O.J. Olatunji, J. Tang, A. Tola, F. Auberon, O. Oluwaniyi, Z. Ouyang. 2018. The genus *Cordyceps*: An extensive review of its traditional uses, phytochemistry and pharmacology. *Fitoterapia* 129: 293–316.

J. Ontyd, J. Schrader. 1984. Measurement of adenosine, inosine, and hypoxanthine in human plasma. *J Chromatogr B Biomed Appl* 307: 404–409.

Y.Y. Ouyang, Z. Zhang, Y.R. Cao, Y.Q. Zhang, Y.Y. Tao, C.H. Liu, … & J.S. Guo (2013). Effects of *Cordyceps* acid and cordycepin on the inflammatory and fibrogenic response of hepatic stellate cells. *Chin J Hepatol* 21(4): 275–278.

A.K. Panda, K.C. Swain. 2011. Traditional uses and medicinal potential of *Cordyceps sinensis* of Sikkim. *J Ayurveda Integr Med* 2(1): 9.

B.T. Park, K.H. Na, E.C. Jung, J.W. Park H.H. Kim. 2009a. Anti-fungal and -cancer activities of a protein from the mushroom *Cordyceps militaris*. *Korean J Physiol Pharmacol* 13: 49–54.

S.E. Park, J. Kim, Y.W. Lee, H.S. Yoo, C.K. Cho. 2009b. Antitumor activity of water extracts from *Cordyceps militaris* in NCI-H460 cell xenografted nude mice. *J Acupunct Meridian Stud* 2: 294–300.

S.E. Park, H.S. Yoo, C.Y. Jin, S.H. Hong, Y.W. Lee, B.W. Kim, S.H. Lee, W.J. Kim, C.K. Cho, Y.H. Choi. 2009c. Induction of apoptosis and inhibition of telomerase activity in human lung carcinoma cells by the water extract of *Cordyceps militaris*. *Food Chem Toxicol* 47: 1667–1675.

J. Peng, X. Li, Q. Feng, L. Chen, L. Xu, Y. Hu. 2013. Anti-fibrotic effect of *Cordyceps sinensis* polysaccharide: Inhibiting HSC activation, TGF-β1/Smad signalling, MMPs and TIMPs. *Exp Biol Med (Maywood)* 238(6): 668–677.

M. Pouliot, T. Treue. 2013. Rural people's reliance on forests and the non-forest environment in west Africa: Evidence from Ghana and Burkina Faso. *World Dev* 43: 180–193.

G.M. Qian, G.F. Pan, J.Y. Guo. (2012). Anti-inflammatory and antinociceptive effects of cordymin, a peptide purified from the medicinal mushroom *Cordyceps sinensis*. *Nat Prod Res*, *26*(24), 2358–2362.

Y.M. Ra, N.S. Hyun, K.M. Young. 2008. Antioxidative and antimutagenic activities of 70%ethanolic extracts from four fungal mycelia-fermented specialty rices. *J Clin Biochem Nutr* 43: 118–125.

I. Rahman. 2003. Oxidative stress, chromatin remodeling and gene transcription in inflammation and chronic lung diseases. *J Biochem Mol Biol* 36: 95–109.

L.M. Renzi-Hammond, E.R. Bovier, L.M. Fletcher, L.S. Miller, C.M. Mewborn, C.A. Lindbergh, J.H. Baxter, B.R. Jr Hammond. 2017. Effects of a lutein and zeaxanthin intervention on cognitive function: A randomized, double-masked, placebo-controlled trial of younger healthy adults. *Nutrients* 9(11): 1246.

M. Romeu, M. Mulero, M. Giralt M. 2002. Parameters related to oxygen free radicals in erythrocytes, plasma and epidermis of the hairless rat. *Life Sci* 71(15): 1739–1749.

V. Rukachaisirikul, S. Chantaruk, C. Tansakul, S. Saithong, L. Chaicharernwimonkoon, C. Pakawatchai, M. Isaka, K. Intereya. 2006. A cyclopeptide from the insect pathogenic fungus *Cordyceps* sp. BCC 1788. *J Nat Prod* 69(2): 305–307.

R. Russell, M. Paterson. 2008. *Cordyceps*—A traditional Chinese medicine and another fungal therapeutic biofactory? *Phytochem* 69: 1469–1495.

L. Shaoping, L. Ping, J. Hui, Z. Quan, T.T. Dong, K.W. Tsim. 2001. The nucleosides contents and their variation in natural *Cordyceps sinensis* and cultured *Cordyceps mycelia*. *J Chin Pharm Sci* 10(4): 175.

W. Shen, D. Song, J. Wu, W. Zhang. 2011. Protective effect of a polysaccharide isolated from a cultivated *Cordyceps mycelia* on hydrogen peroxide-induced oxidative damage in PC12 cells. *Phytother Res* 25(5): 675–680.

L. Sheng, J. Chen, J. Li, W. Zhang. 2011. An exopolysaccharide from cultivated *Cordyceps sinensis* and its effects on cytokine expressions of immunocytes. *Appl Biochem Biotechnol* 163(5): 669–678.

U.B. Shrestha, K.R. Dhital, A.P. Gautam. 2019. Economic dependence of mountain communities on Chinese caterpillar fungus *Ophiocordyceps sinensis* (yarsagumba): A case from western Nepal. *Oryx* 53: 256–264.

U.B. Shrestha, K.S. Bawa. 2014. Economic contribution of chinese caterpillar fungus to the livelihoods of mountain communities in Nepal. *Biol Conser* 177: 194–202.

K.M. Siu, H.F.D. Mak, P.Y. Chiu, K.T.M. Poon, Y. Du, K.M. Ko. 2004. Pharmacological basis of 'Yin-nourishing' and 'Yang-invigorating' actions of *Cordyceps*, a Chinese tonifying herb. *Life Sci* 76: 385–395.

J. Song, Y. Wang, M. Teng, G. Cai, H. Xu, H. Guo, Y. Liu, D. Wang, L. Teng. 2015. Studies on the antifatigue activities of *Cordyceps militaris* fruit body extract in mouse model. *Evid Based Complement Alternat Med* 2015: 1–15.

D. Song, J. Lin, F. Yuan, W. Zhang. 2011. *Ex vivo* stimulation of murine dendritic cells by an exopolysaccharide from one of the anamorphs of *Cordyceps sinensis*. *Cell Biochem Funct* 29(7): 555–561.

B. Stevan. 1999. Angiogenesis and cancer control: from concept to therapeutic trial. *Cancer Control* 6(5): 1–18.

R. Stone. 2008. Last stand for the body snatcher of the Himalayas? *Science* 322(5905): 1182–1182.

N.T. Stringham, P.V. Holmes, J.M. Stringham. 2018. Supplementation with macular carotenoids reduces psychological stress, serum cortisol, and sub-optimal symptoms of physical and emotional health in young adults. *Nutr Neurosci* 21(4): 286–296.

W. Struck-Lewicka, R. Kaliszan, M.J. Markuszewski. 2014. Analysis of urinary nucleosides as potential cancer markers determined using LC–MS technique. *J Pharm Biomed Anal* 101: 50–57.

K. Susztak, A.C. Raff, M. Schiffer, E.P. Bottinger. 2006. Glucose-induced reactive oxygen species cause apoptosis of podocytes and posocyte depletion at the onset of diabetic nephropathy. *Diabetes* 55: 225–233.

K. Suwannasaroj, P. Srimangkornkaew, P. Yottharat, A. Sirimontaporn. 2021. The acute and sub-chronic oral toxicity testing of *Cordyceps militaris* in Wistar Rats. *Bull Dept Med Sci* 63: 628–647.

A.L. Taylor, C.J. Watson, J.A. Bradley. 2005. Immunosuppressive agents in solid organ transplantation: Mechanisms of action and therapeutic efficacy. *Crit Rev Oncol Hematol* 56: 23–46.

J.A. Tobert. 1988. Efficacy and long-term adverse effect pattern of lovastatin. *Am J Cardiol* 62: J28–J34.

P. Trigg, W.E. Gutteridge, J. Williamson. 1971. The effect of Cordycepin on malarial parasites. *T Roy Soc Trop Med H* 65: 514–520.

H.S. Tuli, S.S. Sandhu, A.K. Sharma. 2014. Pharmacological and therapeutic potential of *Cordyceps* with special reference to Cordycepin. *3 Biotech* 4: 1–12.

M. Valko, M. Izakovic, M. Mazur, C.J. Rhodes, J. Telser. 2004. Role of oxygen radicals in DNA damage and cancer incidence. *Mol Cell Biochem* 266: 37–56.

M. Valko, D. Leibfritz, J. Moncol, M.T. Cronin, M. Mazur, J. Telser. 2007. Free radicals and antioxidants in normal physiological functions and human disease. *Int J Biochem Cell Biol* 39: 44–84.

P. Vestergaard, L. Rejnmark, L. Mosekilde. 2009. Diabetes and its complications and their relationship with risk of fractures in type 1 and 2 diabetes. *Calcif Tissue Int*, 84(1): 45–55.

F. Wan, Y. Guo, X. Deng. 1988. Sex hormone like effects of Jin Shui Bao capsule: Pharmacological and clinical studies. *Chinese Traditional Patented Med* 9: 29–31.

B.J. Wang, S.J. Won, Z.R. Yu, C.L. Su. 2005 Free radical scavenging and apoptotic effects of *Cordyceps sinensis* fractionated by supercritical carbon dioxide. *Food Chem Toxicol* 43: 543–552.

L. Wang, N. Xu, J. Zhang, H. Zhao, L. Lin, S. Jia, L. Jia. 2015. Antihyperlipidemic and hepatoprotective activities of residue polysaccharide from *Cordyceps militaris* SU-12. *Carbohydr Polym* 131: 355–362.

C. Wang, Z.P. Mao, L. Wang, F.H. Zhang, G.H. Wu, D.Y. Wang, J.L. Shi. 2017a. Cordycepin inhibits cell growth and induces apoptosis in human cholangiocarcinoma. *Neoplasma* 64: 834–839.

C.W. Wang, W.H. Hsu, C.J. Tai. 2017b. Antimetastatic effects of cordycepin mediated by the inhibition of mitochondrial activity and estrogen-related receptor alpha in human ovarian carcinoma cells. *Oncotarget* 8: 3049–3058.

J. Wang, Y.M. Liu, W. Cao, K.W. Yao, Z.Q. Liu, J.Y. Guo. 2012. Anti-inflammation and antioxidant effect of Cordymin, a peptide purified from the medicinal mushroom *Cordyceps sinensis*, in middle cerebral artery occlusion-induced focal cerebral ischemia in rats. *Metabolic Brain Disease* 27: 159–165.

S. Wangchuk, N. Norbu, N. Sherub. 2012. Impacts of cordyceps collection on livelihoods and alpine ecosystems in Bhutan as ascertained from questionnaire survey of *Cordyceps* collectors. *Royal Government of Bhutan*, UWICE Press, Bumthang.

D. Winkler. 2009. Caterpillar fungus (*Ophiocordyceps sinensis*) production and sustainability on the Tibetan plateau and in the Himalayas. *Asian Med* 5: 291–316.

D. Winkler. 2008. Yartsa Gunbu (*Cordyceps sinensis*) and the fungal commodification of Tibet's rural economy. *Econ Bot* 62: 291–305.

K. Wojcikowski, L. Stevenson, D. Leach, H. Wohlmuth, G. Gobe. 2007. Antioxidant capacity of 55 medicinal herbs traditionally used to treat the urinary system: A comparison using a sequential three-solvent extraction process. *J Altern Complement Med* 13: 103–109.

S.Y. Won, E.H. Park. 2005. Anti-inflammatory and related pharmacological activities of cultured mycelia and fruiting bodies of *Cordyceps militaris*. *J Ethnopharmacol* 96: 555–561.

K.L. Wong, E.C. So, C.C. Chen, R.S. Wu, B.M. Huang. 2007 Regulation of steroidogenesis by *Cordyceps sinensis* mycelium extracted fractions with (hCG) treatment in mouse Leydig cells. *Arch Androl* 53: 75–77.

Y.Y. Wong, A. Moon, R. Duffin, A. Barthet-Barateig, H.A. Meijer, M.J. Clemens, C.H. de Moor. 2010. Cordycepin inhibits protein synthesis and cell adhesion through effects on signal transduction. *J Biol Chem* 285: 2610–2621.

R. Wu, P.A. Yao, H.L. Wang, Y. Gao, H.L. Yu, L. Wang L, X.H. Cui, X. Xu, J.P. Gao. 2018. Effect of fermented *Cordyceps sinensis* on doxorubicin induced cardiotoxicity in rats. *Mol Med Rep* 18: 3229–3241.

J.H. Xiao, J.J. Zhong. 2007. Secondary metabolites from*Cordyceps* species and their antitumor activity studies. *Recent Pat Biotechnol* 1: 123–137.

H.Y. Xu, X. Zheng, C. Xu, Y. Zhao, F. Liu. 2000. The protective effect of *Cordyceps sinensis* on adriamycin-induced myocardial damage. *Acta Chin Med Pharmacol* 28: 64–65.

X.F. Yan, Z.M. Zhang, H.Y. Yao, Y. Guan, J.P. Zhu, L.H. Zhang, Y.L. Jia, R.W. Wang. 2012. Cardiovascular protection and antioxidant activity of the extracts from the mycelia of *Cordyceps sinensis* act partially *via* adenosine receptors. *Phytother Res* 27: 1597–1604.

F.Q. Yang, L. Ge, J.W.H. Yong, S.N. Tan, S.P. Li. 2009. Determination of nucleosides and nucleobases in different species of *Cordyceps* by capillary electrophoresis–mass spectrometry. *J Pharm Biomed Anal* 50: 307–314.

F.Q. Yang, D.Q. Li, K. Feng, D.J. Hu, S.P. Li. 2010. Determination of nucleotides, nucleosides and their transformation products in *Cordyceps* by ion-pairing reversed-phase liquid chromatography–mass spectrometry. *J Chromatogr A* 1217: 5501–5510.

H.Y. Yang, S.F. Leu, Y.K. Wang, C.S. Wu, B.M. Huang. 2006. *Cordyceps sinensis* mycelium induces MA-10 mouse Leydigtumor cell apoptosis by activating the caspase-8 pathway and suppressing the NF-κB pathway. *Arch Androl* 52: 103–110.

L.Y. Yang, A. Chen, Y.C. Kuo, C.Y. Lin. 1999. Efficacy of a pure compound H1-A extracted from *Cordyceps sinensis* on autoimmune disease of MRL lpr/lpr mice. *J Lab Clin Med* 134: 492–500.

L.Y. Yang, W.J. Huang, H.G. Hsieh, C.Y. Lin. 2003. H1-A extracted from *Cordyceps sinensis* suppresses the proliferation of human mesangial cells and promotes apoptosis, probably by inhibiting the tyrosine phosphorylation of Bcl-2 and Bcl-XL. *J Lab Clin Med* 141: 74–83.

M.L. Yang, P.C. Kuo, T.L. Hwang, T.S. Wu. 2011. Anti-inflammatory principles from *Cordyceps sinensis*. *J Nat Prod* 74: 1996–2000.

H.S. Yoo, J.W. Shin, J.H. Cho, C.G. Son, Y.W. Lee, S.Y. Park. 2004. Effects of *Cordyceps militaris* extract on angiogenesis and tumor growth. *Acta Pharm Sinic* 25: 657–665.

N. Yoshikawa, K. Nakamura, Y. Yamaguchi, S. Kagota, K. Shinozuka, M. Kunitomo. 2004. Antitumour activity of cordycepin in mice. *Clin Exp Pharmacol Physiol* 31: S51–S53.

X.M. Yu, S.F. Tan. 2003. Clinical study on the treatment of chronic renal failure with Bailing capsule. *J Trop Med* 3: 203–204.

H.M. Yu, B.S. Wang, S.C. Huang, P.D. Duh. 2006. Comparison of protective effects between cultured *Cordyceps militaris* and natural *Cordyceps sinensis* against oxidative damage. *J Agric Food Chem* 54: 3132–3138.

J.G. Yuan, X.H. Cheng, Y.Q. Hou. 2005. Studies on the components and pharmacological action of polysaccharide from *Cordyceps sinensis*. *Food Drug* 7: 45–48.

Y. Zeng, S. Lian, D. Li, X. Lin, B. Chen, H. Wei, T. Yang. 2017. Antihepatocarcinoma effect of cordycepin against NDEA-induced hepatocellular carcinomas via the PI3K/Akt/mTOR and Nrf2/HO-1/NF-kappaB pathway in mice. *Biomed Pharmacother* 95: 1868–1875.

H. Zhan, C.G. Liu, J.H. Zhou. 1990. The senile changes of male rats. *Acta Physiol Sin* 42: 502–508.

Z. Zhang, W. Huang, S. Liao, L. Lei, J. Lui, F. Leng, W. Gong, H. Zhang, L. Wan, R. Wu, S. Li, H. Luo, F. Zhu. 1995. Clinical and laboratory studies of Jin Shui Bao in scavenging oxygen free radicals in elderly senescent Xu Zheng patients. *J Administration Traditional Chinese Med* 5: 14–18.

Z. Zhang, X. Wang, Y. Zhang, G. Ye. 2011. Effect of *Cordyceps sinensis* on renal function of patients with chronic allograft nephropathy. *Urol Int* 86: 298–301.

S.S. Zhang, D.S. Zhang, T.J. Zhu, X.Y. Chen. 1991. A pharmacological analysis of the amino acid components of *Cordyceps sinensis* Sacc. *Acta Pharm Sin* 26: 326–330.

W. Zhang, J. Yang, J. Chen, Y. Hou, X. Han. 2005. Immunomodulatory and antitumour effects of an exopolysaccharide fraction from cultivated *Cordyceps sinensis* (Chinese caterpillar fungus) on tumour-bearing mice. *Biotechnol Appl Biochem* 42: 9–15.

S.L. Zhao. 2000. Advance of treatment for *Cordyceps* on chronic hepatic diseases. *Shanxi Zhong Yi* 16: 59–60.

J. Zheng, Y. Wang, J. Wang, P. Liu, J. Li, W. Zhu. 2013. Antimicrobial ergosteroids and pyrrole derivatives from halotolerant *Aspergillus flocculosus* PT05-1 cultured in a hypersaline medium. *Extremophiles* 17: 963–971.

X.Q. Zhong. 2006. Oxygen free radicals and disease. *J Shaoguan Univ Nat Sci* 27: 87–90.

X. Zhou, Y. Yao. 2013. Unexpected nephrotoxicity in male ablactated rats induced by *Cordyceps militaris*: The involvement of oxidative changes. *Evid Based Complementary Altern Med* 2013: 1–9.

J.S. Zhou, G. Halpern, K. Jones. 1998. The scientific rediscovery of an ancient Chinese herbal medicine: *Cordyceps sinensis*. *J Altern Complent Med* 4: 429–457.

X. Zhou, Z. Gong, Y. Su, J. Lin, K. Tang. 2009. Cordyceps fungi: natural products, pharmacological functions and developmental products. *J Pharm Pharmacol* 61: 279–291.

J.S. Zhu, J. Rippe. 2004. CordyMax enhances aerobic capability, endurance performance, and exercise metabolism in healthy, mid-age to elderly sedentary humans. In: *Proceedings of the American*

Physiological Society's (APS) Annual Scientific Conference, Experimental Biology. Washington, DC: Convention Center; pp. 28–31.

Z.Y. Zhu, J. Chen, C.L. Si, N. Liu, H.Y. Lian, L. Ding, Y. Liu, Y.M. Zhang. 2012. Immunomodulatory effect of polysaccharides from submerged cultured. *Cordyceps gunnii*, *Pharm Biol* 50: 1103–1110.

J.S. Zhu, G.M. Halpern, K. Jones. 1998. The scientific rediscovery of a precious ancient Chinese herbal regimen: *Cordyceps sinensis* Part II. *J Altern Complement Med* 4: 429–457.

5 *Grifola frondosa* (*Dicks.*) Gray

Barsha Devi
Pandit Deendayal Upadhyaya Adarsha Mahavidyalaya, Tulungia, Assam, India

Nabanita Bhattacharyya
Department of Botany, Gauhati University, Guwahati, Assam, India

CONTENTS

INTRODUCTION

Grifola frondosa (*Dicks.*) Gray is a basidiomycete fungus belonging to the Grifolaceae family and to the Polyporeales order. It is known by various common names in different parts of the world, such as maitake (Japan), gray tree flower (China), hen of the woods (North America), ram's head (Germany), and polypore entouffe (France). The popularity of maitake can be attributed to a variety of factors, including its wonderful taste and texture (similar to chicken) in addition to its medicinal values. Being a good source of protein, carbohydrates, vitamins, and minerals, and with less fat content and calorific value, maitake is considered a healthy food. Besides its nutritional value, *G. frondosa* has been known to possess various pharmacological effects. Here we review the characteristics, distribution, traditional and modern medicinal use, phytochemistry, pharmacological properties as well as toxicity aspects of *G. frondosa,* which will be helpful for nutritionists, food scientists, researchers in particular, and common people in general.

DOI: 10.1201/9781003259763-5

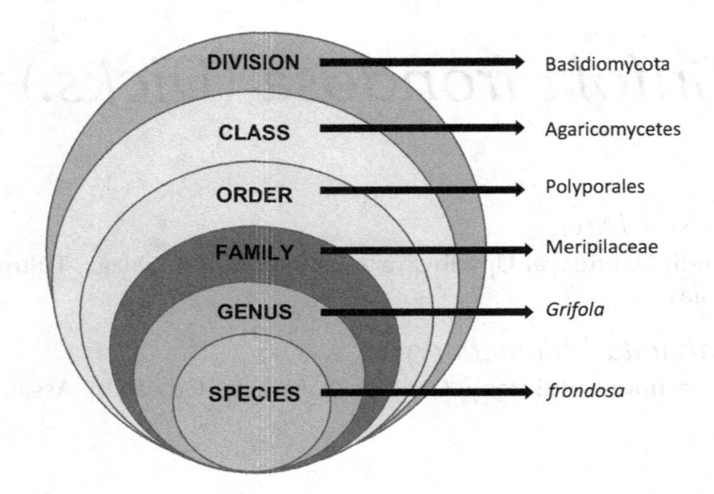

FIGURE 5.1 Systematic botanical classification of *G. frondosa.*

CHARACTERISTICS AND DISTRIBUTION

Maitake is a fleshy polypore fungus that is recognized by its smoked brown overlapping caps (Arora, 1986). The initial color of the mature fruiting body is fleshy dark grayish brown which with age slowly attains a lighter gray color (Stamets, 2000). The mushroom is commonly found at the base of old hardwoods like maples, oaks, and chestnuts (Chen et al., 2000). The fungus is scattered extensively in Japan, China, and other Asian countries including North-east India (Bhattacharjee et al., 2015), and is native to parts of North America and Japan (www.gbif.or). Wide-ranging commercial production of maitake was started in 1981 in Japan followed by the United States and China after ten years (Takama et al., 1981). Figure 5.1 represents the systematic classification of *G. frondose.*

TRADITIONAL AND MODERN MEDICINAL USE

Maitake is best known for cooking and for its healing effects dating back thousands of years. It is mentioned in ancient medical texts as a potent remedy for lungs and liver protection as well as to nourish and enhance "qi," or life force (Zhou et al., 2013). Japan and China have used it in their cuisine and traditional medicine for centuries. Many Japanese herbal books, documented maitake as a medicinal and edible mushroom. Scientists in Japan discovered maitake mushrooms were more powerful than Shiitake, Kawaratake, and Suehirotake, mushrooms, which were used in traditional Asian medicine to stimulate the immune system in the late 1980s (Mayell, 2001). In traditional Chinese medicine, *G. frondosa* is prized as a medicinal mushroom that enhances qi and strengthens the spleen (Herbalism, 1999). As per Traditional Chinese Medicine (TCM), qi is defined as the vital energy circulating in the body that keeps the vital life energy necessary for the body to grow and develop. In Shennong's classic of materia medica, it is prescribed for stomach and spleen ailments, and for calming the nerves and the mind (Mizuno et al., 1995). The Chinese traditionally use this mushroom as an adaptogen – a tonic that provides overall body balance and increases tolerance to stress. In recent studies, extracts of *Grifola* have been shown to have metabolic regulating, antitumor, and immunomodulating properties (He et al., 2017).

PHYTOCHEMISTRY

POLYSACCHARIDES

According to the previous reports, fruiting bodies and the mycelia of maitake contain carbohydrates on an average of 33.53% and 47.84%, respectively (Huang et al., 2011). The polysaccharides are chiefly comprised of glucose (72.2–75.6%), fucose (5.8%), mannose (7.5–7.8%), galactose (10.5–10.9%), and ribose (1.3–2.2%) (Siu et al., 2014; Su et al., 2016). Approximately 3.8% of its dry weight is composed of water-soluble polysaccharides where 13.2% are (1 → 3,1 → 6)-β-D-glucans (Su et al., 2016). α-D glucans are also abundant in maitake but the fruit body contains more α-glucans than the matted mycelium (Ohno et al., 1985).

As of now, more than 47 bioactive polysaccharide fractions have been isolated from maitake employing various isolation techniques. The isolated fractions mostly include MD-fraction (Kodama et al., 2002), SX-fraction (Konno et al., 2013), and X-fraction (Kubo et al., 1994), MZ-fraction (Masuda et al., 2006), and Grifolans (Suzuki et al., 1987). D-fraction and MD-fraction are considered superior among all the fractions and hence used for cancer treatment and immune regulation (Alonso et al., 2017). Nanba and colleagues extracted MD-fraction from dried maitake powder with the help of warm water followed by its isolation with ethanol precipitation. The obtained crude MD-fraction is further purified through various chromatography techniques. Hot water, sodium hydroxide (Iino et al., 1985), and citrate buffer (Adachi et al., 1994), and ultrasound (Ji et al., 2019) are the commonly used solvents for extraction of different polysaccharide fractions from maitake. Table 5.1 summarizes various bioactive polysaccharide fractions isolated from maitake, source, and extraction solvent used along with their biological activity.

Proteins, Peptides, and Enzymes

Apart from polysaccharides, a number of proteins, peptides, and enzymes are present in maitake each having unique medicinal importance (Table 5.2). An *N*-acetylgalactosamine-specific lectin (GFL) was isolated using affinity chromatography techniques. The isolated lectin was found to be cytotoxic against HeLa cells (Kawagishi et al., 1990). GFP, a nonglucan heterodimeric 83 kDa *G. frondosa* protein, was capable of modulating immune response and enhancing antitumor immunity in mice (Tsao et al., 2013). Gu and colleagues found that protein GFAHP from fruiting body of *G. frondosa* has a powerful anti-Herpes simplex virus (HSV) property (Gu et al., 2007). A glycoprotein isolated from cultured mycelia of maitake exhibited its anti-tumor activity (Cui et al., 2013). Zhuang and his team patented a bioactive glycoprotein with antidiabetic, antiobesity, antihypertensive, and antilipidemic effects that could be utilized to treat and prevent these diseases (Zhuang and Kawagishi, 2007). Fruiting bodies of maitake contain highly active proteolytic enzymes. Nishiwaki et al. (2009) isolated protease from *G. Frondosa* (ProGF) which was both an endopeptidase and aminopeptidase. Several peptidases were also extracted from maitake by different researchers. Nishiwaki and Hayashi (2001) purified and characterized an aminopeptidase from maitake fruiting bodies using ammonium sulfate purification and column chromatography. A lysine-specific zinc metalloendopeptidase present in the fruiting bodies of maitake was capable of catalysing acyl-lysine bond cleavage in polypeptides (Nonaka et al., 1995). Another dimeric novel propyl aminopeptidase was isolated by Hiwatash and co-workers in 2004.

Other Biologically Active Substances

Apart from polysaccharides and proteins, there are other components present in maitakes such as fatty acids, phenols, flavonoids, ergosterols, tocopherol, alkaloids, and vitamin C (Table 5.3). A new furanone, Grifolan A was extracted and identified from maitake which displayed antifungal activity against the plant pathogens and opportunistic human pathogen (He et al., 2016). Yaoita et al. (2000) isolated and provided a structural clarification of four new phytosphingosine type ceramides from fruiting bodies of maitake. Chen et al. (2018) obtained pyrrole alkaloids, and ergosterols and identified a new compound pyrrolefronine from GF3 fractions of *G. frondosa*. The compounds successfully displayed anti-α-glucosidase and anti-proliferative activities indicating further utilization of the mushroom as a functional food ingredient. *o*-Orsellinaldehyde, a compound present in the submerged culture of maitake exhibited selective cytotoxicity against Hep 3B

TABLE 5.1
Important Polysaccharide Fractions and Biological Activity from *G. frondosa*

Isolated fractions/ purified polysaccharides	Structural composition	Extraction solvent	Source	Biological activity	References
D-fraction	Isolated beta-glucan polysaccharide compounds (beta-1,6 glucan and beta-1,3 glucan) with protein	Hot water	Fruiting body	Anti-cancer, anti-HIV	Nanba and Kubo (1998), Nanba et al. (2000)
MD-fraction	Purified D-fraction, Range of glucan/protein ratio between 80:20 to 99:1	Hot water	Fruiting body	Anti-tumour	Nanba and Kubo (1998)
Grifolan-7N	(1→3)-linked -D-glucan with a single -D-glucopyrnosyl group attached to sixth position of every third backbone unit	Hot alkali	Fruiting body	Anti-tumour	Iino et al. (1985)
GRN	(1→6) –branched (1→3)-β-D-Glucan	0.5% citrate buffer	Mycelium	Enhanced cytokine production	Adachi et al. (1994)
MZ-fraction	β-1,6 as main chain and a β-1,3 as side chain	Hot water	Fruiting body	enhanced TNF-α and IL-12 productivity, antitumor	Masuda et al. (2006)
X-fraction	B-1,6 glucan with alpha-1,4 branches	hot water	Fruiting body	Anti-diabetic	Kubo et al. (1994)
MZF	heteropolysaccharide consisting of →6)-α-D-Gal*p*-(1 → (36.2%), →3)-α-L-Fuc*p*-(1 → (14.5%), →6)- α -D-Man*p*-(1 → (9.4%), →3)-β-D-Glc*p*-(1→ (10.1%), α -D-Manp-(1 → (23.2%), and →3,6)-β-D-Glcp-(1→ (6.5%)	Hot water, ether, ethanol	Fruiting body	Anti-tumour	Masuda et al. (2009)
GFP-N	consisted of L-arabinose, D-mannose and D-glucose as → 2,6)-α-D-Man*p*-(1 → 4, α-L-Araf-C1→, and →3,6)-β-D-Glc*p*-(1 →	Hot water	Fruiting body	Anti-diabetic	Chen et al. (2019)
GFP	composed of rhamnose, xylose, mannose, and glucose (1.00: 1.04: 1.11: 6.21), (1 → 4)-linked methylation backbone, Glcp as major structural polysaccharide, polysaccharide backbone with GFP every → 3)-Glcp-(1 → and one → 3,4)-Glcp-(1 → connected interval with a small amount of 1→, 1 → 4, 1 → 6 glycosidic linkage.	Hot water	Fruiting body	Immunostimulatory	Meng et al. (2017)

TABLE 5.1 ***(Continued)***
Important Polysaccharide Fractions and Biological Activity from *G. frondosa*

Isolated fractions/ purified polysaccharides	Structural composition	Extraction solvent	Source	Biological activity	References
GFP-22	Backbone composed of 1,4-β-D-Glcp, 1,3-β-D-Glcp, 1,6-α-D-Glcp, 1,6-α-D-Galp, 1,4,6-α-D-Manp and 1,3,6-α-D-Manp units, linear filamentous structure	Hot water	Fruiting body	Immunostimulatory	Li et al. (2018)
GF70-F1	(1→3), (1→6)-β-D-glucan & β-(1→4)-linked backbone with β -(1→6)-linked branches	Hot water	Fruiting body	Immunomodulatory, anti-inflammatory	Su et al. (2020)
GFAP	composed of (1 → 3)-β-D-Glcp and (1 → 3)-α-D-Manp,	Hydrochloric acid	Fruiting body	Anti-tumour	Yu et al. (2020)
Se-GFP-22	1,4-α-D-Glcp units as backbone chain with a branched point at C6 of both 1,3,6-β-D-Manp and 1,4,6-α-D-Galp units	Hot water	Fruiting body	Antioxidant	Li et al. (2017)
MT-α -glucan	Composed of D-glucose,-glucosidic bond	Hot water	Fruiting body	Antidiabetic, antioxidant, immunomodulatory, hypolipidemic	Lei et al. (2013), Lei et al. (2012)
GFP1	possess a 1,6-β-d-glucan backbone with a single 1,3-α-d-fucopyranosyl side-branching unit.	Hot water	mycelium	Antiviral	Zhao et al. (2016)

Cells through apoptosis (Lin and Liu, 2006). Agaricoglycerides of the fermented *G. frondosa* (AGF) have an anti-inflammatory and antinociceptive effect at doses of 500 mg/kg making it an effective alternative medicine for inflammatory pain (Han and Cui, 2012). Extracts of *G. frondosa* contain various antioxidant components such as phenols, flavonoids, α-tocopherol, and ascorbic acid (Yeh et al., 2011). The lipid fractions from the fermented mycelium of *G. frondosa* contain fatty acids and their ester, sterol, sesquiterpenes, squalene, and quinoline, with an unsaturated fatty acid content of 70–80% (Wang et al., 2012).

PHARMACOLOGICAL ACTIVITIES OF *G. FRONDOSA*

Antitumor Activity

Miyazaki et al. (1982) first reported the antitumor activity of *G. frondosa*, afterwards, they examined the chemical structure of glucans extracted from its fruiting bodies. While Hishida et al. (1988) were the first to report D-fraction for the first time from maitake. The D-fraction is a β-glucan complex mainly composed of (1 → 3)-branched (1 → 6)-β-glucans with about 30% protein (Nanba et al.,1987). There were promising prospects for D-fraction as an antitumor drug, as it could be

TABLE 5.2
Important Protein, Peptide, and Enzymes from *G. frondosa* with Biological Activity

Isolated proteins/ enzymes/ peptides	Structural composition/ enzyme specificity	Source	Extraction solvent	Biological activity	References
ProGF	recognizes leucine, phenylalanine, and lysine at the P1" position as an endopeptidase; and an aminopeptidase releasing hydrophobic and aromatic amino acids such as valine, leucine, phenylalanine, and tyrosine.	Fruiting body	water	–	Nishiwaki et al. (2009)
MEP	Single polypeptide; lysine-specific	Fruiting body		–	Nonaka et al. (1995)
prolyl aminopeptidase	Dimer; high activity toward L-proline-p-nitroanilide.	Fruiting body	Sodium phosphate buffer	–	Hiwatashi et al. (2004)
GFL	a glycoprotein comprising 3.3% total sugar, high content of acidic and hydroxy amino acid, and low content of methionine and histidine; GalNAc specific.	Fruiting body	Ethylenediaminetetraacetic acid (EDTA) and 2-mercaptoethanol	cytotoxic against HeLa cells.	Kawagishi et al. (1990)
GFPr	Non-glucan heterodimeric protein with two 41 kDa subunits	Fruiting body	sodium chloride, 2-mercaptoethanol, and Buffer containing acetic acid	Enhances antitumor immunity in mice by activating natural killer and dendritic cells	Tsao et al. (2013)
GFAHP	An 11-amino-acid peptide made up the N-terminal sequence.	Fruiting body	Hot water	anti-inflammatory, antinociceptive	Gu et al. (2007)
GFG-3a	Glycoprotein with O-glycosylation and 6.20% carbohydrate composed of Ara, Fru, Man and Glc in a molar ratio of 1.33:4.51:2.46:1.00; predominantly -sheet glycoprotein with a relatively small -helical content	Mycelium	Water	Anti-tumour	Cui et al. (2013)
Glyco-protein	Protein to saccharide ratio from 75:25 to 90:10. Amino acid composition: Asn, Gln, Ser, Thr, Gly, Ala, Val, Cys, Met, Ile, Leu, Tyr, Phe, Lys. His, Arg and Pro. Monosaccharide composition: Gal, Man, Glc, N-acetylglucosamine and Fuc	Fruiting body	Ethanol, hot water	Antidiabetic, Antihypertensive, Antiobesity, Antihyperlipidemic	Zhuang and Kawagishi (2007)

TABLE 5.3
Other Biologically Active Compounds from *G. frondosa*

Bioactive molecule	Source	Extraction solvent/ technique	Biological activity	References
Ceramide	Fruiting body	Diethyl ether	–	Yaoita et al. (2000)
Ergosterols and pyrrole alkaloids	Fruiting body	Methanol and water	anti-α-glucosidase and anti-proliferative	Chen et al. (2018)
Phenols, flavonoids, α-tocopherol, and ascorbic acid	Fruiting body	Ethanol, cold water, and hot water	antioxidant	Yeh et al. (2011)
fatty acids and coumarins	–	ethanol	alleviate lipid metabolism disorders	Zeng et al. (2021)
Lipids	Mycelium	supercritical flow CO_2	antimicrobial	Wang et al. (2012)
AGF	Mycelium	acetone	anti-inflammatory and antinociceptive	Han and Cui (2012)
Furanone	Fruiting body	Ethyl acetate	antimicrobial	He et al. (2016)
o - Orsellinaldehyde	Mycelium	Ethyl acetate	apoptosis	Lin and Lui (2006)

administered intravenously, intraperitoneally, and orally (Hishida et al., 1988). In terms of effectiveness, *G. frondosa* was more effective in treating liver, lung, and breast cancer than stomach, leukemia, and, bone cancer (Nanba and Kubo, 1998). Likewise, D-fraction was found to be efficacious against mammary tumor cells (Alonso et al., 2017) and human hepatocarcinoma SMMC-7721 cells (Zhao et al., 2017). Additionally, scientists have also reported other polysaccharide fractions and bioactive molecules from maitake that have antitumor activity and are listed in Table 5.4.

Antiviral and Antibacterial Activity

Several antiviral activities from *G. frondosa* have been reported by various researchers. MD-fraction which is a polysaccharide bound to protein with a glucan: protein ratio extending from 80:20 to 99:1, could suppress HIV as mentioned by Nanba et al. (2000). The GFP1 fraction of the mushroom inhibited the pathogen responsible for hand-foot-and-mouth disease, Enterovirus 71 (EV71) (Zhao et al., 2016). By ion-exchange chromatography GFAHP, a novel antiviral protein was purified from maitake fruiting bodies. As described by Gu et al. (2007), the protein was able to inhibit the in vitro replication of herpes simplex virus type 1 (HSV-1) with an IC50 value of 4.1 µg/mL, however, the therapeutic in was >29.3.

In addition to antiviral activity, there are reports on the antibacterial activity of the mushroom. Zhang et al. (2017) reported that *G. frondosa* strain SH-05 potentially inhibited the tested bacteria forming inhibition zones. The antibacterial activity of D-fraction was due to immune-stimulating activity as it could induce cytokine production and enhance the activity of immune-competent cells to kill the target bacteria (Kodama et al., 2001).

Hypolipidemic/Anti-Hyperlipidemic Activity

Abnormal accumulation of high levels of cholesterol, triglycerides, and other lipids in the plasma induces hyperlipidemia. Two novel intracellular polysaccharides (GFP-W1 and GFP-W2) from

TABLE 5.4
Biological Activity of Different Components from *G. frondosa*

Bioactivity	Bioactive components	Isolated fraction	Key findings	References
Anti-tumor	Polysaccharide	D fraction	With 1mg/kg/day for 17 days, 91.3% of the MM46 liver carcinomas were inhibited	Kodama et al. (2005)
		MD-fraction	94.3% inhibition ratio after implanting 0.1 mg/kg of MM46 carcinoma at 10 times	Nanba and Kubo (1998)
		MZ-fraction	70.3% inhibition ratio at 4 mg/kg/day against MM46 carcinoma	Masuda et al. (2006)
		GFP-A	Inhibition of human colon cancer cells at 150 mg/mL for 48 hours (IC50)	Bie et al. (2020)
		GFP-A	Tumor inhibitory rates after daily administration of 50, 100, and 200 mg/kg for 15 days were 17.1%, 28.3%, and 52.2%, respectively	Chen et al. (2020)
		GFAP	tumor inhibitory rates after intragastric administration of 100 and 200 mg/kg for 15 days was 16.36% and 36.72%, respectively	Yu et al. (2020)
		GP11	Cytotoxic against HepG-2 cells in vitro, growth inhibition of Heps cells in vivo	Mao et al. (2015)
	Glycoprotein	GFG-3a	highest antitumor activity on S180 and Bel-7402 cells	Cui et al. (2013)
	Water-soluble extract	-	90% inhibition of gastric cancer cell lines at 10% w/v maitake extract for 3 days.	Shomori et al. (2009)
	o-orsellinaldehyde	-	Apoptosis mediated cytotoxicity against Hep 3B cells.	Lin et al. (2006)
Antiviral	Polysaccharide	GFP1	Inhibited EV71 viral replication, VP1 viral protein expression, and genomic RNA synthesis.	Zhao et al. (2006)
		MD- fraction	Increased CD4+ cell counts to 1.4–1.8 times, decreased viral loads	Nanba et al. (2000)
	Protein	GFAHP	Inhibited in vitro replication of HSV-1 with an IC50 value of 4.1 µg/mL and a therapeutic index >29.3.	Gu et al. (2007)
Antibacterial	Polysaccharide	IZPS	Inhibited the tested pathogens with inhibition zones of 30.0 ± 4.3 mm (*E. coli*), 39.7 ± 2.5 mm (*S. aureus*), 26.3 ± 3.4 mm (*B. megaterium*), and 28.6 ± 3.2 mm (*L. monocytogenes*)	Zhang et al. (2016)
Hypolipidemic	Polysaccharide	GFP-W1, GFP-W2, EP-X1	sodium cholate and sodium glycocholate binding rate of GFP-W1, GFP-W2 and EP-X1 was 30.32%, 37.41%, 47.62%; and 25.08%, 33.68%, 38.70%, respectively.	Yang et al. (2021)
		GFP	Oral administration reduced serum levels of fasting blood glucose (FBG), oral glucose tolerance (OGT), cholesterol (TC), triglyceride, and low-density lipoprotein cholesterol and significantly lowered the hepatic levels of TC, TG and free fatty acids (FFA)	Guo et al. (2020)

TABLE 5.4 *(Continued)*
Biological Activity of Different Components from *G. frondosa*

Bioactivity	Bioactive components	Isolated fraction	Key findings	References
Antidiabetic	Polysaccharide	MT-α-glucan	Increased activity of GSH, SOD and GSHpx and hepatic glycogen content and decreased level of MDA, free fatty acid, triglycerides, serum insulin, and fasting plasma glucose.	Hong et al. (2007)
		F2 and F3	Increased activity and mRNA levels of IR and IRS-1, and decreased levels of fasting serum glucose (FSG) levels, fasting serum insulin (FSI) levels, and a homeostasis model assessment of insulin resistance (HOMA-IR)	Xiao et al. (2015)
	n-Hexane extract	GF-H	Oral administration GF600 (GF at 600mg/kg) lowered average blood glucose, glycated hemoglobin, serum total cholesterol in diabetic mice.	Shen et al. (2015)
	Glycoprotein	SX-fraction	Capable of overcoming the suppressive effects of high glucose	Konno et al. (2013)
	Ergosterol peroxide	–	At 5 μM concentration enhanced glucose uptake, decreased ROS formation, up-regulated the expression of IRS-1, p-IRS-1, PI3K, Akt, p-Akt, and GLUT-4	Wu et al. (2020)
Immunomodulatory	Polysaccharide	D-fraction	Indirect activation of NK cell through IL-12 production by macrophages and dendritic cells.	Kodama et al. (2003)
		MZ-fraction	Enhanced production of TNF-α and IL-12, enhanced antigen presentation of murine macrophage cell line J774	Masuda et al. (2006)
		GFP	Enhanced cytokine and chemokine production	Zhang et al. (2018)
		GRN	Enhanced production of IL-6, IL-1, α (TNFα)	Adachi et al. (1994)
Antioxidant	Polysaccharide	GFP-1, GFP-2, GFP-3	inhibited 1,1-diphenyl-2-picrylhydrazyl (DPPH) radical, hydroxyl radical, and superoxide radical	Chen et al. (2012)
		Se-GFP-22	Possess potent radical scavenging activity (46% scavenging rate)	Li et al. (2017)
	Protein	GFHT-4	Exhibited DPPH scavenging activity, reduces ferrous ion, inhibit autooxidation of linoleic acid	Qi et al. (2015)
	Other compounds	Ergosterol, ergostra-4,6,8(14),22-tetraen-3-one, and 1-oleoyl-2-linoleoyl-3-palmitoylglycerol	At 100 μ/mL, antioxidant activities were 79, 48, and 42%, respectively	Zhang et al. (2002)

G. frondosa in submerged conditions were reported to demonstrate a remarkable hypolipidemic activity (Yang et al., 2021). Polysaccharides from *G. frondosa* prevent hyperlipidemia by-modifying hepatic glycolipid metabolism-related genes in diabetic mice (Guo et al., 2020). When fruiting bodies of maitake were provided as feed, the cholesterol, triglycerides, and other lipid content in the rat serum were suppressed up to 70% (Kubo and Nanba, 1997).

Antidiabetic Activity

Antidiabetic activity from different extracts of *G. frondosa* was thoroughly studied by various researchers. Hypoglycemic activities are mostly linked to insulin activities. Polysaccharide fractions F2 and F3 from maitake are supposed to enhance insulin resistance in diabetic rats by reactivation of insulin receptor (IR) and insulin receptor substrate-1 (IRS-1) (Xiao et al., 2015). Hong and co-workers showed that MT-α-glucan from maitake had an antidiabetic effect on KK-ay mice (Hong et al., 2007). As reported by Konno et al. (2013), SX-fraction could overcome the suppressive effects of high glucose and facilitate glucose uptake and increased insulin secretion. As indicated by Wu et al. (2021), the hypoglycemic effect of maitake could be due to the inhibition of α-glucosidase, an enzyme that hydrolyses starch into disaccharides. Table 5.4 summarizes the antidiabetic effects of various components and their isolated fractions as reported by various researchers.

Immunomodulatory Properties

Among various immunomodulatory components from *G. frondosa*, polysaccharides have been the most widely studied. There is a substantial immunomodulatory effect associated with the polysaccharide fraction D-fraction of *G. frondosa*. As reported by various researchers, maitake could stimulate the production of macrophage-derived IL-12 and activate NK cells (Kodama et al., 2002; Kodama et al., 2003). D-fraction could lessen the dose of mitomycin C (MMC) (a chemotherapeutic agent), by increasing the proliferation, differentiation, and activation of immune cells (Kodama et al., 2005). Table 5.4 shows immunomodulatory activity from different polysaccharide fractions from maitake.

TOXICITY REMARKS

According to the reports available in the literature, maitake has long been regarded as a food that is non-toxic. Maitake polysaccharides were reported to be low in toxicity and have an excellent safety profile. A study conducted by Wang found that β -glucan extracted from mycelium is non-toxic and does not adversely affect growth in mice (Wang et al., 2010). One more study on mice showed that maitake β-glucan was not mutagenic in mice but toxic to germ cells of male mice (Wang et al., 2008). Acute toxicity tests on mice have shown that 3 g/kg of GFP derived from fermentation broth is the maximum safe dose (Zheng et al., 2001). Taking polysaccharides did not cause any bodily discomfort or adverse reactions in a randomized trial involving 28 healthy subjects (Glauco et al., 2004). During the phase I/II trial of breast cancer survivors, no serious adverse effects were reported during the study period (Deng et al., 2009). There was no evidence of cytotoxic, mutagenic, and/or antimutagenic effects from aqueous and methanolic extracts of *G. frondosa*.

ECONOMIC IMPORTANCE

The dried maitake extract is used in a variety of health food products including teas, powders, granules, extracts, and drinks (Figure 5.2). The delicious and unique taste makes it popular as a food ingredient, flavoring agent as well as a health supplement. As indicated by Kim et al. (2007), extracted polysaccharides from *G. frondosa* and mycelia extract can be a potential ingredient for cosmetics. Further, based on the antioxidant activity of polysaccharides extracted from

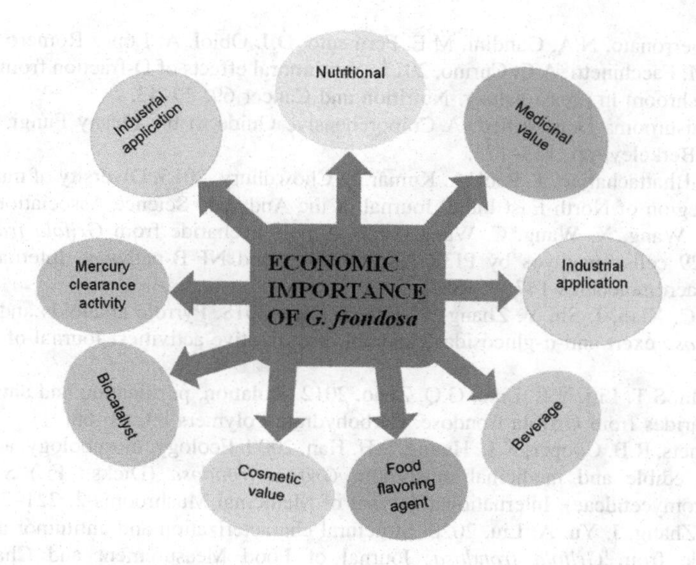

FIGURE 5.2 Various economic applications of *G. frondosa.*

G. frondosa, radical scavenging activity after UV irradiation, fibroblast proliferation, and collagen synthesis, Lee and colleagues suggested that they could be useful in cosmetic applications (Lee et al., 2003). Various compounds isolated from maitake have several industrial applications as well. During growth, this fungus secretes a ligninolytic enzyme laccase into the medium (Xing et al., 2006). Researchers have found that this enzyme could be used for decolorizing synthetic dyes (Nitheranont et al., 2011). Since laccase from maitake is considered safe for use in food processing, this enzyme is well suited for industrial use (Nitheranont et al., 2011). Maitake can also be used for the preparation of beverages. Kameyama and colleagues patented a method of producing beverage by combining roasted, ground maitake with roasted, ground coffee beans to provide a maitake-coffee mixture, which after brewing a maitake-coffee beverage is made (Kameyama et al., 2012). Interestingly, Zhang and colleagues reported mercury clearance activity from maitake for the first time. They isolated a polysaccharide-peptide from dried fruiting bodies which could eliminate the burden of mercury in the liver and kidneys of tested rats and could keep the blood level mercury within a stable range (Zhang et al., 2018)

CONCLUSION

Maitake is a valuable asset for the welfare of human. Emerging studies have confirmed the biological functions of maitake as immune regulator, anti-tumor, anti-ageing, antivirus, and biological response modifier. Polysaccharides are the most significant bioactive components of *G. frondosa* which contribute to its bioactivities as well as health benefits. In the future, they may be used to cure various diseases and be a crucial component of research and development. It has tremendous contribution toward traditional and modern healthcare systems. Therefore, further investigations about its bioactive constituents should be carried out with the aid of modern methods for finding out the huge scope hidden within this majestic wild mushroom.

REFERENCES

Y. Adachi, M. Okazaki, N. Hno, T. Yadomae. 1994. Enhancement of cytokine production by macrophages stimulated with (1→ 3)-β-D-glucan, grifolan (GRN), isolated from *Grifola frondosa*. Biological and Pharmaceutical Bulletin 17: 1554–1560.

E.N. Alonso, M.J. Ferronato, N.A. Gandini, M.E. Fermento, D.J. Obiol, A. López Romero, J. Arévalo, M.E. Villegas, M.M. Facchinetti, A.C. Curino. 2017. Antitumoral effects of D-fraction from *Grifola frondosa* (maitake) mushroom in breast cancer. Nutrition and Cancer 69: 29–43.

D. Arora. 1986. Mushrooms Demystified: A Comprehensive Guide to the Fleshy Fungi, 2nd edition, Ten Speed Press, Berkeley, pp. 163–173.

J. Bhattacharjee, D. Bhattacharjee, T. Paul, A. Kumar, S. Chowdhury. 2015. Diversity of mushrooms in Indo-Bangladesh region of North-East India. Journal of the Andaman Science Association 19: 75–82.

N. Bie, L. Han, Y. Wang, X. Wang, C. Wang. 2020. A polysaccharide from *Grifola frondosa* fruit body induces HT-29 cells apoptosis by PI3K/AKT-MAPKs and NF-B-pathway. International Journal of Biological Macromolecules 147: 79–88.

S. Chen, T. Yong, C. Xiao, J. Su, Y. Zhang, C. Jiao, Y. Xie. 2018. Pyrrole alkaloids and ergosterols from *Grifola frondosa* exert anti-α-glucosidase and anti-proliferative activities. Journal of Functional Foods 43: 196–205.

G.T. Chen, X.M. Ma, S.T. Liu, Y.L. Liao, G.Q. Zhao. 2012. Isolation, purification and antioxidant activities of polysaccharides from Grifola frondose. Carbohydrate Polymers 89: 61–66.

W.A. Chen, P. Stamets, R.B. Cooper, N.L. Huang, S.H. Han. 2000. Ecology, morphology, and morphogenesis in nature of edible and medicinal mushroom *Grifola frondosa* (Dicks.: Fr.) S.F. Gray-Maitake (Aphyllophoromycetideae). International Journal of Medicinal Mushrooms 2: 221–228.

X. Chen, H. Ji, C. Zhang, J. Yu, A. Liu. 2020. Structural characterization and antitumor activity of a novel polysaccharide from *Grifola frondosa*. Journal of Food Measurement and Characterization 14: 272–282.

Y. Chen, D. Liu, D. Wang, S. Lai, R. Zhong, Y. Liu, C. Yang, B. Liu, M.R. Sarker, C. Zhao. 2019. Hypoglycemic activity and gut microbiota regulation of a novel polysaccharide from *Grifola frondosa* in type 2 diabetic mice. Food and Chemical Toxicology 126: 295–302.

F. Cui, X. Zan, Y. Li, Y. Yang, W. Sun, Q. Zhou, Y. Dong. 2013. Purification and partial characterization of a novel anti-tumor glycoprotein from cultured mycelia of *Grifola frondosa*. International Journal of Biological Macromolecules 62: 684–690.

G. Deng, H. Lin, A. Seidman et al. 2009. A phase I/II trial of a polysaccharide extract from *Grifola frondosa* (Maitake mushroom) in breast cancer patients: Immunological effects. Journal of Cancer Research and Clinical Oncology 135: 1215–1221.

S. Glauco, F. Jano, G. Paolo et al. 2004. Safety of maitake D-fraction in healthy patients: Assessment of common hematologic parameters. Alternative and Complementary Therapies 10: 228–230.

C.Q. Gu, J.W. Li, F. Chao, M. In, X.W. Wang, Z.Q. Shen. 2007. Isolation, identification and function of a novel anti-HSV-1 protein from *Grifola frondosa*. Antiviral Research 75: 250–257.

W.L. Guo, J.C. Deng, Y.Y. Pan, J.X. Xu, J.L. Hong, F.F. Shi, X.C. Lv. 2020 Hypoglycemic and hypolipidemic activities of *Grifola frondosa* polysaccharides and their relationships with the modulation of intestinal microflora in diabetic mice induced by high-fat diet and streptozotocin. International Journal of Biological Macromolecules 153: 1231–1240.

C. Han, B. Cui. 2012. Pharmacological and pharmacokinetic studies with agaricoglycerides, extracted from *Grifola frondosa*, in animal models of pain and inflammation. Inflammation 35: 1269–1275.

X. He, X. Du, X. Zang, L. Dong, Z. Gu, L. Cao, D. Chen, N.O. Keyhani, L. Yao, J. Qiu, X. Guan 2016. Extraction, identification and antimicrobial activity of a new furanone, grifolaone A, from Grifola frondosa. Natural Product Research 30: 941–947.

X. He, X. Wang, J. Fang, Y. Chang, N. Ning, H. Guo, Z. Zhao. 2017. Polysaccharides in *Grifola frondosa* mushroom and their health-promoting properties: A review. International Journal of Biological Macromolecules 101: 910–921.

C. Herbalism. 1999. By Chinese Herbalism Editorial Board. State Administration of Traditional Chinese Medicine of the People's Republic of China, Shanghai Scientific and Technology Press, Shanghai, 15: 643.

I. Hishida, H. Nanba, H. Kuroda. 1988. Antitumor activity exhibited by orally administered extract from fruit body of *Grifola frondosa* (Maitake). Chemical and Pharmaceutical Bulletin 36: 1819–1827.

K. Hiwatashi, K. Hori, K. Takahashi, A. Kagaya, S. Inoue, T. Sugiyama, S. Takahashi. 2004. Purification and characterization of a novel prolyl aminopeptidase from Maitake (*Grifola frondosa*). Bioscience, Biotechnology, and Biochemistry 68: 1395–1397. 2: 684–690.

L. Hong, M. Xun, W. Wutong 2007. Anti-diabetic effect of an à-glucan from fruit body of maitake (Grifola frondosa) on KK-Ay mice. Journal of Pharmacy and Pharmacology 59: 575–582.

S.J. Huang, S.Y. Tsai, S.Y. Lin, C.H. Liang, J.L. Mau. 2011. Nonvolatile taste components of culinary-medicinal maitake mushroom, *Grifola frondosa* (Dicks.:Fr.) S.F. gray. International Journal of Medicinal Mushrooms 13: 265–272.

K. Iino, N. Ohno, I. Suzuki, T. Miyazaki, T. Yadomae, S. Oikawa, K. Sato. 1985. Structural characterisation of a neutral antitumour β-d-glucan extracted with hot sodium hydroxide from cultured fruit bodies of *Grifola frondosa*. Carbohydrate Research 141: 111–119.

H.Y. Ji, J. Yu, A.J. Liu. 2019. Structural characterization of a low molecular weight polysaccharide from *Grifola frondosa* and its antitumor activity in H22 tumor-bearing mice. Journal of Functional Foods 61: 103472.

K. Kameyama, M. Tsai, J. Waskiewicz. 2012. Maitake mushroom coffee. US Patent.

H. Kawagishi, A. Nomura, T. Mizuno, A. Kimura, S. Chiba. 1990. Isolation and characterization of a lectin from *Grifola frondosa* fruiting bodies. Biochimica et Biophysica Acta (BBA)-General Subjects 1034: 247–252.

N. Kodama, T. Kakuno, H. Nanba. 2003. Stimulation of the natural immune system in normal mice by polysaccharide from maitake mushroom. Mycoscience 44: 257–261.

S.W. Kim, H.J. Hwang, B.C. Lee, J.W. Yun 2007. Submerged production and characterization of Grifola frondosa polysaccharides-a new application to cosmeceuticals. Food Technology and Biotechnology 45: 295–305.

N. Kodama, K. Komuta, N. Sakai, H. Nanba. 2002. Effects of D-Fraction, a polysaccharide from *Grifola frondosa* on tumor growth involve activation of NK cells. Biological and Pharmaceutical Bulletin 25: 1647–1650.

N. Kodama, Y. Murata, A. Asakawa, A. Inui, M. Hayashi, N. Sakai, H. Nanba. 2005. Maitake D-fraction enhances antitumor effects and reduces immunosuppression by mitomycin-C in tumor-bearing mice. Nutrition 21: 624–629.

N. Kodama, M. Yamada, H. Nanba. 2001. Addition of Maitake D-fraction reduces the effective dosage of vancomycin for the treatment of Listeria-infected mice. Japanese Journal of Pharmacology 87: 327–332.

S. Konno, B. Alexander, J. Zade, M. Choudhury. 2013. Possible hypoglycemic action of SX-fraction targeting insulin signal transduction pathway. International Journal of General Medicine 6: 181–187.

K. Kubo, H. Aoki, H. Nanba, H. 1994. Anti-diabetic activity present in the fruit body of *Grifola frondosa* (Maitake). I. Biological and Pharmaceutical Bulletin 17: 1106–1110.

K. Kubo, H. Nanba. 1997. Anti-hyperliposis effect of Maitake fruit body (*Grifola frondosa*). I. Biological and Pharmaceutical Bulletin 20: 781–785.

B.C. Lee, J.T. Bae, H.B. Pyo, T.B. Choe, S.W. Kim, H.J. Hwang, J.W. Yun. 2003. Biological activities of the polysaccharides produced from submerged culture of the edible Basidiomycete *Grifola frondosa*. Enzym Microbial Technology 32: 574–581.

H. Lei, S. Guo, J. Han, Q. Wang, X. Zhang, W. Wu. 2012. Hypoglycemic and hypolipidemic activities of MT-α-glucan and its effect on immune function of diabetic mice. Carbohydrate polymers 89: 245–250.

H. Lei, W. Wang, Q. Wang, S. Guo, L. Wu. 2013. Antioxidant and immunomodulatory effects of α-glucan from fruit body of maitake (*Grifola frondosa*). Food and Agricultural Immunology 24: 409–418.

Q. Li, F. Zhang, G. Chen, Y. Chen, W. Zhang, G. Mao, T. Zhao, M. Zhang, L. Yang, X. Wu. 2018. Purification, characterization and immunomodulatory activity of a novel polysaccharide from *Grifola frondosa*. International Journal of Biological Macromolecules 111: 1293–1303.

Q. Li, W. Wang, Y. Zhu, Y. Chen, W. Zhang, P. Yu, X. Wu. 2017. Structural elucidation and antioxidant activity: A novel Se-polysaccharide from Se-enriched *Grifola frondosa*. Carbohydrate Polymers 161: 42–52.

J.T. Lin, W.H. Liu. 2006. o-Orsellinaldehyde from the submerged culture of the edible mushroom *Grifola frondosa* exhibits selective cytotoxic effect against Hep 3B cells through apoptosis. Journal of Agricultural and Food Chemistry 54: 7564–7569.

G.H. Mao, Y. Ren, W.W. Feng, Q. Li, H.Y. Wu, T. Zhao, T.X.Y. Wu. 2015. Antitumor and immunomodulatory activity of a water-soluble polysaccharide from *Grifola frondosa*. Carbohydrate Polymers 134: 406–412.

Y. Masuda, N. Kodama, H. Nanba. 2006. Macrophage J774. 1 cell is activated by MZ-Fraction (Klasma-MZ) polysaccharide in *Grifola frondosa*. Mycoscience 47: 360–361.

Y. Masuda, A. Matsumoto, T. Toida, T. Oikawa, K. Ito, H. Nanba. 2009. Characterization and antitumor effect of a novel polysaccharide from *Grifola frondosa*. Journal of Agricultural and Food Chemistry 57: 10143–10149.6.

Y. Masuda, N. Kodama, H. Nanba. 2006. Macrophage J774. 1 cell is activated by MZ-Fraction (Klasma-MZ) polysaccharide in *Grifola frondosa*. Mycoscience 47: 360–366.

M. Mayell. 2001. Maitake extracts and their therapeutic potential. Alternative Medicine Review 6: 48–60. PMID: 11207456.

M. Meng, D. Cheng, L. Han, Y. Chen, C. Wang. 2017. Isolation, purification, structural analysis and immunostimulatory activity of water-soluble polysaccharides from *Grifola frondosa* fruiting body. Carbohydrate Polymer 157: 1134–1143.

T. Miyazaki, T. Yadomae, I. Suzuki, M. Nishijima, S. Yui, S. Oikawa, K. Sato. 1982. Antitumor activity of fruiting bodies of cultured *Grifola frondosa*. Japanese Journal of Medical Mycology 23: 261–263.

T. Mizuno, C. Zhuang. 1995. Maitake, *Grifola frondosa* pharmacological effects. Food Reviews International 11: 135–149.

H. Nanba, A. Hamaguchi, H. Kuroda. 1987. The chemical structure of an antitumorpolysaccharide in fruit bodies of (Maitake). Chemical and Pharmaceutical Bulletin 35: 1162–1168.

H. Nanba, K. Kubo. 1998. Antitumor Substance Extracted from Grifola. U.S. Patent 5,854,404

H. Nanba, N. Kodama, D. Schar, D. Turner. 2000. Effects of maitake (*Grifola frondosa*) glucan in HIV-infected patients. Mycoscience 41: 293–295.

T. Nonaka, H. Ishikawa, Y. Tsumuraya, Y. Hashimoto, N. Dohmae, K. Takio K. 1995. Characterization of a thermostable lysine-specific metalloendopeptidase from the fruiting bodies of a basidiomycete, *Grifola frondosa*. The Journal of Biochemistry 118: 1014–1020.

T. Nishiwaki, S. Asano, T. Ohyama. 2009. Properties and substrate specificities of proteolytic enzymes from the edible basidiomycete *Grifola frondosa*. Journal of Bioscience and bioengineering 107: 605–609.

T. Nishiwaki, K. Hayashi. 2001. Purification and characterization of an aminopeptidase from the edible basidiomycete *Grifola frondosa*. Bioscience, biotechnology, and Biochemistry 65: 424–427.

T. Nitheranont, A. Watanabe, T. Suzuki, T. Katayama, Y. Asada. 2011. Decolorization of synthetic dyes and biodegradation of bisphenol A by laccase from the edible mushroom, *Grifola frondosa*. Bioscience, Biotechnology, and Biochemistry 75: 1845–1847.

N. Ohno, K. Iino, T. Takeyama, I. Suzuki, K. Sato, S. Oikawa, T. Miyazaki, T. Yadomae. 1985. Structural characterization and antitumor activity of the extracts from matted mycelium of cultured *Grifola frondosa*. Chemical and Pharmaceutical Bulletin 33: 3395–3401.

G.D.Y. Qi, Z. Yang, H. Wang, S. Wang, G. Chen G. 2015. Preparation, separation and antioxidant properties of hydrolysates derived from *Grifola frondosa* protein. Czech Journal of Food Sciences 33: 500–506.

K.P. Shen, C.H. Su, T.M. Lu, M.N. Lai, L.T. Ng. 2015. Effects of *Grifola frondosa* non-polar bioactive components on high-fat diet-fed and streptozotocin-induced hyperglycemic mice. Pharmaceutical Biology 53: 705–709

K. Shomori, M. Yamamoto, I. Arifuku, K. Teramachi, H. Ito. 2009. Antitumor effects of a water-soluble extract from Maitake (*Grifola frondosa*) on human gastric cancer cell lines. Oncology Reports 22: 615–620.

K.C. Siu, X. Chen, J.Y. Wu. 2014. Constituents actually responsible for the antioxidant activities of crude polysaccharides isolated from mushrooms. Journal of Functional Foods 11: 548–556.

P. Stamets. 2000. Techniques for the cultivation of the medicinal mushroom royalsun Agaricus? Agaricusblazei Murr. (Agaricomycetideae). International Journal of Medicinal Mushrooms 2: 151–160.

C.H. Su, M.N. Lai, C.C. Lin, L.T. Ng. 2016. Comparative characterization of physicochemical properties and bioactivities of polysaccharides from selected medicinal mushrooms. Applied Microbiology and Biotechnology 100: 4385–4393.

C.H. Su, M.K. Lu, T.J. Lu, M.N. Lai, L.T. Ng. 2020. A (1→ 6)-Branched (1→ 4)-β-D-Glucan from *Grifola frondosa* inhibits lipopolysaccharide-induced cytokine production in RAW264. 7 macrophages by binding to TLR2 rather than Dectin-1 or CR3 receptors. Journal of Natural Products 83: 231–242.

I. Suzuki, T. Takeyama, N. Ohno, S. Oikawa, K. Sato, Y. Suzuki T (1987) Yadomae, antitumor effect of polysaccharide grifolan NMF-5N on syngeneic tumor in mice. Journal of Pharmacobio-Dynamics 10: 72–77.

F. Takama, S. Ninomiya, R. Yoda, H. Ishii, S. Muraki. 1981. Parenchyma cells, chemical components of Maitake mushroom (*Grifola frondosa* S.F. Gray) cultured artificially, and their changes by storage and boiling. Mushroom Science 11: 767–779.

Y.W. Tsao, Y.C. Kuan, J.L. Wang, F. Sheu. 2013. Characterization of a novel maitake (*Grifola frondosa*) protein that activates natural killer and dendritic cells and enhances antitumor immunity in mice. Journal of Agricultural and Food Chemistry 61: 9828–9838.

B.Q. Wang, Z.P. Xu, C.L. Yang. 2012. The chemical compositions of the lipid extracted from fermented mycelium of *Grifola frondosa*. Advanced Materials Research 503: 412–415. Trans Tech Publications Ltd.

Y. Wang, S. Xie, T. Sun T et al. 2010. Inhibitory effect of the polysaccharide of *Grifola frondosa* on carbon tetrachloride induced injury to the liver cell line L02. Journal of Shandong University (Health Science) 48: 32–37.

B. Wang, B. Zhang, Z. Xu et al. 2008. Genotoxicological evaluation of β-glucan from fermented mycelia of *Grifola frondosa*. International Journal of Toxicology 22: 407–408.

S.J. Wu, Y.J. Tung, L.T. Ng. 2020. Anti-diabetic effects of *Grifola frondosa* bioactive compound and its related molecular signaling pathways in palmitate-induced C2C12 cells. Journal of Ethnopharmacology 260: 112962.

Wu, J.-Y. Siu, K.C. Geng, P. 2021. Bioactive ingredients and medicinal values of Grifola frondosa (Maitake). Foods 10: 95.

C. Xiao, Q. Wu, Y. Xie, J. Zhang, J. Tan. 2015. Hypoglycemic effects of *Grifola frondosa* (Maitake) polysaccharides F2 and F3 through improvement of insulin resistance in diabetic rats. Food & Function 6: 3567–3575.

Z.T. Xing, J.H. Cheng, Q. Tan, Y.J. Pan. 2006. Effect of nutritional parameters on laccase production by the culinary and medicinal mushroom, *Grifola frondosa*. World Journal of Microbiology and Biotechnology 22: 799–806.

W. Yang, J. Wu, W. Liu, Z. Ai, Y. Cheng, Z. Wei, L. Yang. 2021. Structural characterization, antioxidant and hypolipidemic activity of *Grifola frondosa* polysaccharides in novel submerged cultivation. Food Bioscience 42: 101187.

Y. Yaoita, T. Ishizuka, R. Kakuda, K. Machida, M. Kikuchi. 2000. Structures of new ceramides from the fruit bodies of *Grifola frondosa*. Chemical and Pharmaceutical Bulletin 48: 1356–1358.

J.Y. Yeh, L.H. Hsieh, K.T. Wu, C.F. Tsai. 2011. Antioxidant properties and antioxidant compounds of various extracts from the edible basidiomycete *Grifola frondosa* (Maitake). Molecules 16: 3197–3211.

J. Yu, H.Y. Ji, C. Liu, A.J. Liu. 2020. The structural characteristics of an acid-soluble polysaccharide from *Grifola frondosa* and its antitumor effects on H22-bearing mice. International Journal of Biological Macromolecules 158: 1288–1298.

F. Zeng, Y. Liu, Y. Pan, J. Xu, X. Ge, H. Zheng, Y. Huang. 2021. Coumarin-rich *Grifola frondosa* ethanol extract alleviate lipid metabolism disorders and modulates intestinal flora compositions of high-fat diet rats. Journal of Functional Foods 85: 104649.

A. Zhang, J. Deng, S. Yu, F. Zhang, R.J. Linhardt, P. Sun. 2018. Purification and structural elucidation of a water-soluble polysaccharide from the fruiting bodies of the *Grifola frondosa*. International Journal of Biological Macromolecules 115: 221–226.

Y. Zhang, G.L. Mills, M.G. Nair M. G. 2002. Cyclooxygenase inhibitory and antioxidant compounds from the mycelia of the edible mushroom *Grifola frondosa*. Journal of Agricultural and Food Chemistry 50: 7581–7585.

C. Zhao, L. Gao, C. Wang, B. Liu, Y. Jin, Z. Xing. 2016. Structural characterization and antiviral activity of a novel heteropolysaccharide isolated from *Grifola frondosa* against enterovirus 71. Carbohydrate Polymers 144: 382–389.

F. Zhao, Y.F. Wang, Y.L. Song, J.X. Jin, Y.Q. Zhang, H.Y. Gan, K.H. Yang. 2017. Synergistic apoptotic effect of d-fraction from *Grifola frondosa* and vitamin C on hepatocellular carcinoma SMMC-7721 cells. Integrative Cancer Therapies 16: 205–214.

C. Zhang, Z. Gao, C. Hu, J. Zhang, X. Sun, C. Rong, L. Jia 2017. Antioxidant, antibacterial and anti-aging activities of intracellular zinc polysaccharides from Grifola frondosa SH-05. International Journal of Biological Macromolecules 95: 778–787.

S. Zheng, S. Chen, W. Gu, et al. 2001. Fractionation and antitumor activity of polysaccharide from *Grifola frondosa*. Pharm Biotechnol 8: 279–283.

C. Zhou, A. Wu, Q. Tang, Y. Liu, S. Zhou, W. Jia, Y. Yang, P. Yan. 2013. Isolation and purification of GFLP, a high-molecule polysaccharide from *Grifola frondosa* fruit bodies, and its effect on immune cells. Acta Edulis Fungi 20: 39–42.

C. Zhuang, H. Kawagishi, H.G. Preuss. 2007. Glycoprotein with antidiabetic, antihypertensive, antiobesity and antihyperlipidemic effects from *Grifola frondosa*, and a method for preparing same. U.S. Patent 7,214,778, 8 May 2007.

6 *Hericium erinaceus* (*Bull.*) and *Hericium coralloides*

Anu Shrivastava and Swati Jain
Department of Food and Nutrition, Lady Irwin College, University of Delhi, India

CONTENTS

INTRODUCTION

Mycotherapy has been used since ancient times for strengthening the body and for long, balanced life (Valu et al., 2021). Mushroom extracts are used for development of miracle medicines and are potential functional food (Ghosh et al., 2021). In recent decades, *Hericium* spp. has gained wide attention owing to their beneficial effects on a broad spectrum of human ailments (Pallua et al., 2012). *H. erinaceus* and *H. coralloides* are Basidiomycetes of the order Russulales and belong to the family Hericiaceae. *H. erinaceus* has a hedgehog-like appearance while *H. coralloides* resembles corals. They are native to North America, Europe, and Asia (Li et al., 2016). These mushrooms have notable edible, nutraceutical, and medicinal values due to which they have been in usage in Chinese traditional medicine since time immemorial and are also consumed as a food supplement (Wabang & Ajungla, 2016; Wittstein et al., 2016). *H. erinaceus* extracts are used for the development of multitarget therapeutics. Mushrooms are a profitable source of drug development (Valu et al., 2021; Ghosh et al., 2021). The mycelium and the fruiting body of *H. erinaceus* contain various bioactive substances that perform different functions of various organ systems via different mechanisms (Jiang et al., 2014).

In many countries, *H. erinaceus* has a rare occurrence and has been red listed in 13 of the 23 European countries due to the destruction of its natural habitat (Thongbai et al., 2015). In India, it

DOI: 10.1201/9781003259763-6

is one of the flagship species for the fungal biodiversity conservation. In the North-Eastern states of India, wild fungi are gathered mainly for usage in food and medicine, and serve as an alternative source of income for the locals. Urgent attention is required along with appropriate management guidelines in order to protect their fast-dwindling population (Wabang & Ajungla, 2016).

BOTANICAL DESCRIPTION

All the species belonging to the Genera *Hericium* have hymenophores consisting of spines of varied lengths and basidiomes adorned with amyloid. With ageing, single clumps of basidiomes develop branches after differentiation from primordia. The tough, watery, and fleshy tubercle of the basidiomes has a seafood flavour similar to that of a crab or lobster. Basidiospores are white, warty, and amyloid measuring 5.5–6.5 × 4.5–5.6 μm and short ellipsoid or subglobose shaped in both species. Their basidia consists of 4 spores measuring 25–40 × 5–7 μm with the sub hymenium giving rise to gloeocystidia which is 7 μm wide. The hyphae of the trama are thick walled which may or may not be inflated measuring 3–20 μm in diameter (Thongbai et al., 2015). *H. erinaceus* and *H. coralloides* are wood saprotrophs which can be found on both fallen as well as standing wood. *H. erinaceus* is mainly found on wounds and stubs of living trees or on fallen wood and fruits on relatively undecayed wood while *H. coralloides* is mainly found on dead trunks and branches with higher levels of underlying decay. The fruiting period of these white rot fungi is generally from August to December (Boddy et al., 2011).

H. erinaceus (Figure 6.1) has a large and irregularly bulbous fruit body, weighing upto 1 kg when fresh and has a diameter of about 40 cm. It is a compact white to creamy-coloured solid plug of tissue which gradually turns yellowish and finally brownish upon ageing thereby leading to a decrease in overall quality (Jiang et al., 2014; Wabang & Ajungla, 2016). A large number of slender and soft spines of upto 5 cm length (which are less than 1 cm in the youngest stage) hang from this tissue arranged in the form of a beard. The spores are nearly round and smooth, subglobose, measuring 4 μm × 6 μm. This species is mostly found during the fall-winter, growing solidary on felled logs or living hardwood (Thongbai et al., 2015; Wabang & Ajungla, 2016).

H. coralloides (Figure 6.2) is a saprobic that forms a bushy fruity body on the dead logs of coniferous or broad-leaved trees (Pala et al., 2013). The hymenial layer of *H. coralloides* is made up of basidioma and basidia intermixed with sterile cystidia that covers the entire surface of basidioma (Pallua et al., 2015). In the hymenium, the mature, spore forming, slightly clavate basidia with generally six surrounding basidioles are regularly distributed. On the sterigmata, the ellipsoid to subglobose basidiospores form the four-spored maturing basidia. The young spores are smooth with a diameter of approximately 2.09 μm which roughen upon maturation. Numerous gloeplerous hyphae originating from trama appear in hymenium as gloeocystidia (Pallua et al., 2012). Its peculiar fruiting bodies resemble white coral (Wittstein et al., 2016). Whitish basidiomata has spines resembling icicles, which is common in all members of this genus and is formed on multiple branching or unbranched basidioma tissue. In a study on *H. coralloides* collected from Austria, the hymenial layer covering the hymenophoral area of basidioma was found to be 50 μm thick. Moisture content was 88.4% while the basidiospore production was 45.1 × 10^7 /gram of the hymenophore (using Micro-CT) (Pallua et al., 2015).

At times, macro morphological differentiation of *H. erinaceus* from some growth forms of *H. coralloides* is difficult mainly due to similar size of basidiospores. Greatly contracted forms of the basidiomes of *H. coralloides* have been found in which the basidiome is a massive body resembling that of *H. erinaceus* rather than a long and graceful branch formation. In such a situation host substrates have been found to be useful for identification since *H. erinaceus* is commonly found on deciduous trees while *H. coralloides* occur on conifers (Thongbai et al., 2015).

FIGURE 6.1 *Hericium erinaceus.*

Source: https://commons.wikimedia.org/wiki/File:Hericium_erinaceus_64176.jpg

TAXONOMICAL CLASSIFICATION

H. erinaceus (Bull.) Persoon

Synonyms: *Clavaria erinaceus* (Bull.) Paulet., *Dryodon erinaceus* (Bull.) P. Karst., *Hericium caput-medusae* (Bull.) Pers., *Hericium echinus* (Scop.) Pers., *Hydnum hystrix* (Pers.) Fr., *Hydnum omasum* Panizzi, *Manina cordiformis* Scop., *Martella hystricinum* (Batsch) Kuntze, *Steccherinum quercinum* Gray (Thongbai et al., 2015).

Common names: Proposed by Bulliard, the Latin word "*erinaceus*" means hedgehog. The species has many common names mainly referring to its uniquely shaped fruiting body, such as monkey head mushroom, white beard or old man's beard, bearded tooth fungi, Lion's mane mushroom, Yambushitake (derived from the word Yambushi meaning mountain priest), hou tou gu, pom-pom mushroom, Shishigashira, Igel-Stachelbart, Bearded Hedgehog, hedgehog mushroom, Bear's head, Hog's head fungus, mountain-hidden mushroom, etc. (Abdullah et al., 2012; Thongbai et al., 2015; Wang et al., 2014).

Its taxonomic classification is as follows,

Division – Basidiomycete
Class – Agaricomycetes
Order – Russulales
Family – Hericiaceae
Genus – *Hericium*
Species – *erinaceus*

FIGURE 6.2 *Hericium coralloides.*

Source: **https://commons.wikimedia.org/wiki/File:2009-09 25_Hericium_coralloides_(Scop.)_Pers_58068.jpg**

H. coralloides (Scop.) Pers.

Synonyms: *H. ramosum*, *H. abietinum* (Wittstein et al., 2016)

Common names: Yaad gab, Comb tooth, Coral fungus (Pala et al., 2013; Bisko et al., 2018; Debnath et al., 2019)

Its taxonomic classification is as follows,

Division – Basidiomycete
Class – Agaricomycetes
Order – Russulales
Family – Hericiaceae
Genus – *Hericium*
Species – *coralloides*

DISTRIBUTION: INDIA AND WORLD

H. erinaceus and *H. coralloides* have been found in many countries in Asia, North America, and Europe with wide variations in their distribution (Boddy et al., 2011). Bear's paws, Shark's fin, trepan, and *H. erinaceus* are considered the "Four famous cuisines" of China. Although it was first described in North America, the species has been extensively used in traditional Chinese and Japanese medicine (Thongbai et al., 2015). Information about this species is also found in

European and South American literature (Chaiyasut & Sivamaruthi, 2017). *H. erinaceus* is mainly found in the northern hemisphere, in North America, Europe, and Asia (Thongbai et al., 2015, Saitsu et al., 2019). In Asia, it is mainly found in the eastern countries such as Japan and China. The Japfer mountains in the Kohima District of Nagaland, Sikkim, etc., in India, host the species, which is generally seen from October to November, at an altitude of about 3000 m (Wabang & Ajungla, 2016). In the year 2003, 13 European countries out of 23 red-listed *H. erinaceus* mainly because of habitat destruction (Thongbai et al., 2015). *H. coralloides* are found in the Himalayan regions of India, like Hipora, Keller, Doodhpathri region of Kashmir Himalayas, and Sikkim. It is also on several Red Data lists in Europe and is classified as "near threatened" (Boddy et al., 2011; Das et al., 2011; Pala et al., 2013).

Spore dispersal is essential for these species in order to establish themselves in new resources. Their spores generally disperse within an area of 1m from the parent fruit body, with very limited spores crossing 100 m area. In a study, the rate of spore germination was found to be less than 1% after 20 weeks. In both species, time for germination is inversely related to the age of the spore with fresh spores taking almost 8 days to germinate while 24-week-old spores taking almost 60 days to germinate (Boddy et al., 2011). The spp. face numerous threats mainly in the form of climate change, overexploitation, habitat destruction, modification or fragmentation and lack of adequate knowledge about their conservation and sustainable use (Wabang & Ajungla, 2016).

PHYTOCHEMICAL COMPOSITION

Mushrooms are known to possess a wide range of nutritional and bioactive compounds, such as polysaccharides, dietary fibre (mainly chitin), proteins and amino acids, essential fatty acids, vitamins, phenolics, organic acids, sterols, alkaloids, and terpenoids (Gąsecka et al., 2020). The species belonging to the genus *Hericium* produce phytochemicals erinacines and hericenones (Tsai et al., 2021). More than 80 structurally diverse bioactive compounds are known to be present in *H. erinaceus* (Gąsecka et al., 2020). Other than erinacines and hericenones, polysaccharides (beta-glucans) and hericerins, sterols like ergosterol, and erinarols G-J, resorcinols, volatile aroma compounds like 2-methyl-3-furanthiol, 2-ethylpyrazine, and 2,6-diethylpyrazine, monoterpenes and diterpenes are also present in *H. erinaceus* (Li et al., 2018). Polysaccharides have been found to be made up of mannose (2.5%), glucuronic acid (1.1%), glucose (60.9%), galactose (28%), and fucose (7.5%) (Sheng et al., 2017). Dominant bioactive compounds present in the fruiting bodies are proteins, lectins, glucan, phenols, and terpenoids (Ghosh et al., 2021; Thongbai et al., 2015). The hericenones are mainly present in the fruit bodies while mycelia has erinacines (Li et al., 2018). 15 erinacines have been identified so far, i.e., erinacine A-K and P-S (Tsai et al., 2021). Some of the hericerins that have been reported from *H. erinaceus* are hericerin A, iso hericenone J, isoericerin, hericerin, N-dephenylethyl isohericerin, hericenone J, hericenols, 4-(30,70-dimethyl-20,60-octadienyl)-2-formyl-3-hydroxy-5-methoxy benzyl alcohol (Chaiyasut & Sivamaruthi, 2017).

Chemical analysis has revealed that 100gm of dried *H. erinaceus* has almost 64 gm carbohydrates, 22 gm of crude protein, 4 gm moisture, and 3 gm crude fat. Glucose, arabitol, mannitol, trehalose and myo-inositol are the soluble sugars present in it, with arabitol being in most abundant quantity, that is, almost 127 mg/gm dry weight. Out of its 16 constituent amino acids, 7 are essential amino acids. The chemical compounds responsible for its specific flavour and odor are 5′-nucleotides and volatile oils (Jiang et al., 2014). In another study, the chemical composition and nutritional content of the dried mycelia and fruit body of *H. erinaceus* were analysed separately (Table 6.1). Bioactive compounds γ-aminobutyric acid (GABA) (42.93 μg/g DW), ergothioneine (629.96 μg/g DW), and lovastatin (4.38 μg/g DW) were found in fruiting body while the mycelial biomass contained only GABA (56μg/g DW) and ergothioneine (149.24 μg/g DW) (Cohen et al., 2014).

Mycochemical analysis of *H. erinaceus* revealed differences in the phytochemical content in the fruiting bodies when analysed using two different solvents. The methanolic extract contained

TABLE 6.1
Chemical Composition and Nutritional Content of the Mycelia and Fruit Body of *H. erinaceus* (Cohen et al., 2014)

Chemical composition					
Nutrient	**Fruiting body**	**Mycelia**	**Nutrient**	**Fruiting body**	**Mycelia**
Carbohydrate	61.1%	42.9%	Protein	20.8%	42.5%
Water	6%	4%	Fat	5.1%	6.3%
Ash	6.8%	4.4%			
Amino acid content					
Amino acid	**Fruiting body** (mg/g DW)	**Mycelia** (mg/g DW)	**Amino Acid**	**Fruiting body** (mg/g DW)	**Mycelia** (mg/g DW)
Cystic acid	0.32	0.62	Aspartic acid	1.30	3.88
Methionine sulfone	0.53	0.77	Threonine	1.18	2.04
Serine	0.83	1.66	Glutamic acid	1.36	4.80
Proline	0.51	1.36	Glycine	0.75	1.63
Alanine	0.89	1.56	Valine	0.90	1.61
Isoleucine	0.77	1.61	Leucine	0.85	2.69
Tyrosine	0.46	1.36	Phenylalanine	0.58	1.64
Lysine	0.97	0.93	Histidine	0.53	0.63
Arginine	1.07	1.86	Total	14.33	30.65
Micro- and macro element content					
Micro- and macro-element	**Fruiting body** (mg/g DW)	**Mycelia** (mg/g DW)	**Micro- and macro-element**	**Fruiting body** (mg/g DW)	**Mycelia** (mg/g DW)
Aluminium	84	62	Arsenic	<0.2	<0.2
Barium	1	11	Beryllium	<0.2	<0.2
Boron	9	9	Cadmium	0.1	0.1
Calcium	395	5526	Chromium	1	1
Cobalt	<0.2	<0.2	Copper	13	8
Iron	270	228	Lead	0.4	0.3
Lithium	<0.2	<0.2	Magnesium	1166	2092
Manganese	11	15	Mercury	<0.5	<0.5
Molybdenum	<0.2	2	Nickel	<0.2	1
Phosphorus	6121	5216	Potassium	29163	6621
Selenium	<0.2	<0.2	Silver	<0.2	<0.2
Sodium	157	116	Sulfur	1932	3285
Strontium	1	9	Tin	1	1
Titanium	1	1	Vanadium	<0.2	<0.2
Zinc	59	38			
Fatty Acid Content					
Fatty acid	**Fruiting body** (mg/g DW)	**Mycelia** (mg/g DW)	**Fatty acid**	**Fruiting body** (mg/g DW)	**Mycelia** (mg/g DW)
Myristic acid	0.3	-	Pentadecanoic acid	1.3	-
Palmitic acid	24.0	14.0	Palmitoleic acid	0.5	-
Stearic acid	8.1	4.1	Trans-Vaccenic acid	4.7	0.8
Oleic acid	33.7	37.2	Linoleic acid	26.9	32.8
Alpha-linolenic acid	0.4	1.0	Arachidonic acid	-	1.5
Eicosenoic acid	-	1.4	Behenic acid	-	4.3
Docosatetraenoic acid	-	0.5	Lignoceric acid	-	1.9
Nervonic acid	-	0.2			

13.5 μg GAE/mg phenol, 0.69 μg QE/mg flavonoid, 2.5 μg/mg ascorbic acid, 0.0174 μg/mg lycopene, and 0.04781 μg/mg β-carotene. The ethanolic extract contained 17.42 μg GAE/mg phenol, 1.68 μg QE/mg flavonoid, 5 μg/mg ascorbic acid, and 0.05045 μg/mg lycopene (Ghosh et al., 2021).

H. coralloides contains almost 2.38 g/100g DW crude fat, 7.25 g/100g DW crude protein, 9.31 g/100g DW ash, 81.06 g/100g DW carbohydrate, 10.79 g/100g DW sugar and 374.67 kcal/100g DW energy. The trace minerals found in it are 134 g/100g DW magnesium, 0.72 g/100g DW copper, 0.31 g/100g DW manganese, 77.96 g/100g DW crude iron and 4.76 g/100g DW zinc (Painuli et al., 2020). The fruiting bodies of *H. coralloides* consist of Corallocin A, B and C along with hericerin (Wittstein et al., 2016). The polysaccharides present in *H. coralloides* are mainly linear starch-type α-1,4 linkages only (which consists of amylose only) with a chain length of nearly 38 glucose molecules (McCracken & Dodd, 1971). Some of the phytochemicals present in *H. erinaceus* and *H. coralloides* have been shown in Figure 6.3.

ETHNOMEDICINAL USES

Medicinal fungi serve as a plentiful source of physiologically active compounds that are beneficial in both prevention and treatment of numerous diseases (Pallua et al., 2012). In traditional Chinese medicine, *H. erinaceus* has been extensively used to cure chronic superficial gastritis as well as oxidative stress-related disorders (Shang et al., 2013; Thongbai et al., 2015). It is widely used as food by locals and in supplements (Adhikari et al., 2005; Anderson & Lake, 2013). In Kashmir, consumption of *H. coralloides* is believed to decrease the chances of developing cancer and heart diseases and is usually recommended by local herbalists for patients suffering from hypertension (Pala et al., 2013). It has been extensively used for nourishing the gut, spleen and for treating cancer in Chinese and Japanese medicinal systems (Elkhateeb et al., 2019).

MEDICINAL PROPERTIES

H. erinaceus and *H. coralloides* exhibit therapeutic effects on physiological systems such as digestive system, immune system, nervous system, etc., via varied mechanisms due to a wide variety of bioactive compounds present in them (Jiang et al., 2014). The antioxidant, antimicrobial, and immunomodulatory properties of many of these substances have been confirmed in several studies (Li et al., 2016). Evidence indicates that polysaccharides obtained from mushrooms have considerable therapeutic value without any harmful side effects (Thongbai et al., 2015).

Antimicrobial Activity and Gut Microbiota Regulation

Mushrooms have aroused the interest of investigators as novel antimicrobial agents amid the rise in multiple drug resistance in recent years (Wong et al., 2009). Ghosh et al. (2021) studied the antibacterial potential of *H. erinaceus* extract using microdilution method against *Bacillus subtilis, Staphylococcus aureus,* and *Escherichia coli*. Different dilutions of methanolic and ethanolic extracts along with 200 μl of nutrient broth and 20 μl of inoculum were used for carrying out reactions in a 96-well plate. The ethanolic extract was capable of inhibiting all three strains with its minimum inhibitory concentration (MIC) being 1,650 μg/ml, 2,750 μg/ml and 1,800 μg/ml for *B. subtilis, S. aureus, E. coli,* respectively, while the methanolic extract was effective against *B. subtilis* and *E. coli*, at MIC 2,100 μg/ml and 1,575 μg/ml respectively (Ghosh et al., 2021). When antimicrobial activities of the mycelium, fresh, oven-dried and freeze-dried fruit body extracts were studied against fungi, gram-positive and gram-negative bacteria, the extracts were discovered to be more beneficial against gram-positive bacteria than gram-negative bacteria. 4 of the 5 gram-positive bacteria and 5 of the 8 gram-negative bacteria were inhibited. This might have been due to different structures of the cell wall of gram-positive and gram-negative bacteria, presence of outer

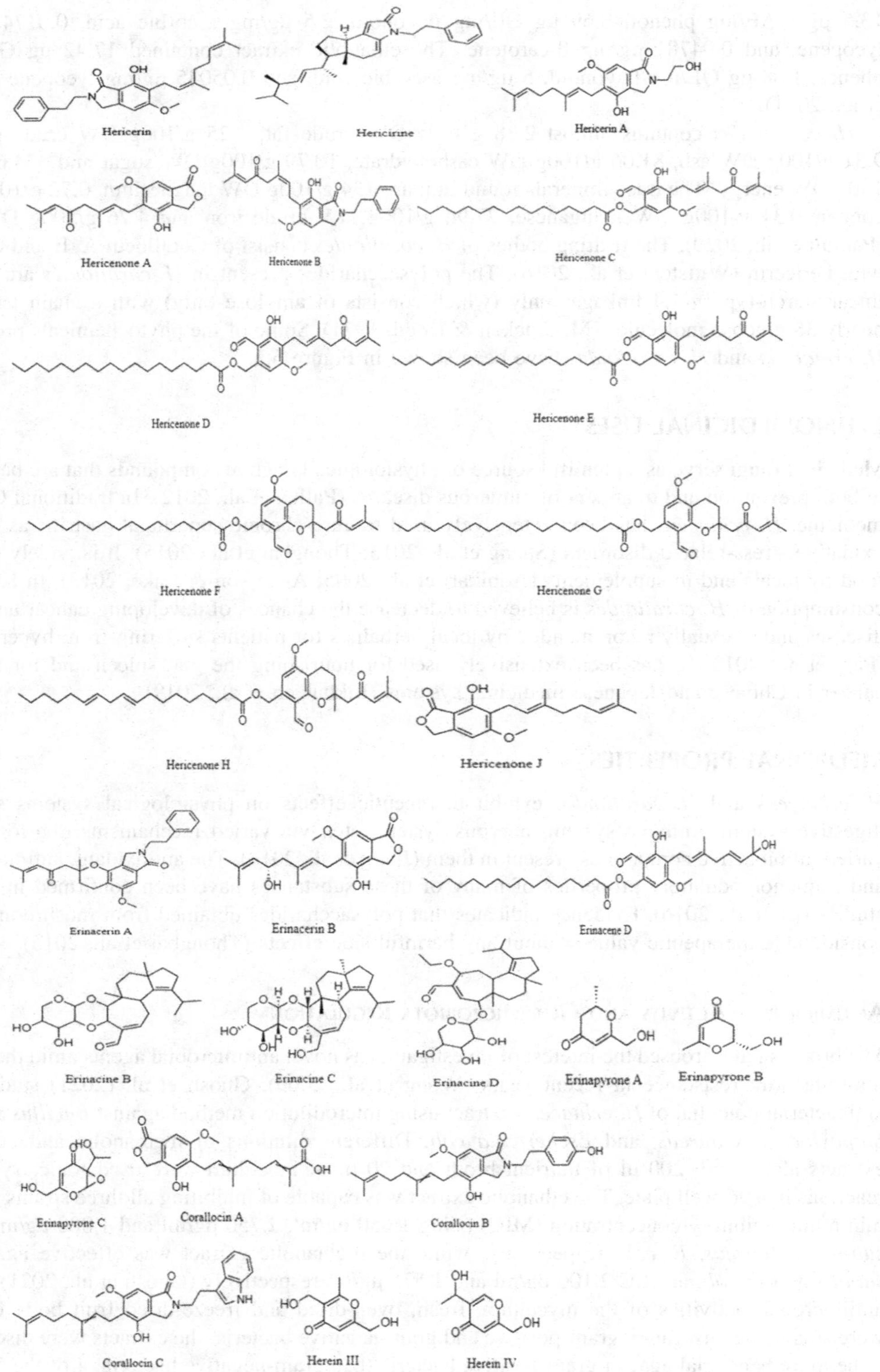

FIGURE 6.3 Some chemical constituents of *H. erinaceus* and *H. coralloides*.

membrane and periplasmic space in the latter which inhibits entry of several environmental substances including antibiotic molecules into the cell. Clear inhibition zone of 8–12 mm diameter was observed in case of *B. cereus* for all 4 types of extracts. In case of gram-positive bacteria the mycelium extract produced clear inhibition zone for *B. cereus, B. subtilis, Enterococcus faecalis* ATCC7080, while freeze-dried extract produced clear inhibition zone for *S. aureus*. The oven-dried extract produced a clear zone only for *B. aureus*. In case of gram-negative bacteria, fresh, dried and mycelium extracts produced a clear zone against *Salmonella sp.* ATCC 13076, *Shigella sp.* and *Plesiomonas shigelloides*. Freeze-dried extract also produced a clear zone against *S. typhimurium* and *P. aeruginosa*. All types of extracts produced hazy inhibition against almost all the bacterial and fungal species (Wong et al., 2009).

H. coralloides tincture exhibits antimicrobial activity against *Listeria innocua, Bacillus cereus, Escherichia coli, Pseudomonas aeruginosa, Staphylococcus aureus, Candida* sp., *Candida albicans* with its MIC range being 8–32 µl/ml (Vamanu & Voica, 2017). Corallocin B exhibits antifungal properties against *Mucor plumbeus* MUCL 49355 (MIC 574 µm) (Wittstein et al., 2016).

Antidiabetic Property

H. erinaceus is beneficial in metabolic diseases due to its high antioxidant potential and bioactive components (Chaiyasut & Sivamaruthi, 2017). In a study by Yi et al. (2015), 40 mg/kg ethanol extract of *H. erinaceus* was administered to alloxan-induced diabetic adult Wistar rats in order to study its effect on neuropathic pain. Thermal hyperalgesia was assessed by exposing a constant beam of radiant light on the hind paw of the rat kept in a plexiglass box. The alloxan rats had reduced pain threshold while the rats in the experimental group administered *H. erinaceus* extract exhibited a significant increase in pain threshold. Similarly, when mechanical hyperalgesia was assessed the alloxan rats had reduced paw pressure withdrawal thresholds which was 68.22 gm as compared to 266.14 gm for control group, while significant and dose-dependent increase in mean paw withdrawal threshold was observed in experimental group. A significant dose-dependent reduction was also observed in blood glucose and urine sugar levels. In order to estimate oxidative stress, some of the endogenous antioxidant enzymes were measured. Significant restoration in glutathione peroxidase (GPx), glutathione reductase (GR), glutathione S transferase (GST), catalase (CAT), and Na+K +ATPase activity was seen for *H. erinaceus* group compared to alloxan group. Serum lactate dehydrogenase level was 130 IU/L in the *H. erinaceus* (40 mg/kg) treated group as compared to 170.2 IU/L in the alloxan group. Glutathione levels were significantly higher in the experimental group i.e., 1.3 nmol GSH/mg protein as compared to 1.011 nmol GSH/mg protein in the alloxan group. Total antioxidant status (TAOS) is an indirect measure of O_2^- and other oxidant species formation. In the control group the TOAS activity was 28.41 µM L-ascorbate, while in the alloxan group it was 80.33 µM L-ascorbate. *H. erinaceus* exhibited significant dose-dependent capacity to attenuate the oxidative stress which reduced from 72.22 µM L-ascorbate in group administered 10 mg/kg *H. erinaceus* extract to 64.30µM L-ascorbate in group administered 20 mg/kg *H. erinaceus* extract and 56.35 µM L-ascorbate in group administered 40 mg/kg *H. erinaceus* extract. Wu and Xu (2015) examined the inhibitory potential of dried fruit bodies of *H. erinaceus* on α-Glycosidase and aldose reductase activity *in vitro*. It was found to have positive α-Glycosidase and aldose reductase inhibitory activity (IC_{50}= 18 mg/ml). When aqueous extract of *H. erinaceus* (AEHE) was administered to streptozotocin-induced diabetic rats for 28 days, the serum glucose levels reduced significantly. In the rats given 100 mg/kg BW AEHE, serum glucose level fell from 286.3 mg/dl on day 0 to 163.2 mg/dl on 28th day and in the rats given 200 mg/kg BW AEHE serum glucose level fell from 291 mg/dl on day 0 to 135.4 mg/dl on 28th day. AEHE also led to significant increase in serum insulin level and in attenuation of liver disorders. Assessment of oxidative stress parameters in the liver revealed restoration of superoxide dismutase (SOD), GSH-Px, glutathione and CAT activity in AEHE-treated diabetic rats which had significantly decreased in diabetic rats. The malondialdehyde level lowered significantly in the liver tissue (Liang et al., 2013).

Wang et al. (2005) used methanol extract of *H. erinaceus* fruiting bodies to study its hypoglycemic effects on streptozotocin-induced diabetic rats. The methanol extract was found to consist of D-threitol, D-arabinitol, and palmitic acid. The food and water intake as well as urine excretion in the group fed *H. erinaceus* extract was significantly lower than the non-*H. erinaceus* group at the dose of 200 mg/kg BW. The loss in body weight was also significantly lower in the group fed 100 mg/kg BW and 200 mg/kg BW *H. erinaceus* extract, i.e., 194.5 g and 197.4 g, respectively, as compared to the diabetic control (187.3 g). Supplementation with fermented juice of *H. erinaceus* had favourable health outcomes for rats with streptozotocin-induced diabetes, which led to an increase in serum insulin level and body mass. A reduction was also observed in fasting plasma glucose and inflammatory markers levels (Chaiyasut & Sivamaruthi, 2017).

Antihypercholesterolemic and Hypolipidemic Properties

The hypolipidemic potential of the exo-biopolymer obtained from *H. erinaceus* mycelial culture has been analysed in rats with dietary-induced hyperlipidemia. There was significant dose-dependent decrease in total cholesterol (TC) and LDL-cholesterol levels in the experimental group administered 50–200 mg/kg/day exo-biopolymer, from 80 mg/dl to 67.5 mg/dl for TC and 56.8 to 40.7 mg/dl for LDL-cholesterol as compared to the control group in which the TC level was 100.6 mg/dl and LDL-cholesterol was 74.5 mg/dl. The observed effect could possibly be either due to the viscous nature of exo-biopolymer which increased with the dose thereby altering intestinal mucosa which would inhibit micelle formation resulting in lower cholesterol and triglyceride (TG) absorption or due to the structure or composition of exo-biopolymer produced by *H. erinaceus*. Increase in HDL cholesterol was also observed from 16.3 mg/dl in the control group to 21.9 mg/dl in mice administered with 200 mg/kg/day exo-biopolymer and a reduction in atherogenic index, plasma TG and phospholipid was also observed, from 5.03 mg/dl, 37.3 mg/dl and 99.1 mg/dl in the control group to 2.08 mg/dl, 24.5 mg/dl and 80.4 mg/dl, respectively, in the group administered 200 mg/kg/day exo-biopolymer. Reduction in liver weight, liver TC and TG was also observed. Reduction in hepatic HMG-CoA reductase activity involved in cholesterol synthesis was noted to be 20.2% in the group administered exo-biopolymer orally which is the primary reason for decrease in TC level of exo-biopolymer (Yang et al., 2003).

When Liang et al. (2013) administered aqueous extract of *H. erinaceus* to streptozotocin-induced diabetic rats, the LDL-cholesterol, TC and TG levels reduced significantly from 2.36 mmol/L, 2.69 mmol/L, and 0.81 mmol/L to 1.71 mmol/l, 1.46 mmol/l, and 0.65 mmol/l, respectively. The HDL-cholesterol level significantly increased from 0.81 mmol/l to 1.91 mmol/l. Methanol extract of the fruiting bodies of *H. erinaceus* has been found to result in significantly lower increase in serum TC than non-*H. erinaceus* fed streptozotocin-induced diabetic rats. On the 20th day, the TC level was 133.5 mg/dl, 123.1 mg/dl, and 120.4 mg/dl when the concentration of *H. erinaceus* extract was 20 mg/kg, 100 mg/kg, and 200 mg/kg, respectively, as opposed to 134.9 mg/dl in the diabetic control group. There was also a significant dose-dependent reduction in the TG levels, which was 156.6 mg/dl for the diabetic control group, and 134.1 mg/dl, 127.8 mg/dl, 125.9 mg/dl for the group fed *H. erinaceus* extract 20 mg/kg BW, 100 mg/kg BW, 200 mg/kg BW, respectively (Wang et al. 2005).

Antioxidant Property

Reactive oxygen species are a leading cause of numerous diseases including CVD, cancer, metabolic syndrome, etc. (Thongbai et al., 2015). The antioxidant potential of a species consists of its ability to scavenge free radicals, chelate metal ions, its reducing potential, etc. It differs according to the part of the mushroom used, method of extraction, solvent used, etc. Therefore, numerous techniques are used to evaluate antioxidant capacity of *H. erinaceus* extracts. Ethanolic extract of the fruiting bodies of *H. erinaceus* has been found to have higher antioxidant potential as compared to its methanolic extract. A dose-dependent radical scavenging pattern was seen when DPPH (1,1-diphenyl-2-picrylhydrazyl) and ABTS (2,2'-azino-bis-(3-ethylbenzothiazoline-6-sulfonic acid)) radical scavenging assays were used. At a concentration of 2000 μg/ml, 68% and 52% radical scavenging activity for ethanolic and methanolic extracts was observed for DPPH assay. The EC_{50} value of the DPPH radical scavenging activity was 1900 μg/ml for methanolic extract and 985 μg/ml for ethanolic extract while for the ABTS radical scavenging activity, EC_{50} value was 1200 μg/ml for methanolic extract and 300 μg/ml for the ethanolic extract. The ethanolic extract had higher chelating ability and chelated 54%, 67%, and 73% ferrous ions at 400 μg/ml, 1200 μg/ml, and 2000 μg/ml, respectively. EC_{50} value for chelating effect was 2900 μg/ml and 358 μg/ml for methanolic and ethanolic extract, respectively. And total antioxidant capacity was 1.42 μg ascorbic acid equivalents (AAE)/ml and 2.17 μg AAE/ml of extract, respectively (Ghosh et al., 2021).

In another study, total phenolic content of *H. erinaceus* extract was found to be 7.18 mg gallic acid equivalents (GAE)/g and total flavonoids content was 0.99 mg catechin equivalent (CE)/g. IC_{50} value for DPPH free radical scavenging activity was 79.5 mg/ml (Wu & Xu, 2015). A reduction in total phenolic content was observed when a fresh sample (3.79 mg/g DW) was dried at 70°C (3.13 mg/g DW) (Gąsecka et al., 2020).

Comparison of the phenolic content and antioxidant activity of the methanolic extract of 3 *Hericium* species, i.e., *H. erinaceus, H. americanum,* and *H. coralloides*, revealed that the antioxidant activity of the species *H. coralloides* was the highest and there are significant differences in the antioxidant properties of different strains of the species *H. erinaceus*. The total phenolic content of the extracts was found to be 3.27 mgGAE/g for *H. coralloides*, 2.31 mgGAE/g for *H. americanum* and it ranged from 2.34 to 3.15 mgGAE/g for the 6 strains of *H. erinaceus* studied. *In vitro* antioxidant potential was assessed using multiple methods. Ferric reducing antioxidant power (FRAP) assay was used to test the reducing power of the *Hericium* spp. For the 6 strains of *H. erinaceus* the FRAP value ranged from 11.1 to 15.7 mmol TE/g, for *H. coralloides* it was 17 mmol TE/g and for *H. americanum*. It was 10.5 mmol TE/g. The scavenging activity was examined using DBTS and DPPH assay. For the DPPH assay the EC_{50} value of *H. erinaceus* strains ranged from 4.27–6.74 mg/ml, for *H. coralloides* it was 4.12 mg/ml and for *H. americanum* it was 7.82 mg/ml. For DBTS, the EC_{50} value of *H. erinaceus* strains ranged from 3.32–4.58 mg/ml, 2.83 mg/ml for *H. coralloides* and 6.36 mg/ml for *H. americanum*. Significant correlation was observed between total phenolic content and antioxidant activities of *Hericium* species. For FRAP, DPPH EC_{50} value and ABTS EC_{50} value the correlation coefficient values were +0.943, −0.969, and −0.855, respectively (Atila, 2019).

Abdullah et al. (2012) examined the antioxidant capacity of *H. erinaceus* extract obtained by boiling *H. erinaceus* in water for 30 minutes and was assessed using 5 different assays (DPPH free radical scavenging activity, β-carotene bleaching assay, inhibition of lipid peroxidation, reducing power ability and cupric ion reducing antioxidant capacity or CUPRAC). Using DPPH assay, the IC_{50} value obtained was 25.451 mg/ml. Using a β-carotene bleaching assay, the IC_{50} value obtained was 8.76 mg/ml. At 10 mg/ml concentration, the ability of the extract to inhibit phospholipid peroxidation was moderate (47.52%) while for the positive controls quercetin, BHA (Butylated Hydroxy Anisole), and ascorbic acid it was 87.35%, 75.13%, and 81.54% respectively. The reducing power of *H. erinaceus* increases with concentration, at 0.05 mg/ml it was 0.017 while at 1 mg/ml it was 0.077. At 0.10 mg/ml concentration, the CUPRAC of *H. erinaceus* extract was 0.173 which increased to 2.442 at 10 mg/ml. Antioxidant index (AI) gives a better picture of the antioxidant potential and is computed by combination of average results of all 5 assays. The AI value of *H. erinaceus* extract is 17.7% which means it has moderate antioxidant potential. As per the results of a study by Mujić et al. (2010), the total phenolic content of *H. erinaceus* ethanol extract is 7.80 mg GAE/g, total flavonoids content is 5.04 mg CE/g. The radical scavenging capacity estimated using DPPH radicals yielded an IC_{50} value of 0.198 mg/ml. The reducing power was examined using the Oyaizu method. At 1 mg/ml concentration, the absorbance of *H. erinaceus* extract was 0.185.

In another study (Wong et al., 2009), methanol extract of mycelium and fresh, oven-dried, freeze-dried fruit bodies of *H. erinaceus* were used. Mycelium extract had the highest total phenolic content (31.20 mg GAE/g), followed by oven dried (2.37 mg GAE/g), freeze-dried (0.78 mg GAE/g), and lowest for fresh fruiting bodies (0.26 mg GAE/g). FRAP was also highest in mycelium extract (21.93 μmol $FeSO_4.7H_2O$ Equivalents/g), oven-dried (13.72 μmol $FeSO_4.7H_2O$ Equivalents/g), freeze-dried (4.06 μmol $FeSO_4.7H_2O$ Equivalents/g), and lowest in fresh fruiting bodies (1.27 μmol $FeSO_4.7H_2O$ Equivalents/g). Fresh fruit bodies had the most potent DPPH activity (EC_{50} value = 3.75 mg/ml) while mycelium extract had the lowest DPPH radical scavenging ability (EC_{50} value = 13.67 mg/ml). At 20 mg/ml concentration, highest antioxidant activity was present in oven-dried extract while the lowest was in mycelium.

The antioxidant potential of four compounds namely hydrospirobenzofuran, sesquibenzopyran, coralcuparene, and spirobenzofuran, extracted from *H. coralloides* was studied against positive controls trolox and BHA. Using ABTS radical-scavenging assay, IC_{50} value of the compounds were found to be 66 μM for hydrospirobenzofuran, 29 μM for sesquibenzopyran, 62.5 μM for coralcuparene, 29.6 μM for spirobenzofuran, 18.5 μM for BHA and 25.2 μM for trolox. For DPPH, the IC_{50} values were 121 μM, 87.3 μM, 118.9 μM, 90.8 μM, 78.3 μM and 55.7 μM, respectively (Kim et al., 2018). When tincture of *H. coralloides* was used to assess antioxidant activity, EC_{50} values were 16.07 mg/ml for DPPH scavenging activity, 21.77 mg/ml for ABTS scavenging activity, 5.98 mg/ml for chelating activity and 0.35 mg/ml *in vitro* antioxidant activity (Vamanu & Voica, 2017).

Antitumor/Anticancer Property

Many primary and secondary metabolites extracted from mushrooms have potential anticancer effects and can serve as multi-targeting anticancer molecules. These include primary metabolites which include, polysaccharides such as β-glucans, krestin, letinan, proteins like lecitins and laccases, fatty acids like linoleic and linolenic acids and secondary metabolites which include alkaloids psilocybin, bufotenin, triterpenes like ergosterol, inotodiol and phenolics like hispidin. β-1,3-glucan and lentinan from mushrooms like *H. erinaceus* and Tremella fuciformis promote proliferation of dendritic cells along with increasing cytokines, interferons and interleukins. Lecitins and Laccases from sources such as *H. coralloides*, *Russula lepida* involve mechanisms such as inactivation of ribosomes in rRNA (Chaitanya et al., 2019).

H. erinaceus extract has been found to be beneficial against numerous organs such as oesophagus, intestines, pancreas, stomach, etc. It has several mechanisms of action which depend upon the solvent or extraction method used, part of the plant used, type of cancer, mode of action and whether it is being used *in vitro* or *in vivo* (Thongbai et al., 2015).

On the lung adenocarcinoma A549 cells, *H. erinaceus* extract exhibits anticancer effect by inhibiting cell growth, affecting morphology of the cells, cell cycle arrest, and preventing colony-forming efficiency. The extract exhibited antiproliferative property in a dose-dependent manner with IC_{50} value of ethanolic extract being 403.12 μg/ml for ethanolic extract and 1.95 mg/ml for methanolic extract while it was 18.26 μg/ml for doxorubicin which is a commercially available anticancer drug. The ethanolic extract also transformed the shape of cells from spindle to distorted longer to round shape. Dose-dependent decrease in colony-forming units and increase in bright-circular dead cells and debris floating was seen. There was an increase in $SubG_0$ phase population, constant accumulation of cells at G_0/G1 phase and decline in number of cells in S & G2/M phases along with increase in dose (Ghosh et al., 2021).

Erinacine A is a bioactive agent with antitumorigenic activity, and inhibits *in vitro* and *in vivo* growth of tumor cells. It has been demonstrated to have a significant dose-dependent reduction in the growth of tumor when male mice with DLD-1 cells grow as xenografts were treated with erinaceus A. It is capable of providing protection against colorectal cancer cells by activation of p70S6K and ROS production (Lu et al., 2016, Tsai et al., 2021). Its role in inducing cancer cell apoptosis in colorectal cancer has also been studied (Lee et al., 2019).

Corallocin B from *H. coralloides* has antiproliferative activity against L929 mouse fibroblast cells, with its IC_{50} value reported to be 14.7μm, and also showed cytotoxicity against HUVEC human cell line with IC_{50} value as 2.1μm, MCF-7 (IC_{50} value 9.2μm) and KB-3-1 (IC_{50} value 11.5 μm) (Wittstein et al., 2016). In a study, the antitumor potential of cytokinins extracted from mycelial biomass of *H. coralloides* was assessed *in vitro* using the cell lines Hela (MTT-assay), T24/83 (viability and level apoptotic cells) and Hep G2 (consumption of glucose). The amount of various cytokinins in the extract was found to be 941.12 ng/g FW Trans-Zeatin, 531.99 ng/g FW Zeatin riboside and 348.60 ng/g FW isopentenyladenine. For T24/83 cell line, under the effect of crude extract, the content of dead cells was 18.7% while for purified fractions it was 7.6%. There were 32.8% apoptotic cells for purified fraction and 27% for crude extracts (Vedenicheva et al., 2020).

Immunomodulatory Property

Mushrooms are known to consist of numerous metabolites among which polysaccharides are the most potent immune-modulators having an impact on both innate and adaptive immunity. Glucans are the most common polysaccharides which are found in the fungal cell wall primarily as cellulose. In *H. erinaceus* heteroglycan peptide is a bioactive polysaccharide with probable immunomodulatory mechanism being regulation of tumor necrosis factor (TNF)-α, interleukin (IL)-12 repression and induction of nitric oxide (NO) production (Chakraborty et al., 2021). Galactoxyloglucan–protein complex is an active immunomodulator in *H. erinaceus* (Jayachandran et al., 2017).

When the effect of *H. erinaceus* was studied on trinitro-benzene-sulfonic acid-induced inflammatory bowel disease (IBD) an improvement was observed in the scores of common morphous and tissue damage index along with reduction in myeloperoxidase (MPO). Differences were also observed in inflammatory factors. There was rise in serum cytokines, Foxp3 and IL-10 while TNF-α and nuclear factor (NF)-κB p65 reduced and there was activation of T-cells. Changes were also seen in the gut microbiota of *H. erinaceus* treated groups, which might be beneficial in promoting healthy gut bacteria along with improving immunity of the host (Diling et al., 2017).

In mice, *H. erinaceus* polysaccharides (HEP) have been found to enhance lymphocyte proliferation stimulated by ConA and can improve serum hemolysin level thereby enhancing innate immunity. On the adaptive immunity *H. erinaceus* polysaccharides work by improving NK cell activity and macrophage phagocytosis. In the group administered *H. erinaceus* polysaccharides an increase in sIgA expression was seen, that protects against foreign antigens (Sheng et al., 2017). In another study, enzymatic hydrolysis product of *H. erinaceus* polysaccharides (EHEP) was used. The EHEP group exhibited more distinct morphological changes consistent with activation of macrophages (RAW264.7). There was dose-dependent stimulation of phagocytic activity of the RAW264.7 cells in the concentration range of 0.391–3.125 μg/ml. Both HEP and EHEP promoted CD40 and CD86 expression in macrophages, both of which are functional molecules on macrophages essential for immune response. On treatment with EHEP, the cells produced a 2.33 times higher amount of nitric oxide compared with the negative control group. Most of these effects were higher for EHEP when compared to HEP, hence suggesting increase in immunomodulatory activity of *H. erinaceus* polysaccharides upon hydrolysis (Liu et al., 2021).

HEP have been found to improve muscovy ducklings' intestinal parameters such as villus height/crypt depth ratio and villus surface area after they were infected with Muscovy duck reovirus (MDRV). In *H. erinaceus* pre-treated ducklings there was significant improvement in injuries to the intestinal mucosa caused due to MDRV in which the villi were longer and thicker, regularly arranged, marrower villus spatium and had complete structure. This group also had increased duodenal villus height/crypt depth (V/C) ratios and villus surface area (VSA) from 3 dpi to 21 dpi. There was an increase in intraepithelial lymphocytes (IELs). Intestinal mucosa is a vital component of the body's defence mechanism against pathogens. HEP can relieve symptoms and decrease mortality in ducklings suffering from MDRV infection. In the upper portion of the epithelial cells, there was a significant increase in IELs which indicates HEP has the potential to stimulate maturation and its migration to the intestinal tract in order to confer immunity by reacting with antigens (Wu et al., 2018).

Neuroprotective Property and Improved Cognitive Function

Much of the neuroprotective and cognitive enhancement property of *H. erinaceus* is attributed to the presence of hericenones and erinacines which have nerve growth factor (NGF) enhancing action (Nagano et al., 2010). Erinacine A is well-known for its neuroprotective properties. NGF is significantly lower in patients suffering from major depressive disorders (MDD) compared to healthy controls (Chen et al., 2015). 8 of the 15 erinacines have been demonstrated to possess neuroprotective properties *in vitro*, which include enhancement of NGF release by erinacines A-I, management of neuropathic pain by erinacine E, reduction in deposition of amyloid-β, etc. Erinacine A and S can cross the blood-brain barrier. Results of a study showed that the brain-to-plasma concentration ratio was near unity with similar influx and efflux clearance rates suggesting that erinacine A is capable of crossing the brain-to-plasma barrier (Tsai et al., 2021).

Erinacine-enriched *H. erinaceus* mycelia can provide protection against numerous neurodegenerative diseases such as ischemic stroke, Parkinson's disease, Alzheimer's disease, depressive symptoms, neuropathic pain, presbycusis by either delaying neuronal cell death or promoting functional recovery and enhancing nerve regeneration. The active components of *H. erinaceus* have beneficial effects on early dementia and age-associated cognitive changes (Li et al., 2018). Erinacine A is capable of increasing NGF and catecholamine (noradrenaline and homovanillic acid) responsible for stimulation of NGF synthesis in the hippocampus and locus coeruleus as demonstrated in study on rats administered erinacine A for 5 weeks since birth, obtained from cultured mycelia of *H. erinaceus* (Shimbo et al., 2005). Erinacine A enriched mycelia can reduce total infarcted volume in cortex and subcortex of transient stroke in Sprague-Dawling rats. In these rats it can also lower the levels of proinflammatory cytokines such as iNOS, TNF-α, IL-6, IL-1β, etc. In APPswe/PS1dE9 transgenic mice, Erinacine A-enriched mycelia can reduce amyloid plaque burden in cerebral cortex and hippocampus, promote hippocampal neurogenesis and increase NGF/pro NGF ratio (Li et al., 2018).

Owing to their small size, both hericenones and erinacines can rapidly cross the blood-brain barrier. Ethanolic extract of *H. erinaceus* is capable of inhibiting scopolamine-induced memory impairment in zebrafish (*Danio rerio*) by inhibiting AChE or reducing oxidative stress (Valu et al., 2021). In a study, supplements prepared using the fruiting bodies of *H. erinaceus* were given to study participants in a randomized, placebo-controlled parallel group comparative study. After 12 weeks the cognitive function was assessed using various tests. The Mini mental state examination (MMSE), which is widely used for diagnosis of dementia, showed significant enhancement in cognitive functions along with preventing deterioration of memory and improvement in cognitive functions hence exhibiting potential to be used for dementia prevention (Saitsu et al., 2019). Assessment of the effect of *H. erinaceus* on depression and sleep quality measured using Center for Epidemiologic Studies Depression Scale (CES-D) and the Pittsburgh Sleep Quality Index (PSQI) showed that the group administered *H. erinaceus* had significantly lower scores after trial as compared to pre-trial (Nagano et al., 2010). Corallocin A-C can exhibit different patterns of neurotrophin expression, induce NGF and brain-derived neurotrophic factor expression in human 1321N1 astrocytes. Corallocin A and C have been shown to significantly increase NGF secretion from 1321N1 astrocytes (Wittstein et al., 2016).

OTHER USES

When *H. erinaceus* extract was applied on wound areas in the posterior neck area of male Sprague-Dawley rats, the results revealed faster recovery, with fewer macrophages, more collagen, and less scar width at wound enclosure when compared to the group treated with distilled water. It is also beneficial in healing gastric ulcers by suppressing neutrophil infiltration and antioxidant activity (Elkhateeb et al., 2019).

Several functional foods are developed using *H. erinaceus* which is also used to enhance food quality. Mushrooms are wild edible delicacies of great importance and are consumed as both a main dish as well as a side dish. In Asian countries, fermented non-alcoholic beverages made from mushrooms are trending nowadays (Chaiyasut & Sivamaruthi, 2017).

CONCLUSION

In recent years, *Hericium* spp. have gained popularity due to their medicinal properties. *H. erinaceus* has a hedgehog like appearance while *H. coralloides* resembles corals. These are rarely found but highly valued mushrooms due to their culinary, nutraceutical, and medicinal properties. These have been part of traditional Chinese and Japanese medicinal systems due to their beneficial effects on various physiological systems of the body, and are mainly used in the treatment of gut and oxidative stress-related disorders. They are also a source of novel phytochemicals such as erinacines, hericenones, hericerins, corallocin, etc., with the potential to treat several present-day ailments of public health concern without any side effects. Numerous *in vitro* and *in vivo* studies have shown their potential antimicrobial, antioxidant, immunomodulatory, antidiabetic, anticholesterolemic, anticancer, and neuroprotective properties. Their fast-deteriorating population points towards an urgent need to strengthen guidelines for their conservation and management.

REFERENCES

Abdullah, N., Ismail, S. M., Aminudin, N., Shuib, A. S., & Lau, B. F. (2012). Evaluation of selected culinary-medicinal mushrooms for antioxidant and ACE inhibitory activities. *Evidence-Based Complementary and Alternative Medicine, 2012*, 1–12.

Adhikari, M. K., Devkota, S., & Tiwari, R. D. (2005). Ethnomycolgical knowledge on uses of wild mushrooms in western and central Nepal. *Our Nature, 3*(1), 13–19.

Anderson, M. K., & Lake, F. K. (2013). California Indian ethnomycology and associated forest management. *Journal of Ethnobiology, 33*(1), 33–85.

Atila, F. (2019). Comparative evaluation of the antioxidant potential of *Hericium erinaceus*, *Hericium americanum* and *Hericium coralloides*. *Acta Scientiarum Polonorum. Hortorum Cultus, 18*(6), 97–106.

Bisko, N. A., Lomberg, M. L., Mykchaylova, O. B., & Mytropolska, N. Y. (2018). Conservation of biotechnological important species diversity and genetic resource of rare and endangered fungi of Ukraine. *Plant & Fungal Research, 1*(1), 18–27.

Boddy, L., Crockatt, M. E., & Ainsworth, A. M. (2011). Ecology of *Hericium cirrhatum*, *H. coralloides* and *H. erinaceus* in the UK. *Fungal Ecology, 4*(2), 163–173.

Chaitanya, M. V. N. L., Jose, A., Ramalingam, P., Mandal, S. C., & Kumar, P. N. (2019). Multi-targeting cytotoxic drug leads from mushrooms. *Asian Pacific Journal of Tropical Medicine, 12*(12), 531.

Chaiyasut, C., & Sivamaruthi, B. S. (2017). Anti-hyperglycemic property of *Hericium erinaceus*–A mini review. *Asian Pacific Journal of Tropical Biomedicine, 7*(11), 1036–1040.

Chakraborty, N., Banerjee, A., Sarkar, A., Ghosh, S., & Acharya, K. (2021). Mushroom polysaccharides: A potent immune-modulator. *Biointerface Research and Applied Chemistry, 11*, 8915–8930.

Chen, Y. W., Lin, P. Y., Tu, K. Y., Cheng, Y. S., Wu, C. K., & Tseng, P. T. (2015). Significantly lower nerve growth factor levels in patients with major depressive disorder than in healthy subjects: A meta-analysis and systematic review. *Neuropsychiatric Disease and Treatment, 11*, 925.

Cohen, N., Cohen, J., Asatiani, M. D., Varshney, V. K., Yu, H. T., Yang, Y. C., Li, Y. H., Mau, J. L. & Wasser, S. P. (2014). Chemical composition and nutritional and medicinal value of fruit bodies and submerged cultured mycelia of culinary-medicinal higher Basidiomycetes mushrooms. *International Journal of Medicinal Mushrooms, 16*(3).

Das, K., Stalpers, J., & Eberhardt, U. (2011). A new species of Hericium from Sikkim Himalaya (India). *Cryptogamie, Mycologie, 32*(3), 285–295.

Debnath, S., Debnath, B., Das, P., & Saha, A. K. (2019). Review on an ethnomedicinal practices of wild mushrooms by the local tribes of India. *Journal of Applied Pharmaceutical Science, 9*(08), 144–156.

Diling, C., Xin, Y., Chaoqun, Z., Jian, Y., Xiaocui, T., Jun, C., … & Yizhen, X. (2017). Extracts from *Hericium erinaceus* relieve inflammatory bowel disease by regulating immunity and gut microbiota. *Oncotarget*, *8*(49), 85838.

Elkhateeb, W. A., Elnahas, M. O., Thomas, P. W., & Daba, G. M. (2019). To heal or not to heal? Medicinal mushrooms wound healing capacities. *ARC Journal of Pharmaceutical Sciences*, *5*(4), 28–35.

Gąsecka, M., Siwulski, M., Magdziak, Z., Budzyńska, S., Stuper-Szablewska, K., Niedzielski, P., & Mleczek, M. (2020). The effect of drying temperature on bioactive compounds and antioxidant activity of*Leccinum scabrum* (Bull.) Gray and*Hericium erinaceus* (Bull.) Pers. *Journal of Food Science and Technology*, *57*(2), 513–525.

Ghosh, S., Nandi, S., Banerjee, A., Sarkar, S., Chakraborty, N., & Acharya, K. (2021). Prospecting medicinal properties of Lion's mane mushroom. *Journal of Food Biochemistry*, *45*(8), e13833.

Jayachandran, M., Xiao, J., & Xu, B. (2017). A critical review on health promoting benefits of edible mushrooms through gut microbiota. *International Journal of Molecular Sciences*, *18*(9), 1934.

Jiang, S., Wang, S., Sun, Y., & Zhang, Q. (2014). Medicinal properties of *Hericium erinaceus* and its potential to formulate novel mushroom-based pharmaceuticals. *Applied Microbiology and Biotechnology*, *98*(18), 7661–7670.

Kim, J. Y., Woo, E., Lee, I. K., & Yun, B. S. (2018). New antioxidants from the culture broth of *Hericium coralloides*. *The Journal of Antibiotics*, *71*(9), 822–825.

Lee, K-C, Lee, K-F, Tung, S-Y, Huang, W-S, Lee, L-Y, Chen, W-P, Chen, C-C, Teng, C-C, Shen, C-H, Hsieh, M-C, & Kuo, H-C (2019). Induction apoptosis of erinacine A in human colorectal cancer cells involving the expression of TNFR, Fas, and Fas Ligand *via* the JNK/p300/p50 signaling pathway with histone acetylation. *Frontiers of Pharmacology*, *10*, 1174. 10.3389/fphar.2019.01174

Li, I., Lee, L. Y., Tzeng, T. T., Chen, W. P., Chen, Y. P., Shiao, Y. J., & Chen, C. C. (2018). Neurohealth properties of *Hericium erinaceus* mycelia enriched with erinacines. *Behavioural Neurology*, *2018*, 1–10.

Li, Q. Z., Wu, D., Zhou, S., Liu, Y. F., Li, Z. P., Feng, J., & Yang, Y. (2016). Structure elucidation of a bioactive polysaccharide from fruiting bodies of *Hericium erinaceus* in different maturation stages. *Carbohydrate Polymers*, *144*, 196–204.

Liang, B., Guo, Z., Xie, F., & Zhao, A. (2013). Antihyperglycemic and antihyperlipidemic activities of aqueous extract of *Hericium erinaceus* in experimental diabetic rats. *BMC Complementary and Alternative Medicine*, *13*(1), 1–7.

Liu, X., Ren, Z., Yu, R., Chen, S., Zhang, J., Xu, Y., Meng, Z., Luo, Y., Zhang, W., Huang, Y. & Qin, T. (2021). Structural characterization of enzymatic modification of *Hericium erinaceus* polysaccharide and its immune-enhancement activity. *International Journal of Biological Macromolecules*, *166*, 1396–1408.

Lu, C. C., Huang, W. S., Lee, K. F., Lee, K. C., Hsieh, M. C., Huang, C. Y., Lee, L. Y., Lee, B. O., Teng, C. C., Shen, C. H. and Tung, S. Y. (2016). Inhibitory effect of erinacines A on the growth of DLD-1 colorectal cancer cells is induced by generation of reactive oxygen species and activation of p70S6K and p21. *Journal of Functional Foods*, *21*, 474–484.

McCracken, D. A., & Dodd, J. L. (1971). Molecular structure of starch-type polysaccharides from *Hericium ramosum* and *Hericium coralloides*. *Science*, *174*(4007), 419–419.

Mujić, I., Zeković, Z., Lepojević, Ž., Vidović, S., & Živković, J. (2010). Antioxidant properties of selected edible mushroom species. *Journal of Central European Agriculture*, *11*(4), 387–392.

Nagano, M., Shimizu, K., Kondo, R., Hayashi, C., Sato, D., Kitagawa, K., & Ohnuki, K. (2010). Reduction of depression and anxiety by 4 weeks *Hericium erinaceus* intake. *Biomedical Research*, *31*(4), 231–237.

Painuli, S., Semwal, P., & Egbuna, C. (2020). Mushroom: Nutraceutical, mineral, proximate constituents and bioactive component. In *Functional Foods and Nutraceuticals* (pp. 307–336). Springer, Cham.

Pala, S. A., Wani, A. H., & Bhat, M. Y. (2013). Ethnomycological studies of some wild medicinal and edible mushrooms in the Kashmir Himalayas (India). *International Journal of Medicinal Mushrooms*, *15*(2).

Pallua, J. D., Kuhn, V., Pallua, A. F., Pfaller, K., Pallua, A. K., Recheis, W., & Pöder, R. (2015). Application of micro-computed tomography to microstructure studies of the medicinal fungus *Hericium coralloides*. *Mycologia*, *107*(1), 227–238.

Pallua, J. D., Recheis, W., Pöder, R., Pfaller, K., Pezzei, C., Hahn, H., Huck-Pezzei, V., Bittner, L. K., Schaefer, G., Steiner, E. & Huck, C. W. (2012). Morphological and tissue characterization of the medicinal fungus *Hericium coralloides* by a structural and molecular imaging platform. *Analyst*, *137*(7), 1584–1595.

Saitsu, Y., Nishide, A., Kikushima, K., Shimizu, K., & Ohnuki, K. (2019). Improvement of cognitive functions by oral intake of *Hericium erinaceus*. *Biomedical Research*, *40*(4), 125–131.

Shang, X., Tan, Q., Liu, R., Yu, K., Li, P., & Zhao, G. P. (2013). In vitro anti-*Helicobacter pylori* effects of medicinal mushroom extracts, with special emphasis on the Lion's Mane mushroom, *Hericium erinaceus* (higher Basidiomycetes). *International Journal of Medicinal Mushrooms*, *15*(2).

Sheng, X., Yan, J., Meng, Y., Kang, Y., Han, Z., Tai, G., Zhou, Y., & Cheng, H. (2017). Immunomodulatory effects of *Hericium erinaceus* derived polysaccharides are mediated by intestinal immunology. *Food & Function*, *8*(3), 1020–1027.

Shimbo, M., Kawagishi, H., & Yokogoshi, H. (2005). Erinacine A increases catecholamine and nerve growth factor content in the central nervous system of rats. *Nutrition Research*, *25*(6), 617–623.

Thongbai, B., Rapior, S., Hyde, K. D., Wittstein, K., & Stadler, M. (2015). *Hericium erinaceus*, an amazing medicinal mushroom. *Mycological Progress*, *14*(10), 1–23.

Tsai, P. C., Wu, Y. K., Hu, J. H., Li, I. C., Lin, T. W., Chen, C. C., & Kuo, C. F. (2021). Preclinical bioavailability, tissue distribution, and protein binding studies of Erinacine A, a bioactive compound from *Hericium erinaceus* mycelia using validated LC-MS/MS method. *Molecules*, *26*(15), 4510.

Valu, M. V., Soare, L. C., Ducu, C., Moga, S., Negrea, D., Vamanu, E., Balseanu, T. A., Carradori, S., Hritcu, L., & Boiangiu, R. S.(2021). *Hericium erinaceus* (Bull.) Pers. ethanolic extract with antioxidant properties on scopolamine-induced memory deficits in a zebrafish model of cognitive impairment. *Journal of Fungi*, *7*(6), 477.

Vamanu, E., & Voica, A. (2017). Total phenolic analysis, antimicrobial and antioxidant activity of some mushroom tinctures from medicinal and edible species, by in vitro and in vivo tests. *Scientific Bulletin. Series F. Biotechnologies*, *21*, 318–324.

Vedenicheva, N., Al-Maali, G., Bisko, N., Kosakivska, I., Garmanchuk, L., & Ostapchenko, L. (2020). Effect of bioactive extracts with high cytokinin content from micelial biomass of *Hericium coralloides* and *Fomitopsis officinalis* on tumor cells in vitro. *Bulletin of Taras Shevchenko National University of Kyiv-Biology*, *79*(3), 31–37.

Wabang, T., & Ajungla, T. (2016) Edible, medicinal and red listed monkey head mushroom *Hericium erinaceus* (Bull.) Pers. from Japfu mountain of Kohima needs immediate protection. *Current Botany*, *7*, 33–35.

Wang, J. C., Hu, S. H., Wang, J. T., Chen, K. S., & Chia, Y. C. (2005). Hypoglycemic effect of extract of *Hericium erinaceus*. *Journal of the Science of Food and Agriculture*, *85*(4), 641–646.

Wang, M., Gao, Y., Xu, D., Konishi, T., & Gao, Q. (2014). *Hericium erinaceus* (Yamabushitake): A unique resource for developing functional foods and medicines. *Food & Function*, *5*(12), 3055–3064.

Wittstein, K., Rascher, M., Rupcic, Z., Löwen, E., Winter, B., Köster, R. W., & Stadler, M. (2016). Corallocins A–C, nerve growth and brain-derived neurotrophic factor inducing metabolites from the mushroom *Hericium coralloides*. *Journal of Natural Products*, *79*(9), 2264–2269.

Wong, K. H., Sabaratnam, V., Abdullah, N., Kuppusamy, U. R., & Naidu, M. (2009). Effects of cultivation techniques and processing on antimicrobial and antioxidant activities of *Hericium erinaceus* (Bull.: Fr.) Pers. extracts. *Food Technology and Biotechnology*, *47*(1), 47–55.

Wu, T., & Xu, B. B. (2015). Antidiabetic and antioxidant activities of eight medicinal mushroom species from China. *International Journal of Medicinal Mushrooms*, *17*(2).

Wu, Y., Jiang, H., Zhu, E., Li, J., Wang, Q., Zhou, W., Qin, T., Wu, X., Wu, B. & Huang, Y. (2018). *Hericium erinaceus* polysaccharide facilitates restoration of injured intestinal mucosal immunity in Muscovy duck reovirus-infected Muscovy ducklings. *International Journal of Biological Macromolecules*, *107*, 1151–1161.

Yang, B. K., Park, J. B., & Song, C. H. (2003). Hypolipidemic effect of an exo-biopolymer produced from a submerged mycelial culture of *Hericium erinaceus*. *Bioscience, Biotechnology, and Biochemistry*, *67*(6), 1292–1298.

Yi, Z., Shao-Long, Y., Ai-Hong, W., Zhi-Chun, S., Ya-Fen, Z., Ye-Ting, X., & Yu-Ling, H. (2015). Protective effect of ethanol extracts of *Hericium erinaceus* on alloxan-induced diabetic neuropathic pain in rats. *Evidence-Based Complementary and Alternative Medicine*, *2015*, 1–5.

7 Hydnum repandum L.

Riya Chugh, Jagdeep Kaur, and Gurmeet Kaur
Department of Chemistry, Chandigarh University, Gharuan, Punjab, India

CONTENTS

INTRODUCTION

Hydnum repandum, ordinarily referred to as the hedgehog mushroom, sweet tooth, wood hedgehog may be a basidiomycetous fungi flora of the Hydnaceae family. Some native names have also been given to this mushroom like: "hedgehog mushroom", "spreading hedgehog", wood urchin, "yellow tooth fungus". sweet tooth" or "pig' trotter" (Kalač, 2013). This mushroom is initially represented in 1753 by plant scientist Linnaeus, *H. repandum* is primarily a European species related to numerous trees, as well as beech and spruce. Various different species of *Hydnum* attributes soft spines on the undersurface of the cap and soft white flesh which Separates the *H. repandum* from other species is now a challenge, however, the moderately massive size, non- or debile blemishing surfaces, pale colors and nearly spherical spores concerning 8 μm long are the vital determiners. This mushroom is well recognized owing to its spore-bearing form that is formed like spine and teeth instead of gills.

In Europe some of its related species consist of *Hydnum magnorufescens*, with hardly bruising surfaces—and numerous species with darker cap colors, for example, *Hydnum ellipsosporum* (darkish orange, flattened spines, ellipsoid spores) *Hydnum rufescens* (slender, yellow-orange).

DOI: 10.1201/9781003259763-7

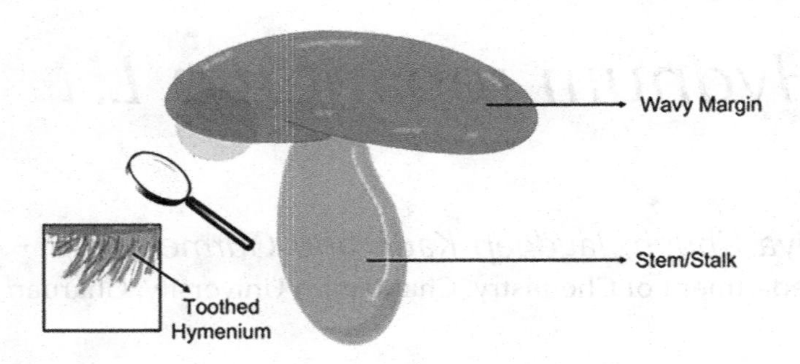

FIGURE 7.1 Illustration represents the general structure of *Hydnum repandum.*

According to current DNA-based definitions, *H. repandum* does no longer arise in North America, despite the fact that this name has been used in North American discipline guides in the past. *Hydnum washingtonianum*, related to conifers and recognized to this point from Labrador, Newfoundland, Washington, and California, is morphologically very comparable and really carefully associated with *H. repandum* (Figure 7.1), as is jap North America's hardwood-related *Hydnum subolympicum*. Various different comparable species in North America consist of *Hydnum subtilior* (smaller and regularly paler) and *Hydnum aerostatisporum* (larger, darker orange). Botanical classification of *H. repandum* is as follows:

Kingdom:	Fungi
Phylum:	Basidiomycota
Class:	Agaricomycetes
Order:	Cantharellales
Family:	Hydnaceae
Genus:	Hydnum
Species:	*Hydnum repandum L.*

HABITAT AND DISTRIBUTION

A mycorrhizal fungi (role in plant's root system) plant life play vital roles in plant nutrition, soil biology, and soil chemistry. This mushroom *H. repandum* is generally distributed in Europe (Swenie, Baroni and Matheny, 2018) wherever it fruits on an individual basis or in shut teams in cone-bearing or deciduous woodland (Feng *et al.*, 2016) this is often an alternative edible mushroom and its specimens can develop a bitter taste. Its non-toxic clone. This mushroom is cosmopolitan in the areas of Europe and North America (Figure 7.2). It is found as single or as in clusters in summer.

It can also form association of Ectomycorrhizal with conifers and hardwoods. This mushroom is also considered as a natural and a major gift within the macro fungi flora of Turkey. It is cosmopolitan within the beech and oak woods of Balikesir, Eskisehir, Adana, Gumushane, Ordu, Manisa, pine, Antalya, Hornbeam, Bolu, Artvin, Giresun, Istanbul, Ordu, Izmir, Samsun and Trabzon Regions and sub-regions of Turkey (*An Antitumor Constituent of the Cultured Mycelia of Hydnum repandum*, no date). It's oversubscribed in the native markets and has a high potential of export (Hong *et al.*, 2019) as a result it is a decent edible and delicious mushroom, having a nutty and sweet taste, delicate smell, and a fresh bright surface.

This mushroom is considered as a mycorrhizal fungi (Petersen, 1977) The fruit of this fungi develops in a scattered or single organizations at the surface or in leaf muddle in the deciduous and coniferous forests (Sterry, 2009; Thi *et al.*, 2015) They can also be developed in the form of closed rings (Dickinson, 1982). The Fruiting process of this fungi mostly takes place in summer till

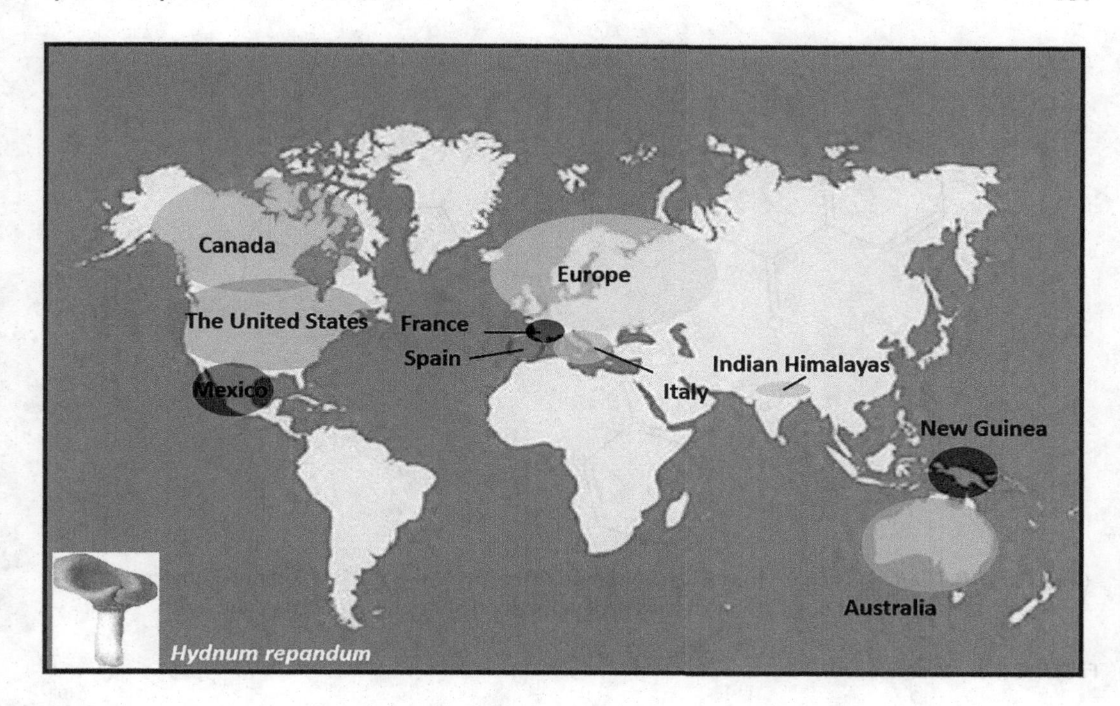

FIGURE 7.2 Distribution of *Hydnum repandum* in various countries.

autumn (Halling, 2004). This fungi is extensively found in Europe, (Swenie, Baroni and Matheny, 2018) and is considered as one of the major enamel fungi (Thi *et al.*, 2015). It is indexed as an inclined species in Europe, as compared to the Red Data Lists of the Germany, Belgium, Sweden, and Netherlands because of having Least Concern (Mu *et al.*, 2020).

Distribution of *H. repandum* in India has been mentioned in the research work done by Prof. R.P. Bhatt on mushrooms of the Uttarakhand Himalayan region. The regions of Uttarakhand where this species has been studied include Adwani, Jageshwar, Phedkhal, Chopta, Tarkeshwar, and Gajar (Bhatt, 2017). In addition, its geographical distribution has been reported in Southeast Asian areas which include the corner of South China, Asia east of India, west of New Guinea, and north of Australia (Thu *et al.*, 2020). The eastern North American region also contains species of *H. repandum* and new six more species of this genus have been discovered (Azeem, Hakeem and Ali, 2020).

GEOGRAPHIC RANGE

H. repandum is mostly widespread in the frigid temperate zones of Europe. The area of occupancy (AOO) of this mushroom is larger than 2,000 km^2 and its extent of occurrence (EOO) is also larger than 20,000 km^2.

H. repandum is often bought with several wild species of fungi in Italy, and in France, it's far one of the formally identified and suitable species for eating often bought in markets (Dickinson, 1982). Also this species is sold under its French name "pied-de-mouton". It is likewise gathered and sells out in nearby markets of Spain, Canada, Mexico, British, and Columbia (Figure 7.3). This edible mushroom (*H. repandum)* is a good source of meal through the pine squirrel (*Sciurus vulgaris*).

MEDICINAL PROPERTIES

Due to its nutritional and medicinal characteristics, *H. repandum* is a vital source of earning as well as a valuable human meal for the rural people. In mice, an extract from culture mycelia

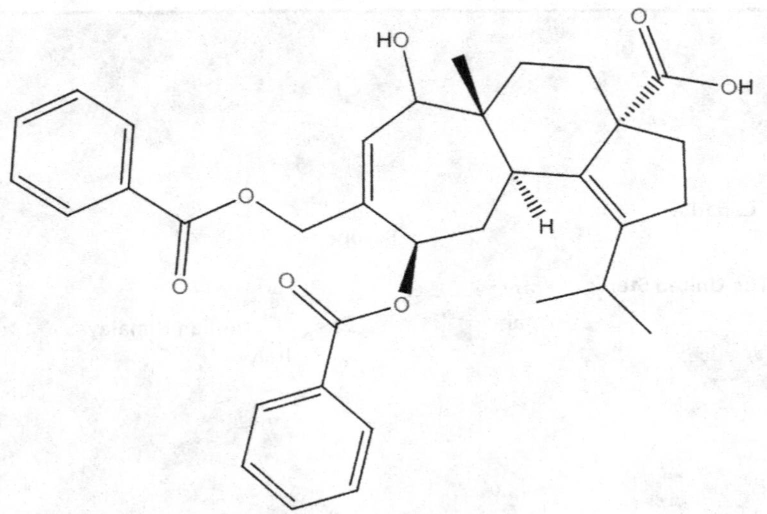

HO O OH H O O O O

(3a*S*,5a*R*,9*R*,10a*R*)-9-(benzoyloxy)-8-((benzoyloxy)methyl)-6-hydroxy-1-isopropyl-5a-methyl-2,3,3a,4,5,5a,6,9,10,10a-decahydrocyclohepta[*e*]indene-3a-carboxylic acid

FIGURE 7.3 Scabronine B.

demonstrated 70% suppression of Sarcoma 180 solid cancer, whereas the fruit bodies extract showed 90% inhibition against both Ehrlich and Sarcoma 180 solid cancer. *H. repandum* variable repandiol had strong cytotoxic activity against a variety of tumor cells. *H. repandum* has a crude ash content of 9.16 percent, a crude protein content of 27.07 percent, a crude cellulose content of 7.60 percent, a crude fat content of 3.16 percent, and a carbohydrate content of 53.01 percent reported that this mushroom has some antioxidant activity.

H. repandum is cheap in calories and fat content, high in protein, and high in a variety of dietary minerals, including iron and manganese, though it's also high in magnesium, calcium, and zinc. Many people enjoy it, and it can be prepared in a variety of ways, making it an appealing sort of healthy cuisine.

Anti-tumor and anti-microbial activities are two potential medical effects, albeit they have yet to be thoroughly investigated. In vitro, Repandiol, a molecule derived from the hedgehog mushroom, displays substantial anti-tumor activity. Several mushroom extracts were also found to be at least partially efficient against test bacteria in culture. Hedgehog is sometimes touted to as an "edible and medicinal mushroom," although it is rarely included in commercially available mushroom supplement mixes because these components have not been investigated in human clinical studies. Now a day, this Repandiol has been also synthesized in labs (Scheme 7.1). ^{1}H-^{13}C-NMR (Table 7.1) of Repandiol is also known.

Anti-tumor Effect

The culture mycelium extract of this mushroom showed inhibition against a solid cancer sarcoma 180 (70%) while the extract from fruit bodies exhibit the inhibition against both the cancers Ehrlich and Sarcoma 180 in mice (90%) (Stachowiak and Reguła, 2012). In addition to this, a major compound Repandiol which is described below exhibits a strong cytotoxic activity against a numerous types of tumor cell in particular colon cancer cells, having their IC50 (Half-maximal inhibitory concentration) is 0.30 (Takahashi, Endo, and NozoE, 1992).

TABLE 7.1
^{1}H-^{13}C NMR Data of Repandiol

Number	^{1}H	^{13}C
1 10	3.44 [each 1H ddd,J=12.7,5.9,4.4] 3.60[each 1H ddd,J=12.7,5.9,2.9]	62.1[each 1C, t]
2 9	3.32[each 1H, br s]	44.1[each 1C,d]
3 8	3.60[each 1H, d, J=2.0]	62.4[each 1C,d]
4 7		77.6[each 1C,s]
5 6		68.8[each1C,S]
O H	4.98[each 1H, t, J=5.9]	

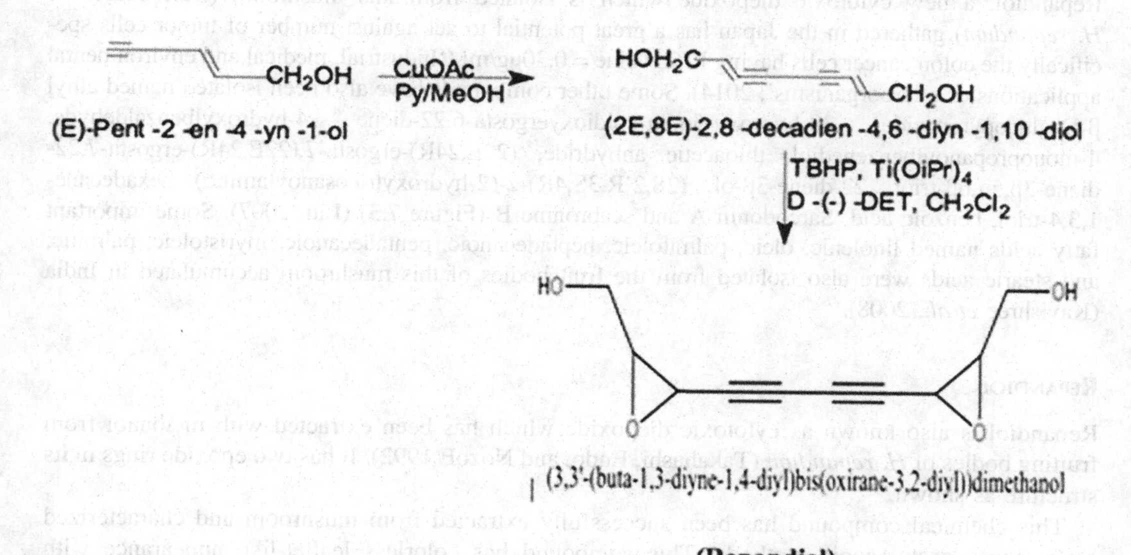

SCHEME 7.1 Chemical synthesis of Repandiol.

Antimicrobial Effect

In a disc diffusion study of antimicrobial activity, a chloroform extract of the hedgehog fungus was found to have minor antibiotic activity against *Staphylococcus epidermidis*, *Staphylococcus aureus*, *Enterobacter aerogenes* and Grass bacillus, but an ethanol extract showed little activity against Grass bacillus (Yamaç and Bilgili, 2006).

H. repandum Mushroom Dosage

Since this mushroom is mainly used as food component, dosage is generally not a problem if Consumed in reasonable amounts, this mushroom is safe. Since its efficacy as a drug has not been established yet, its therapeutic dosage, if any, is unknown.

H. repandum Side Effects, Safety, Dangers, and Warnings

Although *H. repandum* is generally considered safe to eat when cooked, but it is still possible for humans to develop allergies or hypersensitivities—the same caution applies to any food, however.

Like other mushrooms, this mushroom should never be eaten raw. However, its toxicology of this mushroom is well explained in the upcoming paragraph.

PHYTOCHEMICALS

The phytonutrients which are accountable for the fruity smell of *H. repandum* consists of 1-octen-3-ol, (*E*)-2-octenol, and (*E*)-1,3-octadiene which is a 8-C Derivative. After 1986 a study is conducted in Europe, A Nuclear disaster called the Chernobyl disaster occurred in 1986. The fruit bodies showed the high rate of accumulation of the Cesium which is a radioactive isotope.

This edible mushroom(*H. repandum*) which belongs to the Cantharelaceae family (*Edible and Poisonous Mushrooms of the World - SILO.PUB*, no date). Although it is mainly found in Europe but it is also distributed in Portugal, India, China, and Thailand (Kavishree *et al.*, 2008; Kalač, 2013). Repandiol, a new cytotoxic diepoxide which is isolated from this mushroom (fruit bodies of *H. repandum*) gathered in the Japan has a great potential to act against number of tumor cells specifically the colon cancer cells having IC50 value = 0.30µg/ml ('Industrial, medical and environmental applications of microorganisms', 2014). Some other compounds have also been isolated named ethyl β-D-glucopyranoside 3β-hydroxy-5α,8α-epidioxyergosta-6,22-diene 4-hydroxylbenzaldehyde, 4-monopropanoylbenzenediol, thioacetic anhydride, (22E,24R)-ergosta-7,(22E,24R)-ergosta-7,22-diene-3β,5α,6β-triol, 22-diene-3β-ol, (2S,2'R,3S,4R)-2-(2-hydroxytricosanoylamino) hexadecane-1,3,4-triol, benzoic acid, Sarcodonin A and scabronine B (Figure 7.3) (Liu, 2007). Some important fatty acids named linolenic, oleic, palmitoleic, heptadecanoic, pentadecanoic, myristoleic, palmitic, and stearic acids were also isolated from the fruit bodies of this mushroom accumulated in India (Kavishree *et al.*, 2008).

REPANDIOL

Repandiol is also known as cytotoxic diepoxide which has been extracted with methanol from fruiting bodies of *H. repandum* (Takahashi, Endo, and NozoE 1992). It has two epoxide rings in its structure as shown.

This chemical compound has been successfully extracted from mushroom and characterized by various spectroscopic methods. This compound has colorless leaflet-like appearance with 168-degree Celsius melting point and it is optically active compound with formula $C_{10}H_{10}O_4$. It has shown cytotoxic effect for cancer cells especially colon adenocarcinoma cells as Repandiol act as a bifunctional alkylating agent like DAG (dianhydrogalactitol) and WF-3405 which are considered to be potential anti-cancer drugs. The bond formation between two strands of DNA with an external molecule is known as DNA inter strand cross-linking. The significance of forming cross-linking is to inhibit the replication or transcription processes of the DNA. This property can be exploited to combat cancer disease. The diepoxides like DEB (diepoxybutane) and DEO (diepoxyoctane) (Figure 7.4) are compared with present metabolite that is Repandiol. DEB is an active form of drug treosulfan and also it is a potent carcinogen formed from butadiene metabolism by cytochromes in our body (Michaelson-richie *et al.*, 2010). The 1,3-butadiene is a common pollutant in an industrial area. DEB forms cross linking between N-7 positions of deoxyguanosine residues at 5'-GNC sequence (N is any base) particularly on restriction fragments of DNA. Whereas, DEO (diepoxyoctane) having longer chain length than DEN showed cross-linking at 5'-GNnC (n = 1 to 2). It has been observed that there is loss in specificity of binding with the increase in alkyl chain lengths. The nitrogen atoms in a DNA base pairs act as nucleophile which attacks epoxide ring resulting in ring opening and bond formation. Repandiol has complex structure than DEB and DEO. By various studies it has been proved that the Repandiol form cross-links at 5'-GNC and 5'-GNNC sites as DEB and DEO, along with this it form cross-links at additional places. There is a possibility of forming cross-link at 5'-GNT site similar to nitrogen mustards. However, the efficiency of specific binding is low in Repandiol in comparison to DEB and DEO.

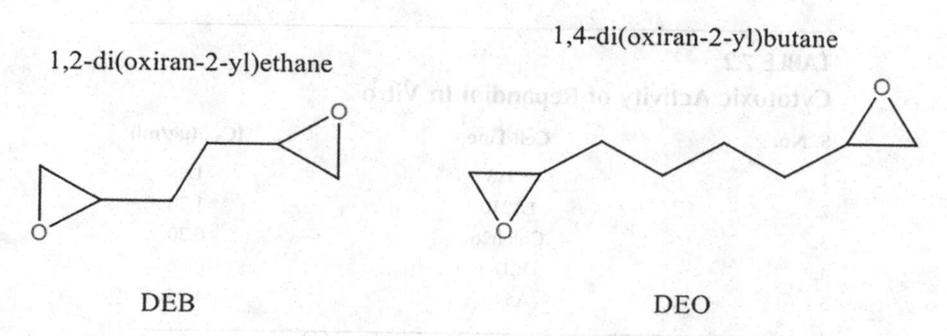

FIGURE 7.4 The chemical structures of DEB and DEO.

Hence, diepoxides form interstrand cross-links in DNA resulting in blocking replication of genetic material and its gene expression (Millard *et al.*, 2004). However, exact role of cross-linking property has not been discovered till now.

The diepoxides like DEB (diepoxybutane) and DEO (diepoxyoctane) [Figure 9] are compared with present metabolite that is Repandiol. DEB is an active form of drug treosulfan and also it is a potent carcinogen formed from butadiene metabolism by cytochromes in our body (Millard *et al.*, 2004).

Vasdekis and Co-workers have reported some of the anti-oxidants, cytotoxic, pro-apoptotic, anti-angiogenic and anti-tumor properties of this mushroom. An important cytotoxicity activity of Repandiol is discussed in (Table 7.2) (IC50 value = 1.0 mg/ml) was showed against A549 cell line and trans-stilbene (a phenolic compound) was also analyzed by (LC-MS) technique (*Fungi for Human Health: Current Knowledge and Future Perspectives - Uzma Azeem, Khalid Rehman Hakeem, M. Ali - Google Books*). The effect of its extract on the development and monogenesis of Ae- penicillium glaucus was also studied in vitro. A Notable decrease is observed in the inhibition and mycelial growth inhibition of an infectious agent and germ sporulation was observed (Florianowicz and Florianowicz, 2000). In vitro anti-bacterial, anti-microbial, and phytochemicals exposure was carried out by many researchers (Yamaç 2006; Heleno *et al.*, 2010; Ozen *et al.*, 2011).

Indole Derivatives

The sample of *H. repandum* was collected from the forests of northern Poland. The components are extracted and concentrated using rota-evaporator. Then the dried mixture is subjected to HPLC for further studies. There are total four Indole compounds that have been detected in this mushroom all of which are anti-oxidant in nature. From Table 7.3, it is vivid that the content of tryptamine is considerably high than others. Tryptamine (Figure 7.5) is generally present in our brain in very trace amounts that acts as neuromodulator or neurotransmitter. The presence of significant amount of tryptamine makes it a good medicinal candidate or a mushroom. The sterols like Ergosterol and Ergocalciferol are not identified in this mushroom(Sułkowska-Ziaja, Muszyńska and Szewczyk, 2015).

DIETARY COMPOSITION

The dried *H. repandum* contains carbohydrates (56%), fat (4%), and protein (20%) (Bofaris and Alzand, 2018). In a reference amount of 100 grams, there are some high-content dietary minerals present in this mushroom, especially manganese and copper. Major Fatty acids are also present including oleic acid (26%), palmitate (16%), linoleic acid (48%), and linolenic acid (20%). *H. repandum* also contains Mycosterol (Sterols obtain from fungi) (*Nutritional value of some wild edible mushrooms from the Black Sea* ..., no date; Kalač, 2009) (Table 7.4).

TABLE 7.2
Cytotoxic Activity of Repandiol In Vitro

S. No.	Cell Line	IC_{50} (μg/ml)
1.	P388	1.88
2.	L1210	1.20
3.	Colon26	0.30
4.	DLD-1	0.66
5.	A549	1.35

TABLE 7.3
Indole Derivatives Content Present in the *Hydnum repandum*

S. No.	Indole compounds	Amount in mg/100 g DW (dried weight)	Structure of the compound
1	Indole	0.13 ± 0.05	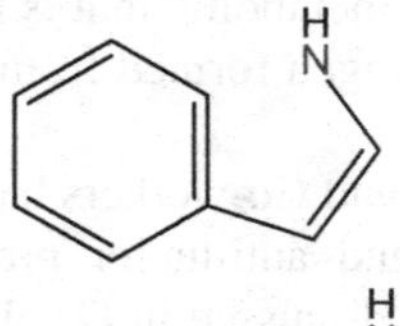
2	Melatonin	0.32 ± 0.28	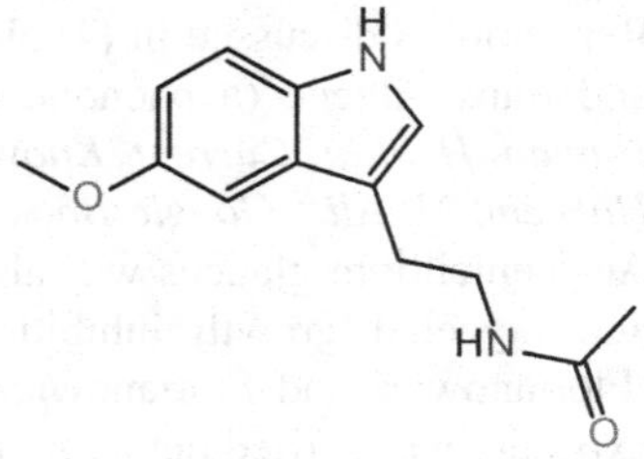
3	L-tryptophan	0.36 ± 0.091	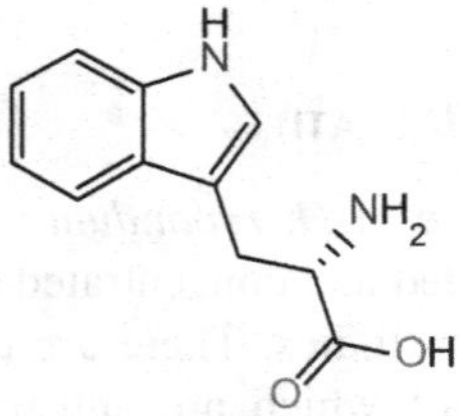
4	Tryptamine	1.46 ± 1.05	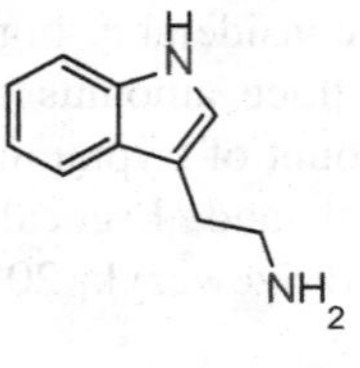

Fats and Fatty Acids

The mushroom is collected from the forest of Himachal Pradesh, India for the study of fatty acids and their amounts. It has been found that a total 4.7% (dry weight) of fat is present in *H. repandum*. Fats are the long chains of fatty acids. It can be saturated (Monounsaturated fatty acid- oleic acid) or unsaturated (Polyunsaturated fatty acid- Linoleic acid) (Figure 7.6). Fat content of *H. repandum*

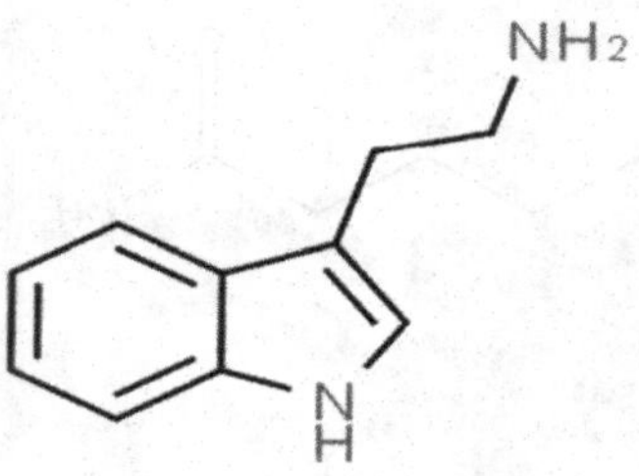

FIGURE 7.5 Tryptamine.

TABLE 7.4
Nutritional Value of *H. repandum* mg/100 g

S. No.	Dietary components	Composition mg	References
1.	Carbohydrates	56.10	Sevindik, 2020
2.	Proteins	19.70	Sevindik, 2020
3.	Fat	4.3	Petersen, 1977
4.	Energy	342 kcal	Stachowiak and Reguła, 2012
5.	Vitamin C	1.1	Stachowiak and Reguła, 2012
6.	Minerals (Copper)	38.9	Petersen, 1977
	(Calcium)	600	Stachowiak and Reguła, 2012
	(Manganese)	23.2	
7.	Ash	9.20	Sevindik, 2020
8.	Moisture	10.65	Sevindik, 2020
9.	Fatty Acids (Oleic acid)	26%	Lucakova, Branyikova and Hayes, 2022
	(Linolenic acid)	20%	Sevindik, 2020
	(Linoleic acid)	48%	
	(Palmitate)	16%	

is extracted with benzene by using soxhlet apparatus. The content undergoes transesterification process catalyzed by boron trifluoride and then it is subjected to GC or gas chromatography for further evaluation of the content. It is clear from Table 7.5 that the content of unsaturated fatty acids is more than the amount of saturated fatty acids. The linoleic acid is known to prevent cardiac issues and also help in lowering cholesterol levels (Kavishree *et al.*, 2008).

PHARMACOLOGY (BIOLOGICAL ACTIVITIES RELATED TO MEDICINAL USES OR CLINICAL EVIDENCE)

Pharmacology refers to the study of action and reaction of drug molecules inside the living organisms (Vallance and Smart, 2006). This mushroom has various medicinal properties such as anti-tumor, anti-oxidant, anti-inflammatory, and other properties that are well described under metabolomics section. Till now Repandiol is one the most studied component of this species. It can act as an anti-cancer drug molecule by interfering DNA replication through the formation of inter-strand cross links at 5' GNC and 5'GNNC (Where, G and C are guanine, and cytosine; N is any nucleotide) sites of the DNA (Takahashi, Endo, and NozoE, 1992). Its detailed study has been mentioned ahead under Repandiol section. Recently, Naz Dizeci *et al.* reported the antioxidant, anticancer, antimicrobial, and antibiofilm properties of *H. repandum* in the *International Journal of Medicinal Mushrooms*. This report demonstrates that *H. repandum* has antibiofilm potential and can combat antibiotic resistance

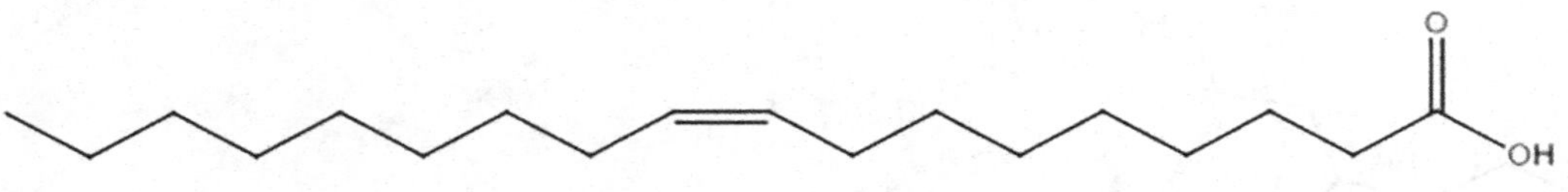

O
OH
Oleic acid

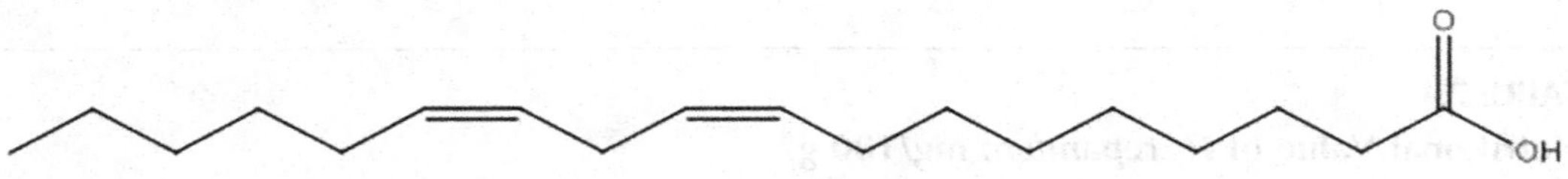

O
OH
linoleic acid

FIGURE 7.6 Oleic acid and linoleic acid.

TABLE 7.5
Fatty Acid Content in *H. repandum*

Fatty acids	Amount (% total fatty acid methyl ester)
Total saturated fatty acids (C14-C18)	24.6
Total mono-unsaturated fatty acids	27.9
Total polyunsaturated fatty acids	47.5
Unsaturated and saturated fatty acid) UFA:SFA	3.06
Linoleic: oleic acid	1.03

in biofilms. The ethanolic extract of *H. repandum* contains a high level of myricetin and apigenin and demonstrated antiproliferative effects on MCF-7 and HT-29 cell lines. Where, MCF-7 is a human breast cancer cell line with estrogen, progesterone, and glucocorticoid receptors (Horwitz, Costlow and McGuire, 1975) and HT-29 is a human colorectal adenocarcinoma cell line (Martínez-Maqueda, Miralles and Recio, 2015). Interestingly, the extract displayed antibacterial and antibiofilm actions against methicillin-resistant *Staphylococcus aureus* ATCC 43300, *S. epidermidis* ATCC 35984, *Escherichia coli* ATCC 25922, and *Pseudomonas aeruginosa* ATCC 27853. When antibiotics like kanamycin and ampicillin are used with mushroom extract, synergetic responses have been observed. *H. repandum* extract showed lesser biofilm eradication concentration than antibiotics (Procházka *et al.*, 2011). *H. repandum* exhibited high cytotoxicity (IC_{50}: 1.0 mg mL^{-1}) and induced apoptosis–necrosis (97.92 and 96.22%) to A549 (lung carcinoma epithelial cells) cells due to the presence of bio-active ingredient piceatannol ((E)- 4-[2-(3,5-dihydroxyphenyl)ethenyl]1,2-benzenediol-3,3′,4,5′-tetrahydroxy-trans-stilbene) an analog of resveratrol (Vasdekis *et al.*, 2018). Its mechanism of action is still undiscovered. Other than Repandiol, a water-soluble protein polysaccharide fraction (WPPF) obtained from mycelium culture of *H. repandum* is also possess anticancer properties. Experiment done on ICR mice implanted with sarcoma 180 revealed 54.3 % inhibition activity against cancer cells. The WPPF consists of 28 % of protein moiety and 42 % polysaccharide moiety which further contains glucose, galactose, mannose, xylose, or fucose as monosaccharide units. Further investigation on the extraction of such anti-tumour compounds is still going on (*An Antitumor Constituent of the Cultured Mycelia of Hydnum repandum*, no date). Free radicals, superoxides, peroxide, and hydrogen peroxide produced in our body is linked with our aging process. Their scavenging activities demonstrating their anti-oxidant property have been studied with the methanol extract of *H. repandum* and are well reported in the paper (Ozen *et al.*, 2011).

METABOLAMICS AND DNA SEQUENCING

Explanation: This ectomycorhizal fungus has a true or well-defined nucleus. Two nuclei per cell hence, kept under sub-kingdom Dikarya. It consists of long filamentous structures called hyphae or has septate mycelium. They show sexual reproduction by the formation of spores through club-shaped bodies called a basidium. In general, the life cycle of this phylum takes place by sexual reproduction. There are no sex bodies yet plasmogamy (fusion of protoplasm) and karyogamy (fusion of nuclei) occur followed by meiosis, a reduction division process in basidium that ultimately forms basidiospores usually four in number (this number varies from species to species). Then, the haploid (n) spores undergo a germination process resulting in the formation of the primary mycelium (filamentous, single nucleus (n), and septate). The mycelium formed by plasmogamy with delayed karyogamy yield (n + n) cells is termed secondary mycelium. This stage is known as the food absorbing stage which grows to form a complete structure. The clamp connections are present that helps in maintaining a bi-nucleate state of the cells. The fruiting bodies are macroscopic in nature. The basidiocarp or sporocarp has a surface called hymenium that bears basidia which forms spores. The fruit body is funnel-shaped or mushroom-shaped and spores are present on gill-like ridges. The sporocarp has toothed hymenium. The generic name represents tooth-like structures instead gills while the species epithet repandum means the wavy margin or bent back structure of the fruiting body. The cell of the fungus contains all essential organelles like Golgi bodies, endoplasmic reticulum, ribosomes, etc., except chloroplast (Figure 7.7) (Swenie, Baroni and Matheny, 2018; Sugawara *et al.*, 2019).

DNA Sequencing

The Phylogenetic studies on *Hydnum* genus are very rare. The DNA or deoxyribonucleic acid is a genetic material present in well-defined nucleus. It consists of de-oxy ribose (5-C) sugar, nitrogen bases (Adenine, Guanine, Cytosine, and Thymine) and phosphate groups. From the database available at GeneBank, it is clear that *H. repandum* has linear DNA. The complete DNA sequence of this species is not available; however, the sequence of nucleotides of some parts of the DNA is given on the GeneBank site (www.ncbi.nlm.nih.gov). The name of the genes and their gene bank identifier number is given below (Table 7.6), and many more are present on the provided site. According to central dogma DNA replicates itself and transcribes to m-RNA (transcription) and m-RNA translates to proteins (translation) that incorporates in cells machinery for healthy functioning of an organism. There are total 87 results on Gene bank site related to DNA sequence of different strains of *H. repandum*. It is cumbersome to mention all of the sequences of different strains here.

METABOLOMICS

It is defined as the scientific study of chemical compounds or metabolites formed by cellular processes in an organism, cells or tissues (Idle and Gonzalez, 2007).

Phenolic Acids or Secondary Metabolites in Acetone Extract of *H. repandum*

This mushroom has various biological properties like anti-cancer, anti-microbial, anti-mutagenic, anti-oxidant, anti-cancer, cytotoxic activity, etc due to the presence of polyphenolic compounds.

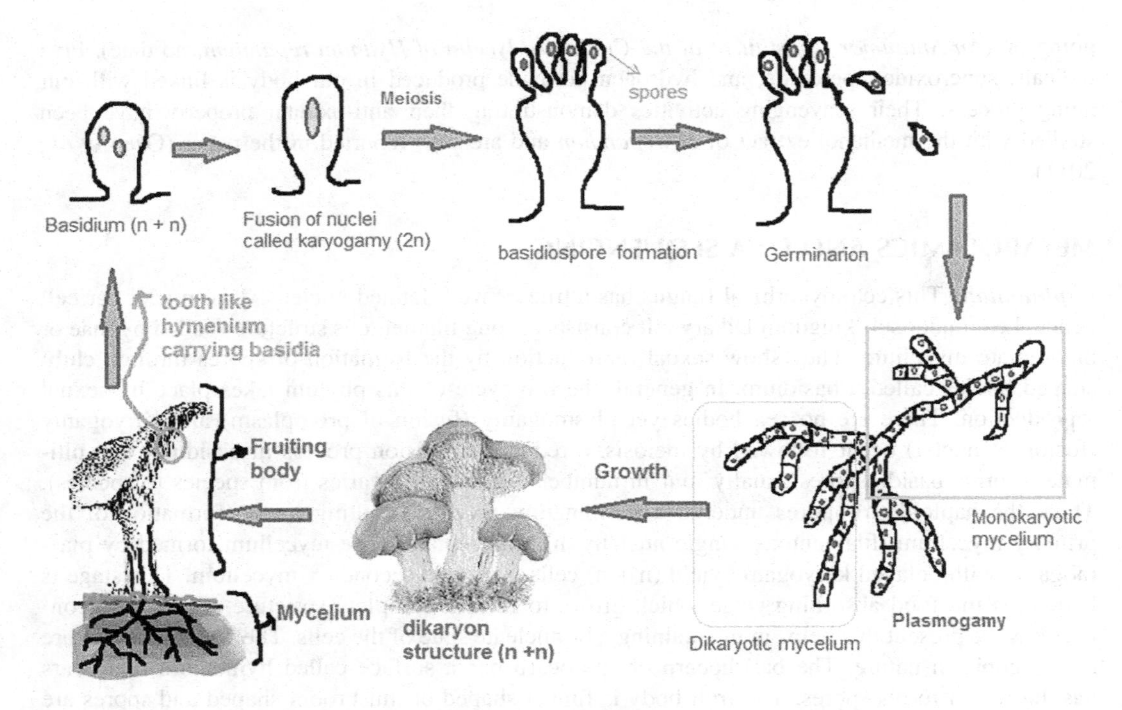

FIGURE 7.7 Diagrammatic representation of a general life cycle of basidiomycetes.

Generally, a mushroom consists of polyphenols, peptides, polysaccharides, vitamins, minerals, iron, zinc, etc as secondary metabolites. The polyphenols are the broadest group among metabolites which further includes phenolic acids, lignins, flavonoids, tannins, stilbenes, oxidized polyphenols, hydroxybenzoic acids etc. The acetone extract of the *H. repandum* was extracted using soxhlet extractor; the obtained extract was filtered and then concentrated using rota-evaporator under low-pressure conditions. The resultant extract was stored at 18 degrees Celsius for further evaluation procedures. The polyphenolic compounds have been identified using HPLC technique. There are five polyphenolic components (syringic acid, p-coumaric acid, chlorogenic acid, gallic acid, and ferulic acid) and three flavonoids (quercetin, rutin, and catechin) that are detected in this mushroom (Table 7.7). The chemical structures of these compounds have shown in Figures 7.8 and 7.9. Moreover, ferulic acid and quercetin are present in more amounts than other members (Tubic *et al.*, 2019).

In addition to the above-mentioned phenolic acids gentistic acid, protocatechuric acid, vanillic acid, and cinnamic acid also have been identified. The protocatechuric acid is present is relatively high amounts (75.23 ± 0.02 mg/100 g DW). All of them have great roles in making human health. Another bioactive compound piceatannol, a stilbenoid is also detected in this mushroom. This compound is known to exhibit anti-proliferative activity. Hence, this mushroom is a promising source of biologically active compounds (Figure 7.10) (Vasdekis *et al.*, 2018).

The phytochemical content of *H. repandum* according to the published data by Tevfic Ozen *et al.* (2011) are given in Table 7.8. The mushroom was acquired from the black sea region of Turkey (Ozen *et al.*, 2011).

ECONOMIC IMPORTANCE

Due to the presence of properties like antioxidant, anti-cancer, anti-inflammatory, etc., it is considered healthy to eat this mushroom. *H. repandum* is collected and marketed in Europe. This

TABLE 7.6
Enlists Different Genes, Their Number of Base Pairs, and its GI Number

S. No.	Gene name	No. of base pairs	GI number
1	18 S ribosomal RNA gene	1,786	2576495
2	Mitochondrial gene for mitochondrion RNA (partial sequence), 12 S ribosomal RNA gene	476	2586038
3	18 S RNA gene	571	6179568
4	5.8 S ribosomal RNA gene, partial sequence; internal transcribed spacer 2, complete sequence; and 28S ribosomal RNA gene, partial sequence	1209	13774444
5	Mitochondrial large subunit ribosomal RNA gene, partial sequence; mitochondrial gene for mitochondrial product	285	15127996
6	DNA-dependent RNA polymerase II second largest subunit gene	2889	45545330
7	18S rRNA gene (partial), 5.8S rRNA gene, 28S rRNA gene (partial), ITS1 and ITS2	611	119507097
8	Internal transcribed spacer 1, partial sequence; 5.8S ribosomal RNA gene, complete sequence; and internal transcribed spacer 2, partial sequence	599	527175461

TABLE 7.7
Polyphenolic Compounds in *H. repandum*

S. No.	Name of polyphenolic compound	Type of compound	Possible activities/uses
1	Gallic acid	Phenolic acid	Anti-oxidant, Anti-inflammatory, anti-cancer, ability to treat cardiovascular disorders, protects nervous system.
2	P-coumaric acid	Phenolic acid	Anti-infective, Anti-inflammatory, Anti-oxidant, etc.
3	Chlorogenic acid	Phenolic acid	Anti-pyretic, anti-bacterial, anti-hypertension, anti-obesity, anti-oxidant, neuroprotective, anti-diabetic, cardioprotective, hepatoprotective
4	Syringic acid	Phenolic acid	Anti-microbial, anti-cancer, anti-diabetic, anti-oxidant, anti-inflammatory.
5	Ferulic acid	Phenolic acid	Anti-aging, anti-oxidant, used in skin lightening
6	Quercetin	Flavonoid	Anti-hypertension, anti-allergy, Anti-inflammatory, Anti-diabetic, anti-oxidant
7	Catechin	Flavonoid	Regulates blood pressure, used in cosmetics, anti-oxidant, anti-obesity, neuroprotective, promote synthesis of collagen protein, etc.
8	Rutin	Flavonoid	Helps to treat autism, osteoarthritis, lymohedema, anti-inflammatory, anti-oxidant

mushroom has a high export potential due to its nutty and sweet taste (*Novel Food Catalogue*, no date). The Eight-carbon derivatives are among the organic volatile compounds accountable for the delicious scent of the mushroom (Fons *et al.*, 2003). It contains 53.01 % carbohydrates, 27.07 % crude protein, 3.16 % crude fat, 7.6 % crude cellulose, and 9.7 % crude ash. Because of its medicinal properties as described earlier in this chapter and nutritional properties, it is a great

gallic acid p-coumaric acid syringic acid

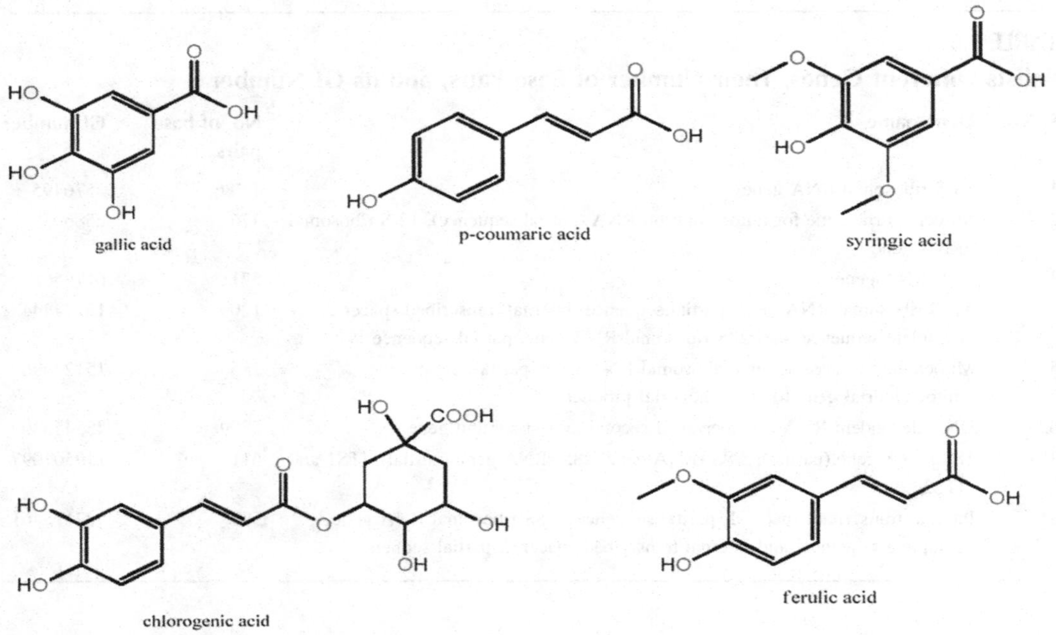

chlorogenic acid ferulic acid

FIGURE 7.8 Chemical structures of phenolic acids.

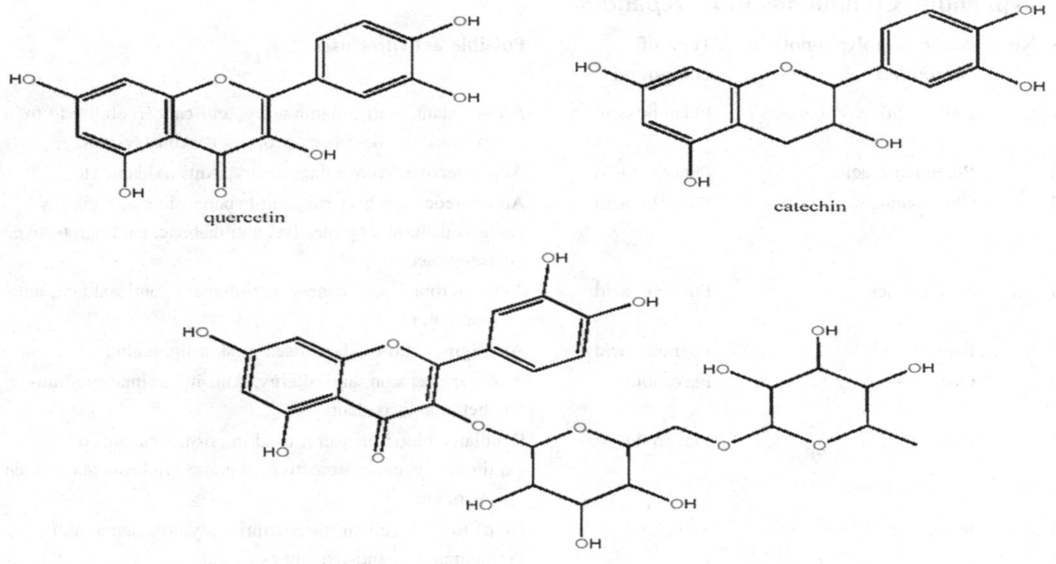

quercetin catechin

FIGURE 7.9 Chemical structures of flavonoids found in *H. repandum.*

source of income especially in Europe and North America where it has been mainly cultivated during the season. For the cultivators, from one study, it has been found that the vegetative growth of the mycelium is efficient at 20–25 degree Celsius temperature, ph 5.5; glucose and mannose are the best carbon source and calcium nitrate as the nitrogen source for the production of vegetative inoculums (Peksen, Kibar and Yakupoglu, 2013).

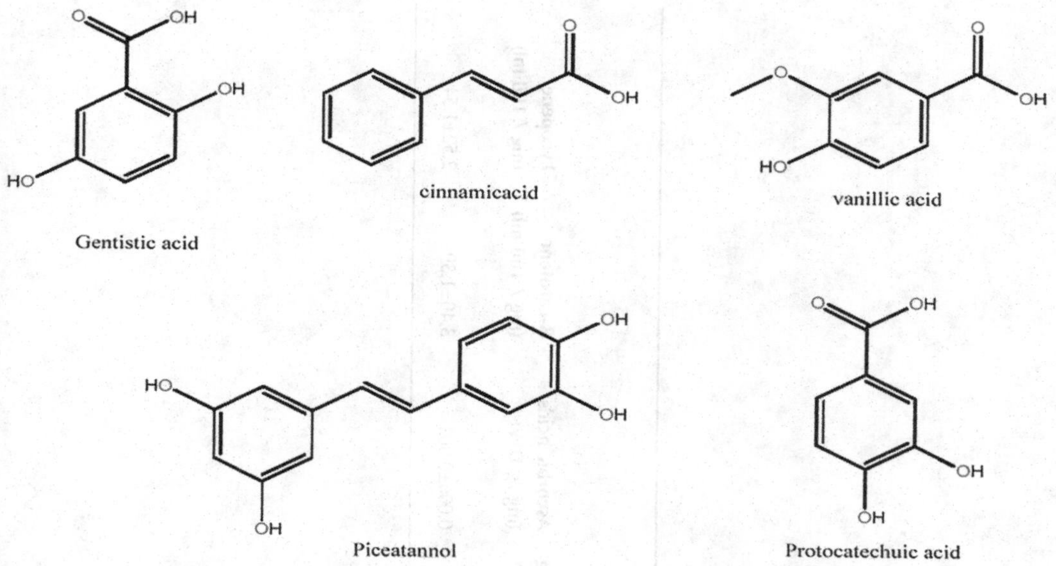

FIGURE 7.10 Biologically active compounds in *Hydnum repandum.*

TOXICOLOGY

The fruiting bodies of this mushroom have a great potential of accumulation of the radioactive isotope (cesium), according to research conducted in Europe after the 1986 Chernobyl tragedy (Stachowiak and Reguła, 2012). The metals or metal salts enter the mushroom through contaminated soil that accumulates in their fruiting bodies. Two metals namely, copper and zinc are found in this species and these two metals are not considered as toxic in nature (Vasdekis *et al.*, 2018). The cadmium, cobalt, and chromium (heavy metals) are also detected in this mushroom. Amount of accumulation of particular metal depends upon the genotype of the mushroom and a mushroom grown in polluted or bad conditions is usually considered as toxic or unsafe to eat (Vázquez *et al.*, 2016). As heavy metals have the tendency to disturb the normal enzymatic reactions in our body or sometimes metals can be carcinogenic in nature. From a study on a Turkish mushroom especially *H. repandum* it has been found that 6.79 mg/kg of Hg (mercury) is present in it by atomic absorption spectroscopy (Demirbasë, 2000). In addition, copper, cadmium, or lead also detected in this mushroom (turkey). The mushrooms have a tendency to pick up these deadly metals from the soil polluted with dangerous metals. This condition would create a great impact on the lives of the consumers. The mushroom itself does not contain any toxic chemical but it can accumulate a toxic substance by absorbing them from the soil. It is advisable to check the quality of soil while growing mushrooms. Rubidium is also detected in it, 1250 mg/kg of its dry matter (Lubomír and Chrastny, 2008).

CONCLUSION AND FUTURE PERSPECTIVE

There is a large number of edible mushrooms containing Hazardous compounds in different proportions. In a present study of an edible mushroom, *H. repandum* clearly demonstrates that it is a rich source of bioactive compounds including anti-microbial, anti-tumor, anti-oxidant, cytotoxic, anti-inflammatory, dietary components, Indole compounds, fatty acids, and polyphenolic compounds. Due to its medicinal characteristics, this mushroom is also used as a therapeutic agent. As, this mushroom contains fatty acids, flavonoids, and some phenolic compounds such as oleic acid, linoleic acid, palmitic and stearic acids, Quercetin, Catechin, Rutin, Gallic acid, Chlorogenic acid,

TABLE 7.8
Phytochemical Content of *Hydnum repandum* According to *Tevfic Ozen* (2011)

Name	Local name	Yield (g)	Phenolics (mg pyrocatechol/g DW)	Flavonoids (mg quercetin/g DW)	Anthocyanins (mg cyaniding-3-glucoside/g DW)	Ascorbic acids (mg /g DW)	B-carotene (mg / 100 ml)	Lycopene (mg / 100 ml)
Hydnum repandum	Beyas post mantan	3.10	3.67±2.21	0.102±0.007	0.62±0.05	0.006±0.002	3.40±1.39	2.52±1.54

Syringic acid which may decrease the risk of various Chronic diseases including atherosclerosis, cancer, diabetes, ageing, cardiac and neuro issues, hypertension, allergy, autism, osteoarthritis, obesity and other degenerative diseases in humans. These Bioactive compounds present in this mushroom can have a good potential as a source of medicinal remedies. The study of its metabolites, DNA sequencing, toxicology and its economic importance has been discussed. It is an edible mushroom with great number of benefits not only for the consumers but also for the economy of our globe. *H. repandum* is a great source of protein However; this edible mushroom is mainly used as a food component and is safe to eat, if consumed in reasonable amount. The toxicology of this mushroom depends upon the type of soil used to grow the mushroom as various heavy metals usually absorbed into it due to the contaminated soil. But, its efficacy as a drug has not been investigated yet and its therapeutic dosage is also unknown. The complete genetic study as well as potency of this mushroom are still unknown or need to be explored yet. Therefore, the clinical and careful investigations are required to identify the various side effects and causes of *H. repandum* and enable their safe consumption.

REFERENCES

Azeem, U., Hakeem, K.R. and Ali, M., 2020. Fungi for Human Health. *Current Knowledge and Future Perspectives*, Publisher Springer, Cham, 5–11, ISBN-10 303058755 :X, ISBN-13 : 978-3030587550.

Bhatt, R.P., Rana, U.S., Stephenson S.I., Uniyal, P. and Mehmood, T., 2017. Wild edible mushrooms from high elevations in the Garhwal Himalaya—II. *Current Research in Environmental & Applied Mycology*, *7*(3), 208–226. ISSN 2229-2225.

Bofaris, M.S.M. and Alzand, K.I., 2018. Chemical composition and nutritional value in turkey species of wild growing edible mushrooms: A review. *World Journal of Pharmaceutical Research*, *8*(3), 63–75.

Demirbaş, A., 2000. Accumulation of heavy metals in some edible mushrooms from Turkey. *Food Chemistry*, *68*(4), 415–419.

Dickinson, C.H. and Lucas, J.A., 1982. *VNR color dictionary of mushrooms*. Publisher Van Nostrand Reinhold, ISBN 10: 0442219989ISBN 13: 9780442219987.

Feng, B., Wang, X.H., Ratkowsky, D., Gates, G., Lee, S.S., Grebenc, T. and Yang, Z.L., 2016. Multilocus phylogenetic analyses reveal unexpected abundant diversity and significant disjunct distribution pattern of the Hedgehog Mushrooms (Hydnum L.). *Scientific Reports*, *6*(1), 1–11.

Florianowicz, T., 2000. Inhibition of growth and sporulation of *Penicillium expansum* by extracts of selected basidiomycetes. *Acta societatis botanicorum Poloniae*, *69*(4), 263–267.

Fons, F. et al., 2003. Les substances volatiles dans les gens *Cantharellus*, Craterellus et Hydnum. *Cryptogamie, Mycologie*, *24*, 367–376.

Halling, R.E., 2004. Edible and Poisonous Mushrooms of the World. *Brittonia*, *56*(2), 150- 150.

Heleno, S.A., Barros, L., Sousa, M.J., Martins, A. and Ferreira, I.C., 2010. Tocopherols composition of Portuguese wild mushrooms with antioxidant capacity. *Food Chemistry*, *119*(4), 1443–1450.

Hong, M.Y., Park, S.W., Kim, D.H., Saysavanh, V. and Lee, J.K., 2019. A Checklist of Mushrooms of Phousabous National Protected Area (PNPA) of Lao PDR. *Journal of Forest and Environmental Science*, *35*(4), 268–271.

Horwitz, K.B., Costlow, M.E., and McGuire, W.L. 1975. MCF-7: A human breast cancer cell line with estrogen, androgen, progesterone, and glucocorticoid receptors. *Steroids*, *26*(6), 785–795. doi: 10.1016/0039-128x(75)90110-5. PMID: 175527.

Idle, J.R. and Gonzalez, F.J., 2007. Metabolomics. *Cell Metabolism*, *6*(5), 348–351.

Kalač, P., 2009. Chemical composition and nutritional value of European species of wild growing mushrooms: A review. *Food Chemistry*, *113*(1), 9–16.

Kalač, P., 2013. A review of chemical composition and nutritional value of wild-growing and cultivated mushrooms. *Journal of the Science of Food and Agriculture*, *93*(2), 209–218.

Kavishree, S., Hemavathy, J., Lokesh, B.R., Shashirekha, M.N. and Rajarathnam, S., 2008. Fat and fatty acids of Indian edible mushrooms. *Food Chemistry*, *106*(2), 597–602.

Liu, J.K., 2007. Secondary metabolites from higher fungi in China and their biological activity. *Drug Discovery Therapy*, *1*(2), 94–103.

Lubomír, S., and Chrastny, V. 2008. Levels of eight trace elements in edible mushrooms from a rural area. *Food Additives and Contaminants*, *25*(1), 51–58.

Lucakova, S., Branyikova, I. and Hayes, M., 2022. Microalgal proteins and bioactives for food, feed, and other applications. *Applied Sciences*, *12*(9), 4402.

Martínez-Maqueda, D., Miralles, B. and Recio, I., 2015. HT29 cell line. *The Impact of Food Bioactives on Health*, 113–124.

Michaelson-Richie, E.D., Loeber, R.L., Codreanu, S.G., Ming, X., Liebler, D.C., Campbell, C. and Tretyakova, N.Y., 2010. DNA – protein cross-linking by 1, 2, 3, 4-diepoxybutane. *Journal of Proteome Research*, *9*(9), 4356–4367.

Millard, J.T., Katz, J.L., Goda, J., Frederick, E.D., Pierce, S.E., Speed, T.J. and Thamattoor, D.M., 2004. DNA interstrand cross-linking by a mycotoxic diepoxide. *Biochimie*, *86*(6), 419–423.

Mu, Y.H., Hu, Y.P., Wei, Y.L. and Yuan, H.S., 2020. Hydnaceous fungi of China 8. Morphological and molecular identification of three new species of Sarcodon and a new record from southwest China. *MycoKeys*, *66*, 83.

Ozen, T., Darcan, C., Aktop, O. and Turkekul, I., 2011. Screening of antioxidant, antimicrobial activities and chemical contents of edible mushrooms wildly grown in the Black Sea region of Turkey. *Combinatorial Chemistry & High Throughput Screening*, *14*(2), 72–84.

Peksen, A., Kibar, B. and Yakupoglu, G., 2013. Favourable culture conditions for mycelial growth of *Hydnum repandum*, a medicinal mushroom. *African Journal of Traditional, Complementary and Alternative Medicines*, *10*(6), 431–434.

Procházka, J., Brom, J., Šťastný, J. and Pecharová, E., 2011. The impact of vegetation cover on temperature and humidity properties in the reclaimed area of a brown coal dump. *International Journal of Mining, Reclamation and Environment*, *25*(4), 350–366.

Petersen, R.H., 1977. The typification of *Hydnum Linn.* per Fries: Time for stability. *Taxon*, 144–146.

Sevindik, M., 2020. Poisonous mushroom (nonedible) as an antioxidant source. *Plant Antioxidants and Health*, 1–25.

Stachowiak, B. and Reguła, J., 2012. Health-promoting potential of edible macromycetes under special consideration of polysaccharides: A review. *European Food Research and Technology*, *234*, 369–380.

Sterry, P., Hughes, B., 2009. *Collins Complete British Mushrooms and Toadstools: The Essential Photograph Guide to Britain's Fungi*. Publisher Collins, ISBN-10: 0007232241, ISBN-13: 978-0007232246.

Sugawara, R., Yamada, A., Kawai, M., Sotome, K., Maekawa, N., Nakagiri, A. and Endo, N., 2019. Establishment of monokaryotic and dikaryotic isolates of Hedgehog mushrooms (*Hydnum repandum* and related species) from basidiospores. *Mycoscience*, *60*(3), 201–209.

Sułkowska-Ziaja, K., Muszyńska, B. and Szewczyk, A., 2015. Antioxidant components of selected indigenous edible mushrooms of the obsolete order Aphyllophorales. *Revista Iberoamericana de Micología*, *32*(2), 99–102.

Swenie, R.A., Baroni, T.J. and Matheny, P.B., 2018. Six new species and reports of *Hydnum* (Cantharellales) from eastern North America. *MycoKeys*, *42*, 35–72.

Takahashi, A., Endo, T. and NozoE, S., 1992. Repandiol, a new cytotoxic diepoxide from the mushrooms *Hydnum repandum* and *H. repandum* var. album. *Chemical and Pharmaceutical Bulletin*, *40*(12), 3181–3184.

Thi, Q.N., Ueda, K., Kihara, J. and Ueno, M., 2015. Inhibition of *Magnaporthe oryzae* by culture filtrates of fungi isolated from wild mushrooms. *Advances in Microbiology*, *5*(10), 686.

Thu, Z.M., Myo, K.K., Aung, H.T., Clericuzio, M., Armijos, C. and Vidari, G., 2020. Bioactive phytochemical constituents of wild edible mushrooms from Southeast Asia. *Molecules*, *25*(8), 1972.

Tubic, J., Grujičić, D., Radovic-Jakovljevic, M., Ranković, B., Kosanic, M., Stanojkovic, T., Ciric, A. and Milošević-Đorđević, O., 2019. Investigation of biological activities and secondary metabolites of *Hydnum repandum* acetone extract. *Farmacia*, 67, 1.

Vallance, P. and Smart, T.G., 2006. The future of pharmacology. *British Journal of Pharmacology*, *147*(Suppl 1), S304.

Vasdekis, E.P., Karkabounas, A., Giannakopoulos, I., Savvas, D. and Lekka, M.E., 2018. Screening of mushrooms bioactivity: Piceatannol was identified as a bioactive ingredient in the order*Cantharellales*. *European Food Research and Technology*, *244*, 861–871.

Vázquez, E.L., García, F.P. and Canales, M.G., 2016. Major and trace minerals present in wild mushrooms. *American-Eurasian Journal of Agricultural Environmental Science*, *16*(6), 1145–1158.

Yamaç, M. and Bilgili, F., 2006. Antimicrobial activities of fruit bodies and/or mycelial cultures of some mushroom isolates. *Pharmaceutical Biology*, *44*(9), 660–667.

8 *Lactarius deliciosus* (L.)

Mandheer Kaur
Chandigarh College of Technology, Landran, Mohali, India

Renuka Sharma
Noida Institute of Engineering & Technology, Noida, India

Harjodh Singh, Palki Sahib Kaur, Vikas Menon, and Aditya Kumar
Chandigarh College of Technology, Landran, Mohali, India

Simranjot Kaur
Sri Guru Granth Sahib World University, Fatehgarh Sahib, India

CONTENTS

DOI: 10.1201/9781003259763-8

INTRODUCTION

Some lower group members of plants known as fungi produce fruiting bodies called mushrooms. Fungi are characterized by the absence of chlorophyll and undifferentiated bodies. They contain a large number of spores that behave similarly to seeds in higher plants for fungus propagation. Mushrooms, which sprout after rains appear in a variety of shapes, sizes and colors. The economic significance of mushrooms stem mostly from their usage as human food. The mushroom's unusual flavor, taste and fleshiness have made it a popular delicacy in human cuisine (Valverde et al., 2015). Mushrooms are a comprehensive, healthful cuisine that is excellent for people of all ages. Mushrooms include a lot of protein, vitamins, minerals and dietary carbohydrates fibres. The nutritional value varies depending on the kind, stage of development and other environmental factors. Because they have a low calorific value due to their low lipid, fat content and high polyunsaturated fatty acid content (PUFAs) (Léon-Guzmán et al., 1997). The protein level of mushrooms varies widely although it is usually high. Mushrooms are high in vitamins C and B (folic acid, riboflavin, niacin, and thiamine) as well as minerals potassium, phosphorus, and sodium. Other important minerals such as Cu, Zn, and Mg are present in trace amounts (Léon-Guzmán et al., 1997). Mushrooms are also recognized to have medical properties since they have been demonstrated to increase immune function, lessen cancer risk, inhibit tumour growth and help the body's detoxification system (Shahidi and Ambigaipalan, 2015). As a result, mushroom farming as a high-quality food has a lot of promise.

Mushrooms are classified as edible or poisonous, implying that they can be consumed or are harmful, toxic and unpleasant to humans. Because they are used by people, edible mushrooms are grown in controlled or natural environments. Due to their low-calorie content and plenty of bio-active chemicals, they are very beneficial to human health. Poisonous mushrooms are known to have chemicals that are harmful to humans if eaten. The majority of these poisonous chemicals have hepatotoxic, nephrotoxic, neurotoxic, and psychotic properties (Flament et al., 2020). Mushrooms have medical value because they contain bioactive substances. These mushrooms have been examined and evaluated as prospective sources of nutraceuticals, but also have potential inhibitors of reactive oxygen species (ROS) and tumour cells Mushrooms are also antimicrobial, working effectively against both (gram-positive and gram-negative bacteria) (Moo-Huchin et al., 2015; Mshandete et al., 2009). The practice of mushroom farming for cultivating medicinal mushrooms for commercial purposes has increased production, which has increased the use of bioactive substances for pharmacological and therapeutic purposes. Mushroom production worldwide and in Indian states are shown in Figure 8.1(a) and (b) depicting that China having maximum pruduction worldwide and Punjab producing miximu in mushroom in India. The nutritional advantages, geographical location (habitat), pharmacological significance of the bio-active chemicals and economic impact of *Lactarius deliciosus* will all be discussed in this chapter.

Although there are several natural molecules that have anti-inflammatory properties, phenolic compounds are the most potent of them all. Several mushroom extracts have been studied for their

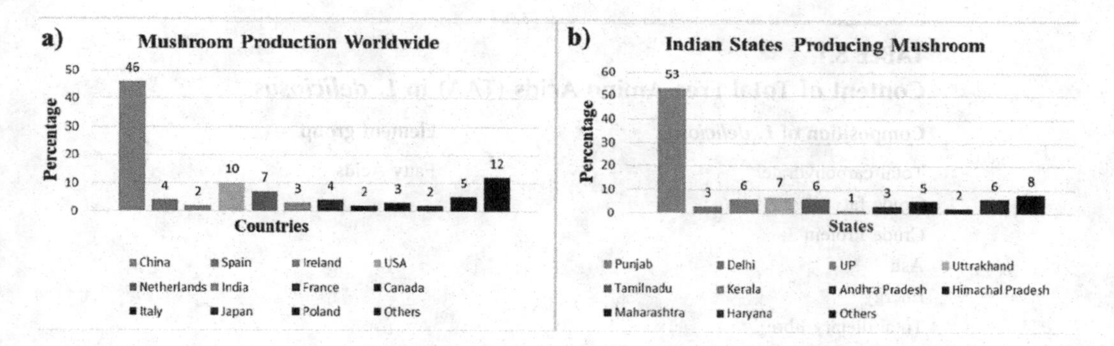

FIGURE 8.1 (a) Mushroom production in the world. (b) Major mushroom-producing states of India.

total phenolic content and anti-inflammatory properties (Fei Pei et al., 2014; Shahidi et al., 2015). Natural antioxidants (phenolic compounds) are a diverse collection of chemicals that come in a variety of shapes and sizes. The biosynthesis of these compounds is influenced by the chemical makeup of the components. Although antioxidants i.e., phenolic acids are the most common chemicals in mushrooms, their content may have nothing to do with the inhibitory potential of the extracts, implying that their role in anti-inflammatory activity is unimportant (Sharma et al., 2021). Several studies were conducted to determine the effect of such compounds on inflammatory effects, and these studies revealed that derivatives from an ethanolic plant extract, such as chlorogenic acid, caffeic acid, ferulic acid, and p-hydroxybenzoic acid, do not have a significant effect on the production of inflammatory agents (Kumar et al., 2019). Pyrogallol, a component of extracts from Bisporus, Cibarius, Cornucopioides, and *L. deliciosus*, may even be responsible for the anti-inflammatory properties, as these extracts have been the most efficient in terms of NO inhibition. Synergies among pyrogallol as well as other bioactive compounds in *L. deliciosus* extracts and other species of mushrooms, such as flavonoids, antioxidants or cinnamic acid and its derivatives, could explain some of the differences in activity observed in the extracts studied (Shahidi et al., 2015).

TAXONOMY AND CLASSIFICATION OF *L. DELICIOSUS*

L. deliciosus, commonly known as traditional mushroom, is a species of the large milk-cap fungi comes under in genus *Lactarius*; order Russulales, and Family Russulaceae. It is also known as the Saffron milk cap. The red pine mushroom is classified as an Agaricomycete in the Basidiomycota Division. It grows in the acidic soil, forming a mycorrhizal relationship with conifers under which they grow.

NUTRITIONAL COMPOSITION OF *L. DELICIOSUS*

Mushrooms are often designated in the category of important health food as they cater to the great proportions of protein, carbohydrate and fat. The composition of all the components found in the *L. deliciosus* is depicted in Table 8.1. It could be deduced that fresh *L. deliciosus* contained minimal dry matter content (8 percent) and was predominately moist in nature. Earlier studies also corroborated that fresh mushrooms were composed generally of dry weight content ranging between 5 and 15 percent (Çayır et al., 2010). Carbohydrate was the most abundant nutrient in the dried fruiting body, accounting for 66.61 g/100 g dw, followed by protein, ash, and fat. Furthermore, *L. deliciosus* has a total dietary fibre content of 31.81 g/100 g, indicating that eating this fungus is an excellent source of dietary fibre. Overall, *L. deliciosus* appears to be a promising meal that may meet the body's low-calorie needs the lipid profile of *L.delicious* revealed as oleic

TABLE 8.1
Content of Total Free Amino Acids (TAA) in *L. deliciosus*

Composition of *L. deliciosus*	Element group
Total carbohydrate	Fatty Acids
Crude fat	
Crude Protein	
Ash	
Energy	
Total dietary fibre	
Insoluble dietary fibre	
Soluble dietary fibre	
C16:0	
C18:0	
C18:1	
C18:2	
Magnesium	Metals
Calcium	
Zinc	
Manganese	
Iron	
Chromium	
Copper	
Arsenic	
Cadmium	
Aspartic acid	Free amino acids
Glutamic acid	
Asparagine	
Serine	
Glutamine	
Histidine	
Glycine	
Threonine	
Citrulline	
Arginine	
Alanine	
Tyrosine	
Valine	
Methionine	
Tryptophan	
Phenylalanine	
Isoleucine	
Leucine	
Lysine	
Proline	
Heptanal	Aroma volatile compounds
Benzaldehyde	
Hexanoic acid	
Octanal	

TABLE 8.1 *(Continued)*
Content of Total Free Amino Acids (TAA) in *L. deliciosus*

Composition of *L. deliciosus*	Element group
Benzyl alcohol	
2-Octenal	
Nonanal	
2-Nonenal	
Dodecane	
Decanal	
2-Decenal	
Undecanal	
2,4-Decadienal	
2-Undecenal	
2-Butyl-2-octenal	
n-Decanoic acid	
Decanoic acid, ethyl ester	
Tetradecane	
Dodecanal	
2-Dodecenal	
Tridecanal	
n-Hexadecanoic acid	
9,12-Octadecadienoic acid	
9-Octadecenoic acid	
Octadecanoic acid	

acid (C18:1) and linoleic acid are examples of unsaturated fatty acids (C18:2) had palmitic acid (C16:0, 5.17 percent) and stearic acid have a larger dominance than saturated fatty acids (C18:0, 16.96 percent). Other edible wild mushrooms have more unsaturated fatty acids (C18:1 and C18:2) than *L. deliciosus* fat (Barros et al., 2007a). Furthermore, a recent study found that ingesting unsaturated fatty acids can lower the risk of cardiovascular disease, cancer and type II diabetes (Orsavova et al., 2015; Calder et al., 2015). As a result, *L. deliciosus* can be considered a nutritious and healthful alternative for combating various disorders.

Because of its nutritional value and culinary value, edible mushrooms are recognised as a valuable food. Non-volatile flavour components such as protein flavour 5'-nucleotides, simple sugars, and acids, among others, contribute to mushroom flavour (Pei et al., 2014). Among the 20 amino acids studied in *L. deliciosus*, glutamic acid, glutamine, histidine, and alanine were identified in quite high concentrations, with a total protein (Total free amino acid) content of 3389.45 mg/100 g dw, as shown in Table 8.1. The total quantity of nine essential amino acids (EAA) in *L. deliciosus* was calculated to be 1026.29 mg/100 g dw, accounting for 30.28 percent of total free amino acids. As a result, *L. deliciosus* can be used as a protein-rich diet replacement. Food flavour is mostly determined by elements such as sweetness, sourness, bitterness, spice, astringency (Yin et al., 2019). The EUC values of mushroom are divided into four categories: >1000 g/100 g dw, 100–1000 g/100 g dw, 10–100 g/100 g dw, and 10–100 g/100 g dw. According to the foregoing characteristics, *L. deliciosus* is a meal that appeals to the senses on the second level.

Minerals have a significant role in maintaining certain physicochemical processes essential for life. *L. deliciosus* includes minerals such as calcium, magnesium, zinc, iron, copper, manganese, chromium, arsenic, and cadmium, according to previous research. Cofactors are indispensable in various biochemical pathways. Even the metal content in mushrooms is greater in comparison to green plants, which is because of the efficient mechanism of easy accumulation of metals from the ecosystem (Yin et al., 2019). As a result, wild edible mushrooms are a good alternative for dietary mineral requirements. Table 8.1 summarizes the mineral content of *L. deliciosus*. The Mg and Ca contents of *L. deliciosus* were 1244.29 and 247.07 mg/kg dw, respectively, according to the findings, which were confirmed by a previous study (Kała et al., 2019), which discovered that calcium and magnesium content were in the same intrinsic range based on the data collected including over 1000 samples of 400 mushroom species. Many studies revealed that Iron content was notably high among all trace elements while copper was lowest as reported in numerous studies (Çayır et al., 2010). In general, 300 g of fresh mushrooms per day is recommended, with 30 g of dry matter (Kalac and Svoboda, 2000). When compared to the Institute of Health and Sciences RDA (recommended dietary allowance)and AI (adequate intake) for females and males (ages 19 to 30), *L. deliciosus* consumption gives a good contribution of magnesium, zinc, manganese, iron, and chromium while being insufficient in providing enough calcium and copper supplementation (Mendil et al., 2004) It was discovered that chromium was present in high amounts in *L. deliciosus*, with a high daily intake value of chromium (up to 344.57 percent for males and 482.40 percent for females of AI percent), but it poses no hazard to humans body when compared to the Institute of Medicine's tolerable upper intake level. *L. deliciosus* was also discovered to contain arsenic, cadmium, and plumbum, all of which are toxic. According to EU scientific committee guidelines, the provisional permissible daily intakes for arsenic, cadmium, and plumbum for adults according to 60 kg of body weight were 0.13, 0.06, and 0.21 mg, respectively. As a result, ingesting harmful metals such as arsenic, cadmium, and plumbum from *L. deliciosus* poses no risk to consumers.

PHYTOCHEMISTRY

Phytochemicals are natural products with a wide range of taxonomic classifications that can be used to develop novel medications based on their pharmacological actions. Glycosides, steroids, alkaloids, saponins, flavonoids, terpenes, tannins and other secondary metabolites are examples (Gomes et al., 2019). These bioactive chemicals are found in plants and operate in conjunction with nutrients and fibers to combat disease. As seen in Figure 8.2, *L. deliciosus* is a plethora of phytochemicals. These data show the *L. deliciosus* phytochemical potential. Even these bioactive substances are responsible for the *L. deliciosus* antioxidant and antibacterial activities.

POLYOLS COMPOUNDS

L. deliciosus fruiting bodies produced two new polyols, methylcyclohexane1, 2,4-triol, and 3-hydroxymethyl-2-methylenepentane-1,4-diol, but also an additional phenylpropanoid glycoside, eugenyl 4″-*O*-acetyl—rutinoside, with seven recognised steroidal hormones. The compositions of these chemicals substances were found through the analysis of spectroscopic data (Mshandete et al., 2009) and these compounds showed basidiomycete activity.

VOLATILE AROMA COMPOUNDS

Although water-soluble non-volatile chemicals are responsible for mushroom flavour, the role of volatile aroma compounds cannot be neglected because they have a direct impact on consumer acceptance. Unsaturated fatty acid oxidation through chemical or enzymatic means, followed through associations with polypeptides, peptides, and free amino acids produced the majority of volatile fragrance molecules in mushrooms The volatile fragrance components of *L. deliciosus* were estimated using gas-solid phase micro-extraction GS-MS integrating library catalogue. Based on qualitative and quantitative tests, acids were shown to be the most essential aromatic volatile

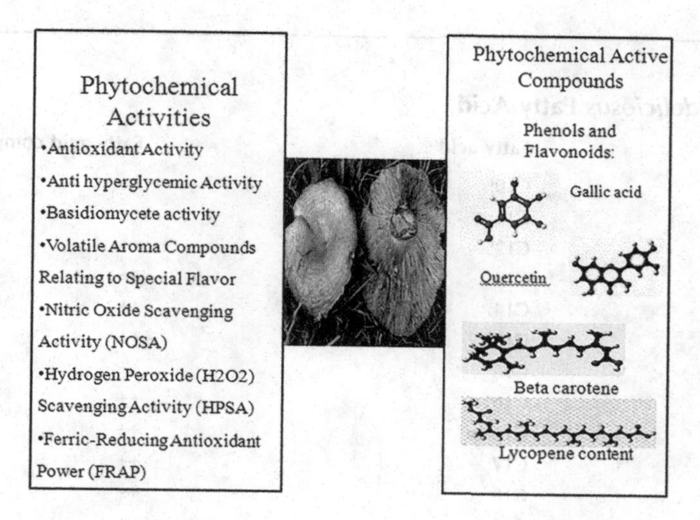

FIGURE 8.2 Phytochemical activities and phytochemical active compounds of *L. deliciosus*.

compounds, accounting for 84.23 percent of total chromatographic area, followed by aldehydes at 14.77 percent. Acids predominated in the aroma volatile components of *L. deliciosus*; this was consistent with a glarrmak's prior research on *L. edodes* and *Pleurotus sajor-caju*. Other research has found that types of alcohol are important aromatic molecules in mushrooms. This might be related to the evaporation, since Tian observed that drying caused a significant drop in alcohols while increasing acids and aldehydes in *L. edodes*. Furthermore, the volatile aroma components of mushrooms were altered by growing circumstances and genetic variations.

Sesqui-terpenoids From the Fruiting Bodies of *L. deliciosus*

Using 1 H NMR-guided extraction of extracts out from edible fungus *L. deliciosus*, researchers discovered two novel azulene-type sesquiterpenoids, 7-isopropenyl-4-methyl-azulene-1-carboxylic acid (Hamid et al., 2017) and 15-hydroxy-3,6-dihydrolactarazulene, as well as seven known compounds. Spectroscopic evidence was used to identify their structures. The 13C NMR of 15-hydroxy-6, 7-dihydrolactarazulene is observed for the first time among the known metabolites. In addition, 7-acetyl-4-methylazulene-1-carbaldehyde demonstrated modest anti-bacterial impact on Staphylococcus aureus (Shahidi and Ambigaipalan et al., 2015).

CLASSES OF COMPOUNDS

L. deliciosus contains a variety of chemicals that contribute to the mushroom's various characteristics. Table 8.2 shows the composition of fatty acid in percentage of *L. deliciosus* from Lesvos Island. At various intervals after grinding, compounds such as oflactarviolin and deterrol, both sesquiterpenoids, were discovered in the hexane extracts of ground fruit-bodies of *L. deliciosus*. Different sterols, phenolic compounds, flavonoids, and terpenoids that contribute to *L. deliciosus* antioxidant potential were tested in substantial amounts.

Orange Pigments

A new study conducted by Feussi et al. (2017) more molecules were discovered to be inseparable orange pigments that decomposed swiftly. Pseudo-molecular ion peaks were discovered using ESI MS, all of which are consistent with the structural formula $C_{15}H_{18}O$. The 1H and 13C NMR

TABLE 8.2
***Lactarius deliciosus* Fatty Acid**

S. No.	Fatty acid	Fatty acid composition %
1	C10:	++
2	C11:	++
3	C12:	++
4	C13:	++
5	C14:	++
6	C15:	++
7	C16	++
8	C161n	++
9	C16:1n	++
10	C17:	++
11	C18	++
12	C18:1trans	++
13	C18:1n	++
14	C18:2n	++
15	C18:3n	++
16	C20:	++
17	CLA	++
18	C20:	++
19	C21:0 nd	++
20	C20:2n	++
21	C20:3n	++
22	C20:4n6 + C22	++
23	C20:5n	++
24	C23:	++
25	C22:2n6	++
26	C24:	++
27	SFA	++
28	MUFA	++
29	PUFA	++
30	n6/n3	++
31	Unsaturation index	++

spectra, on the other hand, indicated that it was a combination of two isomeric molecules 2 and 3 in 1:3 ratios (Figure 8.3).

Total Flavonoids

The extracts' flavonoid concentration of extract was determined using a colorimetric approach modified from. 250 L of mushroom extract were mixed with 1.25 L of deionized water and 75 litres of a 5 percent $NaNO_2$ solution. 150 L of a 10 percent $AlCl_3 \cdot H_2O$ solution will be added after 5 minutes. After 6 minutes, 500 L of 1 M NaOH and 275 L of distilled water were added to the mixture. After completely mixing the solution, the pink colour intensity was measured at 510 nm. The standard calibration curve shows (0.022– 0.34 mM; Y, 0.9629X – 0.0002; R2, 0.9999) was

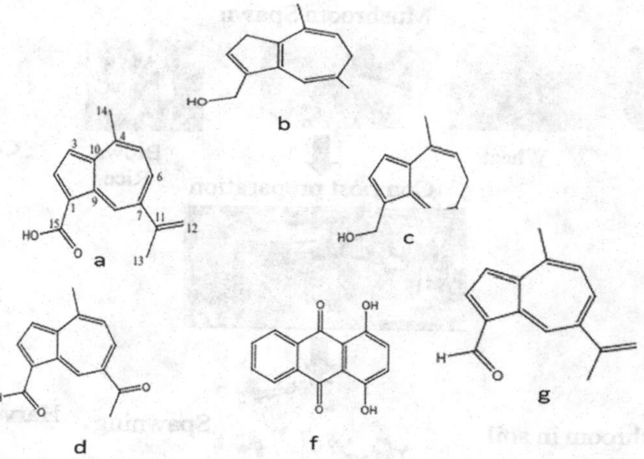

FIGURE 8.3 Different pigments obtained from *Lactarius deliciosus* fruiting bodies.

calculated using (+)-catechin, and the findings were given in milligrams and (+)-chatequin equivalents (CEs) per grams of extract.

CULTIVATION TECHNIQUES

Mushroom cultivation is a technique for cultivating mushrooms from waste from plants, animals, and industry. In a nutshell, it's riches created by repurposing trash technology. Because of the usefulness of dietary fibres and proteins, this method has gained popularity around the world. Mushrooms are fungus that belongs to the basidiomycetes family. Proteins, fibre, vitamins, and minerals abound. There are around 3000 different species of mushrooms. Oyster mushrooms such as (*Pleurotus sp.*), Button mushrooms such as (*Agaricus bisporus*) and paddy straw mushrooms are just a few examples (*Volvariella volvacea*). It takes one to three months to cultivate a plant. The steps of mushroom cultivation are described in detail here (Figure 8.4) (Sharma et al., 2017; Sharma et al., 2020).

COMPOSTING

In the case of button mushrooms, it refers to a material made by blending straw, chicken manure, gypsum, and other ingredients in a specified proportion and fermenting it under aerobic circumstances. Paddy straw is mixed with a variety of organic ingredients, such as cow manure, and inorganic fertilizers to make compost. For one week, it is kept at around 50°C.

The majority of the substrate on which mushrooms live is made up of Plant wastes (cereal straw/ sugarcane bagasse, for example), salts (urea, superphosphate/gypsum, for example), supplements (rice bran/ wheat bran), and water make up the majority of the substrate on which mushrooms grow (Figure 8.4). 220 g of dry base materials are required to generate 1 kilogramme of mushroom. On a dry weight basis, each ton of compost should include 6.6 kg nitrogen, 2.0 kg phosphate, and 5.0 kg potassium (N:P: K- 33: 10:25) enterprises/mushroom-production 2/3, equal to 1.98 percent nitrogen, 0.62 percent phosphate, and 1.5 percent potassium.

SPAWNING

To grasp the concepts of spawn, spawning, and spawn growth, one must first grasp the concept of the mushroom. Mushrooms are made up of two primary parts: the cap and the stem, and they are the

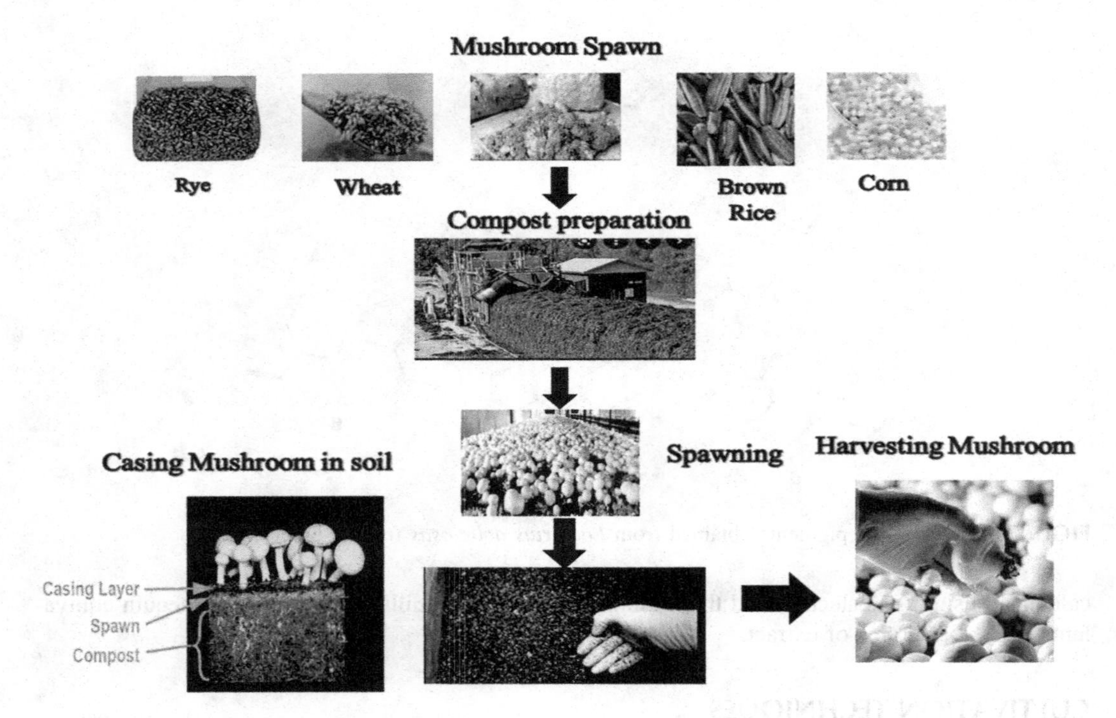

FIGURE 8.4 Mushroom cultivation process.

fruits of the *A. bisporus* fungus. As the mushroom matures, the cap opens, displaying the gills. Spores are formed in the gills of mushrooms. Spores are little spheres around the size of seeds that are found in larger plants. These spores are produced in large quantities by the gills. Each hour, a single 8-cm mushroom can produce up to 40 million spores. Because spores germinate and grow into mycelium in an unpredictable manner, they are not utilised to 'seed' mushroom compost. In laboratories, spores germinate and develop into thread-like mycelium, which is employed in the creation of commercial spawn. The mushroom seed is known as spawn. It's made by inoculating grains with fungal mycelium and growing it under sterile circumstances (Patel, 2014; Sharma et al., 2017).

Mycelium from a mycelial culture is placed onto steam-sterilized grain in the spawn-production technique, and the mycelium grows entirely through the grain over time. Sponge is a grain-and-mycelium mixture that is used to "seed" mushroom compost. Mycelium from a prolonged culture, as opposed to mycelium from a spore, is used to make the majority of spawn. Because each spore is likely to result in the emergence of a new strain with unknown consequences (Sharma et al., 2017).

Casing

The purpose of casing soil is to retain moisture content and pollutant exchange inside the top layer of compost, which aids in the proper development of the mycelium. This casing soil's pH should be 7.5–7.8, and it must be disease-free. Maintain a consistent temperature and humidity level in the mushroom grow chamber. The casing soil can be prepared with a 4 percent formalin solution and layered on the cemented ground. The land has been rotated and will be covered with polythene covering for another 3–4 days. Pasteurization of shell soil for 6–8 hours at 65°C has been demonstrated to be far more effective (Sharma et al., 2017; Patel, 2014)

Once the surface of the compost had been well cleaned, a 3–4 cm thick coating of casing soil was applied (Figure 8.4). A thin layer of soil is applied to the compost. It supports the mushroom's growth, provides humidity, and aids with temperature regulation. (Osdaghi et al., 2019)

Pinning

Mycelium starts to form a little bud, which will eventually become a mushroom. Those small white buds are known as pins. Initials are quite minute; however, they can be seen as rhizomorph outgrowths. Once the structure has quadrupled in size, it is referred to be a "pin". Pins continue to spread and get larger during the button stage, finally converting into a mushroom. Harvestable mushrooms appear after 16 to 18 days. By pumping new air into the growth chamber, the carbon dioxide content in the room air is lowered to 800 ppm or below (depending on the cultivar). The timing of fresh air intake is critical and can only be acquired via experience (Sharma et al., 2017; Sharma et al., 2020;).

Harvesting

Mushrooms thrive at temperatures between 15 and 23°C. They grow 3 cm in a week, which is the standard harvesting size. The first flush mushroom can be harvested in the third week. The cap should be gently twisted off while harvesting. It should be gently grasped with the forefingers, pressed against the dirt, and then twisted off. Chopping off the base of the stalk where mycelial threads and dirt particles cling should be done (Sharma et al., 2017; Sharma et al., 2020; Patel, 2014).

TECHNIQUES FOR ISOLATING PURE CULTURES AND THEIR MAINTENANCE

The spore culture and tissue culture techniques are the two ways to cultivate mushrooms. Mushroom cultures can be obtained from either spores or tissue. Many strains emerge from the germinating spores, some of which are compatible with one another and others which are not. The cultivator preserves the exact genetic character of the contributing mushroom when obtaining a tissue culture (clone) from a living mushroom (Tudses et al., 2016).

Spore Culture

Mushroom cultures can be obtained from either spores or tissue. Many strains emerge from the germinating spores, some of which are compatible with one another and others which are not. The cultivator preserves the exact genetic character of the contributing mushroom when obtaining a tissue culture (clone) from a living mushroom as shown in Figure 8.5. With spores, a single strain must be chosen from among the many that are generated. In both circumstances, the outcome is a network of cells known as the mushroom mycelium (Tudses et al., 2016).

Spore Print

To obtain a spore print or collection of spores, the cap of a disease-free, healthy mushroom was removed; the surface was cleansed with alcohol-soaked cotton swabs, and the cap was placed on

FIGURE 8.5 Spore culture method of mushroom cultivation.

clean sterilized white paper, clean glass plate, or clean glass slides. It's important to clean the environment around your workstation. To prevent airflow, a clean glass jar is placed on the cap surface. After 24–48 hours, spores will fall on the white paper or slide surface, following the radial symmetry of the gills. The spore print can be preserved for a longer amount of time by cutting and folding the paper in half.

Spore Transfer and Germination

Scrape a few spores from the spore print on a paper or glass slide, then streak them aseptically on the agar media to form a pure culture. Clean the scalpel by holding it over a blazing flame for 8–10 seconds until it turns bright red, cool it by dipping it in a sterile medium, scrape a few spores from the spore print on a paper or glass slide, then streak them aseptically on the agar. For each spore print, at least three agar dishes should be infected, and multispore culture is the culture formed following incubation at the appropriate temperature (Kim et al., 2005).

Tissue Culture

A small piece of pileus is cut with a sterile blade or scalpel, cleaned several times in sterilized distilled water, and dried in sterile tissue paper before being inoculated aseptically on a Petri plate or tube with adequate culture material. While the inoculated Petri plates are incubated at 25°C for 6–12 days, mycelial growth is measured at various periods. Only keep the Petri plates or glass tubes with pure growth for future use, and discard any that are contaminated (Tudses et al., 2016)

Sub-culturing

At a low temperature, the pure mushroom culture is retained. Sub-culturing is done by aseptically transferring a small portion of the growing pure culture, as well as the culture media, to test tube slants containing the same or another suitable medium (Tudses et al., 2016)

PHARMACOLOGY

Edible mushrooms are used for both nutritional values and therapeutic benefits. They include a wide variety of fungus species, both wild-harvested and farmed. Depending on the growing climate and harvesting periods, the nutritional value may vary. Medicinal mushroom extracts and essences have been utilized as an alternative medicine in countries like China, Korea, Japan, and Eastern Russia for decades. There is also evidence that certain mushroom species contain compounds that may prevent or improve the immune system, heart disease, diabetes, and viral infections. Bioactive chemicals in mushrooms have been discovered to have a broad connection with the mycelial cell wall, enhancing immunity to carcinogens. Various bioactive compounds with anti-cancer and immunomodulatory properties are increasingly being employed to enhance immune system to function in cancerous patients during radiotherapy and chemotherapy, thereby extending survival time in few types of cancer. With bactericidal, insecticidal, nematicidal, fungicidal and herbicidal actions, mushrooms have also proved their potential as natural biocontrol agents in plant protection (Sharma et al., 2017; Sharma et al., 2020). Many investigations have found *L. deliciosus* to have promising pharmacological action. *L. deliciosus* fungus, which was collected from several locations, was found to have excellent antioxidant and antibacterial properties in one study. This substantial amount of activity is related to the highly presence of phenolic compounds in *L.deliciosus* mushroom, which gives this mushroom a great therapeutic value. This is because the *L. deliciosus* fungus contains both water-soluble phenolic compounds and methanol-soluble lipophilic carotenoids. The antioxidant characteristics of many therapeutic fungi were investigated, and among all the fungi studied, *L. deliciosus* fungi demonstrated substantial antioxidant and

TABLE 8.3
***Lactarius deliciosus*: Bioactive Components and Their Medicinal Effects**

S. No	Medical effects	Bioactive compounds	Diseases	References
1	Antioxidant	Polyphenol and flavonoids are present in fruiting bodies of edible mushrooms, phenolic compounds like protocatechoic,and p-hydroxybenzoic in fruiting bodies of L.volemus.	Reduce the harm of cancer, atherosclerosis, Cardiovascular and other causes of diseases.	Reis et al., 2012
2	Antimicrobial	Effects by phenolic compounds,purines, pyrimidins, quinones, terpenoides and phenyleproponoid, derived sntagonisticsubstanced present in L.delicious.	Bacterial disease	Sagar et al., 2012
3	Antitumor	caused by calvin, volvotoxin, flammutoxin, lentinan and poricin	Oncogenesis metastasis inhibition tumor cell proliferation	Sadi et al., 2016
4	Anticancer	Lectins binds to cancerous cell membrane and Receptors causing cyto-toxicity and apoptosis	Oncogenesis, leukemia	Kosanić et al., 2016; Sadi et al., 2016; Hou et al., 2019.
5	Antihyperglycin	In *L. delicious* alpha amylase and alpha glucosidase key enzyme that catalyze polysaccharides and hydrolysis to enhance blood glucose level.	Blood glucose level	Liu, 2007; Xu et al., 2019
6	Antihypertensive	Have ability to inhibit the angiotensin converting enzyme was reported in edible mushroom	High blood pressure	Ayvaz et al., 2019
7	Antifungal	Alkaloids and necatorin and necotoroun also present in L.necator and L. delicious	Fungal diseases	Sagar et al., 2012
8	Immuno-suppressive	Compounds like flavidulus A present in fruiting bodies of *L. flavidulus*.	inhibits lymphocytes proliferation in the spleen of mice caused by mutagenic factors	Santoyo et al., 2009; Hou et al., 2019.

antibacterial activity of high medical importance (As shown in Table 8.3) (Avci et al., 2019). Gram-positive bacteria were suppressed more successfully than Gram-negative bacteria by phenolic extracts of edible wild fungus species from Portugal (*L. deliciosus*, *Tricholoma portentosum,* and *Sarcodon imbricatus*) (Barros et al., 2007). The IC50 of *L. deliciosus* against HeLa, A549, and LS174 cell lines was found to be 19.01, 33.05, and 74.01 mg/mL, respectively, showing that it had a greater effect on human epithelial cancers (HeLa) cells than on human lung cancer (A549) and human colon cancer (LS174) (Kosanić et al., 2016). This was the first investigation of *L. deliciosus* crude extract's anticancer potential against cell lines, and it paved the door for additional research into this fungus's antitumor activity against a broader spectrum of cell lines. Heteropolysaccharides produced from *L. deliciosus* displayed significant antitumor action in vivo. Anticancer chemicals can serve as an inducer of reactive oxygen species, angiogenesis inhibitors, mitotic kinase inhibitors, topoisomerase inhibitors, antimitotic agents, and inducers of apoptosis, therefore slowing cancer growth (Patel et al., 2012).

The antibacterial, antioxidant, and chemical content of *L. deliciosus* gathered in Turkey's Kastamonu Province were studied, and it was determined that *L. deliciosus* exhibited the greatest inhibitory action against *P. aeruginosa*, with a 3000 mm inhibition zone (Alves et al., 2012). Some studies have also indicated anticancer efficacy of polysaccharides derived from this fungus. A new polysaccharide was isolated from *L. deliciosus* grey, and its structure, characteristics, and anticancer activities were studied. Mice with the S180 tumour were administered the polysaccharide designated as LDG-A to investigate its anticancer effects. According to the findings, LDG-A has the ability to prevent tumour development in a dose-dependent way. The maximum dose of LDG-A tested in mice was 80 mg/kg, with a rate of inhibition of 68.422 percent, indicating the highest rate of inhibition. The isolated polysaccharide's anticancer effect was thought to be due to the activation of the host's cell-mediated immune response (Flament et al., 2020). Table 8.3 summarises the antitumor efficacy of the polysaccharide extracted from *L. deliciosus* by Ding et al. The findings clearly demonstrated that the difference in average liver weight between the test groups was minor, indicating that LDG-A from *L. deliciosus* did not cause substantial liver damage and has the potential to be a promising anticancer agent with a good chemotherapeutic index.

L. deliciosus ethanol and water extracts were used in the α-amylase and α-glucosidase inhibitory studies. The inhibitory activity of ethanol extracts on α-amylase and α-glucosidase were found to be dose-dependent, but aqueous extracts had no inhibitory activity on α-amylase and α-glucosidase. Ethanol extracts reduced α-amylase and α-glucosidase by 29.53 and 52.36 percent, respectively, at 5.0 mg/mL, suggesting that *L. deliciosus* ethanol extracts might be employed as efficient and nontoxic inhibitors of α-amylase and α-glucosidase, which are significant in the treatment of hyperglycemia (Xia et al., 2017).

MYCOLOGY

Fungi are found in 1.5 to 5 million species all over the world. There are 20,000 different types of fruiting mushrooms, but only 200 of them are grown. There are roughly 20 edible mushroom species suitable for large-scale production. When it comes to mushroom-producing fungi, there is lots of potential for discovery in both taxonomy and cultivation (Ergönül et al., 2013).

Fungi are all mushrooms, but not all mushrooms are fungi! The mushroom is a type of fungus that produces fruiting bodies that can be harvested by hand. Mushrooms require a variety of substrates to develop and can be found in a variety of ecological niches. Mushrooms come in a wide variety of colours, forms and morphological traits. Mushrooms have a short life period, making field research challenges. A mushroom is just one part of a larger fungus' life cycle (see Figure 8.6).

Mycelium is the main body portion, which develops in a variety of directions as it looks for food to digest. All fungi require enzymes produced by mycelium to break down their food. Mycelium is particularly sensitive to moisture loss since it only has one cell wall. As a result, in the wild, you must rollover logs or dig through leaves with a greater moisture content to find mycelium.

Spores, like seeds, are the reproductive components of the organism. Small packets of genetic material are disseminated by insects, rain, and wind in the hopes of finding a new food source. Mushrooms release tens of thousands of spores each time they reproduce (Mshandete et al., 2009).

LIFE CYCLE AND GROWTH OF *L. DELICIOSUS*

L. deliciosus, commonly called as the red pine mushroom or saffron milk cap, belongs to the Lactarius genus and the Russulales order. it is one of the well-known member of the huge milk cap mushroom family, which is native to Europe but has been accidentally introduced to other countries (Léon-Guzmán et al., 1997).

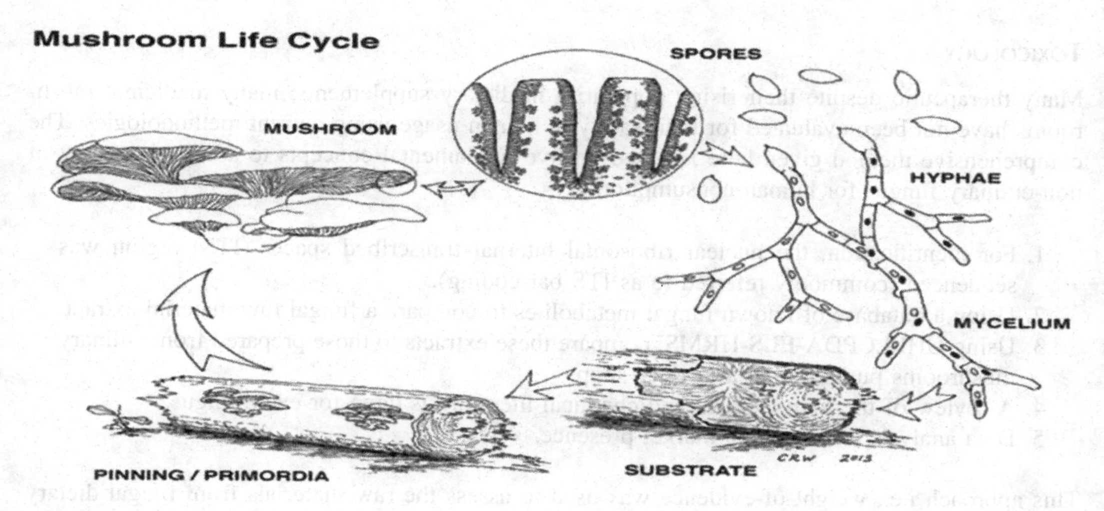

FIGURE 8.6 Life cycle of mushroom.

Following are the growth features of *L. deliciosus*

- On soil in grass or mixed forests, it grows in a scattered to gregarious manner.
- Cap – 4–7 cm in diameter, generally convex with a recessed centre to a shallow tube, bright light orange with a concentric zonal arrangement, incurved to rising border with ageing, slightly sticky touch.
- Gills – Medium orange, sub-distant, medium broad, adnate to slight decurrent.
- Flesh - Medium-thick, white flesh that immediately turns orange due to the orange latex.
- Stalk - central, cylindrical, somewhat mottled, pithy to hollow, tapers downward, 3.5 to 5 cm by 1 cm Pale yellow to buff spore print.
- Spores are around 8–12 × 6.5–10 microns in size, almost spherical, reticulate, and amyloid (Dospatliev, 2017).

L. deliciosus grows in pine woodlands and other similar places where the conditions are good. *L. deliciosus*' crown is carrot orange and convex to vase-shaped. When young, it is in-rolled, with deeper orange lines arranged in concentric rings with a diameter of 4–14 cm (1.5–5.5 inches). The cap is slippery and sticky when wet, but it is normally dry. The gills are densely packed and occur, with a squat hollow orange stripe that measures 3–8 cm (1–3 inches) in length and 1–2 cm (0.5–1 inch) in thickness. While touched, *L. deliciosus* stains a dark green hue; nevertheless, when fresh, the mushroom releases an orange-red "milk" that does not change colour. *L. deliciosus* is a mycorrhizal fungus with hymenium gills. The stipe is bare, the hymenium is decurrent, the spore print is tan, and the biology is mycorrhizal.

It has a carrot-orange head with a smooth, slightly viscid surface that frequently features darker markings and concentric zones. The top and gills can turn greenish with age or damage. The pileus is convex when it is young, with an in-rolling edge and a somewhat depressed centre. It features a squat, hollow orange stripe and packed slightly decurrently orange gills. It has spread to other areas under conifers.

In Girona, this fungus is known as a "pinatell" since it is picked from nearly to natural pine trees; it is often harvested in October following the late-August rain. It demands a premium price due to its scarcity. Anofinic acid, chroman-4-one, 3-hydroxy-acetyl indole, ergosterol, and cyclic dipeptides are produced when the mycelium of this fungus is cultivated in liquid culture (Dospatliev et al., 2017).

Toxicology

Many therapeutic despite their rising popularity in dietary supplements, many medicinal mushrooms have not been evaluated for their safety in human usage using current methodologies. The comprehensive method given here is based on five fundamental concepts to assess the safety of non-culinary fungus for human consumption:

1. For identification, the nuclear ribosomal internal transcribed spacer (ITS) region was sequenced (commonly referred to as ITS bar-coding).
2. Using a database of known fungal metabolites to compare a fungal raw material extract.
3. Using UHPLCPDA-ELS-HRMS, compare these extracts to those prepared from culinary mushrooms purchased at a grocery shop.
4. A review of the toxicological and chemical literature is done for each fungus.
5. Data analysis to determine market presence.

This approach i.e., weight-of-evidence was used to assess the raw materials from fungal dietary components and establish whether or not they were safe for human consumption. To maintain the safety assessment of fungal dietary components, such a method could be a viable alternative to traditional toxicological animal research.

L. deliciosus has a huge number of positive effects in the literature. There have been some investigations on the toxicology of this edible fungus. Regional differences in mushroom preparation practices might lead to disparities in the classification of species as edible or inedible. The well-known milk caps (*Lactarius* spp.) are an example of this. Most milk caps are recommended for ingestion after boiling in Finnish mushroom atlases (Alves et al., 2012). *L. deliciosus* extraxt and compounds were tested for phytotoxicity referred to HydE using Brassica oleracea seeds, but also cytotoxic analysis of this extract on cells of mouse BALB/c monocyte macrophage cell line (J774A.1 cell line) (ATCC TIB-67), and it was associated with improvements for fungal control with no apparent negative effects on mouse BALB/c monocyte-macrophage cell line viability (Nowakowski et al., 2021).

APPLICATION OF *L. DELICIOUS* IN VARIOUS AREAS

L. deliciosus mushroom in numerous studies have been found to be an important medicinal mushroom as well as being an edible mushroom. The edible wild mushrooms are economically valuable all over the world.

Mushroom as Antimicrobial Agent

Sesquiterpenoids, some of which have antimicrobial and anticancer properties, have been found to contribute to the unique colour of *L. deliciosus*. (Kosanić et al., 2016).

Mushroom as an Antitumor Agent

According to Ding et al. and Hou et al., *L. deliciosus* polysaccharides have considerable antitumour action in mice in vivo and immunomodulatory effects in vitro via proliferative proliferation of both B cells and macrophages.

Mushroom as an Anti-inflammatory Agent

Other *L. deliciosus* preparations contain biological activity such as enzyme inhibition, antioxidant, antibacterial, and anti-inflammatory properties.

Mushroom as Antioxidant

Oxidation is necessary in most of the living organisms as it is required for energy production and fueling various biological processes. Sometimes uninhibited production of free radicals containing oxygen-derivates may result in the commencement of numerous diseases like rheumatoid arthritis, cancer and atherosclerosis also degenerative processes like aging (Moo-Huchin et al., 2015). Antioxidants in food or food extracts have long been known to protect cells from free radical damage. DNA mutation, protein degradation, lipid peroxidation, and changes in low-density lipoproteins are all caused by free radicals. Free radicals have also been associated with diabetes, cancer, neurological, and cardiovascular illnesses (Moo-Huchin et al., 2015). However, complex components with many action mechanisms impact the antioxidant potential of food (Ziegler et al., 2015). As a result, Different evaluation methodologies are used concurrently when trying to evaluate the antioxidant properties of food. The antioxidant properties of alcoholic and aqueous extracts of *L. deliciosus* was determined using DPPH and ABTS radical scavenging and ferric-Reducing antioxidant power (FRAP) tests, with aqueous extracts having 2–3 times the antioxidant capacity of ethanol extracts. Trolox equivalent antioxidant capacity measurements were used to express antioxidant capability. The DPPH, ABTS, and FRAP assays were used to evaluate the TEAC values of alcoholic and aqueous extracts. This might be attributable to the fact that the aqueous extract had 3.01 times more phenols than the ethanol extract. Other observations confirmed that *L. deliciosus* extracts had a high antioxidant capacity, implying that it contained a considerable number of antioxidants. (Fereidoon et al., 2015)

Anti-hyperglycemic Activity

α-Amylase and α-glucosidase are digestive enzymes that catalyze carbohydrate breakdown in order to raise blood glucose levels. Hyperglycemia can be treated by blocking carbohydrate-hydrolyzing enzymes, which slows glucose absorption (Bhandari et al., 2008) Effective and nontoxic α-amylase and α-glucosidase inhibitors are required for the treatment of hyperglycemia. Both α-amylase and α-glucosidase inhibitory experiments showed a dose-dependent rise in ethanol extracts of *L. deliciosus,* however, the anti-enzyme activity of ethanol extract was less than that of acarbose. At 5.0 mg/mL, ethanol extracts inhibited α-amylase and α-glucosidase by 29.53 and 52.36 percent, respectively. Aqueous extracts, on the other hand α-amylase and α–glucosidase were not inhibited.

Nitric Oxide Scavenging Activity (NOSA)

Nitric oxide (NO) is a dangerous free radical that is created by nitric-oxide synthase from L-arginine. Hence, preventing excess nitric oxide production is critical. In a previous publication, the NOSA (percent) of Methanolic extracts of *Pleurotus squarrosulus* and farmed *Pleurotus florida* in India were 80 and 81.8 at 1 mg/ml, respectively. The outcomes of this study appeared to be lower in India than *Pleurotus squarrosulus* and farmed *Pleurotus florida* (Bozdoğan et al., 2018).

Hydrogen Peroxide (H_2O_2) Scavenging Activity (HPSA)

Hydrogen peroxide is a moderately stable non-radical oxidant present in tissues during oxidative processes. Hydrogen peroxide (H_2O_2) is produced by plants in their cytosol, cytoplasmic membrane, and extracellular matrix. The electron transport chain associated with the endoplasmic reticulum is the primary generator of H_2O_2 in the cytoplasm. The accumulation of H_2O_2 in plant tissues has been employed as a signalling between cells. It also enhances stress-related proteins such as alternative oxidase catalase, peroxidase, and many genes. H_2O_2 was discovered to induce cancer in any tissue subjected to oxidative stress (Bozdoğan et al., 2018).

Ferric-reducing Antioxidant Power (FRAP)

Ethanolic extracts of *L. deliciosus* (Bolu Turkey) were found to be 0.229 at 250 g/ml and 0.590 at 500 g/ml in Turkey. At 20.0 mg/mL, the FRAP activity of *P. ostreatus* aqueous and MeOH extracts in Thailand (Teekachunhatean, 1997) was 4.38 and 1.61, respectively. Tomato phenolics, lycopene, and carotenoids antioxidant activity and concentration were also shown to be substantially associated. The antioxidant capabilities of *L. deliciosus* and *P. ostreatus* collected from their native habitats in the Amanos Mountains were substantial; nevertheless, additional study is required to isolate and identify the particular compounds that cause the antioxidant properties of the studied species (Xia et al., 2017).

MEDICINAL USES (ANCIENT AND MODERN USES) OF MUSHROOM

For thousands of years, mushrooms are often used in traditional medicine to maintain good health and as therapeutics to treat ailments in practically every corner of the world (as shown in Figure 8.7). Mushrooms have a substantial number of nutrients such as vitamins, proteins, and minerals yet are low in saturated fats, simple carbs, and cholesterol, making them a great diet for diabetics. (De et al., 2012; Lo et al., 2011*). L. deterrimus* and numerous other Lactarius species, including *L. deliciosus, L. salmonicolor, and L. sanguifluus,* have antioxidant, antibacterial, anticancer, and immune-stimulatory qualities in addition to its antidiabetic actions (Hou et al., 2019). Fruit organs of mushrooms of the genus Lactarius (Russulaceae, Basidiomycota) lead to the formation of a juice milky in appearance. In most of the Lactarius species, diverse types of sesquiterpenes play a significant biological role which is responsible for sharpness and bitterness of milky water and for the formation of a defense system against predators. *L. deliciosus* fungus, known as Çıntar mushroom, is widespread in Turkey. *L. deliciosus* is an ectomycorrhizal fungus that grows in coniferous woodlands. *L. deliciosus* mushrooms are high in nutritional value and biological activity (Liu et al., 2007). In this study, biological activities of *L. deliciosus* mushrooms, which are collected intensively from nature and consumed extensively, are reviewed.

After gathering, fresh saffron milk cap is often used in a variety of cuisines, mostly in the fall and winter seasons. Pickling, canning, and drying procedures are used by locals to enjoy this fungus throughout the year. Saffron milk caps are eaten in the form of baked meatballs, scrambled eggs, and pies. It may also be found in rice pilaf and soups. In the Aegean area, *L. deliciosus* is also commonly used as food. Table 8.3 shows the composition and mineral content of saffron milk cap. It has been noted that the chemical composition of mushrooms is affected by the substratum composition, pileus size, harvest period, and mushroom species. Mushrooms have a protein content ranging from 0.8 g/100 g^{-1} fresh matter to 7.6 g/100 g^{-1} dry matter. It has also been discovered that mushrooms have a lower fat content and are hence included in low-calorie diets (Onbaşili et al., 2015). One study found 0.8 percent to 27.5 percent fat content in dried mushrooms (Liu, 2007). The composition of *L. deliciosus* revealed water (91.9 percent), polysaccharides (2.86 percent), protein (0.17 percent), fat (0.32 percent), fibre (4.05 percent), and ash (0.66 percent). The moisture content of Portuguese *L. deliciosus* was 90.05 0.53, total fat 0.22 0.00, crude protein 2.96 0.04, ash 0.51 0.02 and carbs 6.26 0.15 (g/100 g fresh weight) (Bozdoğan et al., 2018; Barros et al., 2007b). *L. deliciosus* from the Turkish region of Kastamonu also had moisture, fat, protein, ash, and dry matter values of 8.750.72, 2.640.16, 75.250.15, 4.610.03, and 89.960.24 percent mg/100 g (percent dry weight) resp. (Onbaşili et al., 2015) *L. deliciosus* is high in dietary fibre, has important physiological effects on glucose and lipid metabolism, and is also helpful to the colon (Reis et al., 2012) revealed that, depending on the mushroom, the primary elements in ash are potassium and phosphorus (Mattila et al., 2002) magnesium in addition to Fe, Ca, Cu and Zn (Guillamón et al., 2010). High content of potassium is distinguishing feature of mushrooms (28.27). Potassium was found to be the most prevalent element in *L. deliciosus* samples (2154 mg/100g fresh weight), followed by phosphorous (36.61.2 mg/100 g fresh weight). *L. deliciosus* was also discovered to be high in Mg, Ca, and Fe, with values of 10.60.1,

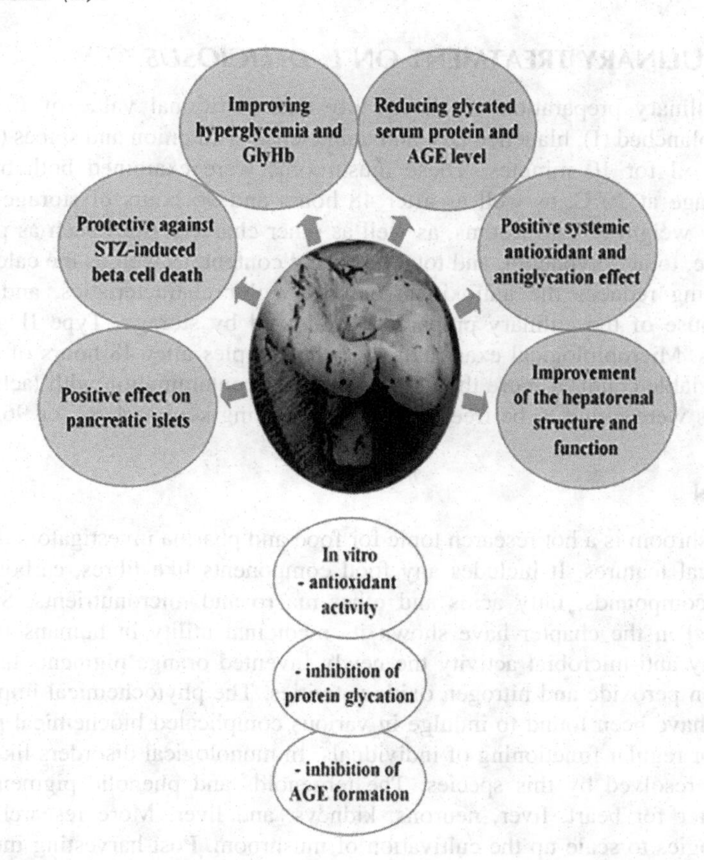

FIGURE 8.7 *Lactarius deliciosus*: Its therapeutics effects.

3.40, and 2.50.6 mg/100 g fresh weight, respectively. Mushrooms are high in vitamins, particularly vitamin B. (Breene, 1990). According to several studies, saffron milk cap is also high in Vitamin B. *L. deliciosus* (1,1-diphenyl-2-picrylhydrazyl radical scavenging effect) activity was 6.43 mg/ml, and (41) determined that the DPPH scavenging activity of *L. deliciosus* methanolic extract was IC50: 17. The DPPH activity of *L. deliciosus* was calculated as 9.9 1/IC50 and 46.4 Mtrolox equivalent /g sample using the TEAC (Trolox Equivalent Antioxidant Capacity/ABTS) technique. The findings of *L. deliciosus* were compared to those of Butylated hydroxyl toluene (BHT), a synthetic antioxidant molecule used as a reference.

ECONOMIC IMPORTANCE

Mushrooms, particularly edible wild mushrooms, are extremely valuable economically all over the world. Around 40 wild mushroom species are consumed at home in countries such as Turkey, while 25 of the species are used for commercial reasons. Saffron milk cap is a remarkable mushroom species because it is widely consumed and traded, particularly in Turkey's Aegean area. *L. deliciosus* is recognised to be important for the people of this region since it is inexpensive and easily available. It is also a fantastic source of revenue for people in the sales industry. The nutritional content revealed that *L. deliciosus* is abundant in vitamin B, numerous minerals, and dietary fibre. According to the TEAC (Trolox Equivalent Antioxidant Capacity/ABTS) technique, the DPPH antioxidant activity of saffron milk caps was 9.9 1/IC50 and 46.4 M trolox equivalent /g sample. *L. deliciosus* was discovered to be a high-quality source of dietary fibre because it has important physiological effects on lipid and glucose metabolism and is also beneficial to the colon.

IMPACT OF CULINARY TREATMENT ON *L. DELICIOSUS*

The effect of culinary preparation on the quality and nutritional value of *L. deliciosus* was investigated. Unblanched (I), blanched (II), and unblanched with onion and spices (III) mushrooms were cooked in oil for 10 minutes. These mushrooms were examined both before and after 48 hours of storage at 20°C, as well as after 48 hours and 96 hours of storage at 4°C. Frying increased the dry weight of mushrooms, as well as other characteristics such as protein, fat, ash, total carbohydrate, total polyphenol, and total flavonoid content, as well as the caloric value of the mushrooms. Frying reduces the antioxidant activity, color characteristics, and texture of the mushroom. Because of the culinary preparation followed by storage, Type II goods displayed notable variances. Microbiological examination of the samples after 48 hours of storage at 20°C revealed a total viable count of more than 106, as well as contamination with lactic acid bacteria. Fried mushrooms were found to be free of germs after being kept at 4 °C for 96 hours.

CONCLUSION

L. deliciosus mushroom is a hot research topic for food and pharma investigators due to its dietary and phytochemical features. It includes any food components like fibres, carbohydrates, amino acids, phenolic compounds, fatty acids and other macro and micronutrients. Several in-house studies mentioned in the chapter have shown its medicinal utility in humans for instance anti cancerous activity anti-microbial activity the newly invented orange pigments have shown anti-oxidant, hydrogen peroxide and nitrogen oxide activities. The phytochemical important activities and compounds have been found to indulge in various complicated biochemical pathways which are mandatory for regular functioning of individuals. Immunological disorders like hyperenstivity are found to be resolved by this species. The terpenoids and phenolic pigments have proven potential supporter for heart, liver, neurons, kidneys, and liver. More research is required to develop technologies to scale up the cultivation of mushroom. Post-harvesting improvements are also a matter of concern. To understand the metabolic pathways there is requirement of more human trials. More detailed statistical data analysis should be done to utilize the mushroom for complete healthcare solutions.

REFERENCES

Alves, M.J., Ferreira, I.C., Martins, A., Pintado, M. 2012. Antimicrobial activity of wild mushroom extracts against clinical isolates resistant to different antibiotics. *Journal of Applied Microbiology* 113(2):466–475.

Avci, E., Avci, G.A. 2019. Antimicrobial and antioxidant activities of medicinally important *Lactarius deliciosus*. *Nutrition* 1:3.

Ayvaz, M.Ç., Aksu, F., Kır, F. 2019. Phenolic profile of three wild edible mushroom extracts from Ordu, Turkey and their antioxidant properties, enzyme inhibitory activities. *British Food Journal* 121(6):1248–1260.

Barros, L., Ferreira, M.J., Queiros, B., Ferreira, I.C., Baptista, P. 2007a. Total phenols, ascorbic acid, β-carotene and lycopene in Portuguese wild edible mushrooms and their antioxidant activities. *Food Chemistry* 103(2):413–419.

Barros, L., Baptista, P., Correia, D.M., Casal, S., Oliveira, B., Ferreira, I.C. 2007b. Fatty acid and sugar compositions, and nutritional value of five wild edible mushrooms from Northeast Portugal. *Food Chemistry* 105(1):140–145.

Bhandari, M.R., Jong-Anurakkun, N., Hong, G., Kawabata, J. 2008. α-Glucosidase and α-amylase inhibitory activities of Nepalee medicinal herb Pakhanbhed (*Bergenia ciliata*, Haw.). *Food Chemistry* 106(1):247–252.

Bozdogan, A., Ulukanlı, Z., Bozok, F., Eker, T., Dogan, H.H., Buyukalaca, S. 2018. Antioxidant potential of *Lactarius deliciosus* and *Pleurotus ostreatus* from Amanos Mountains. *Advancements in Life Sciences* 5(3):114–120.

Breene, W.M. 1990. Nutritional and medicinal value of specialty mushrooms. *Journal of Food Protection*, 53(10):883–895.

Calder, P.C. 2015. Functional roles of fatty acids and their effects on human health. *Journal of Parenteral and Enteral Nutrition* 39:18S–32S.

Çayır, A., Coşkun, M., Coşkun, M. 2010. The heavy metal content of wild edible mushroom samples collected in Canakkale Province, Turkey. *Biological Trace Element Research* 134:212–219.

De Silva, D.D., Rapior, S., Hyde, K.D., Bahkali, A.H. 2012. Medicinal mushrooms in prevention and control of diabetes mellitus. *Fungal Diversity* 56(1):1–29.

Dospatliev, Lilko (2017). Macroelement contents and chemical composition of Lactarius deliciosus (l.) Collected from Batak mountain in Bulgaria. Applied Researches in Technics, Technologies and Education, 5, 165–17010.15547/artte.2017.03.002.

Ergonull, P.G., Sanchez, S. 2013. Evaluation of polycyclic aromatic hydrocarbons content in different types of olive and olive pomace oils produced in Turkey and Spain. *European Journal of Lipid Science and Technology* 115(9):1078–1084.

Feussi Tala, M., Qin, J., Ndongo, J.T., Laatsch, H. 2017. New Azulene-type sesquiterpenoids from the fruiting bodies of *Lactarius deliciosus. Natural Products and Bioprospecting* 7:269–273.

Flament, E., Guitton, J., Gaulier, J.M., Gaillard, Y. 2020. Human poisoning from poisonous higher fungi: Focus on analytical toxicology and case reports in forensic toxicology. *Pharmaceuticals* 13(12):454.

Gomes, D.C., de Alencar, M.V., Dos Reis, A.C., de Lima, R.M., de Oliveira Santos, J.V., da Mata, A.M., Dias, A.C., da Costa Junior, J.S., de Medeiros, M.D., Paz, M.F., de Sousa, J.M. 2019. Antioxidant, anti-inflammatory and cytotoxic/antitumoral bioactives from the phylum Basidiomycota and their possible mechanisms of action. *Biomedicine & Pharmacotherapy* 112:108643.

Guillamon, E., Garcia-Lafuente, A., Lozano, M., Rostagno, M.A., Villares, A., Martínez, J.A. 2010. Edible mushrooms: Role in the prevention of cardiovascular diseases. *Fitoterapia* 81(7):715–723.

Hamid Abd, H., 2017. Between the bioactive extracts of edible mushrooms and pharmacologically important nanoparticles: Need for the investigation of a synergistic combination – A mini review. *Asian Journal of Pharmaceutical and Clinical Research* 10(3):13–24.

Hou, Y., Wang, M., Zhao, D., Liu, L., Ding, X., Hou, W. 2019. Effect on macrophage proliferation of a novel polysaccharide from *Lactarius deliciosus* (L. ex Fr.) Gray. *Oncology Letters* 17(2):2507–2515.

Kała, K., Krakowska, A., Gdula-Argasinska, J., Opoka, W., Muszynska, B. 2019. Assessing the bioavailability of zinc and indole compounds from mycelial cultures of the bay mushroom *Imleria badia* (Agaricomycetes) using in vitro models. *International Journal of Medicinal Mushrooms* 21(4).

Kalac, P., Svoboda, L. 2000. A review of trace element concentrations in edible mushrooms. *Food Chemistry* 69(3):273–281.

Kalac, P. 2009. Chemical composition and nutritional value of European species of wild growing mushrooms: A review. *Food Chemistry* 113(1):9–16.

Kim, H.O., Lim, J.M., Joo, J.H., Kim, S.W., Hwang, H.J., Choi, J.W., Yun, J.W. 2005. Optimization of submerged culture condition for the production of mycelial biomass and exopolysaccharides by *Agrocybe cylindracea. Bioresource Technology* 96(10):1175–1182.

Kosanic, M., Ranković, B., Rančić, A., Stanojkovic, T. 2016. Evaluation of metal concentration and antioxidant, antimicrobial, and anticancer potentials of two edible mushrooms *Lactarius deliciosus* and *Macrolepiota procera. Journal of Food and Drug Analysis* 24(3):477–484.

Kumar, V., Soni, R., Jain, L., Dash, B., Goel, R. 2019. Endophytic fungi: recent advances in identification and explorations. *Advances in Endophytic Fungal Research: Present Status and Future Challenges* 267–281.

Leon-Guzmán, M.F., Silva, I., Lopez, M. 1997. Proximate chemical composition, free amino acid contents, and free fatty acids contents of some wild edible mushrooms from Queretaro, México. *Journal of Agricultural and Food Chemistry* 45:4329–4332.

Liu, J.K. 2007. Secondary metabolites from higher fungi in China and their biological activity. *Drug Discoveries and Therapeutics* 1(2):94–103.

Lo, H.C., Wasser, S.P. 2011. Medicinal mushrooms for glycemic control in diabetes mellitus: history, current status, future perspectives, and unsolved problems. *International Journal of Medicinal Mushrooms* 13(5).

Mattila, P., Salo-Väänänen, P., Könkö, K., Aro, H., Jalava, T. 2002. Basic composition and amino acid contents of mushrooms cultivated in Finland. *Journal of Agricultural and Food Chemistry* 50(22):6419–6422.

Mendil, D., Uluozlu, O.D., Hasdemir, E., Caglar, A. 2004. Determination of trace elements on some wild edible mushroom samples from Kastamonu, Turkey. *Food Chemistry* 88(2):281–285.

Moo-Huchin, V.M., Moo-Huchin, M.I., Estrada-León, R.J., Cuevas-Glory, L., Estrada-Mota, I.A., Ortiz-Vázque, E., Betancur-Ancona, D., Sauri-Duch, E. 2015. Antioxidant compounds, antioxidant activity and phenolic content in peel from three tropical fruits from Yucatan, Mexico. *Food Chemistry* 166:17–22.

Mshandete, A.M., Mgonja, J.R. 2009. Submerged liquid fermentation of some Tanzanian basidiomycetes for the production of mycelial biomass, exopolysaccharides and mycelium protein using wastes peels media. *Journal of Agriculture and Biological Science* 4:1–13.

Nowakowski, P., Markiewicz-Zukowska, R., Gromkowska-Kępka, K., Naliwajko, S.K., Moskwa, J., Bielecka, J., Grabia, M., Borawska, M., Socha, K. 2021. Mushrooms as potential therapeutic agents in the treatment of cancer: Evaluation of anti-glioma effects of *Coprinus comatus*, *Cantharellus cibarius*, *Lycoperdon perlatum* and *Lactarius deliciosus* extracts. *Biomedicine & Pharmacotherapy* 133:111090.

Onbaşili, D.,Celik, G., Katircioglu, H., Narin, I. 2015. Antimicrobial, antioxidant activities and chemical composition of *Lactarius deliciosus* (L.) collected from Kastamonu province of Turkey. *Kastamonu University Journal of Forestry Faculty* 15(1):98–103.

Orsavova, J., Misurcova, L., Vavra Ambrozova, J., Vicha, R., Mlcek, J. 2015. Fatty acids composition of vegetable oils and its contribution to dietary energy intake and dependence of cardiovascular mortality on dietary intake of fatty acids. *International Journal of Molecular Sciences* 16(6):12871–12890.

Osdaghi, E., Martins, S.J., Ramos-Sepulveda, L., Vieira, F.R., Pecchia, J.A., Beyer, D.M., Bell, T.H., Yang, Y., Hockett, K.L., Bull, C.T. 2019. 100 Years since tolaas: Bacterial blotch of mushrooms in the 21st century. *Plant Disease* 103(11):2714–2732.

Patel, S., Goyal, A. 2012. Recent developments in mushrooms as anti-cancer therapeutics: A review. *3 Biotech* 2:1–15.

Patel, S.H. 2014. Review article on mushroom cultivation. *Int. J. Pharm. Res. Technol* 4: 47–59.

Pei, F., Shi, Y., Gao, X., Wu, F., Mariga, A.M., Yang, W., Zhao, L., An, X., Xin, Z., Yang, F., Hu, Q. 2014. Changes in non-volatile taste components of button mushroom (*Agaricus bisporus*) during different stages of freeze drying and freeze drying combined with microwave vacuum drying. *Food Chemistry* 165:547–554.

Reis, F.S., Martins, A., Barros, L., Ferreira, I.C. 2012. Antioxidant properties and phenolic profile of the most widely appreciated cultivated mushrooms: A comparative study between in vivo and in vitro samples. *Food and Chemical Toxicology* 50(5):1201–1207.

Sadi, G., Kaya, A., Yalcin, H.A., Emsen, B., Kocabas, A., Kartal, D.I., Altay, A. 2016. Wild edible mushrooms from Turkey as possible anticancer agents on HepG2 cells together with their antioxidant and antimicrobial properties. *International Journal of Medicinal Mushrooms* 18(1).

Sagar, A., Thakur, 2012. Study on antibacterial activity of *Lactarius deliciosus* (L.) Gray. *Indian Journal of Mushrooms* 30(3):10–14.

Santoyo, S., Ramírez-Anguiano, A.C., Reglero, G., Soler-Rivas, C. 2009. Improvement of the antimicrobial activity of edible mushroom extracts by inhibition of oxidative enzymes. *International Journal of Food Science & Technology* 44(5):1057–1064.

Shahidi, F., Ambigaipalan, P. 2015. Phenolics and polyphenolics in foods, beverages and spices: Antioxidant activity and health effects – A review. *Journal of Functional Foods* 18(15):820–897.

Sharma, Nitin, Kumar, Vikas, & Gupta, Nidhi (2021). Phytochemical analysis, antimicrobial and antioxidant activity of methanolic extract of Cuscuta reflexa stem and its fractions. Vegetos, 34, 876–88110.1007/s42535-021-00249-3.

Sharma, V.P., Annepu, S.K., Gautam, Y., Singh, M., Kamal, S. 2017. Status of mushroom production in India. *Mushroom Research* 26(2):111–120.

Sharma, V.P., Heera, G., Kumar, S., Nath, M. 2020. Development and identification of high yielding strain of *Calocybe indica* based on the multilocation trials. *Mushroom Research* 29(1):47–50.

Teekachunhatean, T. 1997. Cultivation of shiitake mushroom (*Lentinula edodes* (Berk) Sing.) in Nakhon Ratchasima. *Warasan Technology Suranaree*.

Tudses, N. 2016. Isolation and mycelial growth of mushrooms on different yam-based culture media. *Journal of Applied Biology and Biotechnology* 4(5):033–036.

Valverde, M.E., Hernández-Perez, T., Paredes-López, O. 2015. Edible mushrooms: Improving human health and promoting quality life. *International Journal of Microbiology* 376387. doi: 10.1155/2015/376387. Epub 2015 Jan 20. PMID: 25685150; PMCID: PMC4320875.

Xia, Q., Wang, L., Xu, C., Mei, J., Li, Y. 2017. Effects of germination and high hydrostatic pressure processing on mineral elements, amino acids and antioxidants in vitro bioaccessibility, as well as starch digestibility in brown rice (*Oryza sativa* L.). *Food Chemistry* 214:533–542.

Xu, Z., Fu, L., Feng, S., Yuan, M., Huang, Y., Liao, J., Zhou, L., Yang, H., Ding, C. 2019. Chemical composition, antioxidant and antihyperglycemic activities of the wild *Lactarius deliciosus* from China. *Molecules* 24(7): 1357.

Yang, X.L., Luo, D.Q., Liu, J.K. 2006. A new pigment from the fruiting bodies of the basidiomycete *Lactarius deliciosus*. *Zeitschrift für Naturforschung B* 61(9): 1180–1182.

Yin, C., Fan, X., Fan, Z., Shi, D., Yao, F., Gao, H. 2019. Comparison of non-volatile and volatile flavor compounds in six Pleurotus mushrooms. *Journal of the Science of Food and Agriculture* 99(4):1691–1699.

Ziegler, D.V., Wiley, C.D., Velarde, M.C. 2015. Mitochondrial effectors of cellular senescence: beyond the free radical theory of aging. *Aging Cell* 14(1):1–7.

9 *Laetiporus sulphureus* (Bull.) Murrill

Kanika Dulta
Department of Food Technology, School of Applied and Life Science, Uttaranchal University, Dehradun, Uttarakhand, India

Keshav Kumar
Department of Biotechnology, Dr YS Parmar University of Horticulture and Forestry, Nauni, Solan, HP, India

Arti Thakur
Department of Botany, Shoolini Institute of Life Sciences and Business Management, Solan, Himachal Pradesh India

Somvir Singh
University Institute of Biotechnology, Department of Biosciences, Chandigarh University, Mohali, India

Gözde Koşarsoy Ağçeli
Hacettepe University, Department of Biology/Biotechnology, Ankara, Turkey

Divyanshi Singh
Department of Food Technology, Shoolini University, Solan, Himachal Pradesh, India

CONTENTS

DOI: 10.1201/9781003259763-9

INTRODUCTION

For centuries, mushrooms have been used as a traditional medicine. Secondary metabolites originate in the cultured mycelium, fruit bodies, and cultured broth of mushrooms, which are higher Basidiomycetes and Ascomycetes. For many years, mushrooms have been used in a variety of human activities. Mushrooms are consumed as a delicacy in many countries, particularly for their flavour. Researchers have recently lengthened their research into supplementary uses of mushrooms, particularly for food-preserving and medicinal purposes. Because of their various ecological, physiological, and morphological characteristics, which likewise contribute to their diversity, some of these mushrooms have been dubbed medicinal mushrooms. There are 16,000 species of mushrooms worldwide, with over 2000 of them being safe.

Eatable mushrooms have been recognized to be a suitable candidate for both nutritional elements and bioactive chemicals, and they can be used to create modern anticancer and antibacterial compounds (Fontana et al., 2014; Xu et al., 2014). Many mushroom species contain antibacterial and anticancer compounds in different parts of the world (Vamanu, 2012). Organic and aqueous preparations of *Ganoderma lucidum* fruiting bodies have antibacterial action (Ofodile et al., 2005). Further studies show that Plectasin, derived from *Pseudoplectania nigrella*, is an effective gram-positive bacteria inhibitor, and 2-aminoquinoline, derived from *Leucopaxillus albissimus*, is an effective negative bacteria inhibitor. *Laetiporus* is derived from two Latin words, "laeti" and "por," and alludes to a hymenial layer as well as the size of the specially organized fruiting bodies (Alves et al., 2012). The Latin adjective "sulphureus" refers to the colour of the fruiting bodies. Except for Antarctica, *L. sulphureus* is a global species that can be found on all continents. It is readily accessible throughout Europe and North America. *L. sulphureus* is also known by the common names Chicken Mushroom and Chicken-of-the-Woods, (Bull.: Fr.) Murrill (= Polyporus sulphureus (Bull.) Fr). (Fomitopsidaceae, Polyporales). W. A. Murrill, an American mycologist, named the genus *Laetiporus* in 1904 and added *Polyporus sulphureus* to it in 1920. Brown rot of wood is linked to *L. sulphureus*, which can be found in the roots, butte as well as rotting logs; its preferred host/substrate is *Quercus* spp (Zjawiony, 2004). Decadent species such as *Aesculus* sp. and *Populus* sp. are attacked and colonised by it, while coniferous species like *Larix* sp. and *Taxus* sp. are less frequently affected. Sulfur-yellow fleshy semicircular hats, known as fruiting bodies, begin to grow rapidly in May and continue to be produced until the end of the season. *L. sulphureus* is classified as a conditionally edible fungus since only immature fruiting bodies can be consumed, but raw fruiting bodies can be hazardous (Gumińska & Wojewoda, 1985). *Laetiporus* species is medicinal fungi that is traditionally used by Europeans to treat pyretic disorders, rheumatism, cough and stomach cancer (Rios et al., 2012). As a traditional food and herbal medicine in Asia for thousands of years, studies have revealed that it has antibacterial properties (Turkoglu et al., 2007). Emphasis is being placed on identifying *L. sulphureus* fractions that are beneficial to human health, such as glucan and other saccharides, pigments, and phenolic chemicals that have antibacterial and anticancer properties (AM, 2017; Kolundzic et al., 2016). Researchers have recently discovered the chemical composition, which exposed occurrence of numerous bioactive components, including polysaccharides, vitamins, lectins, sterols, minerals, laetiporic acid, polyphenols, and so on. Consequently, this mushroom found headed for medically energetic in extensive range of treatments, comprising antibacterial, anti-inflammatory, antioxidant, antitumor, antithrombotic, antiviral, immune-modulating therapies, and antiproliferative (Luangharn et al., 2014; Petrović, Papandreou et al., 2014). To provide researchers and food industry professionals with current scientific information about fungus as valuable health-oriented medicinal food, the review seeks to provide a comprehensive worldwide overview.

TAXONOMY

Brown rot is caused by the fungus *Laetiporus* Murrill, which is found in living hardwoods and conifers (Murrill, 1904). The genus contains many pathogens that can affect forests; however, some species are edible and have medicinal properties. Previous research has accepted 15 species in the genus globally (Banik et al., 2012; Song et al., 2014; Song & Cui, 2017). A characteristic of *L. sulphureus* complex is its annual basidiocarps, soft, fleshy environment, and dimitic hyphal system, composed of simple septate hyphae that produce material and a hyphal system that binds it (Ota et al., 2009).

Species Taxonomy

Eukaryota; Opisthokonta; Fungi; Dikarya; Basidiomycota; Agaricomycotina; Agaricomycetes; incertae sedis; Polyporales; Laetiporaceae; *Laetiporus*; *Laetiporus sulphureus* ((Bull.) Murrill, 1920).

Scientific Classification of *Laetiporus sulphureus*

Kingdom: Fungi
Division: Basidiomycota
Class: Agaricomycetes
Order: Polyporales
Family: Fomitopsidaceae
Genus: *Laetiporus*
Species: *sulphureus*

RESOURCE AVAILABILITY

The taxon is classified as part of the Basidiomycotina division, Agaricomycotina subdivision, Agaricomycetes class, Aphyllophorales order, and Polyporaceae family. The mushroom can be found all over the world. From May to October, it can be found growing alone or in enormous clusters of 5–50 on the stems of many surviving deciduous and coniferous trees in a semi-circular pattern. In the subalpine zone, the fungus can be found on live Larix decidua or Acacia species, while in the montane zone, it can be found on broadleaf trees (Kovács & Vetter, 2015). This mushroom is recognized easily in forests and cities by its brilliant sulfur-yellow to orange porous basidiocarps and great size up to 40 cm across (Smith et al., 1980). Sulphur shelf and sulphur polypore, the fungus's common names are derived from its distinctive hue, as are its specific epithet, *Laetiporus*, which signifies a fungus with bright pores and sulphureus, which means sulphur in colour. In addition to the fruit body's large, bracket shape, fleshy margin, and tubular hymenopores, this species has an unusually large mass of wet weight of over 40 kg (Wiater et al., 2012).

TRADITIONAL USES

Apart from Antarctica, *L. sulphureus* is a global species that can be found on all continents. This mushroom has become a vital feature of certain country cuisines over the generations due to its flavour. In addition to appearance, *L. sulphureus* has a distinct taste, texture, and odour. The taste has been compared to chicken, crab, or lobster meat, while the odour has been described as somewhat pleasant to highly fungal (Bulam et al., 2019; Khatua et al., 2017; Petrović et al., 2014; Rapior et al., 2000). Young fruit bodies can be eaten and have a crab or lobster flavour. As a result, it's also called "crab of the woods," "chicken of the woods," "chicken polypore," or "chicken mushroom," and it's a vegetarian alternative to chicken. For many years, the distinct aroma and texture of young fruit bodies have inspired the acceptance of young fruit bodies as an important food source in oriental civilizations. Mushrooms are unique among polypores because of their great history of use, specifically in Japan, Thailand and North America, where they are well-thought-out delicacies. It is considered a delicacy in some regions of these continents and can also be used as a vegetarian substitute for chicken. Some researchers in Turkey have already identified this species in various locations. This species is available for purchase at local markets and enjoyed in local cuisines throughout Turkey and the world (Pekşen et al., 2016; Sułkowska-Ziaja et al., 2018). Asian folk medicine has used the fruit bodies of *L. sulphureus* for centuries for numerous purposes, not only as a food, but also for the regulation of the body, the improvement of health, and the prevention of chronic ailments (Klaus et al., 2013; Mandic et al., 2018; Sesli, 2007; Sesli & Denchev, 2008). Several scientists from around the world have addressed this ethno-mycological component in recent decades. Multiple studies have shown that the mushroom has a number of important nutritional components which could provide a wonderful opportunity to enhance health equity. Nutritional studies have recently verified that it is a sustainable food supply for the world's rising population due to its high carbohydrate, protein, mineral, vitamin, polyunsaturated fatty acid, and fibre content. However, after the successful biosynthesis of eburicoic acid in the 1960s, the mushroom gained prominence in the scientific community (Bulam et al., 2019; Pekşen et al., 2016).

BIOACTIVE PHYTOCHEMICALS

L. sulphureus not only provides important macro and micronutrients to the body, but it also contains bioactive compounds that have health advantages. The greatest significant therapeutically active primary metabolites such as polysaccharide-protein complexes, polysaccharides and proteins. Aside from them, secondary metabolites are responsible for a considerable portion of known biological activities. Triterpenoids of various structures represent approximately 75% of secondary metabolites, with other secondary metabolite groups producing a lesser amount. Organic acids are the second most abundant secondary metabolite found in mushrooms (14%). Flavonoids, benzofurans, N-containing compounds and coumarins are among other chemicals found in *L. sulphureus* (Grienke et al., 2014; Khatua et al., 2017). Polysaccharides are a well-known class of metabolites found in fruiting bodies. Exopolysaccharides found in the fruiting bodies of that species included linear water-insoluble -1,3-glucan-laminaran and fuco-galactomannan. At positions 2 and 3, these residues are substituted by L-fucopyranosyl, 3-O-D-L-mannopyranosyl and -D-mannopyranosyl, -moieties (Alquini et al., 2004). (Olennikov et al., 2009) isolated an alkali-soluble polysaccharide, latiglucan I, but were unable to identify any biological activity. Also extracted from mushroom cultures was a phenolic acid derivative, whose structure was determined using spectroscopic data. 6-((2E, 6E)-3, 7-dimethyldeca-2, 6-dienyl)-7-hydroxy-5-methoxy-4-methylphtanlan-1-one, component was identified, but it is still not known if it has any therapeutic uses (Sun et al., 2014).

Phenolic compounds of the benzofuran lignans type egonol, demethoxyegonol, and egonol glucoside, mycophenolic acid and its derivatives, and laetirobin, which can inhibit the division of cancer cells, are capable of inhibiting the growth of tumour cells (Deol et al., 1978; Lear et al., 2009; Rios et al., 2012; Sugimoto et al., 2009; Yin et al., 2015; Yoshikawa et al., 2001). Four lanostane-type triterpenes, including three novel compounds: 15a-hydroxy-3,4-secolanosta-4(28),8,24-triene-3,21-dioic acid, 5a-hydroxy-3,4-seco-lanosta-4(28),8,24-triene-3,21-dioic acid 3-methyl ester, and 15a-acetoxyltrametenolic acid (3). Triterpenes, including laetiposides A–D, laetiporins A–B, sulphurenic acid, and dehydroeburicoic acid, have been isolated from *Laetiporus* spp. in the past (Chepkirui et al., 2017; Hassan et al., 2021). Triterpene chemicals found in *L. sulphureus* fruiting bodies are lanostane derivatives: 3-oxosulfurenic acid, 15-hydroxy-trametenolic acid, fomefficinic acid, trametenolic acid, and sulfurenic acid were discovered among the triterpene acids (Kariyone & Kurono, 1940; Wu et al., 2004; Yoshikawa et al., 2001). Dehydrotrametenolic acid and etenolic acid, which are typically detected in wood-rotting fungi, were also discovered in fruiting bodies (León et al., 2004). Two novel triterpenes isolated from the fruiting body of *L. sulphureus* were identified as laetiporins C and D (Hassan et al., 2021). Sterols including ergost-7,22-dien-3-ol, ergosterol, ergost-7-en-3-ol, and 24-ethylcholestan-3-ol, benzofuran glycoside, acetylenic acids, and laetiporic acids; the latter give the fungus its distinctive colour (Ericsson & Ivonne, 2009). Figure 9.1. represents some key bioactive compounds of *L. sulphureus.*

NUTRITIONAL VALUE

The nutritional importance of mushrooms was highlighted during the 1940s as a result of wartime food shortages. Since then, scientists have investigated the nutritional composition of edible macrofungi in an effort to discover alternative food sources, especially in cases of protein deficiency. The Recommended Dietary Allowance (RDA) states that 300 grams of carbs, 50 grams of protein, 65 grams of fat, and 25 grams of dietary fibre are enough to meet the nutritional needs of practically healthy people. As a result, the carbohydrate, protein, and fibre content of mushrooms has been determined to range from 51 to 88%, 19 to 35% of the recommended daily amount, and 4 to 20% of the recommended daily allowance, respectively. As a result, *L. sulphureus* has emerged as an outstanding nutrient supply, influencing substantial research in various laboratories. However, statistics vary significantly since a variety of factors influence nutritional value, including growth place, substrate type, pileus size, developmental phases, and the portion of fungal samples examined (Bernaś et al., 2006; Chan et al., 2013; Valverde et al., 2015).

There has been little research into the individual profiles of phenolic and flavonoids chemicals in wild edible mushrooms. In terms of substance, cultivated mushroom species are better known; nonetheless, wild edible mushrooms have been little studied and, to the best of our knowledge, documented in Tables 9.1 and 9.2, whereas Table 9.3 presents the composition of Macronutrients and energy content of *L. sulphureus.*

ORGANIC ACIDS

Organic acids are organic carboxylic acids with the structural formula R-COOH (Table 9.4).

MINERAL COMPOSITION OF *L. SULPHUREUS*

An elemental mineral is an inorganic compound that plays a crucial role in bone and tooth formation, oxygen transfer, and the normal function of enzymes, hormones, neurons, and muscles, among other things. Macro minerals are demanded in higher concentrations (>100 mg/dl), but micro minerals are sufficient in lower concentrations (100 mg/dl), as their name suggests. The characteristics of a wood-rotting mushroom are listed in Table 9.5.

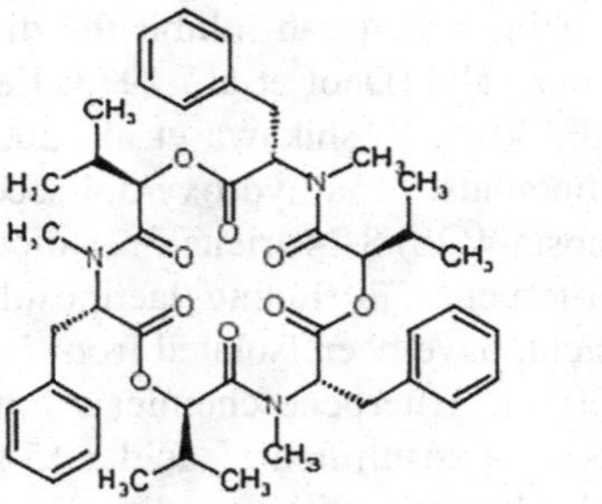

Beauvericin

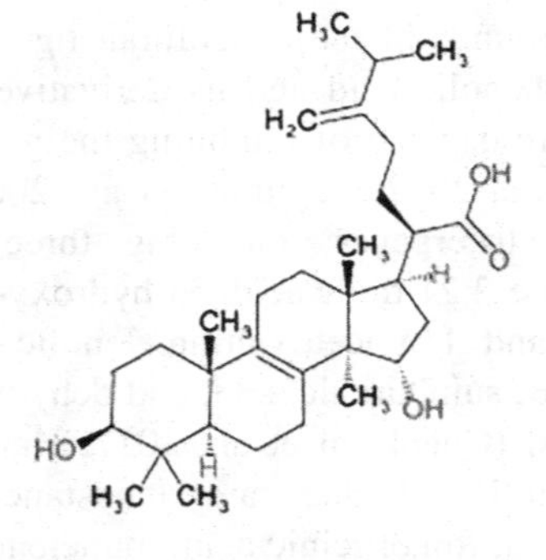

Sulphurenic acid (15α-hydroxyeburicoic acid)

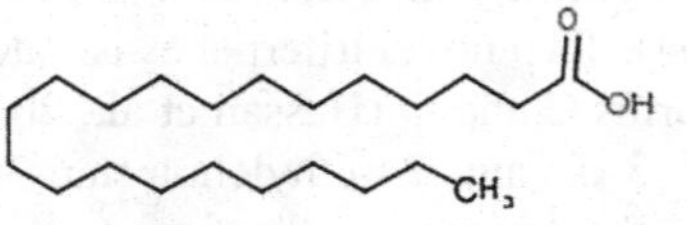

Masutakic acid

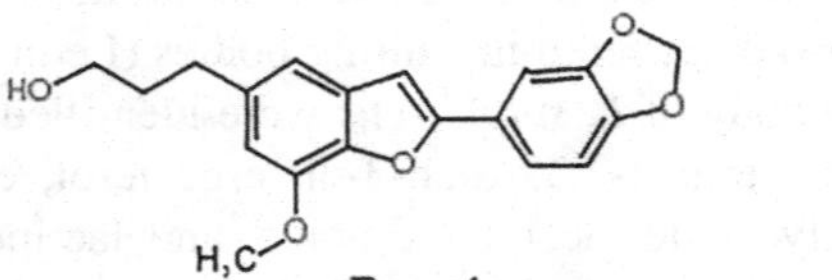

Egonol

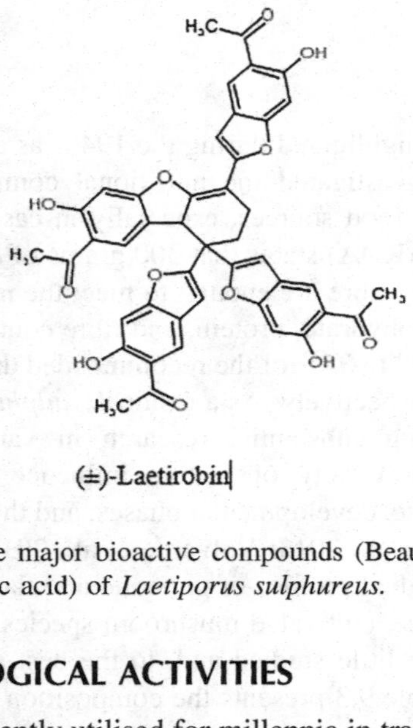

(±)-Laetirobin

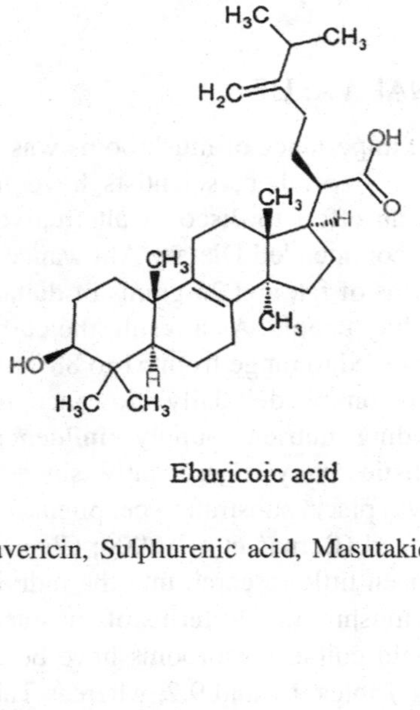

Eburicoic acid

FIGURE 9.1 Some major bioactive compounds (Beauvericin, Sulphurenic acid, Masutakic acid, Egonol, Laetirobin, Eburicoic acid) of *Laetiporus sulphureus.*

PHARMACOLOGICAL ACTIVITIES

L. sulphureus anciently utilised for millennia in traditional medicine in several European countries due to its numerous pharmacological effects. Examining its extracts and individual constituents extensively confirms the known, traditional uses while also revealing new pharmacological activity profiles. Some of the medicinal properties of *L. sulphureus* are as follows.

ANTIBACTERIAL PROPERTIES

Both gram-negative and gram-positive bacteria, including methicillin-resistant *S. aureus* and glycopeptide-resistant *Leuconostoc mesenteroides* strains, are susceptible to *L. sulphureus* (Elu et al., 2003). Fruiting body extracts were found to be effective against the following strains: *Bacillus cereus, B. subtilis Enterobacter cloacae, Escherichia coli, Micrococcus flavus, Listeria monocytogenes, Pseudomonas aeruginosa, Salmonella typhimurium, Staphylococcus aureus* (DEMİR & YAMAÇ, 2008; Šiljegović et al., 2011).

TABLE 9.1
Qualitative Screening of *Laetiporus sulphureus* (Chatterjee & Acharya, 2016)

S. No	Name of Phytochemical	Reagent used	Result
1	Alkaloid	Dragendroff's	+++
2	Carbohydrates	–	–
3	Phenol	Folin-Ciocalteu	+++
4	Flavonoid	Ferric chloride	++
5	Cardiac glycoside	Keller-Killani	+++
6	Steroids	Limbermann-Buchard	+++
7	Terpenoids	Acetic anhydride with sulphuric acid	+++
8	Saponin	Foam	++
9	Tanins	–	–
10	Fixed oils and fats	–	–

+++ strongly present, ++ moderately present

TABLE 9.2
Phenolic, Flavonoid Contents, Polysaccharides and Free Radical Scavenging Activity of *Laetiporus sulphureus*

S. No	Components (mg g^{-1} d.m.)	Quantity	References
1	Total phenolic content	6.07±0.50	(Kovács & Vetter, 2015)
2	Total flavonoid content	0.312±0.018	
3	Polysaccharides	24.93±1.15	
4	Free radical scavenging activity	159.8±6.94	

TABLE 9.3
Macronutrients and Energy Content of *Laetiporus sulphureus*

S. No	Macro-constituents	Quantity	References
1	Moisture (%)	66.67–77.69	(Khatua et al., 2017)
2	Ash (g/100 g DW)	4.00–9.03	(Kovács & Vetter, 2015)
3	Carbohydrate (g/100 g DW)	64.90–74.47	(Kovács & Vetter, 2015)
4	Fibre (g/100 g DW)	5.55–7.57	(Luangharn et al., 2014)
5	Protein (g/100 g DW)	10.61–21.00	(Saha et al., 2014)
6	Fat (g/100 g DW)	2.96–4.50	(Palazzolo et al., 2012)
7	Energy (kcal/100 g DW)	341.06–375.62	(Petrović, Papandreou, et al., 2014)

Antifungal Properties

Like a few other fungus species, ethanolic preparations of *L. sulphureus* fruiting bodies have high antifungal activity against *Candida albicans* (Turkoglu et al., 2007). Furthermore, the water-ethanol extract was found to be antifungal against *Botrytis cinerea, Aspergillus niger and Fusarium*

TABLE 9.4
Organic Acids and Amino Acids in *Laetiporus sulphureus*

S. No	Name of organic acid	Quantity	References
1	Ascorbic acid	0.006	(Khatua et al., 2017)
2	Citric acid	1.24	
3	Malic acid	1.12	
4	Malonic acid	0.32	
5	Tartaric acid	1.39	
6	Succinic acid	0.45	
7	Quinic acid	0.16	
8	Fumaric acid	0.25	
9	Oxalic acid	2.66	
Amino acids (%)			
1	Arginine	0.47	(Agafonova et al., 2007)
2	Histidine	0.40	
3	Isoleucine	0.11	
4	Leucine	0.52	
5	Lysine	0.28	
6	Methionine	0.18	
7	Threonine	0.24	
8	Tryptophan	0.79	

oxysporum. Minimum concentration required to stop these microbes from growing was comparable to fluconazole, a well-known antifungal (Pârvu et al., 2010). *Fusarium tricinctum, Penicillium griseofulvum, Aspergillus wentii, Alternaria alternata,* and *Microsporum gypseum* have all been demonstrated inhibition the presence of extracts from the mycelium of *L. sulphureus* (Sakeyan, 2006).

Antioxidant Properties

An ethanol extract of *L. sulphureus* inhibited DPPH free radical scavenging activity by 14, 26, 55, and 86% at doses of 100, 200, 400, and 800 mg/mL, respectively (Turkoglu et al., 2007). At a dosage of 0.5 mg/mL, the extract of *L. sulphureus* reduced over 60% of the radicals in the ABTS radical scavenging experiment (Khatua et al., 2017).

Anti-inflammatory Properties

Eburicoic acid is the primary bioactive metabolite in *L. sulphureus* in a mouse model of stomach ulcers (Wang et al., 2015). Inflammation is a generalised immune response that results in changes in flow of blood, vascular permeability, and tissue damage. Pro-inflammatory cells begin the inflammatory process after being exposed to immune stimulants by producing biomarkers such as IL-1, IL-6, IL-8, nuclear factor-B, TNF-, prostaglandin E2, and COX-2. New anti-inflammatory therapeutic compounds must be created because current treatments are unable to cure these disorders (Elsayed et al., 2014). LSMH7 was isolated from fruiting bodies in order to identify active anti-inflammatory combinations from wood-rotting fungi. Using sophisticated tools, acetyl eburicoic acid was identified as LSM-7's molecular structure. The findings revealed that LSM-H7 suppressed NO generation dose-dependently, but had no adverse effects across the entire dosage range (12.5, 25, 50, and 100 g/mL). These findings suggested that LSM-H7 plays a role in *L. sulphureus* anti-inflammatory activity (Saba et al., 2015).

TABLE 9.5
Minerals Content of *Laetiporus sulphureus*

S. No	Mineral	Quantity (mg kg^{-1} DM)	References
1	Aluminium (Al)	53.9	(Ayaz et al., 2011; Doğan et al., 2006; Durkan et al., 2011; Khatua et al., 2017; Kovács & Vetter, 2015)
2	Arsenic (As)	ND	
3	Boron (Al)	16.4	
4	Barium (Al)	1.70	
5	Calcium (Ca)	4 200	
6	Copper (Cu)	22.7	
7	Cadmium (Cd)	0.68	
8	Cobalt (Co)	1.2	
9	Chromium (Cr)	58.3	
10	Iron (Fe)	5.09	
11	Lead (Pb)	24.5	
12	Potassium (K)	18 500	
13	Molybdenum (Mo)	0.07	
14	Manganese (Mn)	30.7	
15	Magnesium (Mg)	2 100	
16	Nickel (Ni)	22.7	
17	Sodium (Na)	285.0	
18	Silicon (Si)	230	
19	Strontium (Sr)	*6.95*	
20	Silver (Ag)	0.26	
21	Titanium (Ti)	*ND*	
22	Tin (Sn)	4.5	
23	Zinc (Zn)	5.65	

ND = Not Detected

Hypoglycemic Action

In vivo, exopolysaccharides were discovered to have a hypoglycemic effect. When given to rats 48 hours after a streptozotocin treatment, it reduced plasma glucose concentration by 43.5% and cholesterol and triglyceride levels to near normal levels, compared to the control group. EPS stimulated pancreatic islet cell proliferation and regeneration while also increasing the activity of antioxidant enzymes such as glutathione peroxidase, superoxide dismutase, and catalase (Hwang & Yun, 2010). Anti-diabetic properties of dehydrotrametenolic acid extracted from fruiting bodies This triterpene molecule has a similar biological activity to PPAR-receptor agonists-thiazolidinediones. In vitro, it promotes adipocyte differentiation and reduces hyperglycemia in mice with non-insulin-dependent diabetes mellitus. As a result, dehydrotrametenolic acid is a hypoglycemic molecule with a mechanism that includes tissues that respond to insulin (Sato et al., 2002).

Cytotoxic Action

Triterpene byproducts, such as lanostan extracted from fruiting bodies, and their semi-synthetic derivatives, have been shown to be cytotoxic. The most potent action was well-known for the acetyl derivative of eburic acid, which induces apoptosis by activating caspase-3 and degrading poly (ADP-ribose) polymerase (PARP), one of the enzymes that repairs DNA damage Acetyl-eburic acid is a crucial component that could lead to the creation of novel anticancer drugs (León et al., 2004). Egonol, demethoxyegonol, and egonol glucoside extracted from *L. sulphureus* var. *miniatur* were found to have in vitro cytotoxic activity against the human gastric cancer cell line KATO III (Yoshikawa et al., 2001). Letirobin has been discovered to have cytotoxic effect, and its mode of action varies from that of antimitotic drugs previously identified. It penetrates cancer cells quickly in vitro, decreased their proliferation in late mitosis, and triggered apoptosis (Lear et al., 2009). Carboxymethyl derivatives of -(13)- -D-glucans isolated from the fruiting bodies of *L. sulphureus* inhibit tumour cell line metabolism but not normal cell metabolism (Wiater et al., 2011).

Seven triterpenoids isolated from *L. sulphureus* were examined for and it was discovered that they inhibited the proliferation of HL-60 myeloid leukemia cells in a dose-dependent manner (León et al., 2004). The results of quantitative fluorescence microscopy, DNA fragmentation analysis, caspase-3 activity, and cytochrome C release show that these triterpenoids induce apoptosis by activating caspase-3 and releasing cytochrome C. Caspase-3 activity was found to be at its peak when cells were preserved with acetyl eburicoic acid, making it a likely model chemical for the development of anticancer medicines (Radic & Injac, 2009).

Anticancer Properties

TACA antigens that are specific to lung, thyroid, and breast cancer cells can be detected using LSL (*L. sulphureus* Lectin). Because they have hemolytic and hemagglutination structures, as well as an action similar to that of poisonous lectins in that they have enzymatic properties (RNA--glycosidase) and inhibit protein combination by deactivating ribosomes, *L. sulphureus* (Końska et al., 2008) lectins can be utilized to treat a variety of ailments.

Anti-HIV

The 57 species of wood-eating fungi extract of methanol and dichloromethane found effective against HIV–1 reverse transcriptase was investigated. *L. sulphureus* was the most active member, with a repressive potential of 90%. Those who performed dichloromethane formulation found that Methanolic extract was superior. In addition, the strong fraction was subjected to an initial HPLC analysis in which five fractions were created and tested for repressing activity. Using thin-layer chromatography (TLC), the most active fraction, fraction 1, was analysed, and the results revealed the presence of an amino group and an acidic chemical (MLINARIC et al., 2005).

Anti-malarial Activity

Malaria is one of the most lethal infections seen in tropical areas, causing millions of illnesses and deaths annually. Parasitic resistance to medications has rendered several antimalarial drugs ineffective, necessitating the development of new drugs, a novel type of antimalarial compound discovered in *L. sulphureus*, shown efficacy against the HBO strain of *Plasmodium falciparum*. DMSO and ethanol, were made from the fruiting bodies showed inhibition rates of 0.6% and 2.4%, respectively. As a result, the mushroom exhibited a promising trait for anti-malarial medications (Lovy et al., 2000).

Anti-diabetic Activity

It is a chronic metabolic disorder characterized by an increase in blood sugar due to defective insulin function. In diabetic rats induced by Streptozotocin (STZ), the glucose-lowering impact of crude extracellular polysaccharides generated from *L. sulphureus* var. *miniatus* submerged mycelial culture was investigated., and the results revealed its excellent hypoglycemic action. After 48 hours of STZ treatment, the polysaccharide extract was orally administered, resulting in plasma concentrations of glucose, total cholesterol, and triglycerides that were close to normal. In addition, immunohistochemical staining of pancreatic tissues revealed that the treatment increased the insulin antigenicity of diabetic islet -cells, suggesting the possibility of -cell proliferation or regeneration. The polysaccharide fraction may aid in the treatment and management of type 2 diabetes (Hwang & Yun, 2010; Zjawiony, 2004). Several polypores, most notably *L. sulphureus*, contain dehydrotrametenolic acid. As an insulin sensitizer, it decreases hyperglycemia in mice with noninsulin-dependent diabetes during glucose tolerance tests.

Anti–thrombin

(Okamura et al., 2000) identified *L. sulphureus* as one of the most active Basidiomycetes after screening for anti–thrombin activity in higher Basidiomycetes. The researchers used protoplast fusion between *Hypsizygus marmoreaus* and *L. sulphureus* a commonly cultivated Basidiomycete, to avoid cultivation of the fungus. In contrast to other synthetic drugs, cultivable fusants generated a harmless anti–thrombin compound. Thus, consuming these fusant mushrooms may prevent future thrombosis.

Immunomodulatory

Immunomodulators are potent medications that stimulate leukocytes and macrophages to modulate the immune system. It has been suggested that these components be regularly added to feed to prevent diseases caused by immunological deficiencies and other immune deficiency states. Several immunomodulatory compounds, including polysaccharides (particularly -D-glucans), polysaccharopeptides, polysaccharide proteins, and proteins, have been extracted from mushrooms. Consequently, these metabolites may be utilised to treat immunodeficiency diseases (Seo et al., 2011). Mycelial cultures of *L. sulphureus* var. *miniatus* yielded an extracellular polysaccharide with a molecular weight of 6.95 kDa. A polysaccharide was used in culture to study the effect and Bax and Bad protein levels in cells treated with the pure component were nearly double those of control cells. The findings could point to a link between the polysaccharide and the activation of immunomodulatory mediators (Singdevsachan et al., 2016).

Hepatoprotective

Numerous xenobiotics, including alcohol, numerous medications, malnutrition, infection, and anaemia, can cause liver damage, a prevalent disease. The hepatic damage causes an upsurge in the aspartate aminotransferase (AST), serum concentration of aminotransferases: and alanine aminotransferase (ALT) (Soares et al., 2013). Although the fact that synthetic medications are readily, available, long-term usage of these substances may have negative consequences. As a result, substantial research is required to find a medicine that is both safe and cost-effective (Chatterjee & Acharya, 2016). (Sun et al., 2014) discovered that eburicoic acid and trametenolic acid B have hepatoprotective properties against CCl_4-induced hepatic fibrosis (Table 9.6).

CULTIVATION OF *L. SULPHUREUS* FRUITING BODY

Production of *L. sulphureus* fruiting bodies on a vast scale in an artificial medium A few *L. Sulphureus* isolates were cultivated in various media. In the instance of two strains, the primordia appeared 5–6 days after initiation, and the fruiting bodies appeared 2 days later. This experiment's findings pave the door for commercial preparation of these lucrative fruiting bodies (Pleszczyńska et al., 2013). Many attempts have been made in the past to cultivate *L. sulphureus* under various

TABLE 9.6
Key Pharmacological Properties of *Laetiporus sulphureus* and Responsible Bioactive Compounds

Pharmacological activity	Active component	References
Antibacterial activity	Lipids such as phosphatidylcholine, phosphatidylethanolamine	(Sinanoglou et al., 2015)
Antioxidant activity	*p*-coumaric acid, quercetin, kaempferol, caffeic acid, catechin, gallic acid and 5-caffeyl quinic acid, Laetiporan A, oxalic acid	(Petrović, Papandreou et al., 2014)
Anti-inflammatory	Egzopolysaccharides	(Jayasooriya et al., 2011)
Antifungal activity	Hexane extracts	(Sinanoglou et al., 2015)
Hypoglycemic effect	Dehydrotrametenolic acid	(Sato et al., 2002)
Cytotoxic activity	Demethoxyegonol	(León et al., 2004)
Anticancer activity	Lectins	(Końska et al., 2008)
Anti- HIV	Methanolic extract	(MLINARIC et al., 2005)
Anti-malaria	Extract of dimethyl sulfoxide and ethanol fractions	(Lovy et al., 2000)
Anti-diabetic	Dehydrotrametenolic acid	(Zjawiony, 2004)
Immunomodulatory	Polysaccharide, Bax and Bad proteins	(Singdevsachan et al., 2016)
Hepatoprotective	Eburicoic acid and trametenolic acid	(Sun et al., 2014)

conditions. In the vast majority of cases, the fruiting bodies produced in this manner did not reach the required size, which was satisfactory to the industry. In this invention's method of breeding fruiting bodies, the substrate is placed in a specialised, hermetically sealed container. In 6–12 days, these developed fruiting bodies reach a weight of 250–300 g, which is an ideal bulk for extracting a high number of essential metabolites.

TOXICITY

A 6-year-old child experienced hallucinations and ataxia after consuming a slight amount of raw *L. sulphureus* fruit body in August 1987, despite the fact that *L. sulphureus* is generally considered safe for human consumption. The mushroom tested negative for recognised hallucinogens in a gas chromatographic study. Several factors, including the patient's age, the quantity consumed, and the fact that it was consumed uncooked, contributed to the negative outcomes in this instance (Appleton et al., 1988). Zebrafish (Danio rerio) as a preclinical animal model to show that ethanol extract of *L. sulphureus* is not harmful at high doses up to 400–500 g/mL while efficiently inhibiting melanogenesis in a dose-dependent manner. The extracts suppressed tyrosinase activity and melanin formation in zebrafish skin melanocytes without causing melanocytotoxicity, inflammation, or immunosuppression, making them ideal items for topical application or usage as functional additions in cosmetic compositions (Pavic et al., 2021).

MANAGEMENT

Trees suspected of being invaded by members of the *L. sulphureus* species complex in an urban area should be evaluated and rated as potential danger trees by an ISA-certified arborist. Because members of the *L. sulphureus* species complex might weaken a tree, it may need to be removed, although an arborist can assist with risk assessment. Because *Laetiporus* and other wood-decay fungus can be vital components of our native ecosystems, it may be more preferable to leave the tree alone in a natural context. Once the fruiting bodies of a *Laetiporus* are seen on a tree, it is too late to prevent the fungus from establishing itself. To avoid further injury to a decaying tree, we propose following the advice given by Downer and Perry (Downer & Perry, 2019). (1) prune branches properly to promote balanced branching structure, (2) remove dead or diseased limbs (especially if they potentially endanger persons or structures), and (3) avoid injury to the bark or branches (Benitez et al., 2020).

CONCLUSION

We can draw conclusions about the various benefits of mushrooms for humans from the words of Hippocrates, the father of medicine, who said, "Let food be your medicine, and medicine be your food." This proverb pertains to mushrooms, which have high medical, pharmacological, and mineral values. Mushrooms occur naturally in many regions of the world and are also commercially grown. Mushrooms are a nutritious health food that has been utilised medicinally for millennia in various regions of the world. Although the nutritional information and culinary uses of mushrooms are well known, the fungi's therapeutic properties have yet to be widely accepted. As a result, they are a vital value to human welfare. In the future, *L. sulphureus* may be used to cure numerous diseases and may play a crucial role or act as a backbone in the study sector. They serve as a guiding principle for several studies in pharmacology, pharmacognosy, microbiology, and biotechnology.

REFERENCES

Agafonova, S. V., Olennikov, D. N., Borovskii, G. B., & Penzina, T. A. (2007). Chemical composition of fruiting bodies from two strains of *Laetiporus sulphureus*. *Chemistry of Natural Compounds*, *6*(43), 687–688.

Alquini, G., Carbonero, E. R., Rosado, F. R., Cosentino, C., & Iacomini, M. (2004). Polysaccharides from the fruit bodies of the basidiomycete Laetiporus sulphureus (Bull.: Fr.) Murr. *FEMS Microbiology Letters*, *230*(1), 47–52.

Alves, M. J., Ferreira, I. C. F. R., Dias, J., Teixeira, V., Martins, A., & Pintado, M. (2012). A review on antimicrobial activity of mushroom (Basidiomycetes) extracts and isolated compounds. *Planta Medica*, *78*(16), 1707–1718.

AM, Y. (2017). Anticancer potential of *Hericium erinaceus* extracts against particular human cancer cell lines. *Microbial Biosystems*, *2*(1), 9–20.

Appleton, R. E., Jan, J. E., & Kroeger, P. D. (1988). *Laetiporus sulphureus* causing visual hallucinations and ataxia in a child. *CMAJ: Canadian Medical Association Journal*, *139*(1), 48.

Ayaz, F. A., Torun, H., Özel, A., Col, M., Duran, C., Sesli, E., & Colak, A. (2011). Nutritional value of some wild edible mushrooms from Black Sea Region (Turkey). *Turkish Journal of Biochemistry/Turk Biyokimya Dergisi*, *36*(3).

Banik, M. T., Lindner, D. L., Ortiz-Santana, B., & Lodge, D. J. (2012). A new species of *Laetiporus* (Basidiomycota, Polyporales) from the Caribbean basin. *Kurtziana*, *37*(1), 15–21.

Benitez, B., Paez, C., Smith, M., & Smith, J. A. (2020). Chicken of the woods (*Laetiporus sulphureus* species complex): PP358, 10/2020. *EDIS*, *2020*(5).

Bernaś, E., Jaworska, G., & Lisiewska, Z. (2006). Edible mushrooms as a source of valuable nutritive constituents. *Acta Scientiarum Polonorum Technologia Alimentaria*, *5*(1), 5–20.

Bulam, S., Üstün, N. Ş., & Pekşen, A. (2019). Nutraceutical and food preserving importance of *Laetiporus sulphureus*. *Turkish Journal of Agriculture-Food Science and Technology*, *7*(sp1), 94–100.

Chan, P.-M., Kanagasabapathy, G., Tan, Y.-S., Sabaratnam, V., & Kuppusamy, U. R. (2013). *Amauroderma rugosum* (Blume & T. Nees) Torrend: Nutritional composition and antioxidant and potential anti-inflammatory properties. *Evidence-Based Complementary and Alternative Medicine*, *2013*, 806180.

Chatterjee, A., & Acharya, K. (2016). Include mushroom in daily diet—A strategy for better hepatic health. *Food Reviews International*, *32*(1), 68–97.

Chepkirui, C., Matasyoh, J. C., Decock, C., & Stadler, M. (2017). Two cytotoxic triterpenes from cultures of a Kenyan *Laetiporus* sp.(Basidiomycota). *Phytochemistry Letters*, *20*, 106–110.

Demir, M. S., & Yamaç, M. (2008). Antimicrobial activities of basidiocarp, submerged mycelium and exopolysaccharide of some native basidiomycetes strains. *Journal of Applied Biological Sciences*, *2*(3), 89–93.

Deol, B. S., Ridley, D. D., & Singh, P. (1978). Isolation of cyclodepsipeptides from plant pathogenic fungi. *Australian Journal of Chemistry*, *31*(6), 1397–1399.

Doğan, H. H., Şanda, M. A., Uyanöz, R., Öztürk, C., & Çetin, Ü. (2006). Contents of metals in some wild mushrooms: Its impact in human health. *Biological Trace Element Research*, *110*, 79–94.

Downer, A. J., & Perry, E. J. (2019). Wood decay fungi in landscape trees. *UC IPM Pest Notes; UC ANR Publication: Oakland, CA, USA*, *74109*, 6.

Durkan, N., Ugulu, I., Unver, M. C., Dogan, Y., & Baslar, S. (2011). Concentrations of trace elements aluminum, boron, cobalt and tin in various wild edible mushroom species from Buyuk Menderes River Basin of Turkey by ICP-OES. *Trace Elements and Electrolytes*, *28*(4), 242.

EIu, E., Tikhonova, O. V., Lur'e, L. M., Efremenkova, O. V., & Kamzolkina, O. V. (2003). Antimicrobial activity of *Laetiporus sulphureus* strains grown in submerged culture. *Antibiotiki i Khimioterapiia= Antibiotics and Chemoterapy [Sic]*, *48*(1), 18–22.

Elsayed, E. A., el Enshasy, H., Wadaan, M. A. M., & Aziz, R. (2014). Mushrooms: A potential natural source of anti-inflammatory compounds for medical applications. *Mediators of Inflammation*, *2014*, 805841.

Ericsson, D. C. B., & Ivonne, J. N. R. (2009). Sterol composition of the macromycete fungus Laetiporus sulphureus. *Chemistry of Natural Compounds*, *45*, 193–196.

Fontana, S., Flugy, A., Schillaci, O., Cannizzaro, A., Gargano, M. L., Saitta, A., de Leo, G., Venturella, G., & Alessandro, R. (2014). In vitro antitumor effects of the cold-water extracts of Mediterranean species of genus *Pleurotus* (higher Basidiomycetes) on human colon cancer cells. *International Journal of Medicinal Mushrooms*, *16*(1).

Grienke, U., Zöll, M., Peintner, U., & Rollinger, J. M. (2014). European medicinal polypores–A modern view on traditional uses. *Journal of Ethnopharmacology*, *154*(3), 564–583.

Gumińska, B., & Wojewoda, W. (1985). *Grzyby i ich oznaczanie*. Państwowe Wydawnicto Rolnicze i Leśne.

Hassan, K., Matio Kemkuignou, B., & Stadler, M. (2021). Two new triterpenes from basidiomata of the medicinal and edible mushroom, *Laetiporus sulphureus*. *Molecules*, *26*(23), 7090.

Hwang, H. S., & Yun, J. W. (2010). Hypoglycemic effect of polysaccharides produced by submerged mycelial culture of *Laetiporus sulphureus* on streptozotocin induced diabetic rats. *Biotechnology and Bioprocess Engineering*, *15*, 173–181.

Jayasooriya, R., Kang, C.-H., Seo, M.-J., Choi, Y. H., Jeong, Y.-K., & Kim, G.-Y. (2011). Exopolysaccharide of *Laetiporus sulphureus* var. miniatus downregulates LPS-induced production of NO, PGE2, and TNF-α in BV2 microglia cells via suppression of the NF-κB pathway. *Food and Chemical Toxicology*, *49*(11), 2758–2764.

Kariyone, T., & Kurono, G. (1940). Constituents of Fomes officinalis Fr. *Journal of Pharmaceutical Society of Japan*, *60*, 318.

Khatua, S., Ghosh, S., & Acharya, K. (2017). *Laetiporus sulphureus* (Bull.: Fr.) Murr. as food as medicine. *Pharmacognosy Journal*, *9*(6s).

Klaus, A., Kozarski, M., Niksic, M., Jakovljevic, D., Todorovic, N., Stefanoska, I., & van Griensven, L. J. L. D. (2013). The edible mushroom *Laetiporus sulphureus* as potential source of natural antioxidants. *International Journal of Food Sciences and Nutrition*, *64*(5), 599–610.

Kolundzic, M. D., Grozdanic, N. O., Stanojkovic, T. P., Milenkovic, M. T., Dinic, M. R., Golic, N. E., Kojic, M., & Kundakovic, T. D. (2016). Antimicrobial and cytotoxic activities of the sulphur shelf medicinal mushroom, *Laetiporus sulphureus* (Agaricomycetes), from Serbia. *International Journal of Medicinal Mushrooms*, *18*(6).

Końska, G., Wójtowicz, U., & Pituch-Noworolska, A. (2008). Possible application of lectins in diagnostics and therapy. Part I. Diagnostic application. *Przeglad Lekarski*, *65*(4), 189–194.

Kovács, D., & Vetter, J. (2015). Chemical composition of the mushroom *Laetiporus sulphureus* (Bull.) Murill. *Acta Alimentaria*, *44*(1), 104–110.

Lear, M. J., Simon, O., Foley, T. L., Burkart, M. D., Baiga, T. J., Noel, J. P., DiPasquale, A. G., Rheingold, A. L., & la Clair, J. J. (2009). Laetirobin from the parasitic growth of *Laetiporus sulphureus* on *Robinia pseudoacacia*. *Journal of Natural Products*, *72*(11), 1980–1987.

León, F., Quintana, J., Rivera, A., Estévez, F., & Bermejo, J. (2004). Lanostanoid Triterpenes from *Laetiporus sulphureus* and apoptosis induction on HL-60 human myeloid leukemia cells. *Journal of Natural Products*, *67*(12), 2008–2011.

Lovy, A., Knowles, B., Labbe, R., & Nolan, L. (2000). Activity of edible mushrooms against the growth of human T4 leukemic cancer cells, HeLa cervical cancer cells, and *Plasmodium falciparum*. *Journal of Herbs, Spices & Medicinal Plants*, *6*(4), 49–57.

Luangharn, T., Hyde, K. D., & Chukeatirote, E. (2014). Proximate analysis and mineral content of *Laetiporus sulphureus* strain MFLUCC 12-0546 from northern Thailand. *Chiang Mai Journal of Science*, *41*(4), 765–770.

Mandic, R., Mesud, A., & Marjanović, Ž. (2018). Conservation and trade of wild edible mushrooms of Serbia-history, state of the art and perspectives. *Nature Conservation-Bulgaria*, *25*, 31–53.

Mlinaric, A., Kac, J., & Pohleven, F. (2005). Iskanje inhibitorjev HIV-1 reverzne transkriptaze v lesnih glivah. *Acta Pharmaceutica*, *55*(1), 69–79.

Murrill, W. A. (1904). The polyporaceae of North America-IX. Inonotus, sesia and monotypic genera. *Bulletin of the Torrey Botanical Club*, *31*(11), 593–610.

Ofodile, L. N., Uma, N. U., Kokubun, T., Grayer, R. J., Ogundipe, O. T., & Simmonds, M. S. J. (2005). Antimicrobial activity of some Ganoderma species from Nigeria. *Phytotherapy Research: An International Journal Devoted to Pharmacological and Toxicological Evaluation of Natural Product Derivatives*, *19*(4), 310–313.

Okamura, T., Takeno, T., Dohi, M., Yasumasa, I., Hayashi, T., Toyoda, M., Noda, H., Fukuda, S., Horie, N., & Ohsugi, M. (2000). Development of mushrooms for thrombosis prevention by protoplast fusion. *Journal of Bioscience and Bioengineering*, *89*(5), 474–478.

Olennikov, D. N., Agafonova, S. V., Borovskii, G. B., Penzina, T. A., & Rokhin, A. V. (2009). Alkali-soluble polysaccharides of *Laetiporus sulphureus* (Bull.: Fr.) Murr fruit bodies. *Applied Biochemistry and Microbiology*, *45*, 626–630.

Ota, Y., Hattori, T., Banik, M. T., Hagedorn, G., Sotome, K., Tokuda, S., & Abe, Y. (2009). The genus *Laetiporus* (Basidiomycota, Polyporales) in East Asia. *Mycological Research*, *113*(11), 1283–1300.

Palazzolo, E., Letizia Gargano, M., & Venturella, G. (2012). The nutritional composition of selected wild edible mushrooms from Sicily (southern Italy). *International Journal of Food Sciences and Nutrition*, *63*(1), 79–83.

Pârvu, M., Andrei, A.-Ş., & Roşca-Casian, O. (2010). Antifungal activity of *Laetiporus sulphureus* mushroom extract. *Contributii Botanice*, *45*.

Pavic, A., Ilic-Tomic, T., & Glamočlija, J. (2021). Unravelling anti-melanogenic potency of edible mushrooms *Laetiporus sulphureus* and *Agaricus silvaticus* in vivo using the zebrafish model. *Journal of Fungi*, *7*(10), 834.

Pekşen, A., Bulam, S., & Üstün, N. Ş. (2016). Edible wild mushrooms sold in Giresun local markets. *1st International Mediterranean Science and Engineering Congress (IMSEC 2016) Çukurova University, Congress Center*, 3358–3362.

Petrović, J., Papandreou, M., Glamočlija, J., Ćirić, A., Baskakis, C., Proestos, C., Lamari, F., Zoumpoulakis, P., & Soković, M. (2014). Different extraction methodologies and their influence on the bioactivity of the wild edible mushroom *Laetiporus sulphureus* (Bull.) Murrill. *Food & Function*, *5*(11), 2948–2960.

Petrović, J., Stojković, D., Reis, F. S., Barros, L., Glamočlija, J., Ćirić, A., Ferreira, I. C. F. R., & Soković, M. (2014). Study on chemical, bioactive and food preserving properties of *Laetiporus sulphureus* (Bull.: Fr.) Murr. *Food & Function*, *5*(7), 1441–1451.

Pleszczyńska, M., Wiater, A., Siwulski, M., & Szczodrak, J. (2013). Successful large-scale production of fruiting bodies of *Laetiporus sulphureus* (Bull.: Fr.) Murrill on an artificial substrate. *World Journal of Microbiology and Biotechnology*, *29*, 753–758.

Radic, N., & Injac, R. (2009). Sulphur tuft culinary-medicinal mushroom, *Laetiporus sulphureus* (Bull.: Fr.) Murrill (Aphyllophoromycetideae): bioactive compounds and pharmaceutical effects. *International Journal of Medicinal Mushrooms*, *11*(2).

Rapior, S., Konska, G., Guillot, J., Andary, C., & Bessiere, J.-M. (2000). Volatile composition of *Laetiporus sulphureus*. *Cryptogamie Mycologie*, *21*(1), 67–72.

Rios, J.-L., Andujar, I., Recio, M.-C., & Giner, R.-M. (2012). Lanostanoids from fungi: A group of potential anticancer compounds. *Journal of Natural Products*, *75*(11), 2016–2044.

Saba, E., Son, Y., Jeon, B. R., Kim, S.-E., Lee, I.-K., Yun, B.-S., & Rhee, M. H. (2015). Acetyl eburicoic acid from *Laetiporus sulphureus* var. miniatus suppresses inflammation in murine macrophage RAW 264.7 cells. *Mycobiology*, *43*(2), 131–136.

Saha, D., Sundriyal, M., & Sundriyal, R. C. (2014). *Diversity of food composition and nutritive analysis of edible wild plants in a multi-ethnic tribal land, Northeast India: an important facet for food supply.*

Sakeyan, C. Z. (2006). Antifungal activity of several xylotrophic medicinal mushrooms against filamentous fungi-potentially pathogenic for humans and animals. *Electronic Journal of Natural Sciences*, *6*(1).

Sato, M., Tai, T., Nunoura, Y., Yajima, Y., Kawashima, S., & Tanaka, K. (2002). Dehydrotrametenolic acid induces preadipocyte differentiation and sensitizes animal models of noninsulin-dependent diabetes mellitus to insulin. *Biological and Pharmaceutical Bulletin*, *25*(1), 81–86.

Seo, M.-J., Kang, B.-W., Park, J.-U., Kim, M.-J., Lee, H.-H., Choi, Y.-H., & Jeong, Y.-K. (2011). Biochemical characterization of the exopolysaccharide purified from *Laetiporus sulphureus* mycelia. *Journal of Microbiology and Biotechnology, 21*(12), 1287–1293.

Sesli, E. (2007). Preliminary checklist of the macromycetes of the East and Middle Black Sea Regions of Turkey. *Mycotaxon, 99*, 71–74.

Sesli, E., & Denchev, C. M. (2008). Checklists of the myxomycetes, larger ascomycetes, and larger basidiomycetes in Turkey. *Mycotaxon, 106*(2008), 65.

Šiljegović, J. D., Stojković, D. S., Nikolić, M. M., Glamočlija, J. M., Soković, M. D., & Ćirić, A. M. (2011). Antimicrobial activity of aqueous extract of *Laetiporus sulphureus* (Bull.: Fr.) Murill. *Zbornik Matice Srpske Za Prirodne Nauke, 120*, 299–305.

Sinanoglou, V. J., Zoumpoulakis, P., Heropoulos, G., Proestos, C., Ćirić, A., Petrovic, J., Glamoclija, J., & Sokovic, M. (2015). Lipid and fatty acid profile of the edible fungus*Laetiporus sulphurous*. Antifungal and antibacterial properties. *Journal of Food Science and Technology, 52*, 3264–3272.

Singdevsachan, S. K., Auroshree, P., Mishra, J., Baliyarsingh, B., Tayung, K., & Thatoi, H. (2016). Mushroom polysaccharides as potential prebiotics with their antitumor and immunomodulating properties: A review. *Bioactive Carbohydrates and Dietary Fibre, 7*(1), 1–14.

Smith, A. H., Weber, N. S., & Weber, N. S. (1980). *The mushroom hunter's field guide*. University of Michigan Press.

Soares, A. A., de Sá-Nakanishi, A. B., Bracht, A., da Costa, S. M. G., Koehnlein, E. A., de Souza, C. G. M., & Peralta, R. M. (2013). Hepatoprotective effects of mushrooms. *Molecules, 18*(7), 7609–7630.

Song, J., Chen, Y., Cui, B., Liu, H., & Wang, Y. (2014). Morphological and molecular evidence for two new species of *Laetiporus* (Basidiomycota, Polyporales) from southwestern China. *Mycologia, 106*(5), 1039–1050.

Song, J., & Cui, B.-K. (2017). Phylogeny, divergence time and historical biogeography of *Laetiporus* (Basidiomycota, Polyporales). *BMC Evolutionary Biology, 17*, 1–12.

Sugimoto, S., Chi, G., Kato, Y., Nakamura, S., Matsuda, H., & Yoshikawa, M. (2009). Medicinal Flowers. XXVI. Structures of acylated oleanane-type triterpene oligoglycosides, Yuchasaponins A, B, C, and D, from the flower buds of *Camellia oleifera*—Gastroprotective, aldose reductase inhibitory, and radical scavenging effects. *Chemical and Pharmaceutical Bulletin, 57*(3), 269–275.

Sułkowska-Ziaja, K., Muszyńska, B., Gawalska, A., & Sałaciak, K. (2018). *Laetiporus sulphureus*: Chemical composition and medicinal value. *Acta Scientiarum Polonorum. Hortorum Cultus, 17*(1).

Sun, W. J., He, H. B., Wang, J. Z., Wu, L., Cheng, F., & Deng, Z. S. (2014). The main components analysis of *Laetiporus sulphureus* crude extract and its hepatoprotective effect on carbon tetrachloride-induced hepatic fibrosis in rats. *Applied Mechanics and Materials, 568*, 1934–1939.

Turkoglu, A., Duru, M. E., Mercan, N., Kivrak, I., & Gezer, K. (2007). Antioxidant and antimicrobial activities of *Laetiporus sulphureus* (Bull.) Murrill. *Food Chemistry, 101*(1), 267–273.

Valverde, M. E., Hernández-Pérez, T., & Paredes-López, O. (2015). Edible mushrooms: improving human health and promoting quality life. *International Journal of Microbiology, 2015*.

Vamanu, E. (2012). In vitro antimicrobial and antioxidant activities of ethanolic extract of lyophilized mycelium of Pleurotus ostreatus PQMZ91109. *Molecules, 17*(4), 3653–3671.

Wang, J., Sun, W., Luo, H., He, H., Deng, W.-Q., Zou, K., Liu, C., Song, J., & Huang, W. (2015). Protective effect of eburicoic acid of the chicken of the woods mushroom,*Laetiporus sulphureus* (higher Basidiomycetes), against gastric ulcers in mice. *International Journal of Medicinal Mushrooms, 17*(7).

Wiater, A., Paduch, R., Pleszczyńska, M., Próchniak, K., Choma, A., Kandefer-Szerszeń, M., & Szczodrak, J. (2011). α-(1→ 3)-d-Glucans from fruiting bodies of selected macromycetes fungi and the biological activity of their carboxymethylated products. *Biotechnology Letters, 33*, 787–795.

Wiater, A., Pleszczyńska, M., Szczodrak, J., & Janusz, G. (2012). Comparative studies on the induction of *Trichoderma harzianum* mutanase by α-(1→ 3)-glucan-rich fruiting bodies and mycelia of *Laetiporus sulphureus*. *International Journal of Molecular Sciences, 13*(8), 9584–9598.

Wu, X., Yang, J., Zhou, L., & Dong, Y. (2004). New lanostane-type triterpenes from *Fomes officinalis*. *Chemical and Pharmaceutical Bulletin, 52*(11), 1375–1377.

Xu, B. B., Li, C., & Sung, C. (2014). Telomerase inhibitory effects of medicinal mushrooms and lichens, and their anticancer activity. *International Journal of Medicinal Mushrooms, 16*(1).

Yin, X., Li, Z.-H., Li, Y., Feng, T., & Liu, J.-K. (2015). Four lanostane-type triterpenes from the fruiting bodies of mushroom *Laetiporus sulphureus* var. miniatus. *Journal of Asian Natural Products Research, 17*(8), 793–799.

Yoshikawa, K., Bando, S., Arihara, S., Matsumura, E., & Katayama, S. (2001). A benzofuran glycoside and an acetylenic acid from the fungus *Laetiporus sulphureus* var. miniatus. *Chemical and Pharmaceutical Bulletin*, *49*(3), 327–329.

Zjawiony, J. K. (2004). Biologically active compounds from Aphyllophorales (polypore) fungi. *Journal of Natural Products*, *67*(2), 300–310.

10 *Morchella esculenta* (Gucchi)/ *Morchella conica* – Krombh/ *Morchella elata* Fr., Family – Morchellaceae

Pooja Kapoor, Priyanka Kumari, Satyaranjan Padhiary, Abhishek Katoch, and Jeewan Tamang
University Institute of Agriculture Science, Chandigarh University, Gharuan, Mohali, India

CONTENTS

INTRODUCTION

India has a vast and diverse geographical distribution. The northern greater range of the Himalayas, extending from the northeast to the far northwest, constitutes an area of approximately 0.53 million km2, equivalent to 16% of the total land coverage. Over 50% of green forests, including some national parks and preserve areas, come under this belt, which comprises 40% of endemic species (Kandari et al., 2012). Numerous wild mushroom species with nutritional and therapeutic benefits can be found in abundance in these areas. Humans place great value on wild mushrooms because of their delectability, high nutritional content, and medicinal properties (Raman et al., 2018). Most

DOI: 10.1201/9781003259763-10

mushrooms are classified as Basidiomycota or Ascomycota. They are described as a fungus that often has an expanded fleshy fruiting body that resembles an umbrella on a stalk grown above the ground, known as a pileus (Chang and Miles, 2004). Bio-macromolecules like polysaccharides, polynucleotides, and proteins, found in mushrooms, have been shown to help treat diabetes, heart disease, and even cancer (Thakur et al., 2021). *Morchella* spp., often known as morels or true morels, is one of the most well-known, expensive, and nutrient-dense mushrooms (Sayeed et al., 2018). *Morchella* species frequently encountered include *M. rufobrunnea, M. dunesis, M. importuna, M. dispaeilis, M. esculenta, M. semilibera, M. arbutiphila, and M. steppicola* (Loizides et al., 2017). These species are found in the Northern Hemisphere's temperate zones, primarily in Asia and parts of North America and Asia. Currently, 32 species of the genus Morchella are known in India. Out of these, six in the Himalayas are mentioned below.

1. *M. crassipes* (Broad-stemmed)
2. *M. deliciosa* (Delectable)
3. *M. esculenta* (True-Morels)
4. *M. conica*
5. *M. angusticeps* (Dark Moth)
6. *M. hybrida* (Cross Moth) (Yoon et al., 1990; Lakhanpal et al., 2010).

As per Dorfelt (Table 10.1), *Morchella esculenta*, which has a delicate flavor and an alluring look, is the most significant and economically beneficial wild mushroom in the genus *Morchella*. In different places, it is called yellow morel, true morel, sponge morel, and common morel. In Kashmir, it is called Kann Gitch, which appears ear-shaped (Sayeed et al., 2018). Its remarkable bee comb appearance and web of elevated and concave holes give the pileus, the edible fruiting body of morels, a sponge-like appearance (Longley et al., 2019). Despite being widely distributed worldwide, *M. esculenta* is most frequently found in the northwest Himalayan region, particularly in J&K, H.P., and Uttarakhand (Manikandan et al., 2011). It can develop "mycorrhizal" associations with broadleaf, coniferous, and hardwood trees. It is thought to be found near wild strawberry plants, ferns, ash trees, and elm trees (Wagay and Vyas, 2011).

A wide variety of active compounds, such as polysaccharides, carotenoids, multiple phenolic compounds, tocopherols, and organic acids, are abundant in the fruiting bodies of *M. esculenta* and have pharmacological and therapeutic properties. The mushroom extract is antibacterial, anti-inflammatory, antioxidant, immune-stimulatory, and fights tumors (Tietel and Masaphy, 2018). Many of the morel's elements, namely the fruiting body, mycelium, are used as an antiseptic for wound healing and as a powder to treat stomach issues. *Morchella* species have been used as traditional medicine by people who have lived at high altitudes in the Himalayas for the past 2,000 years (Ajmal et al., 2015). This has helped them eat and stay healthy in harsh weather conditions (Ajmal et al., 2015). Cancer, chronic arthritis, asthma, and indigestion are just some diseases it is often used to treat. Proteins, carbs, fibers, all three necessary vitamins (A, B, and C), minerals, and aromatic chemicals like ketone, aldehyde, esters, and terpenes are all found in them. Because of its unique aroma, taste, and texture, it is found in various types of food around the globe (Raman et al., 2018).

Notably, the commercial demand for *M. esculenta* has continuously risen in fancy restaurants and drug companies. The world's most expensive mushroom is a rare wild mushroom found only in a few spots that involve challenging harvesting. Due to this, it is renowned as the "growing gold of mountains," which significantly promotes the economic progress of any nation (Taşkın et al., 2012). Various researchers have spent considerable time and energy hoping to culture this rare and precious fungus artificially. Since then, much work has been done to make its culture marketable (Thakur et al., 2021). This review paper talks about the current state of *M. esculenta* production in the Central Himalayas and several research and development projects.

TABLE 10.1
Vernacular Names of *M. esculenta* (Ajmal et al., 2015)

Region/ Language	Vernacular Name	References
French	Morille	(Roody 2003)
Germany	Speisemorchel	
Italian	spugnola bruna	
Spanish	Colmenilla	
Nepal	Guchi chyau	
India	Guchhi	(Paliwal et al., 2013)
Pakistan	Gujae	(Razaq et al., 2010)
	Guchhi	(Mahmood et al., 2011)
	Spina Guchhi	(Sher et al., 2014)
	Kerkichoke	(Gilani et al., 2003)
	Khosay	(Hassan et al., 2015)

TAXONOMY

True morels are treasured wild edible mushrooms that are regarded as a wonderful gift to humanity, given their curative, dietary, and economic utility (Laalaa et al., 2019). Morels are typically polymorphic, ranging in dimension, morphology, flavor, texture, aroma, and genetic complexity among their species. The existing polymorphism has sparked taxonomical disagreements for over a century (Irfan et al., 2017). Morels are currently recognized as constituents of the Discomycetes group, order Pezizales, and family *Morchellaceae* of Ascomycota based on the use of molecular phylogenetic approaches (Liu et al., 2017). In several parts of the globe, morels are widespread and recognized by different names, namely Morille-France, Colmenilla-Spain, Guchi-chyau-Nepal, and Guchhi-Hisdustan (Ajmal et al., 2015) (Table 10.1).

Taxonomy of Morchella esculenta

Kingdom: Fungi
Phylum: Ascomycota
Class: Discomycetes
Order: Pezizales
Family: *Morchellaceae*
Genus: *Morchella*
Species: *Morchella escuelnta (L.) Pers*

FEATURES OF MORELS

The color, texture, form, and size of true morel fruiting bodies can vary as they are polymorphic (Masaphy et al., 2010). Hyphae masses are the constant form of *Morchella esculenta* mycelium. It grows in humus soil and has discrete, branching, filamentous, septate hyphae. The morel's fruiting body is recognized as "pileus," mainly composed of the ascus, and every ascus comprises eight circular ascospores (Kanwal et al., 2012). They have a 5–12 cm-high, cone-shaped pileus, porous on its inner side and masked by random holes and furrows, creating a peculiar bee comb pattern. The tone's appearance varies from whitish to yellow-brownish.

Further, it gets darker with age (Reilly 2016). The base part or stem is generally cream-white to grayish in appearance. Subsequently, it again develops a grey-brownish color, whereas the cap is yellowish in appearance, comprising 70%–80% of the whole mass (Longley et al., 2017). The soil containing a pH value of (4.7–6.9), enriched in organic matter, and has a temperature of (4–5.5)°C is the requirement of *Morchella esculenta* growth and development in shades of evergreen trees such as oak, maples, etc. It needs chilly temperatures as it primarily thrives in regions having hot conditions, especially in the spring (Kakakhel 2020). It creates symbiotic partnerships with coniferous plants by fusing mycelium and roots. This symbiotic relationship is referred to as the ectomycorrhizal association (Kuo et al., 2008). It contains various bioactive compounds that enhance the food's taste, aroma, delectability, nutritional value, and medicinal properties. Local cultivators often perceive fake morels as they mimic Morchella species (true morels). Based on the chemical gyromitrin, which is quickly converted to the compound monomethyl hydrazine, fake morels are potentially toxic (MMH). This compound elevates the risk of cancer and digestive issues, and in extreme cases, it becomes fatal. False morels encompass species like *Gyromitra esculenta, G. caroliniana, G. infula, and Verpa conica. Gyromitra esculenta* comprises a reddish-brown cap that resembles the brain and is packed with cotton-like hairs. This structural recognition, along with breaking the cap into half portions, aids in differentiating between false morels and actual morels (Lagrange and Vernoux, 2020) (Figure 10.1).

SETTLEMENT OF MORELS, AND ITS BIOGEOGRAPHICAL DISTRIBUTION

Morels grow in the Northern Hemisphere, in the Mediterranean basin and subtropical areas like the western states of the United States, the Mexican mountain range, Egypt and Arab countries, eastern Asia, and the northern sides of India, Pakistan, and Nepal (Loizides et al., 2017). In contrast, locations such as the northern side of the USA, Canada, some mountainous areas, and several European countries have a habitat of *Morchella esculenta* that is bulky and dense (Figure 10.2). In Indian sub-continents, places like the far northwest include J & K (Jammu and Kashmir), and northwest states such as Himachal and UP (hilly regions) are the dominant states (Lakhanpal and Shad, 1986). In addition, a small portion is identified in the northeast and seven sister states (Ghosh and Pathak, 1962). The species prefer a chilly, temperate climate and thrive in loamy soil enriched in organic matter and nitrogen. It relies on other core components such as soil containing a moisture level of 20%, a temperature of (11–21)°C on the surface and (9–20)°C beneath, and a pH value of 7.5 above (Kaul et al., 1981). It is commonly prevalent in evergreen woods, broad-leaved forests, apple orchards, meadows, pastures, paddy cultivation bunds, and corn fields in Solan (H.P.) and occasionally in burnt regions. It usually thrives in mountainous areas, roughly at an elevation of 2500–2000 meters in evergreen woods (Ajmal et al., 2015). Morels are primarily prevalent in the spring, between March and May, when the snowfall is melting, followed by precipitation. In addition, some Morchella species have been discovered throughout the monsoon season (Shad et al., 1989). Therefore, chilly, damp weather with an R.H. range of 60–70% and an average temperature of 60–70 degrees Fahrenheit is ideal for developing morels (Manikandan et al., 2011).

DIVERSITY OF SPECIES

The genus Morels has about 80 species, which are categorized into three independent clades: the Rufobrunnea clade, which is entirely made up of white morel species; the Esculenta clade- yellow variant, Elata clade- Dark variant species (Loizides et al., 2018) (Figure 10.3) (Table 10.2). Rufobrunnea clades include *M. anatolica* and *M. rufobrunnea* (Isiloglu et al., 2010). The Esculenta clade of Morchellaincludes *M. americana* (Clowez et al., 2012), *M. dunensis* (Taşkın et al., 2010), *M. esculenta* (Du et al., 2012a), *M. fluvialis,* and *M. steppicola* (Yatsiuk et al., 2016). Furthermore, the Elata clade also contains *M. arbutiphila* (Taşkın et al., 2010, 2012), *M. deliciosa* (Jacquetant et al., 1984; O'Donnell et al., 2011), *M. disparilis* (O'Donnell et al., 2011), *M. dunaii* (Bouldier et al., 1909), *M. eximia* (Bouldier et al., 1909), *M. exuberans* (Boul). Roughly about 32 species are

FIGURE 10.1 Identification of fake and real i.e (A) Real morels with hollow in its inner surface (B) cut (section of true morels (C) & (D) denotes Gyromitra esculenta, fake morel with a whitish soft smooth fiber on the inner surface).

believed to exist in India, of which 6 members of the genus Morchella are identified in the Himalayan region. The above six species encompass *M. angusticeps* (black morel*), M. crassipes* (thick stem variety), *M. esculenta* (common variety), *M. semilibera* (hybrid variety), *M. conica* (conical morel), and *M. deliciosa* (delicious morel) (Prasad et al., 2002; Lakhanpal et al., 2010).

SYSTEMATIC DEVELOPMENT OF MORELS OVER THE PERIOD OF TIME

Evolution of Morchella Species

The existence of the true Morels began during the Cretaceous period approximately 129 million years ago (O'Donnell et al., 2011). Then, the economically significant genus *Morchella* has over

FIGURE 10.2 Global presence of Morels (Lladó et al, 2017).

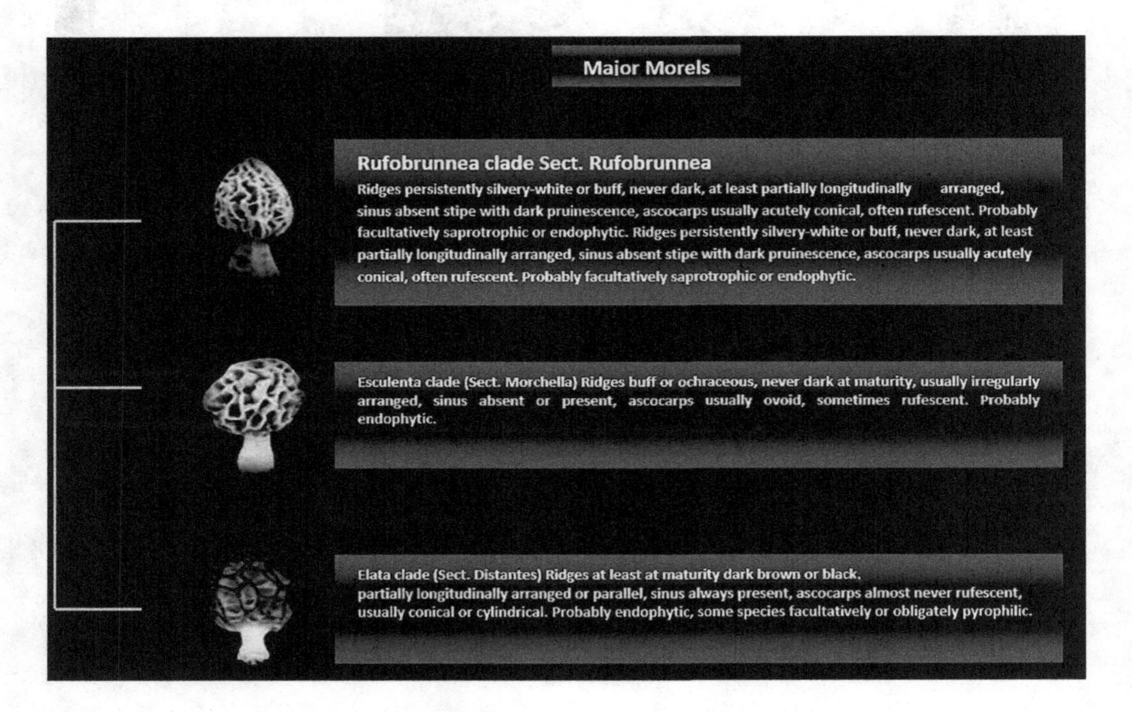

FIGURE 10.3 Different Morel-clades along it's their unique features (Loizides, 2018).

80 species classified and dispersed over several geographical locations (Baroni et al., 2018). The species of *Morchella*, such as *M. Anatolia* and *M. rufobrunnea*, comprise on first diverging species and are significant for evaluating the Morchella's background, ecology, race of evolution, etc. (Loizides et al., 2021). The three clades that make up the genus *Morchella* are *Morchella elata* (late-diverging black morels), *M. esculenta* (yellow morels), and *M. rufobrunnea* (white morels). Several Morchella species are genetically and physically unique (Loizides et al., 2021).

Regarding economic, sociological, and ethnomycological considerations, the real morel is one of these three clades and is one of the significant fungi; its consumption is increasing significantly across all nations. However, morels are only available in a few places, which drives costs (Thakur et al., 2021). The native morels available in nature are being picked from their original habitat and exported to countries like China, the USA, and Turkey to meet the expanding demand. (Taşkın

TABLE 10.2
Current Global Distribution of Morels Along with Its Environment (Loizides et al., 2018)

Rufobrunnea clade

Species	Ecology/trees association	Biogeographical distribution	References
Morchella anatolica	Uncertain, perhaps saprotrophic	Spain, Turkey	(Isiloglu et al. 2010; Clowez et al., 2012; Richard et al., 2015)
M. rufobrunnea	Saprotrophic, but probably also facultatively biotrophic or endophytic, found on disturbed ground, wood mulch and under Olea	Australia, Cyprus, Israel, Mexico, USA	Guzmän and Tapia 1998; Kuo et al., 2008; Masaphy et al., 2009; Loizides et al., 2011; O'Donnell et al., 2011; Kuo et al., 2012; Elliott et al., 2014; Loizides et al., 2011; 2015; 2016)

Esculenta-clade

Species	Ecology/trees association	Biogeographical distribution	References
Morchella americana	Probably biotrophic or endophytic, associated with Fraxinus, Ulmus, Populus, Quercus and Acer	Canada, France, Germany, Turkey, USA	(Taşkın et al., 2010; 2012; O'Donnell et al., 2011; Clowez et al., 2012; Kuo et al., 2012; Richard et al., 2015)
M. castanea	Reported with Castanea, Fraxinus and Populus	Spain	(Clowez et al., 2012; Richard et al., 2015)
M. dunensis	Probably saprotrophic as well as facultatively biotrophic or endophytic, found on sand dunes and under Malus	Cyprus, Spain, Turkey	Castañera and Moreno 1996; Taşkın et al., 2010; 2012; Clowez, 1997; 2012; Loizides et al., 2011; 2016)
M. esculenta	Mostly associated with Fraxinus, Malus, Populus, Quercus and Ulmus	Belgium, China, Czech Republic, France, Netherlands, Norway, Spain, Sweden, Switzerland, Turkey	Persoon et al., 1794; Fries 1822; Krombholz 1834; Boudier 1897; Jacquetant 1984; Taşkin et al., 2010; 2012; O'Donnell et al., 2011; Du et al., 2012a; 2012b; Clowez, 1997; 2012; Loizides et al., 2011; 2016; Richard et al., 2015)
M. fluvialis	Found in riparian forests of Fraxinus, Alnus and Ulmus	Spain, Turkey	(Taşkin et al., 2010; 2012; Clowez, 1997; Clowez 2012; Richard et al., 2015)
M. galilaea	Uncertain, reported under a wide diversity of vegetation, including greenhouses	Africa, China, Hawaii, India, Israel, Java, New Zealand, Turkey	(Masaphy et al., 2009; O'Donnell et al., 2011; Taşkin et al., 2010; 2012; Clowez et al., 2012; Du et al., 2012a; 2012b; Richard et al., 2015; Loizides et al., 2016)
M. palazonii	Reported under Quercus ilex and Fraxinus angustifolia	Spain	(Richard et al., 2015)
M. steppicola	Probably saprotrophic or endophytic, found in steppic meadows and dry grasslands	Germany, Hungary, Serbia, Slovakia, Ukraine, Uzbekistan	(Zerova 1941; O'Donnell et al., 2011; Kuo et al.,

(*Continued*)

TABLE 10.2 ***(Continued)***
Current Global Distribution of Morels Along with Its Environment (Loizides et al., 2018)

Rufobrunnea clade

Species	Ecology/trees association	Biogeographical distribution	References
			2012; Richard et al., 2015; Yatsiuk et al., 2016)
M. vulgaris	Likely biotrophic or endophytic, associated with Acer, Crataegus, Fraxinus, Sorbus and Ulmus	France, Spain	(Persoon 1801; Fries 1822; Gray 1821; Clowez 2012; Richard et al., 2015)

Elata-clade

Species	Ecology/trees association	Biogeographical distribution	References
Morchella arbutiphila	So far strictly associated with Arbutus andrachne	Cyprus, Turkey	(Taşkin et al., 2010; 2012; Loizides et al., 2016)
M. conifericola	Reported with Pinus, Cedrus and Abies	Turkey	(Taşkin et al., 2012; 2016)
M. deliciosa	Associated with conifers, mostly Larix, Picea and Pinus	France, Sweden, Turkey	(Taşkin et al., 2010; 2012; Clowez 2012; Richard et al., 2015)
M. disparilis	Probably biotrophic or endophytic, found under Cupressus and Arbutus	Cyprus	(Loizides et al., 2016)
M. dunaii	Associated with Pinus brutia and evergreen Quercus spp., rarely also with Cistus	Cyprus, France, Spain, Turkey	(Boudier 1897; Taşkin et al., 2010; 2012; Loizides et al., 2011; 2016; Clowez 2012; Richard et al., 2015; Loizides 2016)
M. eohespera	Uncertain, reported under mixed vegetation	France, Canada, China, Germany, Sweden, Switzerland, JUSA	(Richard et al., 2015; Voitk et al., 2016)
M. eximia	Strictly pyrophilic, found in 1–2-year-old burned coniferous forests	Argentina, Australia, Canada, China, Cyprus, France, Spain, Turkey, USA	(Boudier 1909; Jacquetant 1984; Taşkin et al., 2010; 2012; O'Donnell et al., 2011; Kuo et al., 2012; Du et al., 2012b; Richard et al., 2015; Loizides et al., 2016)
M. eximiodes	Uncertain, originally reported under deciduous trees	China, Norway	(Jacquetant 1984; Du et al., 2012b; Richard et al., 2015)
M. exuberans	Strictly pyrophilic, found in 1–2-year-old burned coniferous forests	China, Cyprus, Sweden, Turkey, USA	(Taşkın et al., 2010; 2012; O'Donnell et al., 2011; Du et al., 2012a; 2012b; Loizides et al., 2016)
M. fekeensis	Mostly associated with Pinus	Turkey	(Taşkin et al., 2010; 2012; 2016)
M. importuna	Facultatively pyrophilic, found in 1–2-year-old burned forests, disturbed ground, wood mulch, but also under Malus	Canada, China, Cyprus, Finland, France, Germany, Spain,Switzerland, Turkey, USA	(Taşkın et al., 2010; 2012; O'Donnell et al., 2011; Kuo et al., 2012; Du et al., 2012a; 2012b; O'Donnell 2014; Loizides 2016
M. kakiicolor	Associated with Castanea sativa	Spain	O'Donnell et al., 2011; Clowez 2012; Richard

TABLE 10.2 *(Continued)*

Current Global Distribution of Morels Along with Its Environment (Loizides et al., 2018)

Rufobrunnea clade

Species	Ecology/trees association	Biogeographical distribution	References
			et al., 2015; Loizides et al., 2015)
M. magnispora	Mostly reported from coniferous forests	Turkey	(Taşkin et al., 2010; 2012; 2016)
M. mediterraneensis	Mostly reported from coniferous forests	Spain, Turkey	(Taşkin et al., 2010; 2012; 2016)
M. pulchella	Uncertain, reported with both broadleaved and coniferous trees	China, France, Turkey	(Taşkın et al., 2010; 2012; Clowez 2012; Du et al., 2012b; Richard et al., 2015)
M. purpurascens	Associated with Pinus and other conifers	China, Denmark, France, Norway, Sweden, Turkey	(Krombholz 1834; Boudier 1897; Jacquetant 1984; Jacquetant and Bon 1984; Taşkın et al., 2010; 2012; Clowez 2012; Du et al., 2012b; Richard et al., 2015)
M. semilibera	Reported with Castanea, Fraxinus, Malus and Ranunculus	Czech Republic, France, Germany, India, Italy, Netherland, Turkey	(Taşkın et al., 2010; 2012; O'Donnell et al., 2011; Clowez 2012; Du et al., 2012a; Richard et al., 2015)
M. tridentina	Most likely biotrophic or endophytic, associated with Abies, Arbutus, Castanea, Corylus, Fraxinus, Olea, Pinus, Pseudotsuga, Quercus, etc.	Argentina, Armenia, Chile, Cyprus, France, India, Italy, Spain, Turkey, USA	(Bresadola 1898; Jacquetant 1984; Röllin and Anthoine 2001; Barseghyan et al., 2012; Taşkın et al., 2010; 2012; O'Donnell et al., 2011; Clowez 2012; Kuo et al., 2012; Richard et al., 2015; Loizides et al., 2015; 2016; Loizides 2016)

et al., 2012) says that overharvesting for profit has made it harder to keep their wild population and keep their biological and genetic variations safe.

Research and Development of Morchella species

Morel mushroom production has attracted people for 130 years. Both indoor and outdoor attempts to produce morel mushrooms have yet to be successful since morels are challenging to grow for commercial gain. Through scientific intervention, the cultivation of the Guchhi has made progress. It is said that apple compost was used to help it grow, but it didn't make much of a difference (Thakur et al., 2021). In 1982 (Ower, 1982), the first successful attempt to grow morel mushrooms in a lab was made. However, indoor cultivation on a large scale failed. Chinese researchers put forward considerable effort and developed *M. importuna* (Zhu et al., 2011). Since the inception of the ICAR-Directorate of Mushroom Research, Solan, several attempts to domesticate the Guchhi mushroom have been undertaken in India (H.P.). However, no noteworthy outcome was discovered. Indian and American scholars successfully identified and domesticated *Morchella importuna* strain ANI1 of Morel under controlled media (Kumar et al., 2022).

LIFE CYCLE

Before learning how to grow Morels in a lab, it is important to know how they live (Liu et al., 2017). By doing so, we may better grasp the phases and develop appropriately *in vitro* cultures.

Morels has several uses demonstrating different phases (Volks and Leonard, 1990) during the 1990s with a well-explained lifecycle (Castillo et al., 2014).

The latest one illustrates the morphological and physiological stages with cellular phases, whereas the older one did it relatively broadly, despite being fundamentally the same (Castillo et al., 2014). Morels start their life cycle once mature adult ascocarps, which comprise eight ascospores, develop new haploid ascospores, which are released for dispersion (Castillo et al., 2014). When adverse circumstances first appear, the mycelium that the released ascospores generate might turn into sclerotia (Liu et al., 2017). When circumstances are suitable, ascocarps generate a germinative tube that lengthens and thickens to produce a haploid hypha, which then transforms into a multicellular mycelium known as the main mycelium (Castillo et al., 2014). A heterokaryotic mycelium, also known as secondary mycelium, is created when two healthy main mycelia are in contact with each other. Recurrent branching and plasmogamy of the secondary mycelium produce masses that eventually evolve into sclerotia, a transitional phase. Chlamydospores may also develop as a result of this. The sclerotia evolve into a vegetative, mycelogenic, or fruiting carpogenic phase by obtaining the proper stimulus, which is often an externally adverse situation, giving birth to ascocarps (Castillo et al., 2014). The existence of nodes and pinheads, which continue to develop and distinguish into adult ascocarps for fruitification, is a unique characteristic of the carpogenic mycelial phase. Despite the complexity of the morel's life cycle and its ongoing research, the studies conducted with it have produced more or less identical results and conclusions that remain understandable (Figure 10.4).

NUTRIENTS PRESENT IN *MORCHELLA ESCULENTA*

Morchella has a wide range of extremely important nutrients to human health. It is a rich source of minerals like iron (195 mg/g), zinc (62.6 mg/g), phosphorus (3.49 mg/g), manganese (54.7 mg/g), potassium (23.5 mg/g), calcium (0.85 mg/g), potassium (23.5 mg/g), calcium (0.85 mg/g), sodium (0.18 mg/g), fats-2%, ash-9.7%, and fiber-17.6%, (Sud & Sud, 2017). It is also a good source of these nutrients, making them regarded as nutritious as meat or fish, and they are frequently used with leafy green vegetables, along with different cuisines (Ajmal et al., 2015).

COMPOSITION OF BIOACTIVE COMPOUNDS IN *MORCHELLA ESCULENTA*

Global public health has been supported and preserved by *Morchella esculenta* as it contains multiple significant bioactive compounds (Figure 10.5). Cheung (2010) states that it is an extremely good source of lectins, polysaccharides, proteases, and ribonucleases. Polysaccharides are effective against inflammation, microbial growth, and carcinoma. In addition to these, *Morchella esculenta* ascocarp contains bioactive substances such as phenolic compounds, tocopherols, carotenoids, and organic acids (Table 10.3). Protocatechuic acid, p-hydroxybenzoic acid, and coumaric acid are the three most prevalent phenolic substances discovered. Yellow morels contain aromatic substances such as aldehydes, ketones, terpenes, and esters (Thakur et al., 2021). With a 50.9 percent alcohol content, phenol is the most prevalent aromatic element, trailed by ester (15.6 percent) & carbamic acid (11.6 percent). Examples of organic acids include citric, malic, oxalic, fumaric, and quinic. Additionally, it was discovered that *M. esculenta's* mycelium extracts in methanol, ethanol, and chloroform had antibacterial effects against bacteria, including Listeria monocytogenes, *Salmonella typhimurium*, *E. cloacae*, Staph, & *Staphylococcus aureus* (Raman et al., 2018).

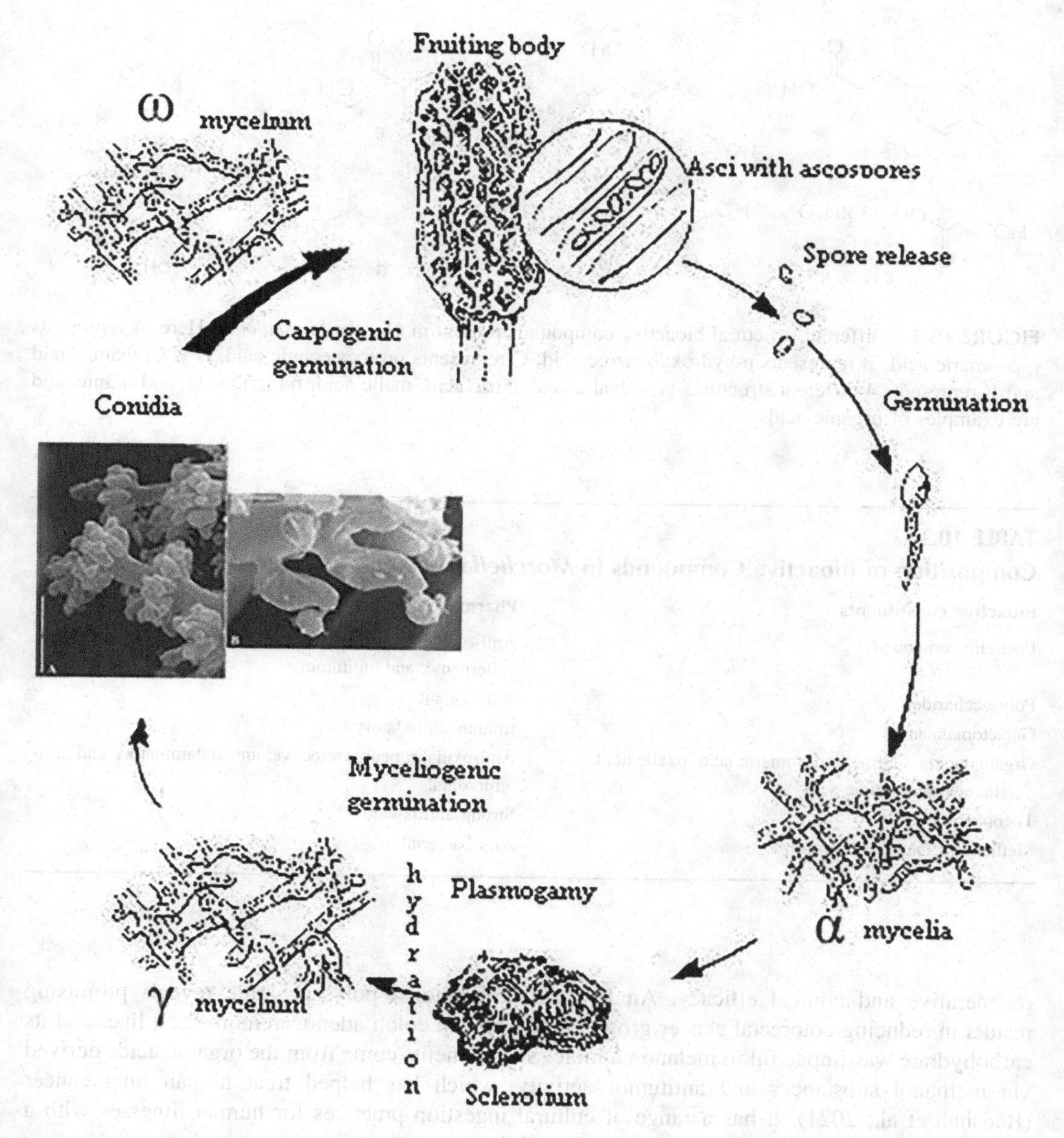

FIGURE 10.4 Life cycle of *Morchella* (Liu et al., 2017).

MEDICINAL PROPERTIES OF *M. ESCULENTA*

Morels have been utilized as remedies (Table 10.4) for more than 2000 years in China, Japan, and Malaysia because of the continued availability of their nutritional and biological elements (Ajmal et al., 2015). In nature, polysaccharides, proteases, ribonucleases, polysaccharide-peptides, polysaccharide-proteins, and lectins are the chemical building blocks of many compounds found in morels, including anti-oxidants, antimicrobials, antiinflammatories, antitumor, hypocholesterolemia, and immunosuppressives (Thakur et al., 2021). It is frequently described as "mushroom therapeutical" owing to the medicinal features of metabolites like phenolic, alkaloids, terpenoids, glycoproteins, lactones, oxidative enzymes, peptides, lipids, and hydrolytic that have substantial

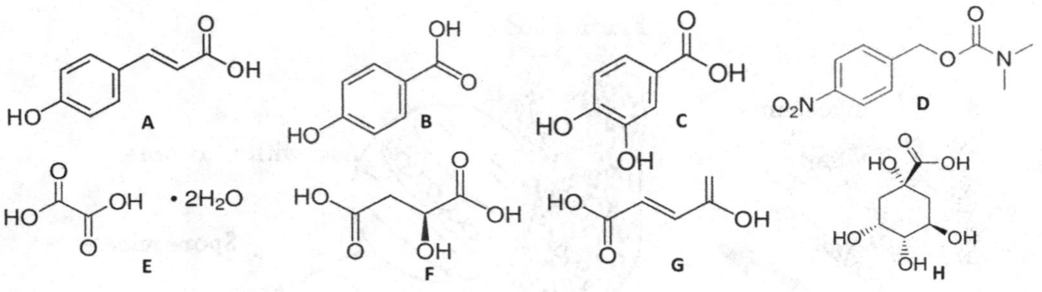

FIGURE 10.5 Different structural bioactive compounds present in *Morchella esculenta.* Here, **A** represents p-coumaric acid, **B** represents p-hydroxybenzoic acid**, C** represents protocatechuic acid, **D** is Carbamic acid and **E** represents 4 different structures *viz.* Oxalic acid, citric acid, malic acid, fumaric acid, and quinic acid are examples of organic acids.

TABLE 10.3
Composition of Bioactive Compounds in *Morchella esculenta* (Raman et al., 2018)

Bioactive constituents	Pharmacological properties
Phenolic compounds	Anti-oxidant, antimicrobial, anti-inflammatory, anti-allergenic, and antitumor
Polysaccharides	Antioxidant
Galactomannan	Immunostimulatory
Organic acids – citric acid, fumaric acid, oxalic acid, malic acid and quinic acid	Anti-oxidant, neuroprotective, anti-inflammatory and anti-microbial
Tocopherols	Strong antioxidant
Methanol, Ethanol, Chloroform	Anti-bacterial

regenerative and clinical efficacy. An 81 kDa (kilodaltons) polysaccharide reveals promising results in reducing colorectal cancer growth in the human colon adenocarcinoma cell line, and its carbohydrate was impactful to melanoma mole. Such benefits come from the organic acids derived via methanol substances and antitumor activity, which has helped treat human lung cancer (Badshah et al., 2021). It has a range of cultural ingestion practices for human illnesses with a variation of dosages or intake (Thakur et al., 2021).

PROCESSING POTENTIAL OF *MORCHELLA ESCULENTA*

Morels are applied for manufacturing and utilized in food and medicine. The efficient fermentation requirements, which included a brewing process temp around 22.6°C, 0.2 percent $MgSO_4$ $7H_2O$, 1.5 percent Ammonium sulfate, 75 percent H_2O, & 4 percent of sugar, were stabilized by *M. esculenta* and led to the yield of 95.82 1.37 mg/g of crude polysaccharides via soybean curd. These generate unrefined polysaccharides in food manufacturing and drug companies (Li et al., 2019). As *Morchellas* were introduced to crystalline-forming cornmeal to lower its carbohydrate presence, it was discovered that the carbohydrate (starch) content declined from 64.5% to 23.5%, and the degree of carbohydrate decomposition increased to 74.8 percent (Zhang et al., 2010).

TABLE 10.4
Traditionally Utilization of Methodology of Consuming *Morchella* as a Clinical Remedy (Thakur et al. 2021)

S. No.	Aliment	Methods	Uses
1	Stomach pain	Aqueous extracts of the fruiting bodies are given to the patient, or fruiting bodies are ground in water or raw milk It Relieves pain	Relieves from pain
2	Pneumonia	Pneumonia decoction is prepared of old fruiting bodies of Guchhi as "Garam kandha" is given to the patients.	Cures congestion
3	Dehydration/ blood stools	Aqueous extract of fruiting bodies by adding little salt/sugar.	Recovers water loss from the body
4	Respiratory problem	A decoction is prepared by adding 2/3 fruiting bodies to three glasses of water. Boil till water remained ¼.	Help in reducing respiratory problems.
5	Fever, cough, and cold	Boiled soup or raw Guchhi is eaten by adding a little salt and desi butter.	Heavy sweating is there which lowers the body temperature.
6	Healing of wounds	The paste is prepared by grinding the fruiting bodies with desi butter and then applied on wounds.	Heals the wounds as it provides warmness to the Wound
7	Pregnant women/ Lactating Mothers	Soup, dry vegetables (cooked) given to the ladies.	Provides strength, warmness to the body and is considered very nutritious. It provides energy to the body, which is indirectly given to the body.
8	1–12 months kids	The fruiting body is ground with honey/ raw milk -3 spoons/day; soup once per day.	Growth of kids
9	Asthmatic patients	Simple frying of fresh and rehydrated fruiting bodies is given to the patients.	Helps in proper functioning of lungs.
10	In acne	The paste is prepared while grinding in water	Due to warmth, it will dry acne and cures it
11	Sex appeal	A decoction is prepared in apple wine	Provides warmness to the body
12	Weakness	The decoction is prepared by adding milk and while serving add two spoons of honey.	Provides strength to the body.

THE NECESSITY OF IN VITRO CULTURE OF MORELS

Morchella has emerged as the most well-liked and valuable mushroom on the market due to its gourmet flavor, nutrient benefits, and medicinal potential. It is quite tasty and is comparable to fish or meat in terms of nutrition. Morel mushrooms include a variety of bioactive substances that promote their use in preserving global health across the globe. They contain various saccharides, lectins, proteases, and ribonucleases, and they exhibit properties that are effective against inflammatory, anti-microbial, antioxidant, immune suppressive, and anti-tumor (Xia et al., 2010). Locals in the Himalayan area utilize this fungus and its components for medical reasons based on folklore. The old practices of using Morels as a remedy for extreme ailments follow Table 10.4. (Lakhanpal et al., 2010). The growing interval of *Morchella* is brief, and they are often only sold for a limited time in the markets, primarily in the spring. Due to their limited availability and high cost, medicinal mushrooms are now accessible to all members of society. Despite its high cost, there is a rising global commercial demand for morel (Tan et al., 2019). Various wild morels are collected, dried, and graded annually before being sent to other markets. Overharvesting, primarily in the Himalayas, could result in its extinction because it remains a natural resource (Sayeed, 2018). There have been no conservation initiatives for morels so far. This fungus is both aesthetically and economically important, but it could become extinct before the growing process is standardized. In order to provide an appropriate substrate for morel cultivation in a regulated setting, effective preservation and scientific assistance are indeed required (Sayeed et al., 2018).

VARIOUS INITIATION FOR CULTURING OF MORELS

A. To suit the rising commercial demand on the global market, numerous efforts have been undertaken over the past two decades by various workers for the outdoor cultivation of morels (Liu et al., 2018). However, the invention of a sufficient organic substrate contained in an exogenous nutrition bag has resulted in the most significant improvement in the cultivation of morels (ENB). This technique led to a substantial advancement in the morel mushroom industry, especially in China's production of *M. importuna* (black morel mushrooms) (Tan 2019). Exogenous nutrition is added to the mushroom bed after successful spawning, followed by control of the fruiting body, harvest, and storage. Spawning is the first step in the exogenous nutrition bag technique of culture. For more cultivation, spawn quality is essential since Morel's development was impacted by different elements, including heat, humidity, soil moisture and pH, and many more (Liu et al., 2017). The combination of 85% wheat and 15% rice bran in a sterilized thermal plastic bag is the most popular ENB formulation for spawn generation. The formula comprising 85 percent of brewed corn, 10 percent wheat bran, 4 percent grass-ash, 1 percent of gypsum, and 60 percent H_2O is another blend of raw substrate materials for ENB. The following phase is land preparedness, which begins with ploughing the soil to change the pH and eliminate the pest. The loose soil is frequently treated with quick lime before 0.8–1 m wide, 0.2–0.3 m deep mushroom beds are constructed. The processed morel spawn is planted immediately into the mushroom beds at a depth of about 1 inch after several days once the temperature is below 20°C (Liu et al., 2018). When the right humidity and temperature conditions are met, like 20°C and 50–70%, soil humidity, morel mycelia spawn and begin to invade the soil. After 10 to 15 days, a whitish appears on the mushroom beds' cover, which is known as the "powdery mildew stage," and it is at this point that exogenous nutrition bags can be applied to the mushroom beds. Then, ENB bags are positioned over the black morel-inoculated mushroom beds at a spacing of 50 cm.
B. However, descending openings must first be formed to allow the mycelia to invade and draw nutrients via ENB (Tan et al., 2019). The black Morel can end its life cycle also in regulated ecological systems with the aid of ENB since it provides critical organic components like enough nitrogen and carbon for the growth of morel mycelium. The bags are taken out of the mushroom beds after 40 to 45 days since the nutrient bag will run out entirely. Following that, fruiting maintenance is needed, and pest control from pests including *Trichoderma, Aspergillus, Rhizopus, mucor,* and bacteria is crucial. Biological and physical pest-effective strategies are allowed. However, chemical pesticides are prohibited. When the ascocarps reach a size of 10–15 cm and have pits and ridges visible, the fruiting body is harvested as the final step. The moisture is removed from the mushroom and stored longer (Liu et al., 2017).
C. ***Indoor culture:*** More and more scientists and farmers are becoming interested in artificial cultivation. The efficient indoor culture of *Morchella* and the duplication of its life cycle led to the development of distinctive ascocarps (Ower, 1982; Qizheng et al., 2017).

He cultivated *M. rufobrunnea* in a jar, promoting the formation of fruiting bodies by encouraging the germination of pseudo-sclerotia (Ge, 2019). According to his study, morel mushrooms can be cultivated at the appropriate temperatures, humidity, and ventilation (Pilz et al., 2007). To isolate pure colonies, several mycologists utilized tetracycline-laced malt extract agar media (Arkan and Guler, 1992). At 20°C incubation, they noticed the development of pure mycelia, which grew to 90 mm. The Morchella spp. mycelium grows the quickest among other mushrooms and possesses the same flavor and nutritional value as the ascocarp. Sclerotia grow for about 12 days and are influenced by the in vitro temp and media contents (Baran and Boron, 2017). The most often used nutritive medium for *in-vitro* culture is Potato Dextrose Agar (PDA), Malt Extract Agar, and Yeast Extract Agar. Casein, casamino acid, peptone, and sodium nitrate were included in the complete yeast medium that Giieler and Arkan used to grow *Morchella* ascocarps (Baran and Boron, 2017). Through *in-vitro* culture, several researchers have recently produced *Morchella esculenta* with effectiveness. With the aid of three nutrient-rich mediums, including soil, Luria agar, and soil extract agar media, one of the *in vitro* research projects was carried out in Pakistan. Although mycelial development was encouraged by each of these media, the LA medium was the best among all. Numerous hyphal structures, including the rhizomorph and the dense bulk of prosenchyma and pseudo-parenchyma tissue, were visible under the microscope. Secondary mycelium is most likely causing the pigmented mycelium forms that have been observed frequently. These produce ascogenous hyphae, which produce ascogonium and antheridium with dilated tips. (Kanwal et al., 2016)

Hand collection is typically used to gather morels, usually in the spring or early summer (Ajmal et al., 2015). Choosing the right mushrooms to pick manually entails separating the stem with a surgical blade and a pair of scissors (Keefer et al., 2010). Because moisture content reduces shelf life, drying is required during the rainy season. This must be done before using the proper storing technique. It takes roughly 10–15 days to gather morels in favorable weather (Thakur et al., 2021). An ordinary person might gather 1–3 kg of morel mushrooms (Kumar and Sharma, 2010). The ascocarp is dried by hanging it near a fire, usually near Chula, which is a traditional cooking fireplace in its natural environment, mostly in Himachal Pradesh (Thakur et al., 2021). Making ascocarp necklaces and placing them on the walls is yet another method. A little air during drying aids in better mushroom preservation (Ajmal et al., 2015). They shouldn't be kept in sealed spaces since they can begin to collect moisture and begin to rot. Morels need to be fumigated with insecticide pills to keep insects away (Hamayun and Khan, 2003). Paper bags or polythene bags are used to store the ascocarps that are intended for marketing (Sayeed et al., 2018).

OBSTACLES

Morels have been successfully grown outdoors in several nations, including India. Even *Morchella importuna* and *M. rufobrunnea* indoor growing methods are practical. However, *M.esculenta in-vitro* cultivation is still hindered (Tan et al., 2019). It produces spores (conidia & sclerotia) in order to reproduce. The various phenomena that occur during Morels' journey and reproduction period are critical to achieving an invitro culture (Liu et al., 2017). Because of the *M. esculenta* life cycle's complexity, it has been unable to fully understand how different phases develop, which has prevented successful *in vitro* production. Sclerotia that grew inappropriately in a controlled setting occasionally turned into fake mushrooms, making it difficult for it to grow commercially and to produce (Ge, 2019). It takes certain growth conditions for *Morchella* to be successfully cultivated inside. Mycelial growth and the development of pseudosclerotia are significantly influenced by pH, light, temperature, the makeup of the medium, and water potential (Ge, 2019). The infection of *Morchella* spp. under artificial culture by bacteria and other pests, such as the rots of *M. importuna*, which is most frequently brought on by *diplosporus* is another significant problem (Peixin et al.,

2014). Additionally, because of its decline in yield attributed to various pests and bacterial infections, the *in vitro* culture of *M. rufobrunnea* across the USA was entirely disrupted during 2008 (Tan et al., 2019).

MARKETS, YIELD, AND IMPORT-EXPORTS STATUS OF *MORCHELLA ESCULENTA*

Asian nations like India, Pakistan, the Peoples' Republic of China, and nations from North America and the Mediterranean, produce and trade a significant share of the world's largest wild morels (Raut et al., 2019). Around 1.5 million tons of mushrooms are generated globally (Ajmal et al., 2015). Morels can be sold for as much as 10,000–30,000 rupees per kilogram on the open market in India, where intermediaries buy them from local farmers for a pitiful price of 2000–3000 rupees per kilogram (Kumar et al., 2022). Because of their astronomical price, morels are known as the "growing gold of mountains" (Ajmal et al., 2015). China exports roughly 900 tonnes of dried morels annually, compared to Indian exporters accounting for 70 tonnes annually (Negi, 2006; Sayeed et al., 2018). This is a relatively small amount compared to China. Disorganized commerce is the main cause of the export rate since locals gather morels and sell them privately, making it impossible for a formal trade route to be followed (Kumar and Sharma, 2010). A barrier is a difficulty in developing methods for the artificial growing of morels (Kumar et al., 2022), which prevents effective production volume and quality from being reached in comparison to a nation like China with a morel trade surplus. The creation of an organized trading system and new ways to farm could increase both exports and output. The supply chain includes everyone from harvesters and farmers to local traders, district traders, regional traders, and the international market. Another format includes regional traders procuring from farmers via road head traders (Raut et al., 2019).

IMPACTS OF TOXICITY OF *MORCHELLA ESCULENTA*

The species has a positive effect on medicine and health in many ways, but it has also been said to have some bad effects in some situations. In particular, the French Poison Control Center's data reveals that *M. esculenta* intake caused neurological and gastrointestinal symptoms in 67% of patient cases (Saviuc et al., 2010). Six to twelve hours after eating *M. esculenta*, there were signs of brain effects (Thakur et al., 2021). After eating partially cooked or raw morels, some people develop the syndrome, which includes gait ataxia or postural instability, tremor, and gastro-intestinal or hemolytic syndrome (Saviuc et al., 2010; Lagrange and Vernoux, 2020). Monomethylhydrazine (MMH), which is found in morel mushrooms, has been linked to food poisoning, stomach problems, weakness, loss of coordination, jaundice, and even coma. In extreme situations, it can result in death (Thakur et al., 2021).

Usually mistakenly identified and ingested as Morels, False Morels (*Gyromitra* spp.), which do not belong to the Morchallacae family but have a look remarkably similar to True Morels (*M. esculenta*), present serious health risks (Lagrange and Vernoux, 2020). *Gyromitrin*, a hydrazine derivative that makes up the main toxin in fake morels, causes nausea, vomiting, and stomach pain. Seizures, vertigo, stupor, delirium, and coma are occasionally caused by epilogenetic neurotoxic effects that follow. The hydrolysis of gyromitrin brings these symptoms into monomethyl hydrazine (MMH), degrading the glutamic acid decor (Lagrange and Vernoux, 2020). According to the Finnish Food Authority, this substance is not advised since it has been hard to determine its dangerous status and because it has been extremely harmful to children, pregnant women, and nursing mothers (Lagrange and Venoux, 2020).

A few misconceptions and presumptions about the species exist globally. For example, villagers in Pakistan consider that those who find them are the luckiest or most prosperous. At the same time, those in Himachal Pradesh think that intense lightning and thunder have an impact on the expansion and prosperity of their region (Thakur et al., 2021; Ajmal et al., 2015). As immuno-regulatory, hallucinogenic, immunostimulant, and laxatives, as healing properties once fried in butter, as a therapy for chronic arthritis (Ajmal et al., 2015), and as a culinary ingredient in dried vegetables, vegetables with gravy, and pulao, these mushrooms are utilized in many different ways by the locals. The Kullu district residents of Himachal Pradesh cooked the ascocarps of *M. esculenta* with milk before eating them (Thakur et al., 2021). The variety in eating patterns and culinary customs points to the significance of morels in sociobiology and ethnobotany. Since it is a wild and scarce mushroom and there are currently no preservation initiatives. Overexploitation of morels should be avoided to prevent their decline or extinction in the worst-case scenario. This is because of the need for more advanced technical assistance for building revolutionary systems instead of constantly failing (Sayeed et al., 2018). Even though there is traditional knowledge as well as methodologies, like leaving 1–2 ascocarps beside native mushroom hoarders, abandoning the ground and buried mycelium that serves as a source of inoculum for the remainder of the season, which in turn will result in the growth in their inhabitants, it's also presumed more by scientists and researchers to pursue these methodologies to keep the *Morchella* inhabitants (Sayeed et al., 2018).

CONCLUSION

The morel mushroom, *M. esculenta*, also known as Guchhi, is well-known all over the globe because of its delicious flavour and aroma. Some of the chemically active chemicals created from these morel extracts can be utilized to create a wide range of delectable meals and medicines for treating various ailments. Due to several qualities, including dietary, immune-stimulatory, antioxidative, and anti-inflammatory qualities, *M. esculenta* is recognized as one of the world's most significant wild and cultivated fungi. Morels' outstanding qualities drew much attention, spurring global demand and finally elevating it to a highly sought-after mushroom. Due to its peculiar texture, a rarity in nature, considerable interaction with living molecules, quick commercial culture, and bulk storage of organic and native species, a biotechnological way to cultivate morels in a controlled setting is now required. It is important to note that implementing conservation measures for morels in their native habitat is conceivable.

REFERENCES

M. Ajmal, A. Akram, A. Ara, S. Akhund, B.G. Nayyar. 2015. Morchella esculenta: An edible and health beneficial mushroom. *Pak J Food Sci* 25(2): 71–78.

O. Arkan, P. Güler. 1992. The sclerotia formations of Morchella conica Pers., Doğa. Tr. *J. Biol*. 16: 217–226.

S.L. Badshah, A. Riaz, A. Muhammad, G. Tel Çayan, F. Çayan, M. Duru Emin, N. Ahmad, A.H. Emwas, M. Jaremko. 2021. Isolation, characterization, and medicinal potential of polysaccharides of *Morchella esculenta*. *Molecules* 26(5): 1459.

J. Baran, P. Boroń. 2017. Two species of true morels (the genus *Morchella*, Ascomycota) recorded in the Ojców National Park (south Poland). *Acta Mycologica* 52(1).

G.S. Barseghyan, A. Kosakyan, O.S. Isikhuemhe, M. Didukh, S.P. Wasser. 2012. Phylogenetic analysis within genera Morchella (Ascomycota, Pezizales) and Macrolepiota (Basidiomycota, Agaricales) inferred from nrDNA ITS and EF-1a sequences. *Systematics and evolution of fungi*. Jersey: Science Publishers. 159–205.

E. Boudier. 1897. Révision analytique des morilles de France (Taxonomic revision of the morels of France). *Bulletin de la Société Mycologique de France (Bull French Mycol Soc)* 13:129–153.

E. Boudier. 1909. Icones mycologicae ou Iconographie Des champignons de France principalement Discomycètes avec texte descriptif. *Librairie des Sciences Naturelles, Paris* 2: 194–421.

G. Bresadola. 1898. *Fungi Tridentini* 2 (11–13):47–81, pl. 151–195.

V. Castañera, G. Moreno. 1996. Una Morchella (Morchella esculenta forma dunensis f. nov.) frecuente en las dunas de Cantabria. *Yesca. Revista Sociedad Micólogica de Cántabria*. 8:27.

A.G. Castillo, G. Mata, C.W. Sangabriel. 2014. Understanding the life cycle of morels (Morchella spp.). *Revista Mexicana de Micología* [en linea] 40: 47–50.

S.T. Chang, P.G. Miles. 2004. Mushrooms cultivation, nutritional value, medicinal effect, and environmental impact. *CRC Press* 480.

P.C.K. Cheung. 2010. The nutritional and health benefits of mushrooms. *Nutrition Bulletin* 35(4): 292–299.

P. Clowez. 1997. Morchella dunensis (Boud.) Clowez (stat. et comb. nov. ad. int.). Une bonne espèce pour un mystère nomenclatura. *Documents Mycologiques 26*(104): 13–20.

P. Clowez. 2012. Les morilles: A new global approach to the genus *Morchella. Quarterly Bulletin of the Mycological Society of France* 126(3–4): 199–376.

X.H. Du, Q. Zhao, K. O'Donnell, A.P. Rooney, Z.L. Yang. 2012b. Multigene molecular phylogenetics reveals true morels (Morchella) are especially species-rich in China. *Fungal Genetics and Biology.* 49(6):455–469.

X.H. Du, Q. Zhao, Z.L. Yang, K. Hansen, H. Taşkin, S. Büyükalaca, D. Dewsbury, J.M. Moncalvo, G.W. Douhan, V.A. Robert, P.W. Crous. 2012a. How well do ITS rDNA sequences differentiate species of true morels (*Morchella*)? *Mycologia* 104(6): 1351–1368.

T.F. Elliott, N.L. Bougher, K. O'Donnell, J.M. Trappe. 2014. Morchella australiana sp. nov., an apparent Australian endemic from New South Wales and Victoria. *Mycologia.* 106(1):113–118.

E.M. Fries. 1822. Morchella, Helvella. *Systema Mycologicum.* 2:5–18.

S. Ge. 2019. Studies on the molecular identification. *Biological Characteristics and Indoor Cultivation of Morchella spp.*

R.N. Ghosh, N.C. Pathak. 1962. Fungi of India-1 *Morchella*, Verpa, and Helvella. *Bulletin Nat Bot Gardens (Lucknow)* 71: 1–19.

S.S. Gilani, S.Q. Abbas, Z.K. Shinwari, F. Hussain, K. Nargis. 2003. Ethnobotanical studies of Khurram agency, Pakistan through rural community participation. *Pak. J. Biol. Sci.* 6 (15): 1368–1375.

S.F. Gray. 1821. Natural Arrangement of British Plants: According to Their Relations to Each Other, as Pointed Out by Jussieu, De Candolle, Brown &c.: Including Those Cultivated for Use with an Introduction to Botany, in which the Terms Newly Introduced are Explained. Baldwin, Cradock and Joy.

G. Guzmán, F. Tapia. 1998. The known morels in Mexico, a description of a new blushing species, Morchella rufobrunnea, and new data on M. guatemalensis. *Mycologia.* 90(4):705–714.

M. Hamayun, M.A. Khan. 2003. Studies on collection and marketing of Morchella (Morels) of Utror-Gabral Valleys, District Swat, Pakistan. *Ethnobot Leafl* 1(6).

H. Hassan, W. Murad, N. Ahmad, A. Tariq, I. Khan, N. Akhtar, S. Jan. 2015. Idigenous uses of the Plants of Malakand Valley, District Dir (Lower), Khyber Pakhtunkhwa, Pakistan. *Pak. J. Weed Sci. Res.* 21(1): 83–99.

M. Irfan, S. Yang, L. Yuxin, J.X. Sun. 2017. Genetic diversity analysis of *Morchella* sp. By RAPD. *Mol Biol Res Commun* 6(1): 27.

M. Isiloglu, H. Alli, B.M. Spooner, M.H. Solak. 2010. *Morchella anatolica* (Ascomycota): A new species from southwestern Anatolia, Turkey. *Mycologia* 102(2): 455–458.

E. Jacquetant. 1984. Les Morilles. *Paris, La Bibliothèque des Arts* 114.

E. Jacquetant, M. Bon. 1984. Typifications et mises au point nomenclaturales dans l'ouvrage Les Morilles (de E. Jacquetant), Nature-Piantanida. *Documents Mycologiques.* 14:1.

S.F. Kakakhel. 2020. True morels *(Morchella spp.)* and community livelihood improvement in Mankial Valley, District Swat, Khyber Pakhtunkhwa, Pakistan. JBM *Journal of Bioresource Management* 7(3): 5.

L.S. Kandari, P.C. Phondani, K.C. Payal, K.S. Rao, R.K. Maikhuri. 2012. Ethnobotanical study towards conservation of medicinal and aromatic plants in upper catchments of Dhauli Ganga in the central Himalaya. *J Mt Sci* 9: 286–296.

H.K. Kanwal, M.S. Reddy. 2012.The effect of carbon and nitrogen sources on the formation of sclerotia in Morchella spp. *Annals of Microbiology.* 62(1):165–168.

N. Kanwal, K. William, K. Sultana. 2016. In vitro propagation of Morchella esculenta and study of its life cycle. *Journal of Bioresource Management.* 3(1):6.

T.N. Kaul, M.L. Khurana, J.L. Kachroo, A. Krishna, C.K. Atal. 1981. Myco-ecological studies on morel bearing sites in Kashmir Mush. *Sci.* 11: 789–797.

M. Keefer, R. Winder, T. Hobby. 2010. Commercial development morels in the East Kootenay, British Columbia. *J Ecosystems Manage* 11(1–2).

J.V. Krombholz. 1834. Naturgetreue Abbildungen und Beschreibungen der essbaren, schädlichen und verdächtigen Schwämme. *In Commission in der JG CALVE'schen Buchhandlung.*

S. Kumar, Y.P. Sharma. 2010. Morel trade in Jammu and Kashmir-Need for organized commercialisation. *Everyman's Science* 45(2): 12–113.

A. Kumar, V.P. Sharma, S. Kumar. 2022. First report from India on identification and domestication of *Morchella importuna* strain ANI1. Research square In Press (DOI: 10.21203/rs.3.rs-1814026/v1)

M. Kuo. 2008. *Morchella tomentosa,* a new species from western North America, and notes on *M. rufobrunnea. Mycotaxon* 105: 441–446.

M. Kuo, D.R. Dewsbury, K. O'Donnell, M.C. Carter, S.A. Rehner, J.D. Moore, J.M. Moncalvo, S.A. Canfield, S.L. Stephenson, A.S. Methven, T.J. Volk. 2012. Taxonomic revision of true morels (Morchella) in Canada and the United States. *Mycologia* 104(5): 1159–1177.

G. Laalaa, M.U. Raja, S.R.A. Gardezi, G. Irshad, A. Akram, I. Bodlah. 2019. Study of macro-fungi belonging to order Agaricales of Poonch District Azad Jammu and Kashmir (AJK). *Pure and Applied Biology,* 8(1): 27–33.

E. Lagrange, J.P. Vernoux. 2020. Warning on false or true morels and button mushrooms with potential toxicity linked to hydrazinic toxins: An update. *Toxins* 2(8): 482.

T.N. Lakhanpal, O.S. Shad. 1986. Studies on wild edible Mushrooms of Himachal Pradesh (N.W. Himalayas)-I. Ethnomycology, Production and Trade of *Morchella* species. *Indian J. Mush* 12–13: 5–13.

T.N. Lakhanpal, O. Shad, M. Rana. 2010. *Biology of Indian Morels.* I. K. International Publishing House, New Delhi. p. 39 (Link: *Biology of Indian Morels – Google Books).*

S. Li, D. Tang, R. Wei, S. Zhao, W. Mu, S. Qiang, Z. Zhang, Y. Chen. 2019. Polysaccharides production from soybean curd residue via Morchella esculenta. *J. Food Biochem.* 43(4): e12791.

Q. Liu, H. Ma, Y. Zhand, C. Dong. 2017. Artificial cultivation of true morels: Current state, issues and perspectives, ISSN: 0738-8551 (Print) 1549-7801 (Online) Journal homepage: http://www.tandfonline.com/loi/ibty20

Q. Liu, H. Ma, Y. Zhang, C. Dong. 2018. Artificial cultivation of true morels: Current state, issues and perspectives. *Crit. Rev. Biotechnol.* 38(2): 259–271.

S. Lladó, R. López-Mondéjar, P. Baldrian. 2017. Forest soil bacteria: diversity, involvement in ecosystem processes, and response to global change. *Microbiol. Mol. Biol. Rev.* 81(2): e00063- 16.

M. Loizides. 2016. Macromycetes within Cistaceae- dominated ecosystems in Cyprus. *Mycotaxon* 131(1): 255–256.

M. Loizides. 2017. Morels: the story so far. *Field Mycol.* 18: 42–53. Doi:10.1016/j.dmyc.2017.04.004

M. Loizides, P. Alvarado, P. Clowez, P.A. Moreau, L. Romero, A. Palazón. 2015. *Morchella tridentina, M. rufobrunnea* and *M. kakiicolor*: A study of three poorly known Mediterranean morels, with nomenclatural updates in section Distantes. *Mycol. Prog.* 14: 13.

M. Loizides, J.M. Bellanger, P. Clowez, P. Richard, P.A. Moreau. 2016. Combined phylogenetic and morphological studies of true morels (Pezizales, Ascomycota) in Cyprus reveal significant diversity, including *Morchella arbutiphila* and *M. disparilis* spp. nov. *Mycol. Prog.* 15: 1–28.

M. Loizides, J.M. Bellanger, Y. Yiangou, P.A. Moreau. 2018. Preliminary phylogenetic investigations into the genus Amanita (Agaricales) in Cyprus, with a review of previous records and poisoning incidents. *Documents Mycologiques.* 37:201–218.

M. Loizides, T. Kyriakou, A. Tziakouris. 2011. Edible and toxic fungi of Cyprus. *Manitari, Greece.*

M. Loizides, Z.G. Zagou, G. Fransuas, P. Drakopoulos, C. Sammut, A. Martinis, J.M. Bellanger. 2021. Extended phylogeography of the ancestral *Morchella anatolica* supports preglacial presence in Europe and Mediterranean origin of morels. *Mycologia* 113(3): 559–573.

R. Longley, G.M. Benucci, G. Mills, G. Bonito. 2019. Fungal and bacterial community dynamics in substrates during the cultivation of morels (*Morchella rufobrunnea*) indoors. *FEMS Microbiology Letters* 366(17): fnz215.

A. Mahmood, R.N. Malik, Z.K. Shinwari, A. Mahmood. 2011. Ethnobotanical survey of plants from Neelum, Azad jammu & kashmir, Pakistan. *Pak. J. Bot.* 43: 105–110.

K. Manikandan, V.P. Sharma, S. Kumar, S. Kamal, M. Shirur. 2011. Edaphic conditions of natural Sites of Morchella and Phellorinia. *Mushroom Res.* 20(2): 117–120.

S. Masaphy. 2010. Biotechnology of morel mushrooms: successful fruiting body formation and development in a soilless system. *Biotechnol. Lett.* 32: 1523–1527.

S. Masaphy, L. Zabari, D. Goldberg. 2009. New long-season ecotype of Morchella rufobrunnea from northern Israel. *Micologia Aplicada International.* 21(2):45–55.

C.S. Negi. 2006. Morels (*Morchella* spp.) in Kumaun Himalayas, Natural Product. *Radiance* 5(4): 306–310.

K. O'Donnell. 2014. A preliminary assessment of the true morels (Morchella) in Newfoundland and Labrador. *Omphalina.* 5:3–6.

K. O'Donnell, A.P. Rooney, G.L. Mills, M. Kuo, N.S. Weber, S.A. Rehner. 2011. Phylogeny and historical biogeography of true morels (*Morchella*) reveals an early cretaceous origin and high continental endemism and provincialism in the holarctic. *Fungal Genet. Biol.* 48(3): 252–265.

R. Ower. 1982. Notes on the development of the morel ascocarp: *Morchella esculenta*. *Mycologia* 74(1): 142–144.

A. Paliwal, U. Bohra, D.K. Purohit Pillai. 2013. First report of Morchella – An edible morel from mount Abu, Rajasthan. *Middle East J. Sci. Res.* 18(3): 327–329.

H.E. Peixin, L.I. Wei, H.E. Xin-sheng, G.E. Lu-jing. 2014. Diversity of endophytic fungi in ascocarps of Morchella crassipes. Journal of Zhengzhou University of Light Industry, *Natural Science Edition*. 29(3):1.

C.H. Persoon. 1794. Disposita methodical fungorum. *Romers Neues Mag Bot.* 1:81–128.

C.H. Persoon. 1801. Synopsis methodica fungorum: sistens enumeratione omnium huc usque detectarum specierum, cum brevibus descriptionibus nec non synonymis et observationibus selectis. *Apud Henricum Dieterich.*

D. Pilz, R. McLain, S. Alexander, L. Villarreal-Ruiz, S. Berch., T.L. Wurtz,... & J.E. Smith. 2007. Ecology and management of morels harvested from the forests of western North America. United States Department of Agriculture Forest Service Pacific Nortwest Research Station. *Gen. Tech. Rep. PNW-GTR-710.*

P. Prasad, K. Chauhan, L.S. Kandari, R.K. Maikhuri. 2002. *Morchella esculenta* (Guchhi): Need for scientific intervention for its cultivation in Central Himalaya. *Curr. Sci* 82(9):1098–1100.

L. Qizheng, M. Husheng, Z. Ya, D. Caiong. 2017. Artificial cultivation of true morels: Current state, issues and perspectives. *Critical Reviews of Biotechnology* 2017:1–13.

V.K. Raman, M. Saini, A. Sharma, B. Parashar. 2018. *Morchella esculenta*: A herbal boon to pharmacology. *Int. J. Dev. Res.* 8(3): 19660–19665.

J.K. Raut, J. Upadhyaya, V. Raghavan, M. Adhikari, S. Bhushal, P.S. Sainju, C.M. Gurmachhan, A. Giri, L.R. Bhatt. 2019. Trade and conservation of morel mushrooms in Nepal. *International Journal of Natural Resource Ecology and Management* 4(6): 183–187.

Razaq, A. Rashid, H. Ali, H. Ahmad, M. Islam. 2010. Ethnomedicinal Potential of Plants of Changa Valley District Shangla, Pakistan. *Pak. J. Bot.* 42(5): 3463–3475.

P. Reilly. 2016. *Fascinated by fungi : Exploring the majesty and mystery, facts and fantasy of the quirkiest kingdom of life on earth.* First Nature. pp. 448.

F. Richard, J.M. Bellanger, P. Clowez, K. Hansen, K. O'Donnell, A. Urban, M. Sauve, R. Courtecuisse, P.A. Moreau. 2015. True morels (Morchella, Pezizales) of Europe and North America: Evolutionary relationships inferred from multilocus data and a unified taxonomy. *Mycologia*. 107(2):359–382.

O. Röllin, A. Anthoine. 2001. Les morilles noires du Chablais savoyard genre Morchella, section Distantes. 1-Remarques préliminaires et présentation de Morchella elata Fr. et Morchella tridentina Bres. *Bull Trimest Féd Mycol Dauphiné-Savoie*. 161:7–12.

W.C. Roody. 2003. *Mushrooms of West Virginia and the Central Appalachians*. Lexington, Kentucky: University Press of Kentucky. pp. 485.

P. Saviuc, P. Harry, C. Pulce, R. Garnier, A. Cochet. 2010. Can morels (*Morchella* sp.) induce a toxic neurological syndrome. *Clin. Toxicol.* 48(4): 365–372.

R. Sayeed, S. Kausar, M. Thakur. 2018. Morchella esculenta Fr: Biodiversity, sustainable conservation, marketing and ethno-mycological studies on medicinal fungus from Kashmir Himalayas. *India. Mushroom Res* 27(1): 77–86.

O.S. Shad. 1989. Biological studies on Morchella species (morels) of Himachal Himalayas (Ph.D. Thesis). *Himachal Pradesh University, Shimla, India.*

H. Sher, A. Aldosari, A. Ali, H.J. de Boer. 2014. Economic benefits of high value medicinal plants to Pakistani communities: An analysis of current practice and potential. *J. Ethnobiol. Ethnomed.* 10(71): 1–16.

V.S. Sud, V.K. Sud. 2017. A review of toxic effects and aphrodisiac action of *Morchella esculenta* (Wild Morel-Guchhi Mushroom): A Himalayan Delight. *European Journal of Pharmaceutical and Medical Research* 4(8): 726–730.

H. Tan, A. Kohler, R. Miao, T. Liu, Q. Zhang, B. Zhang, L. Jiang, Y. Wang, L. Xie, J. Tang, X. Li. 2019. Multi-omic analyses of exogenous nutrient bag decomposition by the black morel *Morchella importuna* reveal sustained carbon acquisition and transferring. *Environ. Microbiol.* 21(10): 3909–3926.

H. Taşkın, S. Büyükalaca, H.H. Doğan, S.A. Rehner, K. O'Donnell. 2010. A multigene molecular phylogenetic assessment of true morels (*Morchella*) in Turkey. *Fungal Genet. Biol.* 47(8): 672–682.

H. Taşkın, S. Büyükalaca, K. Hansen, K. O'Donnell. 2012. Multilocus phylogenetic analysis of true morels (Morchella) reveals high levels of endemics in Turkey relative to other regions of Europe. *Mycologia.* 104(2): 446–461.

H. Taşkın, H. Doğan, S. Büyükalaca, P. Clowez, P-A. Moreau, K. O'Donnell. 2016. Four new morel (Morchella) species in the elata subclade (M. sect. Distantes) from Turkey. *Mycotaxon.* 131(2): 467–482.

M. Thakur, I. Sharma, A. Tripathi. 2021. Ethnomedicinal aspects of morels with special reference to *Morchella esculenta* (Guchhi) in Himachal Pradesh (India): A Review. *Fungal Biol.* 11(1): 284–293.

Z. Tietel, S. Masaphy. 2018. True morels (*Morchella*)—nutritional and phytochemical composition, health benefits and flavor: A review. *Crit Rev Food Sci Nutr.* 58(11): 1888–1901.

Timothy, J. Baroni, W. Michael, A. Sharon, A. Teresa, I. Teresa, L. Thomas, E. María, M. Frank, O. Quispe, D. Jean Lodge, O. Kerry. 2018. Four new species of *Morchella* from the Americas. *Mycologia* 1–17.

T.J. Volk, T.J. Leonard. 1990. Cytology of the life cycle of *Morchella. Mycol. Res.* 94(3): 399–406.

A. Voitk, M.V. Beug, K. O'Donnell, M. Burzynski. 2016. Two new species of true morels from Newfoundland and Labrador: Cosmopolitan Morchella eohespera and parochial M. laurentiana. *Mycologia.* 108(1): 31–37.

J.A. Wagay, D. Vyas. 2011. Phenolic quantification and antioxidant activity of *Morchella esculenta. Int. J. Pharma Bio Sci.* 2(1): 188–197.

D. Wipf, S. Koschinsky, P. Clowez, J.C. Munch, B. Botton, F. Buscot. 1997. Recent advances in ecology and systematics of morels. *Cryptogamie. Mycologie.* 18(2):95–109.

H. Xia, W.K. Cheung, L. Sze, G. Lu, S. Jiang, H. Yao, X.W. Bian, W.S. Poon, H.F. Kung, M.C. Lin MC. 2010. miR-200a regulates epithelial-mesenchymal to stem-like transition via ZEB2 and β-catenin signaling. *Journal of Biological Chemistry* 285(47):36995–7004.

I. Yatsiuk, I. Saar, K. Kalamees, S. Sulaymonov, Y. Gafforov, K. O'Donnell. 2016. Epitypification of *Morchella steppicola* (Morchellaceae, Pezizales), a morphologically, phylogenetically and biogeographically distinct member of the *Esculenta clade. Phytotaxa* 284(1): 31–40.

C.S. Yoon, R.V. Gessner, M.A. Romano. 1990. Population genetics and systematics of the *Morchella esculenta* complex. *Mycologia* 82 (2): 227–235.

M. Zerova. 1941. A new morel from the virgin steppe (Morchella steppicola Zerova sp. nov.) *Ukrayins' kyi Botanicnyi Zhurnal.* 2(1):155–159.

G.P. Zhang, F. Zhang, W.M. Ru, J.R. Han. 2010. Solid-state fermentation of cornmeal with the ascomycete *Morchella esculenta* for degrading starch and upgrading nutritional value. *World J. Microbiol. Biotechnol.* 26(1): 15–20.

Y. Zhu, S. Du, J. Che, X. Chang, W. Jiang, C. Wang. 2011. Effect of different carbon sources and Nitrogen sources on the mycelial growth o*f Morchella* spp. (Chinese). *Journal of Northwest A and F University (Natural Science Edition)* 39(3): 113–118.

11 *Pleurotus ostreatus (Jacq.) P. Kumm*

Sugandha Sharma, Shiv Kumar, and Rekha Kaushik
MMICTBM (HM), Maharishi Markandeshwar (Deemed to be University), Mullana, Ambala, Haryana, India

Poonam Baniwal
Department of Quality Control Food Corporation of India, New Delhi, India

Rahul Mehra
MMICTBM (HM), Maharishi Markandeshwar (Deemed to be University), Mullana, Ambala, Haryana, India

Yogender Singh Yadav
Department of Dairy Engineering, College of Dairy Science and Technology, Lala Lajpat Rai University of Veterinary and Animal Sciences, Hisar, Haryana, India

CONTENTS

DOI: 10.1201/9781003259763-11

INTRODUCTION

The most widely grown variety of mushrooms worldwide is the oyster mushroom (*Pleurotus ostreatus*). Africa, Europe, India, and Asia are its second-largest markets for production. For *Pleurotus* species, the term "oyster mushroom" is appropriate (Patel et al. 2012). In order to distinguish it from other species in the genus, *P. ostreatus* is known as the oyster mushroom tree or the grey mushroom. Many benefits come with it. When the various enzymes are secreted and the pH is between 6 and 8, the growth of the organism is accelerated. They have the ability to degrade lignocellulosic biomass, call for environmental regulation, do not require substrate compositing, may occupy the substrate for a shorter period of time, and have significant therapeutic and dietary value (Ejigu et al. 2022). These substrates are utilised for culture, therefore sterilising, which can be an expensive process, is not necessary. Pets and illnesses do not damage its fruiting structure. They are grown in an easy and affordable manner. For the cultivation of mushrooms, lignin, cellulose, and hemicellulose are appropriate substrates (Jennifer and Devi 2020). Typically, lignocellulosic decomposing fungus and wood make up these mushrooms. They flourish on lignocellulosic agricultural wastes including teff straw, cotton waste, sugarcane bagasse, wheat straw, paddy straw, and coffee pulp. The expense of the substrate used to grow mushrooms is frequently a barrier in the industry (Shen et al. 2014).

TAXONOMY

It's a member of the Pleurotaceae and Agaricales families. These are classified based on their habitat and colour. The cap of the mushroom is fan, board, or oyster-shaped, and it ranges in colour from white to grey in natural specimens or from tan to black brow. When they are young, their borders are wavy, enrolled, lobed, or smooth (Leo et al. 2019). The flesh is solid, white, and varies in thickness according onto how the stripes are arranged. The gills of the mushroom range in colour from white to cream and, if present, descend the stem. If it is, the stipe is laterally attached to the wood and off-center. The mushroom's spore print ranges from white to lilac-grey. However, the mushroom's stipe is missing (Nhi and Hung 2012). Most of them are saprophytic, and just a few are parasitic. The morphological characteristics of this species are changeable because of the diverse substrate and fluctuating environmental circumstances. This species' phytogenic and taxonomy identification is a little complicated, which causes misidentification. Numerous molecular and biochemical methods are used to examine the taxonomical hierarchy and phylogenetic connection (Ho et al. 2020). The methods used to identify the DNA of this species include random amplified polymorphic DNA (RAPD), internal transcribed spacer region identification, restriction fragment length polymorphism (RFLP), and amplified fragment length polymorphism (AFLP).

BOTANICAL CLASSIFICATION (ELATTAR ET AL. 2019)

Kingdom	Fungi
Division	Basidiomycota
Class	Agaricomycetes
Order	Agaricales
Family	Pleurotaceae
Genus	*Pleurotus*
Species	*P. ostreatus*

CHARACTERISTICS AND DISTRIBUTION OF OYSTER MUSHROOM

Oyster mushrooms come in a variety of varieties, including pink oyster mushrooms (*P. flabellatus*), white oyster mushrooms (*P. ostreatus*), abalone oyster mushrooms (*P. cystidiosus*), and grey oyster mushrooms (*P. sajorcaju*) (Muswati et al. 2021). The Indonesian population grows (Onchonga et al. 2013). It grows saprophytically at temperatures between 12 to 32 degrees Celsius and is found all throughout the world, from temperate to tropical locations. These have over 40 species and range in colour from white to a variety of colours, are sessile or stalked, grow underground or above ground, and occasionally parasitize other organisms (Bandopadhyay 2013). These produce oyster mushrooms that are incredibly nutritious and full of proteins, vitamins, and minerals when they grow on various types of lignocellulosic composted agricultural waste. Additionally, it contains very low amounts of sugars, carbohydrates, and very less cholesterol. Each *Pleurotus* species' active mycelia and fruiting bodies have a unique medicinal characteristics, such as anti-inflammatory, immunomodulatory, ribonuclease activity, anticancer activity, etc (Mukhopadhyay 2019). According to Shirmila et al. (2018), the different biological compounds discovered in mushrooms belong to various classes of natural products, including proteins, complex starches, polysaccharides, triterpenoids, glycoproteins, peptides, lipids, lipopolysaccharides, hemicelluloses, nucleosides, and lectins. Oyster mushrooms, however, rank third in importance among mushrooms of edible value. In addition to being well-known for its therapeutic and nutritional benefits, *Pleurotus* species have bio potentialities that include the bioconversion of lignocellulosic wastes, recycling of agricultural wastes, production of improved animal feed, degradation of industrial dyes, bioremediation and degradation of xenobiotics, bioremediation, enzyme production, and degradation of xenobiotics for bioremediation (Tesfaw et al. 2015). It is one of the varieties of mushrooms that is grown the most widely. In particular, in India, Africa, Europe, and South Asia, it is the commercially produced mushroom that is the second biggest after the Agaricus bisporus. It has several benefits (Ejigu et al. 2022). It can thrive in a wide variety of pH (6–8) and temperature (10–30°C) conditions. Oyster mushrooms are often lignocellulosic decomposing fungus. They are developing a variety of lignocellulosic agricultural wastes, including cotton waste, coffee pulp, wheat straw, paddy straw, teff straw, and sugarcane bagasse (Dubey et al. 2019).

As a result, all of the substrates are quite expensive and in great demand. The expensive substrates on which mushrooms are cultivated provide a number of difficulties for the industry. When it comes to nutritional value, production, and financial return, the best substrates are employed for oyster mushroom growing (Fikadu et al. 2019). In various regions of Europe, *Pleurotusostreatus* are found, including all of Ireland and Britain. It is widespread in North America, Japan, and Asia. A unique feature of this mushroom is to feed and trap worms by using 'lassos' which is made up of hypae (Correa et al. 2016).

MEDICAL USES

It provides the necessary vitamins and minerals that aid in the treatment of disease and malnutrition. The low-fat content of mushrooms is crucial for heart patients (Koutrotsios et al. 2017). It has every vital mineral that the human body needs, including calcium, iron, and phosphorus. Folic acid is abundant in this food because it both prevents and treats anaemia. Oyster mushrooms aid with weight reduction, muscle relaxation, blood vessel leak prevention, and tendons and veins strengthening (Taofiq et al. 2015). It is crucial in preventing health issues that can lead to death, such as obesity, diabetes, and cardiovascular disease (Facchini et al. 2014).It is essential for preventing and treating a wide range of illnesses, including hepatitis B, chronic fatigue syndrome, microbial infections, diabetes, hypertension, heart disease, high blood cholesterol, kidney difficulties, gastric cancer, impaired immune response, and others.

Oyster mushrooms are high in vitamins (B, C, D, and K), minerals, and trace elements (Cr, Na, I, Cu, Zn, and Se), and low in fatty acids. When consumption is higher than average, a large number of fibres and some naturally occurring compounds that reduce cholesterol formation have a good impact on lipid profiles (Younis et al. 2015). The enzymes Alanine transaminase in plasma and Aspartate transaminase, which are produced by hepatocytes, are decreased by oyster mushrooms. Numerous studies have found that oyster mushroom extract decreases liver damage-inducing enzymes such as glutathione (GSH), superoxide dismutase (SOD), catalase (CAT), carbon tetra chloride (CCl_4), and glutathione peroxidase (Gpx) and increases the production of antioxidant enzymes from the liver (Cao et al. 2015). The cholesterol level is reduced by this mushroom. Lovastatin, a naturally occurring substance in mushrooms that passes the placental and blood-brain barriers, is present in them. By preventing the HMG-CoA reductase enzyme from producing cholesterol in the liver, statins reduce the amount (Pasnik 2017). Beta-1,3-D-glucan and pectin, two compounds that gel when exposed to water, are found in oyster mushrooms. The bile acids and join together. As a result, the development of cholesterol-bile micelles and cholesterol absorption is inhibited.

Oyster mushrooms are high in fibre because they bind cholesterol in the intestines, which helps the body expel waste products. Fibers raise the size of stools, shorten the period of evacuation, and speed up bowel movements (Golak-Siwulska et al. 2018). Reduce gastrointestinal tract disorders and colon cancer as a result. Beta-glucan is found in the cell walls of oyster mushrooms. Pathogen-associated molecular patterns is a class of immunostimulants (PAMP). Oyster mushroom beta-glucan strengthens the immune system and aids in the fight against aberrant cells (Abidin et al. 2017). Chemotherapy and radiation treatments are used to combat the tumour cells (Sharma et al. 2018). Additionally, it serves as an aphrodisiac and prevents constipation, obesity, and hangovers (Aishah and Wan Rosli 2013).

PHYTOCHEMISTRY

In light of *Pleurotus ostreatus's* bioactivity, several studies have been conducted on bioactive substances. A variety of distinct chemical components were present in different extracts (Abdullah et al. 2013). Figure 2 illustrates the significant bioactive chemicals. The abundance of nutraceuticals and bioactive components in oyster mushrooms lends them a wide range of medicinal

TABLE 11.1
Functions of Bioactive Compounds

Bioactive compounds	Functions	References
α-glucan, β-glucans, proteins	Inhibit cancer	(Chanakya et al. 2015; Raman et al. 2021)
Polysaccharides, lectin, proteoglycans	Protect from tumour	(Correa et al. 2015)
Polysaccharides, heteroglycan	Helps in immunomodulatory	(El-Batal et al. 2015)
Lovastatin, ergosterol	Prevent from hypercholesterolemic	(Facchini et al. 2014)
Chrysin, lovastatin	Protect from atherogenic	(Fernandes et al. 2015)
Resveratrol, *Pleurotusostreatus* mycelium polysaccharides 2 (POMP2), Cibacron blue affinity purified protein; *Pleurotusostreatus* polysaccharides -1(POPS-1).	Act as anti-cancer	(Huang et al. 2015; Inacio et al. 2015)
β-glucans, vitamin C, and phenolic components	Act as anti-arthritic, Anti-nociceptive	(Feeney et al. 2014)
Polysaccharides	Work as an anti-oxidant, anti-inflammatory	(Golak-Siwulska et al. 2018)
Laccase	Inhibition of hepatoma and hepatitis C virus	(Bernardi et al. 2013)
Unspecified	Work as Anticataractogenic,	(Jegadeesh et al. 2018)
Astrans-2-decenedioic acid	Help in the activity of Nematicidal	(Ahmed et al. 2016)
Cibacron blue affinity purified protein, Concanavalin A	Work as immunomodulating and Antitumour effects	(Khatun et al. 2015)
Ubiquitin-like protein	Inhibitory activity against antiviral effects and HIV-1 reverse transcriptase	(Khan and Tania 2012)
Water soluble proteoglycans	Arresting the tumour cell growth	(Jafarpour and Eghbalsaeed 2012)
Pleuran, α-tocopherol; Ergosta-7,22-dienol; Ergosta-7-enol; Ergosta-5,7-dienol; Ergosterol	Anti-atopic dermatitis potential	(Jatwa et al. 2016)
p-Hydroxybenzoic acid, 3-(2-aminopheny1thio)-3-hydroxypropanoic acid, 14, 17- Octadecadienoic acid, methyl ester (E,E); Protocatechuic acid, Syringic acid, Caffeic acid, Vanillic acid, Cinnamic acid, Syringaldazine, 4- hydroxy-benzoic Acid, O- dianisidine, 2,6-dimethoxyphenol, 1-octen-3-ol, 3-octanone, Benzoic acid, 3-octanol; N-8-Guanidino- spermidine; Protocatechuic acid; 9,12-Octadecadienoic acid, methyl ester (E,E); p-anisaldehyde (4- methoxybenzaldehyde); Chlorogenic acid	Antimicrobial activity	(Barh et al. 2019; He 2017)

capabilities. When it comes to treating various diseases including diabetes, cancer, liver, and cardiovascular issues, they are quite successful. In light of this, anti-microbial and antioxidant substances have anti-aging, immuno-modulating, and anti-microbial actions (Atri et al. 2013). In Table 11.1, the health advantages of several natural compounds present in oyster mushrooms (Figure 11.1) are listed.

Due to its ability to combat oxidation, polyphenol is the most significant component (Baggio et al. 2012). The primary purpose of phenolic compounds is to stabilise the oxidation of lipids, which protects people from oxidative damage and prevents oxidative stress. *Pleurotusostreatus*

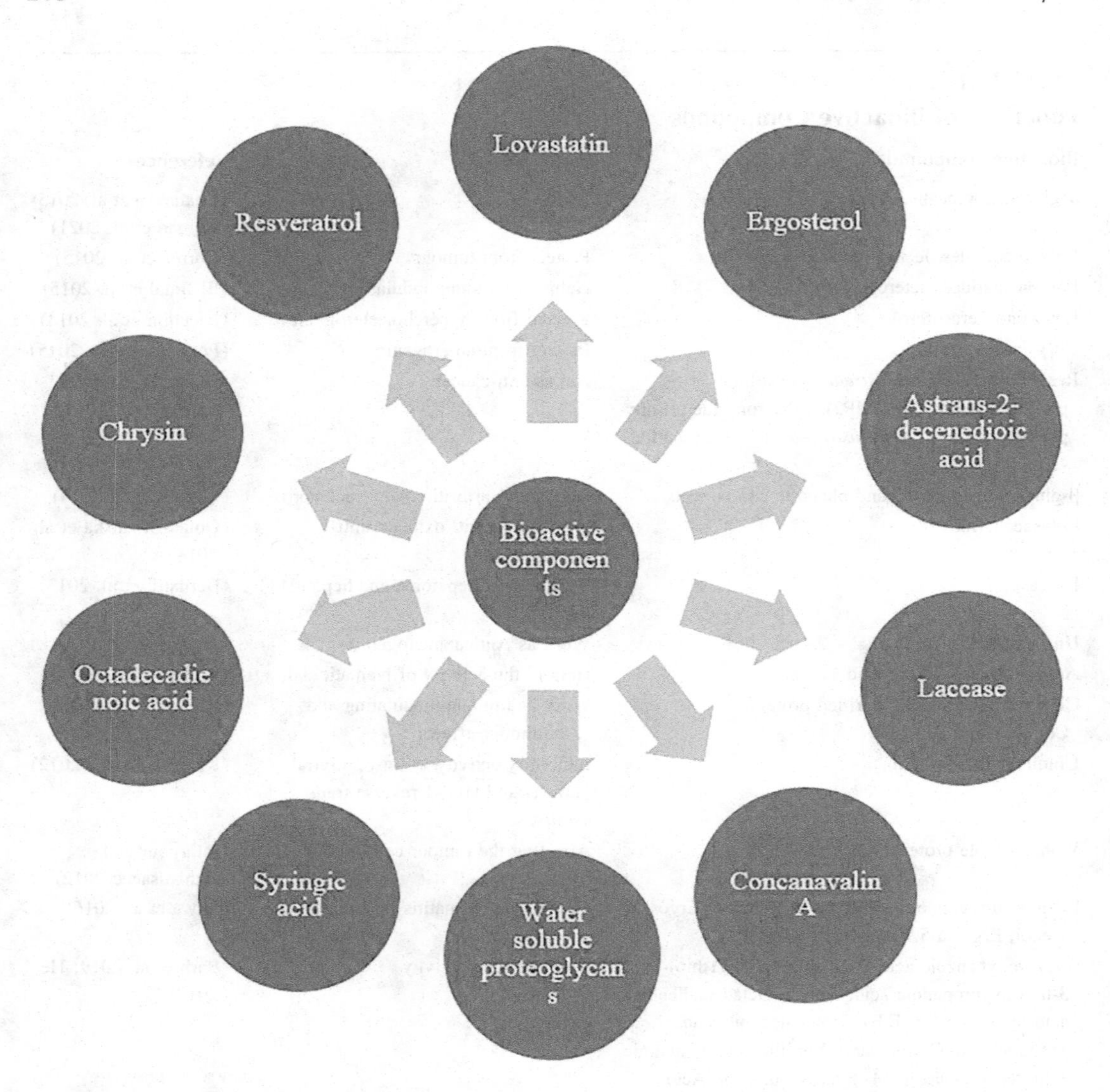

FIGURE 11.1 Bioactive compounds found in Oyster mushroom (dos Santos Bazanella et al. 2013).

contains phenolics as well as flavonoids (Chen et al. 2012). The presence of phenolic compounds in mushrooms gives them their well-known anti-microbial properties as well as other therapeutic advantages. According to Correa et al. (2016), benzoic acids are the primary phenolic chemical derivatives detected in *Pleurotus* species. The key phenolic elements of the *Pleurotus* family include quercetin, catechin, and chrysin (Facchini et al. 2014; Fernandes et al. 2015).

A major supply of ergothioneine, an amino acid with thiols, is found in mushrooms and has a number of antioxidant effects. Nitric oxide causes oxidative stress, which is primarily what it protects the cells against (Correa et al. 2015; Gąsecka et al. 2016). Oyster mushrooms have the highest ergothioneine concentration. Atherosclerosis and oxidative stress are treated and prevented by ergothioneine, which is present in *Pleurotus*. A lack of these nutrients can cause oxidative stress-induced lipid and protein oxidation as well as DNA damage. Specifically in the *Pleurotus* genus of mushrooms, ergothioneine is predominantly excreted. They are located in the mushroom's mycelia and fruiting bodies. It has anti-oxidant qualities in addition to other advantages (Lavelli et al. 2018).

DIETARY COMPOSITION

The nutritional qualities of mushrooms are highly valued. Protein (28.6–15.4%), carbs (84.1–61.3%), and dietary fibre (3–33.3%) are all present in significant amounts in these foods. Its protein concentration is higher than that of vegetables, but lower than that of milk and meat. Proteins, minerals (Na, Fe, Ca, P, and K), and vitamins (Vitamin C and B complex) are abundant in *Pleurotus* species (Royse et al. 2014). The umami flavour of *Pleurotus* mushrooms may be enhanced by the mushroom's nutritional value. Dietary fibres, which are fibre-rich and indigestible by humans, may make up a percentage of the carbs. All of the essential amino acids are present, along with a minimal amount of sulfur-containing cysteine, and methionine.

They include lipids such as free fatty acids, mono, di, triglycerides, phospholipids, and sterol esters (Wendiro et al. 2019). Along with these active ingredients, it also contains minerals including zinc, copper, selenium, iodine, polysaccharides, triterpenoids, oligosaccharides, proteins, peptides, phenols, alcohols, vitamins, and amino acids. They operate as hepatoprotectors, immune system boosters, and hypercholesteremic agents. They also have anti-cancer capabilities. Due to the high potassium and salt content, they are an excellent diet for those with hypertension and cardiovascular disease (Jang et al. 2016).

PROTEINS AND AMINO ACIDS

A wonderful source of edible protein, especially for vegans, may be found in oyster mushrooms. Chitin, amino acids and nucleic acids are examples of nitrogen-containing compounds that are not proteins. Species differences in mushroom protein content. The agro-climatic conditions and innate variables affect the protein in mushrooms (Adebayo and Martinez-Carrera 2015). The use of nitrogen-rich substrates that are supplemented by the nitrogen source may increase the protein content of the mushroom. The total amino acid makeup of a mushroom can be used to determine its quality (Gąsecka et al. 2016). The mushroom flavour is deliciously enhanced by the amino acids. With their antioxidant properties, amino acids including cysteine, tryptophan, glycine, and alanine have a synergistic impact with vitamins E and C. These species are known as a source of amino acid and protein content. They have high amount of orni- thine and c-aminobutyric acid (GABA) (Sardar et al. 2017). GABA is required for mental activity and brain functioning and for treatment of the wasting muscles after post-operative care and illness.

CARBOHYDRATES

Except for sugar-free components, which are crucial in maintaining the high osmotic and releasing energy owing to a rapid metabolic rate, carbohydrates contained in mushrooms mostly contribute to the structural makeup. It has a significant quantity of carbs in it, ranging between 24.95 and 75.88%. (Sekan et al. 2019). The mushroom's two main nutrients are chitin and carbohydrates. There is a high concentration of celluloids compounds, or rich dietary fibre, which acts as a high therapeutic diet and low-calorie diet for diabetes patients, reduces gastrointestinal ulcers, as well as reduces obesity. Depending on the type of mushroom, the range of carbs (Raman et al. 2021).

DIETARY FIBRE

Consuming foods high in dietary fibre, including mushrooms, is crucial for maintaining good health. Polysaccharides and chitin make up the mushroom's cell wall, which contains dietary fibre (Golak-Siwulska et al. 2018). Due to the fact that they cannot be digested by humans, these fibres provide a variety of physiological and nutritional advantages. These fibres reduce both the body's blood glucose and cholesterol levels. B-glucans, chitin, and polysaccharide protein are all dietary mushrooms that may be found in mushrooms (Girmay et al. 2016). Increased anti-tumour and immunomodulatory actions may result from consuming a lot of mushrooms. It could also lessen other illnesses. But the *Pleurotus* species are regarded as being high in dietary fibre (Hsu et al. 2018).

Lipids

Mushrooms include lipids that are excellent for patients who experience plaque buildup in their blood vessels. It consists of lipids, mono-, di-, triglycerides, fatty acids, sterols, and sterol esters (Lee et al. 2018). The species of *Pleurotus* are good suppliers of oleic and linoleic acid and have low lipid levels. In terms of human health, they are beneficial for hypocholesterolaemia and anti-inflammation. They lower the body's cholesterol levels and prevent the buildup of VLDL and LDL (very low-density lipoprotein and low-density lipoproteins) (Zhang et al. 2019).

Vitamins and Minerals

Because folic acid cannot be produced by the human body but must be obtained through diet, *Pleurotus* mushrooms are extremely rich in it (Tolera and Abera 2017). It has a low salt content and all the necessary nutrients, including phosphate and potassium. The iron included in mushrooms is 90% easily absorbed. Potassium aids in preserving the (Owaid et al. 2015), heart rhythm, blood cholesterol levels, fluid balance, nerve function, and blood pressure. The *Pleurotus* mushrooms are also rich in zinc which contributes to the nutrition to human health. High blood pressure can be prevented when there is a balance between sodium and potassium levels (Lavelli et al. 2018). The trace element selenium work as a cofactor in antioxidants which is mainly the potential source of nutrients. Selenium also reduces cancer if consumed in higher doses. Many of the bioactive compounds found in *Pleurotus* species are fully investigated. The frequent intake of *Pleurotus* species may improve immunity and reduce the level of cholesterol in humans (Smiderle et al. 2012).

ISOLATION TECHNIQUES FOR NUTRIENTS FROM DIFFERENT MUSHROOMS

Mushrooms are the richest source of nutrients like polysaccharides, polyphenolic components, proteins, lipids, micronutrients and other vitamins. Industries those process mushrooms produce a large variety of by-products during pickling, canning, and processing (Morris et al. 2017). Organic waste produced by this sector is dangerous to the environment and has the efficiency to exploit a bioresource in bioactive extraction from mushrooms. The chemicals of bioactive extract are hired in the pharmaceutical and food sector as nutritional and nutraceutical supplements (Park et al. 2016). They have plenty of B-group vitamins which are good source for vegetarians. There are many isolation techniques which are shown in Figure 11.2.

EXTRACTION OF BIOACTIVE COMPONENTS BY CONVENTIONAL TECHNIQUE

When extracting bioactive components from various bioresources, this method employs either water or organic solvents. Valuable components from mushrooms are extracted using the conventional technique, known as solvent-assisted extraction. There is no specific equipment needed for this procedure, making it affordable (Vital et al. 2015). Because it takes 1.5 to 5 hours and reaches high temperatures of 50 to 80 °C, it is the most time-consuming process. The thermolabile components, which are found in mushrooms, are degraded by the prolonged high temperature. Temperatures between 25°C and 60°C are needed for hydro-alcoholic extractions, which also need high concentrations (30 to 98.6%) and time periods of 1 to 24 hours. In order to extract the components of nutraceuticals from plant bioresources, organic solvents such percolation, 2:1 (v/v) chloroform-methanol, pressured liquid extraction, maceration (soaking), and Soxhlet technique are frequently utilised (Piska et al. 2017). Solvents are typically used in this approach because it causes

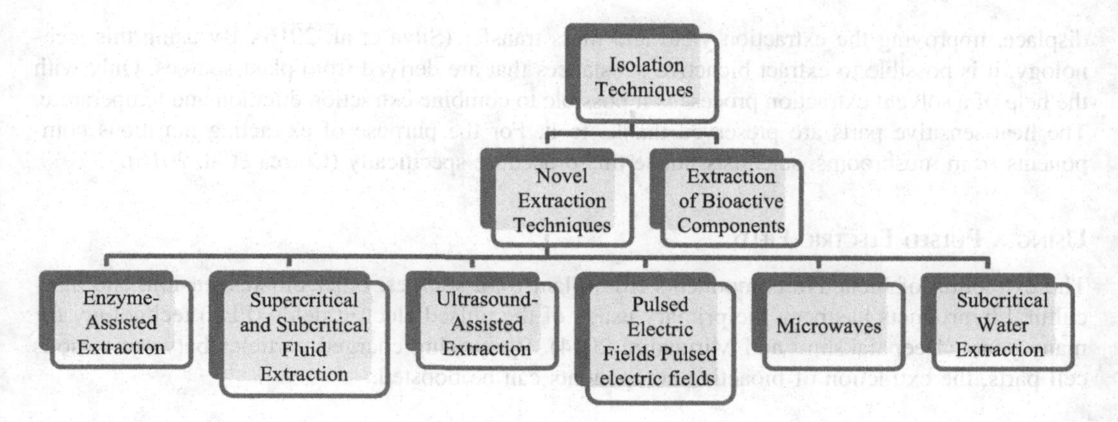

FIGURE 11.2 Isolation (Jesenak et al. 2013).

heat-labile components to degrade. In order to extract useful components from mushrooms, the food industry is increasing its hunt for innovative sustainable extraction methods.

Novel Extraction Uses Technique

There are many techniques which are used for the bioactive components extraction from plants which includes (Ravi et al. 2013), Extraction by Enzyme assisted, Supercritical and subcritical fluid extraction, Using a pulsed electric field, Ultrasound-assisted, Microwave-assisted and by Subcritical water, etc.

Enzyme-assisted Extraction Technique

This processing is carried out in the food business at a low temperature for a shorter amount of time with a high yield (Suseem and Saral 2013). Hydrolytic enzymes like chitinase and glucanase hydrolyze polysaccharides like glucans and chitin found in the cell wall of mushrooms. These hydrolytic enzymes are found in the cell walls of mushrooms. Consequently, these enzymes are beneficial for improving the extraction of bioactive components (Taofiq et al. 2016).

Subcritical and Supercritical Fluid Extraction

The traditional extraction method was recently superseded by this one. Scientists are drawn to this method because it is environmentally benign and has a larger capacity for extracting bioactive compounds from a variety of sources, including mushrooms (Abidin et al. 2017). The technique through which the fluid's characteristics are sandwiched between the liquid and the gas is known as supercritical fluid. The critical point where the liquid and gas phases do not exist is subjected to extremely high temperatures and pressures. Subcritical carbon dioxide and rich linoleic meals both produce higher extraction yields as compared to supercritical carbon dioxide (Oloke 2017). Soe and Lee use the subcritical water extraction (SWE) technique which recovers the bioactive compounds of oyster mushrooms with different pressure and high temperatures (50–300°C). It was seen that the extract which are obtained have higher antioxidant activities (Tiram et al. 2013).

Ultrasound-assisted Extraction (UAE)

Due to its numerous advantages, this approach has found widespread use and has caught the interest of food scientists. The cavitation bubbles created by this procedure cause the mushroom cells to displace, improving the extraction yield and mass transfer (Silva et al. 2016). By using this technology, it is possible to extract bioactive substances that are derived from plant sources. Only with the help of a solvent extraction process is it possible to combine extraction duration and temperature. The heat-sensitive parts are preserved thanks to it. For the purpose of extracting nutritious components from mushrooms, scientists utilise this procedure specifically (Correa et al. 2016).

Using a Pulsed Electric Field

The extraction of bioactive components from plant food sources, other bio-suspension, and agricultural byproducts has been the primary usage of the pulsed electric fields (PEF) technology for many years (Deepalakshmi and Mirunalini 2014). By moving charged particles between various cell parts, the extraction of bioactive components can be boosted.

Microwave-assisted Extraction

This method includes heating the food and allowing the moisture to evaporate, which causes the cells to rupture and creates a lot of pressure. as a result, encourages the release of desirable components from plant cells (Elisashvili 2012). The frequency of radio waves and infrared light range from 300 MHz to 300 GHz depending on the scientific and industrial applications. Microwaves mostly use the 915 and 2450 MHz heating frequencies. These microwaves cause food components that are based on dielectric materials to heat up. This method is far more efficient than any other method since it releases the methanolic extract (Hassan and Medany 2014).

Subcritical Water Extraction

To keep the water in a liquid form under high pressure, this technique makes use of hot water that is heated to a temperature between 100 and 374°C. The process is known as hot pressurized water extraction (Jaworska et al. 2015). High-polarity compounds can dissolve readily in water when the temperature is low.

PHARMACOLOGY

Oyster mushrooms are well-liked because of their abilities as functional foods and nutraceuticals. Due to their low-fat content, excellent protein quality, all of the vitamins, including riboflavin, thiamine, niacin, ergosterol, and ascorbic acid, as well as several minerals, including phosphorus and iron, they contain a variety of vital nutritional features (Ritota and Manzi 2019). In addition to providing nutritional benefits, they also exhibit therapeutic effects against several lifestyle-related diseases like cancer, hypertension, diabetes, obesity, and hypercholesterolemia (Kortei and Wiafe-Kwagyan 2015). Vitamins, polysaccharide-peptides, lectins, polysaccharides, and minerals are among the nutritive components that give foods their antioxidant, immunomodulatory, anticancer, antidiabetic, antihypercholesteremic, and antibacterial properties (Figure 11.3).

Antimicrobial

It has been discovered that the oyster mushroom is the most versatile and straightforward drug resistance for *S. laureus*, *E. coli*, species of Candida, *Staphylococcus epidermidis*, *Enterococcus*, and *Streptococcus*. The growth of *Klebsiella pneumoniae*, *C. glabrata*, species of Trichophyton, *C. albicans*, and *Bacillus megaterium* was thought to be inhibited by methanolic extract from *Pleurotus* species (Koutrotsios et al. 2014). Depending on the nature and kind of the solvent, oyster mushrooms have different antifungal and antibacterial properties.*E. coli*, *S. cerevisiae,* and *B. subtilis* are all successfully combatted by the acetone and ether extract of oyster mushrooms (Salami et al. 2017).

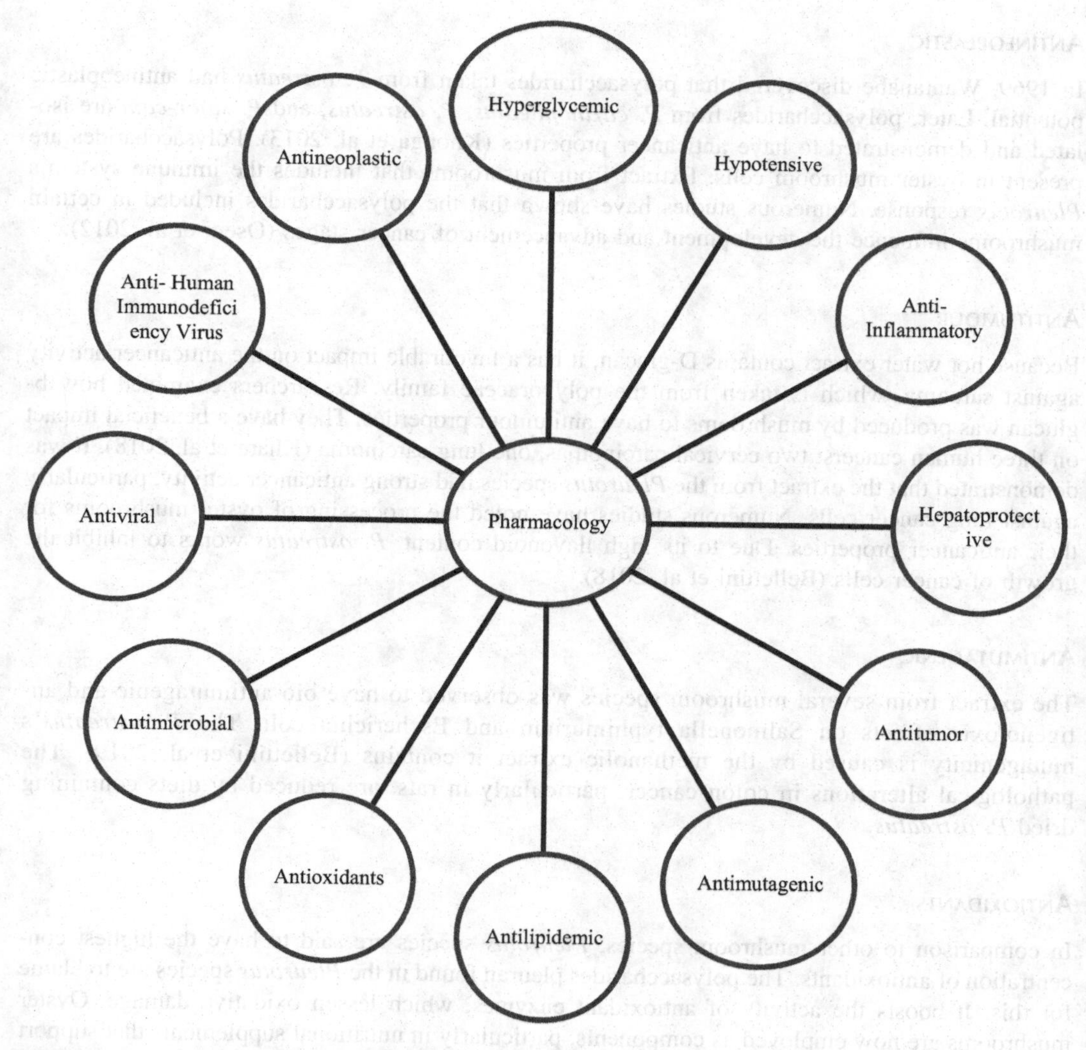

FIGURE 11.3 Pharmacology of oyster mushroom (Yehia 2012).

ANTIVIRAL

Numerous compounds found in *Pleurotus* mushrooms have both direct and indirect antiviral actions and consequent immune-stimulating activities. The oyster mushroom's ubiquitin, an antiviral protein, was discovered and isolated (Xu et al. 2012). The antibacterial action of the numerous virus particles is induced by the binding of the sulphated glucans, which prevents the infection of the host cells. The *Pleurotusostreatus* contains polysaccharides that have an immune-suppressing impact (Okwulehie and Nosike 2015).

ANTI-HUMAN IMMUNODEFICIENCY VIRUS (HIV) POTENTIAL

Ribonucleases have been identified and isolated from *Pleurotusostreatus*, and they have the capacity to neutralize HIV by degrading genetic material. In contrast, ribonuclease was characterized and isolated from sclerotia which showed the maximum ribonucleolytic activity for 30 minutes at 100°C. (Nongthombam et al. 2021). It was discovered that hemolysin, a monomeric protein, exhibited anti-HIV-1 action in cell culture.

Antineoplastic

In 1969, Wantanabe discovered that polysaccharides taken from *P. ostreatus* had antineoplastic potential. Later, polysaccharides from *P. citrinopileatus*, *P. ostreatus*, and *P. sajor-caju* are isolated and demonstrated to have anticancer properties (Khonga et al. 2013). Polysaccharides are present in oyster mushroom cells. Extract from mushrooms that includes the immune system's *Pleurotus* response. Numerous studies have shown that the polysaccharides included in certain mushrooms influence the development and advancement of cancer stages (Oseni et al. 2012).

Antitumour

Because hot water extract contains D-glucan, it has a favourable impact on the anticancer activity against sarcoma, which is taken from the polyporaceae family. Researchers examined how b-glucan was produced by mushrooms to have antitumour properties. They have a beneficial impact on three human cancers: two cervical carcinomas, one lung carcinoma (Khare et al. 2018). It was demonstrated that the extract from the *Pleurotus* species had strong anticancer activity, particularly against lung cancer cells. Numerous studies have noted the processing of oyster mushrooms for their anticancer properties. Due to its high flavonoid content, *P. ostreatus* works to inhibit the growth of cancer cells (Bellettini et al. 2018).

Antimutagenic

The extract from several mushroom species was observed to have bio-antimutagenic and antigenotoxic effects on Salmonella typhimurium and Escherichia coli. The *P. ostreatus*'s mutagenicity is caused by the methanolic extract it contains (Bellettini et al. 2019). The pathological alterations in colon cancer, particularly in rats, are reduced by diets containing dried *P. ostreatus*.

Antioxidants

In comparison to other mushroom species, *Pleurotus* species are said to have the highest concentration of antioxidants. The polysaccharides pleuran found in the *Pleurotus* species are to blame for this. It boosts the activity of antioxidant enzymes, which lessen oxidative damage. Oyster mushrooms are now employed as components, particularly in nutritional supplements that support health and ward off various ailments (Sharma et al. 2013). The oyster mushroom's capacity to scavenge free radicals is influenced by the colour of the fruiting bodies. The oyster mushroom has a significant amount of phenols. These substances have the ability to scavenge free radicals, which may lessen the impact of mutagens and carcinogens (Nongthombam et al. 2021).

Antilipidemic

Hyperlipidaemia is a risk factor for the disease i.e atherosclerosis. Taking of mushroom may increase the total lipids excretion and cholesterol. Mevinolin present in the oyster mushroom lead to the anti-hypocholesterolaemia activities (Bellettini et al. 2015). *P. pulmonarius* gives the antihyperglycemic effect when it is combined with glyburide.

Hyperglycemic

The compound guanide which is corelated to the bi-guanide is isolated with the species of *Pleurotus* that have anti-hypoglycaemic effect. High protein and fibre and low fat in edible mushroom make the great food for diabetic patient. *Pleurotus* extract decrease the level of serum glucose (Bellettini et al. 2016). Polysaccharides which are extracted from the *Pleurotus* improve anti-hyperglycaemic effect by increasing the activity of glutathione peroxidase.

Hypotensive

The property of antihypertensive property depends on the variety of mushroom with their combinations. Oyster mushroom maintain the blood pressure. Died fine particles and extract of fruiting bodies of *Pleurotus* lead to hypertension and congestive heart failure prevention (Hatvani et al. 2012). It also inhibits the enzyme which converts angiotensin-I that leads to the blood vessels concentration therefore increasing the blood pressure.

Anti-Inflammatory

The oyster mushroom's pleura, which is the fruiting body, processes the anti-inflammatory action. Both chronic and acute inflammation responds less to *Pleurotus* extract. Leukocyte migration to acetic acid-damaged tissue is inhibited by an isolated b-glucan related to *P. ostreatus*. The anti-inflammatory properties of oyster mushrooms were shown to be processed via inhibiting AP-1 and NF-kB signalling (Oloke 2017).

Hepatoprotective

Chronic liver injury damages the liver, causing fibrosis, which then sets the foundation for the development of cirrhosis and hepato-carcinoma. The active ingredients found in several *Pleurotus* species, including as phenol, vitamin C, and b-glucan, boost the activity of various antioxidant enzymes (Ahmed et al. 2013). They are in charge of reducing hepatic cell necrosis. This kind of mushroom's hepatoprotective effects caused a rise in blood aminotransferase enzyme levels. The polysaccharopeptides from *P. ostreatus*' fruiting body reduced inflammation, steatosis, thioacetamide-induced changes, fibrosis, and necrosis in the treatment protocol, particularly (Avagyan et al. 2013).

MYCOLOGY

Mushrooms have many medical uses as well as they are broadly utilized by humans for their food at the immortal time. In order to see the hereditary variety in *Pleurotus* types of mushrooms utilizing the unequal enhance polymorphic DNA (RAPD) markers and morphological, all around seven species were collected (Chen et al. 2011). The mushroom has fan, oyster-shaped cap spanning 5–25 cm and board. It ranges from tan to dark brown or white to grey. When young, margin is enrolled and smooth and somewhat wavy or lobed. The flesh is firm, white and differs in thickness due to the arrangement of stripes (Deepalakshmi and Sankaran 2014). The mushroom gills are cream to white and descend if stalks are present. The strip are off-centre with the attachment of lateral to wood. The mushroom spore print is white to lilac grey and dark background. The stipe mushroom is often absent. If present, it is thick and short (El Enshasy and Hatti-Kaul 2013).

Metabolomics and DNA Sequencing

The life cycle of the organisms are linked with the breeding techniques. *Pleurotus* species are the heterothallic homobasidiomycete. The bifactorial tetrapolar genetic system controls mating. The fungus lifecycle is started by the haploid basidiospores germination. The basidiospores which are germinated developed into the haploid mycelia (Kibar 2021). The haploid mycelia which are compatible fuse and form into the dikaryotic mycelium which are having two unfused nuclei. Karyogamy is the two haploid nuclei fusion that occurs in the body of the fruit in basidium (basidiocarp). This develop the four basidiospore haploid which enter in the same cycle again.

Monokaryotic hyphae lack in clamp connections just as the basidiomycetes (Krakowska et al. 2020). Due to the presence of clamp connections, Dikaryotic mycelium can be classified easily.

Techniques of Mushroom Breeding in *Pleurotus* species

There are three principal methods through which breeding approaches bring some variations for crop improvement (Lam and Okello 2015). These are recombinant breeding, transgenic breeding and mutation breeding.

Recombination breeding is breeding in which the new varieties and strains developed are there by conventional breeding such as hybridization/ matting and selection (Lin et al. 2013). The selection is utilizers for the primary improvement of *Pleurotus* species. Selection uses direct selection by germplasm collected from cultures or during fungal forays by multispore germination. Mating and selecting genetically diverse parents is the other approach for the exploit heterosis by hybridization (Mahalakshmi et al. 2019).

Multispore Selection

The various spore collected are germinated which are intermate in the culture media plates from the particular fruiting body. The dikaryotic mycelium in this process is selected on the basis of natural selection in this process (Mishra et al. 2016). According to many researchers, the new varieties are selected by the multispore cultures screening.

Selection of Hybridization and Oarents (Dual Culture Technique/ Two- Point Technique)

The genetic makeup of strains affects agricultural yield and quality. By using phenotypic and hybridization selection for mushroom quality and yield during breeding, any trait may be genetically improved (Mutukwa et al. 2019). For mushroom breeders, the idea of hybridization is new, and it became more common after 1983. The agronomically adapted strain receives the necessary features by hybridization (Nolle et al. 2017). For the hybridization of *Pleurotus*, the following three techniques are used: The genetic makeup of strains affects agricultural yield and quality. By using phenotypic and hybridization selection for mushroom quality and yield during breeding, any trait may be genetically improved (Mutukwa et al. 2019). For mushroom breeders, the idea of hybridization is new, and it became more common after 1983. The agronomically adapted strain receives the necessary features by hybridization (Nolle et al. 2017). For the hybridization of *Pleurotus*, the following three techniques are used: Intraspecific hybridization, Interspecific hybridization, and Intergeneric hybridization

Intraspecific Hybridization

Intraspecific hybridization, sometimes referred to as mono-mono or mon-mon hybridization, is the intermating of compatible monokaryons from the same species but from different strains to produce a dikaryon (Palacios et al. 2011). The monokaryotic mycelia emerged from different fruit bodies as a result of several allelism and incompatibility factors. Dikaryotization is the process by which compatible homokaryon fuses with other dikaryotic mycelium or undergoes a change from homokaryon to dikaryon (Patel et al. 2012). Regardless of the form of mating, the hybridization that develops between monokaryon and dikaryon is now known as di-mon mating. This is the most suitable technique for hybridization. The wide range of parents chosen for the hybridization that produces basidiospores is essential. Single spore isolates (SSIs) selection is another factor for economically significant characteristics, such as yield (Piska et al. 2017).

Interspecific Hybridization

This hybridization occurs between suitable monokaryons from two different types of organisms. Given that their alleles at the two loci differ, they are compatible. The compatible Gasteromycetes, Angiosperms, and Hymenomycetes all have a large population of this system (Piskov et al. 2020). The conditions for hybridization and the enhancement of many commercial features are made similar by this method.

Intergeneric Hybridization

This hybridization involves the union of two distinct monokaryon genera. It is used to increase certain features or properties, such as dye production and media use, or to combine traits that are either not present in the genus (Rout et al. 2015). Due to monokaryotization and sterility, the main drawback is the lack of fruit body production. The oyster mushroom has commercial value as a food for human consumption. It has the highest levels of protein, which range from 1.6 to 2.5%, and is the richest source of vitamins B and B complex (Samsudin and Abdullah 2019).

Folic acid is a nutrient that can help treat anaemia. Additionally, it is excellent for those with diabetes, hypertension, and obesity. Since oyster mushrooms have a lower ratio of sodium to potassium, fat to starch, and caloric value (Srikram and Supapvanich 2016). Mushrooms' high fibre and alkaline ash content help to reduce hyperacidity and constipation. Antibiotic capabilities are brought about by a substance called polycyclic aromatic. Following the addition with rice or wheat barn, the stem straw is recycled for the development of oyster mushrooms. It is employed to prepare compost for white button mushrooms after being properly supplemented with chicken or nitrogen-rich horse dung (sun-dried before use). The spent straw is used for cattle feed pr for biogas production (Tan et al. 2015).

TOXICOLOGY

Consuming oyster mushrooms is safe. At the commercial level, however, it causes allergies to the pores. The way that oyster mushrooms affect people varies from person to person. A few people get digestive discomfort after ingesting it (Kumar 2020). In this situation, the person should get in touch with the nearest poison control centre. These facilities offer professional, no-cost medical advice every day of the week, 24 hours a day. If you can, keep some leftover food that contains mushrooms so you can confirm the identification (Correa et al. 2016).

CONCLUSION AND FUTURE PROSPECTIVE

Oyster mushrooms are an innovative edible fungus with significant medicinal and dietary value, it has been determined. It has a large number of bioactive ingredients, including ascorbic acid, phenolic compounds, polysaccharides, protein, beta glucan, lectins, and antioxidants, making it useful for nutritionally enhancing mushrooms (Patel et al. 2012). These mushrooms help fight malnutrition illnesses as well as nutritional deficiencies. Because mushrooms are produced in enormous quantities, much of them are lost due to ineffective processing methods. Therefore, there is a huge need to create processing technology to stop perishable. The bioactive ingredients from mushrooms were isolated and capsuled to maximise the medicinal advantages (Jesenak et al. 2013). The immune-modulating polysaccharides can be used with health-related vitamins. The extraction of secondary metabolites will be discovered in the following years. The key focus in research into the future is on the extraction of beneficial bioactive components, diverse edible mushroom extraction, and subsequently inclusion of these components in functional foods that lead to the treatment of various health issues.

REFERENCES

Abdullah, N., Ismail, R., Johari, N. M. K., & Annuar, M. S. M. (2013). Production of liquid spawn of an edible grey oyster mushroom, *Pleurotuspulmonarius* (Fr.) Quél by submerged fermentation and sporophore yield on rubber wood sawdust. *Scientia Horticulturae*, *161*, 65–69.

Abidin, M. H. Z., Abdullah, N., & Abidin, N. Z. (2017). Therapeutic properties of *Pleurotus* species (oyster mushrooms) for atherosclerosis: A review. *International Journal of Food Properties*, *20*(6), 1251–1261.

Adebayo, E. A., & Martinez-Carrera, D. (2015). Oyster mushrooms (*Pleurotus*) are useful for utilizing lignocellulosic biomass. *African Journal of Biotechnology*, *14*(1), 52–67.

Ahmed, M., Abdullah, N., & Nuruddin, M. M. (2016). Yield and nutritional composition of oyster mushrooms: An alternative nutritional source for rural people. *Sains Malaysiana*, *45*(11), 1609–1615.

Ahmed, M., Abdullah, N., Ahmed, K. U., & Bhuyan, M. H. M. (2013). Yield and nutritional composition of oyster mushroom strains newly introduced in Bangladesh. *PesquisaAgropecuáriaBrasileira*, *48*, 197–202.

Aishah, M. S., & Wan Rosli, W. I. (2013). The effect of addition of oyster mushroom (*Pleurotussajor-caju*) on nutrient composition and sensory acceptation of selected wheat-and rice-based products. *International Food Research Journal*, *20*(1).

Atri, N. S., Sharma, S. K., Joshi, R., Gulati, A., & Gulati, A. (2013). Nutritional and neutraceutical composition of five wild culinary-medicinal species of genus *Pleurotus* (higher Basidiomycetes) from northwest India. *International Journal of Medicinal Mushrooms*, *15*(1).

Avagyan, I. A., Nanagulyan, S. G., & Minasbekyan, L. A. (2013). Increasing of Fermentative and antiinflamatory activity of the *Pleurotus ostreatus* (Jacq.: Fr.) Kumm. Culture by modification of growth conditions by MM-waves. *Progress In Electromagnetics Research*, *1659*.

Baggio, C. H., Freitas, C. S., Marcon, R., de Paula Werner, M. F., Rae, G. A., Smiderle, F. R., ... & Santos, A. R. S. (2012). Antinociception of β-d-glucan from *Pleurotuspulmonariusis* possibly related to protein kinase C inhibition. *International Journal of Biological Macromolecules*, *50*(3), 872–877.

Bandopadhyay, S. (2013). Effect of supplementing rice straw with water hyacinth on the yield and nutritional qualities of oyster mushrooms (*Pleurotus* spp.). *MicologíaAplicada International*, *25*(2), 15–21.

Barh, A., Sharma, V. P., Annepu, S. K., Kamal, S., Sharma, S., & Bhatt, P. (2019). Genetic improvement in *Pleurotus* (oyster mushroom): A review. *3 Biotech*, *9*(9), 1–14.

Bellettini, M. B., & Fiorda, F. A. (2016). Production pests and diseases in mushroom *Pleurotus* spp crops. *Apprehendere, Guarapuava*, *152*.

Bellettini, M. B., Bellettini, S., Fiorda, F. A., Pedro, A. C., Bach, F., Fabela-Morón, M. F., & Hoffmann-Ribani, R. (2018). Diseases and pests noxious to *Pleurotus* spp. mushroom crops. *Revista Argentina de microbiologia*, *50*(2), 216–226.

Bellettini, M. B., Fiorda, F. A., & Bellettini, S. (2015). Aspectosgerais do cultivo de cogumeloPleurotusostreatus e djamor pela técnica Jun–Cao. *Apprehendere, Guarapuava (in Portuguese)*.

Bellettini, M. B., Fiorda, F. A., Maieves, H. A., Teixeira, G. L., Ávila, S., Hornung, P. S., ... & Ribani, R. H. (2019). Factors affecting mushroom spp. *Saudi Journal of Biological Sciences*, *26*(4), 633–646.

Bernardi, E., Minotto, E., & Nascimento, J. S. D. (2013). Evaluation of growth and production of *Pleurotus* sp. in sterilized substrates. *Arquivos do Instituto Biológico*, *80*, 318–324.

Cao, X. Y., Liu, J. L., Yang, W., Hou, X., & Li, Q. J. (2015). Antitumor activity of polysaccharide extracted from *Pleurotusostreatus mycelia* against gastric cancer in vitro and in vivo. *Molecular Medicine Reports*, *12*(2), 2383–2389.

Chanakya, H. N., Malayil, S., & Vijayalakshmi, C. (2015). Cultivation of *Pleurotus* spp. on a combination of anaerobically digested plant material and various agro-residues. *Energy for Sustainable Development*, *27*, 84–92.

Chen, J. N., De Mejia, E. G., & Wu, J. S. B. (2011). Inhibitory effect of a glycoprotein isolated from golden oyster mushroom (*Pleurotuscitrinopileatus*) on the lipopolysaccharide-induced inflammatory reaction in RAW 264.7 macrophage. *Journal of Agricultural and Food Chemistry*, *59*(13), 7092–7097.

Chen, S. Y., Ho, K. J., Hsieh, Y. J., Wang, L. T., & Mau, J. L. (2012). Contents of lovastatin, γ-aminobutyric acid and ergothioneine in mushroom fruiting bodies and mycelia. *Lwt*, *47*(2), 274–278.

Correa, R. C. G., Brugnari, T., Bracht, A., Peralta, R. M., & Ferreira, I. C. (2016). Biotechnological, nutritional and therapeutic uses of *Pleurotus* spp. (Oyster mushroom) related with its chemical composition: A review on the past decade findings. *Trends in Food Science & Technology*, *50*, 103–117.

Correa, R. C. G., de Souza, A. H. P., Calhelha, R. C., Barros, L., Glamoclija, J., Sokovic, M., & Ferreira, I. C. (2015). Bioactive formulations prepared from fruiting bodies and submerged culture mycelia of the Brazilian edible mushroom*Pleurotusostreatoroseus* Singer. *Food & Function*, *6*(7), 2155–2164.

Deepalakshmi, K., & Mirunalini, S. (2014). Toxicological assessment of *Pleurotusostreatus* in Sprague Dawley rats. *International Journal of Nutrition, Pharmacology, Neurological Diseases*, *4*(3), 139.

Dos Santos Bazanella, G. C., de Souza, D. F., Castoldi, R., Oliveira, R. F., Bracht, A., & Peralta, R. M. (2013). Production of laccase and manganese peroxidase by *Pleurotuspulmonarius* in solid-state cultures and application in dye decolorization. *Folia microbiologica*, *58*(6), 641–647.

Dubey, D., Dhakal, B., Dhami, K., Sapkota, P., Rana, M., Poudel, N. S., & Aryal, L. (2019). Comparative study on effect of different substrates on yield performance of oyster mushroom. *Global Journal of Biology, Agriculture, Health Sciences*, *8*(1).

Ejigu, N., Sitotaw, B., Girmay, S., & Assaye, H. (2022). Evaluation of oyster mushroom (*Pleurotusostreatus*) production using water hyacinth (*Eichhornia crassipes*) biomass supplemented with agricultural wastes. *International Journal of Food Science*, *2022*.

El Enshasy, H. A., & Hatti-Kaul, R. (2013). Mushroom immunomodulators: Unique molecules with unlimited applications. *Trends in Biotechnology*, *31*(12), 668–677.

El-Batal, A. I., ElKenawy, N. M., Yassin, A. S., & Amin, M. A. (2015). Laccase production by *Pleurotusostreatus* and its application in synthesis of gold nanoparticles. *Biotechnology Reports*, *5*, 31–39.

Elattar, A. M., Hassan, S., & Awd-Allah, S. F. (2019). Evaluation of oyster mushroom (*Pleurotusostreatus*) cultivation using different organic substrates. *Alexandria Science Exchange Journal*, *40*(July-September), 427–440.

Elisashvili, V. I. (2012). Submerged cultivation of medicinal mushrooms: Bioprocesses and products. *International Journal of Medicinal Mushrooms*, *14*(3).

Facchini, J. M., Alves, E. P., Aguilera, C., Gern, R. M. M., Silveira, M. L. L., Wisbeck, E., & Furlan, S. A. (2014). Antitumor activity of *Pleurotusostreatus* polysaccharide fractions on Ehrlich tumor and Sarcoma 180. *International Journal of Biological Macromolecules*, *68*, 72–77.

Feeney, M. J., Dwyer, J., Hasler-Lewis, C. M., Milner, J. A., Noakes, M., Rowe, S., & Wu, D. (2014). Mushrooms and health summit proceedings. *The Journal of Nutrition*, *144*(7), 1128S–1136S.

Fernandes, A., Barros, L., Martins, A., Herbert, P., & Ferreira, I. C. (2015). Nutritional characterisation of *Pleurotusostreatus* (Jacq. ex Fr.) P. Kumm. produced using paper scraps as substrate. *Food Chemistry*, *169*, 396–400.

Fikadu, A., Wedu, T. D., & Derseh, E. (2019). Review on economics of teff in Ethiopia. *Open Access Biostatistics & Bioinformatics*, *2*(3), 1–8.

Gąsecka, M., Mleczek, M., Siwulski, M., & Niedzielski, P. (2016). Phenolic composition and antioxidant properties of *Pleurotusostreatus* and *Pleurotuseryngii* enriched with selenium and zinc. *European Food Research and Technology*, *242*(5), 723–732.

Girmay, Z., Gorems, W., Birhanu, G., & Zewdie, S. (2016). Growth and yield performance of *Pleurotusostreatus* (Jacq. Fr.) *Kumm* (oyster mushroom) on different substrates. *Amb Express*, *6*(1), 1–7.

Golak-Siwulska, I., Kałużewicz, A., Spiżewski, T., Siwulski, M., & Sobieralski, K. (2018). Bioactive compounds and medicinal properties of Oyster mushrooms (sp.). *Folia Horticulturae*, *30*(2), 191–201.

Hassan, F. R., & Medany, G. M. (2014). Effect of pretreatments and drying temperatures on the quality of dried *Pleurotus* Mushroom Spp. *Egyptian Journal of Agricultural Research*, *92*(3), 1009–1023.

Hatvani, L., Sabolić, P., Kocsubé, S., Kredics, L., Czifra, D., Vágvölgyi, C., ... & Kosalec, I. (2012). The first report on mushroom green mould disease in Croatia. *Archives of Industrial Hygiene and Toxicology*, *63*(4), 481–487.

He, X. L., Li, Q., Peng, W. H., Zhou, J., Cao, X. L., Wang, D., ... & Gan, B. C. (2017). Intra-and inter-isolate variation of ribosomal and protein-coding genes in *Pleurotus*: Implications for molecular identification and phylogeny on fungal groups. *BMC Microbiology*, *17*(1), 1–9.

Ho, L. H., Zulkifli, N. A., & Tan, T. C. (2020). Edible mushroom: Nutritional properties, potential nutraceutical values, and its utilisation in food product development. *An introduction to mushroom*, *10*.

Hsu, C. M., Hameed, K., Cotter, V. T., & Liao, H. L. (2018). Isolation of mother cultures and preparation of spawn for oyster mushroom cultivation: SL449/SS663, 1/2018. *EDIS*, *2018*(1).

Huang, S. J., Lin, C. P., & Tsai, S. Y. (2015). Vitamin D2 content and antioxidant properties of fruit body and mycelia of edible mushrooms by UV-B irradiation. *Journal of Food Composition and Analysis*, *42*, 38–45.

Inacio, F. D., Ferreira, R. O., Araujo, C. A. V. D., Brugnari, T., Castoldi, R., Peralta, R. M., & Souza, C. G. M. D. (2015). Proteases of wood rot fungi with emphasis on the genus*Pleurotus*. *BioMed Research International*, *2015*.

Jafarpour, M., & Eghbalsaeed, S. (2012). High protein complementation with high fiber substrates for oyster mushroom cultures. *African Journal of Biotechnology*, *11*(14), 3284–3289.

Jang, K. Y., Oh, Y. L., Oh, M., Woo, S. I., Shin, P. G., Im, J. H., & Kong, W. S. (2016). Introduction of the representative mushroom cultivars and groundbreaking cultivation techniques in Korea. *Journal of Mushroom*, *14*(4), 136–141.

Jatwa, T. K., Apet, K. T., Wagh, S. S., Sayyed, K. S., Rudrappa, K. B., & Sornapriya, S. P. (2016). Evaluation of Various agro-wastes for production of *Pleurotus* spp. (*P. florida*, *P. sajor-caju* and *P. eous*). *Journal of Pure and Applied Microbiology*, *10*, 2783–2792.

Jaworska, G., Pogoń, K., Bernaś, E., & Duda-Chodak, A. (2015). Nutraceuticals and antioxidant activity of prepared for consumption commercial mushrooms *A. garicusbisporus* and *P. leurotusostreatus*. *Journal of Food Quality*, *38*(2), 111–122.

Jegadeesh, R., Lakshmanan, H., Kab-Yeul, J., Sabaratnam, V., & Raaman, N., (2018). Cultivation of pink oyster mushroom *Pleurotusdjamor* var. roseus on various agro-residues by low cost technique. *Journal of Mycopathology Research*, *56*(3), 213–220.

Jennifer, O., & Devi, L. J. (2020). Bioconversion of selected solid wastes by control cultivation of oyster mushroom, *Pleurotusflorida* and its nutrient analysis. *International Journal of Research in Engineering, Science and Management*, *3*(12), 34–36.

Jesenak, M., Majtan, J., Rennerova, Z., Kyselovic, J., Banovcin, P., & Hrubisko, M. (2013). Immunomodulatory effect of pleuran (β-glucan from *Pleurotusostreatus*) in children with recurrent respiratory tract infections. *International Immunopharmacology*, *15*(2), 395–399.

Khan, M. A., & Tania, M. (2012). Nutritional and medicinal importance of *Pleurotus* mushrooms: An overview. *Food Reviews International*, *28*(3), 313–329.

Khare, K. B., Jongman, M., & Loeto, D. (2018). Oyster mushroom cultivation at different production systems: A review, *5*(5), 72–79.

Khatun, S., Islam, A., Cakilcioglu, U., Guler, P., & Chatterjee, N. C. (2015). Nutritional qualities and antioxidant activity of three edible oyster mushrooms (*Pleurotus* spp.). *NJAS-Wageningen Journal of Life Sciences*, *72*, 1–5.

Khonga, E. B., Khare, K. B., & Jongman, M. (2013). Effect of different grain spawns and substrate sterilization methods on yield of oyster mushroom in Botswana, *2*(10), 1308–1311.

Kibar, B. (2021). Influence of different drying methods and cold storage treatments on the postharvestquality and nutritional properties of *P. ostreatus* mushroom. *Turkish Journal of Agriculture and Forestry*, *45*(5), 565–579.

Kortei, N. K., & Wiafe-Kwagyan, M. (2015). Comparative appraisal of the total phenolic content, flavonoids, free radical scavenging activity and nutritional qualities of *Pleurotusostreatus* (EM-1) and *Pleurotuseous* (P-31) cultivated on rice (*Oryzae sativa*) straw in Ghana. *Journal of Advances in Biology and Biotechnology*, *3*(4), 153–164.

Koutrotsios, G., Kalogeropoulos, N., Stathopoulos, P., Kaliora, A. C., & Zervakis, G. I. (2017). Bioactive compounds and antioxidant activity exhibit high intraspecific variability in *Pleurotusostreatus* mushrooms and correlate well with cultivation performance parameters. *World Journal of Microbiology and Biotechnology*, *33*(5), 1–14.

Koutrotsios, G., Mountzouris, K. C., Chatzipavlidis, I., & Zervakis, G. I. (2014). Bioconversion of lignocellulosic residues by *Agrocybecylindracea* and *Pleurotusostreatus* mushroom fungi–assessment of their effect on the final product and spent substrate properties. *Food chemistry*, *161*, 127–135.

Krakowska, A., Zięba, P., Włodarczyk, A., Kała, K., Sułkowska-Ziaja, K., Bernaś, E., ... & Muszyńska, B. (2020). Selected edible medicinal mushrooms from *Pleurotus* genus as an answer for human civilization diseases. *Food Chemistry*, *327*, 127084.

Kumar, K. (2020). Nutraceutical potential and processing aspects of oyster mushrooms (*Pleurotus* species). *Current Nutrition & Food Science*, *16*(1), 3–14.

Lam, Y. S., & Okello, E. J. (2015). Determination of lovastatin, β-glucan, total polyphenols, and antioxidant activity in raw and processed oyster culinary-medicinal mushroom, *Pleurotusostreatus* (higher Basidiomycetes). *International Journal of Medicinal Mushrooms*, *17*(2).

Lavelli, V., Proserpio, C., Gallotti, F., Laureati, M., & Pagliarini, E. (2018). Circular reuse of bio-resources: The role of *Pleurotus* spp. in the development of functional foods. *Food & Function*, *9*(3), 1353–1372.

Lee, S. J., Kim, H. H., Kim, S. H., Kim, I. S., & Sung, N. J. (2018). Culture conditions of liquid spawn and the growth characteristics of *Pleurotusostreatus*. *Journal of Mushroom*, *16*(3), 162–170.

Leo, V. V., Passari, A. K., Muniraj, I. K., Uthandi, S., Hashem, A., Abd_Allah, E. F., ... & Singh, B. P. (2019). Elevated levels of laccase synthesis by *Pleurotuspulmonarius* BPSM10 and its potential as a dye decolorizing agent. *Saudi Journal of Biological Sciences*, *26*(3), 464–468.

Lin, S. Y., Chen, Y. K., Yu, H. T., Barseghyan, G. S., Asatiani, M. D., Wasser, S. P., & Mau, J. L. (2013). Comparative study of contents of several bioactive components in fruiting bodies and mycelia of culinary-medicinal mushrooms. *International Journal of Medicinal Mushrooms*, *15*(3).

Mahalakshmi, A., Suresh, M., & Rajendran, S. (2019). Cultivation of oyster mushroom (*Pleurotusflorida*) in various seasons on paddy straw. *Research Journal of Life Sciences, Bioinformatics, Pharmaceutical and Chemical Sciences*, *5*(6), 79–86.

Mishra, K. K., Pal, R. S., Mishra, P. K., & JC, B. (2016). Antioxidant activities and mineral composition of oyster mushroom (*Pleurotussajor-caju*) as influenced by different drying methods, *28*(9), 2025–2030.

Morris, H., Beltrán, Y., Llauradó, G., Batista, P., Perraud, I. G., García, N., ... & Diez, J. (2017). Mycelia from *Pleurotus* sp. (oyster mushroom): a new wave of antimicrobials, anticancer and antioxidant bio-ingredients. *International Journal of Phytocosmetics and Natural Ingredients*, *4*(1), 3.

Mukhopadhyay, S. B. (2019). Oyster mushroom cultivation on water hyacinth biomass: Assessment of yield performances, nutrient, and toxic element contents of mushrooms. In *An Introduction to Mushroom.* IntechOpen.

Muswati, C., Simango, K., Tapfumaneyi, L., Mutetwa, M., & Ngezimana, W. (2021). The effects of different substrate combinations on growth and yield of oyster mushroom (*Pleurotusostreatus*). *International Journal of Agronomy*, *2021*.

Mutukwa, I. B., Hall III, C. A., Cihacek, L., & Lee, C. W. (2019). Evaluation of drying method and pre-treatment effects on the nutritional and antioxidant properties of oyster mushroom (*Pleurotusostreatus*). *Journal of Food Processing and Preservation*, *43*(4), e13910.

Nhi, N. N. Y., & Hung, P. V. (2012). Nutritional composition and antioxidant capacity of several edible mushrooms grown in the Southern Vietnam, *19*(2), 611–615.

Nolle, N., Argyropoulos, D., Ambacher, S., Müller, J., & Biesalski, H. K. (2017). Vitamin D2 enrichment in mushrooms by natural or artificial UV-light during drying. *LWT-Food Science and Technology*, *85*, 400–404.

Nongthombam, J., Kumar, A., Ladli, B., Madhushekhar, M., & Patidar, S. (2021). A review on study of growth and cultivation of oyster mushroom. *Plant Cell Biotechnology and Molecular Biology*, *22*(5&6), 55–65.

Oloke, J. K. (2017). Oyster mushroom (*Pleurotus* species); a natural functional food. *The Journal of Microbiology, Biotechnology and Food Sciences*, *7*(3), 254.

Onchonga, N. A., Oima, D., & Oginda, M. (2013). Utilization of water hyacinth as an alternative substrate for mushroom farming: A study of vihiga mushroom project in western Kenya. *International Journal of Educational Research*, *1*, 1–10.

Oseni, T. O., Dlamini, S. O., Earnshaw, D. M., & T MASARIRAMBI, M. (2012). Effect of substrate pre-treatment methods on oyster mushroom (*Pleurotusostreatus*) production. *International Journal of Agriculture and Biology*, *14*(2).

Owaid, M. N., Abed, A. M., & Nassar, B. M. (2015). Recycling cardboard wastes to produce blue oyster mushroom *Pleurotusostreatus* in Iraq. *Emirates Journal of Food and Agriculture*, 537–541.

Palacios, I., Lozano, M., Moro, C., Arrigo, D., Rostagno, M. A., Martinez, J. A., Garcia–Lafuente, A., Quillamon, E., & Villares, A. (2011). Antioxidant properties of phenolic compounds occuring in edible mushrooms. *Food Chemistry*, *128*, 674–678.

Park, K. H., Lee, E. S., Jin, Y. I., Myung, K. S., Park, H. W., Park, C. G., ... & Kim, Y. O. (2016). Inhibitory effect of Panax ginseng and *Pleurotusosteratus* complex on expression of cytokine genes induced by extract of Dermatophagoidespteronissinus in human monocytic THP-1 and EoL-1 cells. *Journal of Mushroom*, *14*(4), 155–161.

Pasnik, J. (2017). Preventive effect of pleuran (β-glucan from *Pleurotusostreatus*) in children with recurrent respiratory tract infections-open-label prospective study. *Current Pediatric Research*, *21*(1), 99–104.

Patel, Y., Naraian, R., & Singh, V. K. (2012). Medicinal properties of *Pleurotus* species (oyster mushroom): A review. *World Journal of Fungal and Plant Biology*, *3*(1), 1–12.

Piska, K., Sułkowska-Ziaja, K., & Muszyńska, B. (2017). Edible mushroom *Pleurotusostreatus* (oyster mushroom): its dietary significance and biological activity. *Acta ScientiarumPolonorum. Hortorum Cultus*, *16*(1).

Piskov, S., Timchenko, L., Grimm, W. D., Rzhepakovsky, I., Avanesyan, S., Sizonenko, M., & Kurchenko, V. (2020). Effects of various drying methods on some physico-chemical properties and the antioxidant profile and ACE inhibition activity of oyster mushrooms (*Pleurotusostreatus*). *Foods*, *9*(2), 160.

Raman, J., Jang, K. Y., Oh, Y. L., Oh, M., Im, J. H., Lakshmanan, H., & Sabaratnam, V. (2021). Cultivation and nutritional value of prominent *Pleurotus* spp.: An overview. *Mycobiology*, *49*(1), 1–14.

Ravi, B., Renitta, R. E., Prabha, M. L., Issac, R., & Naidu, S. (2013). Evaluation of antidiabetic potential of oyster mushroom (*Pleurotusostreatus*) in alloxan-induced diabetic mice. *Immunopharmacology and Immunotoxicology*, *35*(1), 101–109.

Ritota, M., & Manzi, P. (2019). *Pleurotus* spp. cultivation on different agri-food by-products: Example of biotechnological application. *Sustainability*, *11*(18), 5049.

Rout, M. K., Mohapatra, K. B., Mohanty, P., & Chandan, S. S. (2015). Studies on effect of incubation temperature and light intensity on mycelial growth of oyster species. *Journal Crop and Weed*, *11*(2), 44–46.

Royse, D. J. (2014, November). A global perspective on the high five: Agaricus,*Pleurotus*, Lentinula, Auricularia &Flammulina. In *Proceedings of the 8th International Conference on Mushroom Biology and Mushroom Products (ICMBMP8)* (Vol. 1, pp. 1–6).

Salami, A. O., Bankole, F. A., & Salako, Y. A. (2017). Nutrient and mineral content of oyster mushroom (*Pleurotusflorida*) grown on selected lignocellulosic substrates. *Journal of Advances in Biology & Biotechnology*, *15*(1), 1–7.

Samsudin, N. I. P., & Abdullah, N. (2019). Edible mushrooms from Malaysia; a literature review on their nutritional and medicinal properties. *International Food Research Journal*, *26*(1), 11–31.

Sardar, H., Ali, M. A., Anjum, M. A., Nawaz, F., Hussain, S., Naz, S., & Karimi, S. M. (2017). Agro-industrial residues influence mineral elements accumulation and nutritional composition of king oyster mushroom (*Pleurotuseryngii*). *Scientia Horticulturae*, *225*, 327–334.

Sekan, A. S., Myronycheva, O. S., Karlsson, O., Gryganskyi, A. P., & Blume, Y. (2019). Green potential of *Pleurotus* spp. in biotechnology. *PeerJ*, *7*, e6664.

Sharma, S., Yadav, R. K. P., & Pokhrel, C. P. (2013). Growth and yield of oyster mushroom (*Pleurotusostreatus*) on different substrates. *Journal on New Biological Reports*, *2*(1), 03–08.

Sharma, D., Saha, A. K., & Datta, B. K. (2018). Bioactive compounds with special references to anticancer property of oyster mushroom *Pleurotusostreatus*. *J PharmacognPhytochem*, *7*(4), 2694–2698.

Shen, Y., Gu, M., Jin, Q., Fan, L., Feng, W., Song, T., ... & Cai, W. (2014). Effects of cold stimulation on primordial initiation and yield of *Pleurotuspulmonarius*. *Scientia Horticulturae*, *167*, 100–106.

Shirmila, J. G., Athira, S. V., Devika, A. V., Aishwarya, N. A. M., Parvathy, A. M., & Merry, D. M. (2018). Bioconversion of water hyacinth as an alternate substrate for mushroom (*Pleurotuseous*) cultivation. *Trends in Biosciences*, *11*(7), 1447–1451.

Silva, P. P., Andrade, C. L., Junior, J. C. B., Magalhaes, B. G., Melo, B. F., & y Garcia, A. G. (2016). Response of tropical maize to supplemental irrigation strategies. In *2016 ASABE Annual International Meeting* (p. 1). American Society of Agricultural and Biological Engineers.

Smiderle, F. R., Olsen, L. M., Ruthes, A. C., Czelusniak, P. A., Santana-Filho, A. P., Sassaki, G. L., ... & Iacomini, M. (2012). Exopolysaccharides, proteins and lipids in *Pleurotuspulmonarius* submerged culture using different carbon sources. *Carbohydrate Polymers*, *87*(1), 368–376.

Srikram, A., & Supapvanich, S. (2016). Proximate compositions and bioactive compounds of edible wild and cultivated mushrooms from Northeast Thailand. *Agriculture and Natural Resources*, 50, 432–436.

Suseem, S. R., & Saral, A. M. (2013). Analysis on essential fatty acid esters of mushroom *Pleurotuseous* and its antibacterial activity. *Asian Journal of Pharmaceutical Clinical Research*, *6*(1), 188–191.

Tan, Y. S., Baskaran, A., Nallathamby, N., Chua, K. H., Kuppusamy, U. R., & Sabaratnam, V. (2015). Influence of customized cooking methods on the phenolic contents and antioxidant activities of selected species of oyster mushrooms (*Pleurotus* spp.). *Journal of Food Science and Technology*, *52*(5), 3058–3064.

Taofiq, O., Calhelha, R. C., Heleno, S., Barros, L., Martins, A., Santos-Buelga, C., ... & Ferreira, I. C. (2015). The contribution of phenolic acids to the anti-inflammatory activity of mushrooms: Screening in phenolic extracts, individual parent molecules and synthesized glucuronated and methylated derivatives. *Food Research International*, *76*, 821–827.

Taofiq, O., González-Paramás, A. M., Martins, A., Barreiro, M. F., & Ferreira, I. C. (2016). Mushrooms extracts and compounds in cosmetics, cosmeceuticals and nutricosmetics—A review. *Industrial Crops and Products*, *90*, 38–48.

Tesfaw, A., Tadesse, A., & Kiros, G. (2015). Optimization of oyster (*Pleurotusostreatus*) mushroom cultivation using locally available substrates and materials in Debre Berhan, Ethiopia. *Journal of Applied Biology and Biotechnology*, *3*(1), 0–2.

Tiram, C. (2013). Effect of different drying techniques on the nutritional values of oyster mushroom (*Pleurotussajor-caju*). *Sains Malaysiana*, *42*(7), 937–941.

Tolera, K. D., & Abera, S. (2017). Nutritional quality of Oyster Mushroom (*Pleurotusostreatus*) as affected by osmotic pretreatments and drying methods. *Food Science & Nutrition*, *5*(5), 989–996.

Vital, A. C. P., Goto, P. A., Hanai, L. N., Gomes-da-Costa, S. M., de Abreu Filho, B. A., Nakamura, C. V., & Matumoto-Pintro, P. T. (2015). Microbiological, functional and rheological properties of low fat yogurt supplemented with *Pleurotusostreatus* aqueous extract. *LWT-Food Science and Technology*, *64*(2), 1028–1035.

Wendiro, D., Wacoo, A. P., & Wise, G. (2019). Identifying indigenous practices for cultivation of wild saprophytic mushrooms: Responding to the need for sustainable utilization of natural resources. *Journal of Ethnobiology and Ethnomedicine*, *15*(1), 1–15.

Xu, W., Huang, J. J. H., & Cheung, P. C. K. (2012). Extract of *Pleurotuspulmonarius* suppresses liver cancer development and progression through inhibition of VEGF-induced PI3K/AKT signaling pathway. *PLoS One*, *7*(3), e34406.

Yehia, R. S. (2012). Nutritional value and biomass yield of the edible mushroom *Pleurotusostreatus* cultivated on different wastes in Egypt. *Innovative Romanian Food Biotechnology*, *11*, 9.

Younis, A. M., Wu, F. S., & El Shikh, H. H. (2015). Antimicrobial activity of extracts of the oyster culinary medicinal mushroom *Pleurotusostreatus* (higher basidiomycetes) and identification of a new antimicrobial compound. *International journal of medicinal mushrooms*, *17*(6).

Zhang, W. R., Liu, S. R., Kuang, Y. B., & Zheng, S. Z. (2019). Development of a novel spawn (block spawn) of an edible mushroom, *Pleurotusostreatus*, in liquid culture and its cultivation evaluation. *Mycobiology*, *47*(1), 97–104.

12 *Pycnoporus cinnabarinus*

Ali Ikram
University Institute of Food Science and Technology, The University of Lahore, Pakistan

Farhan Saeed, Muhammad Afzaal, and Amna Saleem
Department of Food Sciences, Government College University Faisalabad, Pakistan

Shahid Bashir, Muhammad Zia Shahid, and Awais Raza
University Institute of Food Science and Technology, The University of Lahore, Pakistan

CONTENTS

DOI: 10.1201/9781003259763-12

INTRODUCTION

Mushrooms are high in protein, vitamins, and minerals, and are commonly employed in both food and medicine. Saprophytes are commonly found on soil, urban fields, garden soil, trees, and roadside vegetation. White fungi belong to the Polyporales group and are connected to the *Pycnoporus* species (Sevindik et al. 2020). They are assumed to be related to the trametes type. *Pycnoporus* is a genus of white fungi that is related to Agaricomycetes and Polyporaceos (Hibbett et al. 2007). Pycnoporus is a tetrapolar connective tissue heterothallic homobasidiomycetes. The Size, shape, color, texture, unity, hyphalic system, pores, tubes, and chemical reactions are only a few of the genetic alterations that Pycnoporus-bearing organisms go through. Melzer solution with phenol is used for 5% KOH (Gong et al. 2020). Every year, this type of genetically created genus emerges, with traits such as flat, frequently flattened, and scaly skin with wide pores, and occasionally no well-marked or spotted. This layer is white or silver in matured cases (Guzmán 2004). Isodiametric perforations varying in size from small to medium adorn the lower section. Whether little or huge, the subject is the same hue as the pile. Genes with a solid wall, hyaline, and septic generative attachment to a solid bone wall identify the hyphal dimitic or trimitic type. The pigment phenoxazinone, which contains cinabarin, cinnabaric acid, and eventually tramesanguine, is responsible for the hue (Hyder and Dutta 2021).

Pycnoporus fungi, which are saprotrophic homobasidiomycetes, are recognized for having lignocellolytic capabilities (Zhou et al. 2020; Lomascolo et al. 2011). Pycnoporus food is also of significant interest in a range of applications, including textile manufacturing, whitening, agricultural product recycling, phenolic compound removal, wastewater treatment, and industrial (Levasseur et al. 2014). Pycnoporus is found in four different types of geographical areas: New Caledonia, Africa, and India, to name a few (Levasseur et al. 2014). *Pycnoporus cinnabarinus*, commonly called polypore cinnabar, is a saprophytic white matter. The body of the fruit is orange. It can be found in different parts of the world. It is not a feast. (Angelini et al. 2020).

Cinnabarine has also been shown to have antiviral and antibacterial activity against Pycnoporus fungi and harmful microorganisms (Smânia et al. 2003) as well as infectious viruses such as *Klebsiella pneumoniae* and *Salmonella typhi* (Kuang et al. 2021). Basidiocarps of various ages, sessile, single-celled, or intervertebral, skin or anus, 70 × 130 × 140 mm (diameter, width, and thickness). The ocher gray-orange stem is semi-oblong or oblong. As the sample ages, the color deepens, the mature or dry-type whitens, and the whiteness-type lightens. It grows slowly on top of the cracks at an early age, and is usually fast or hardy, with or without a lattice pattern, with or without a well-developed, sharp, smooth, or flat surface, poor (Li et al. 2021). It is robust and fibrous, 10–15 mm thick, and has a hue that is usually more durable than loose. A 1503 tube diskoration tube in childhood, up to a depth of 7 mm; hymenophore pores, from coral to red; pores (2–3 mm each). It causes rot on dead plants, especially Alnus, Corylus, Betula, Fraxinus, *Sorbus aucuparia*, Malus, Quercus, Populus, Salix and Prunus. Macrochemical reactions: Adding KOH to the dye causes it to break down quickly in all directions. Microscopic information: Tritical hypal system with hyphae, thin walls, broad hyalinfibules, 1.6–2.6 microns, hyaline skeletal hyphae, cylindrical spores 5-6 × 2-3 m, flat, hyaline, non-amyloid Disclosure: The presence of red fruit bodies in flat piles is a reliable indicator of their qualities (Krupodorova et al. 2020). In some places of the northern hemisphere, this kind is more common. The presence of *P. cinnabarinus* in the hilly environment is thought to evoke tropical monuments (Guzmán 2004).

Common name: Cinnabar-red polypore
Scientific name: *Pycnoporus cinnabarinus*

CHARACTERISTICS OF *PYCNOPORUS CINNABARINUS*

The vivid red color of the fruit is an important feature of every Pycnoporus species, while the color vibrations may wane over time. When Pycnoporus' foot touches with potassium hydroxide (KOH),

TABLE 12.1
Identification of *Pycnoporus cinnabarinus*

1	**Fruit body**	The whole fruit body is bright orange; up to 10 cm across and when in bracket form projecting typically 4 to 6 cm from the substrate; usually between 1 and 2 cm thick; the upper (infertile) surface is rough or wrinkled, orange-red, fading with age; margins are rounded (left) in young specimens, which are downy or finely hairy on the upper surface; margin becomes more acute as fruit body ages; lower (fertile) surface with tubes.
2	**Tubes**	Pale orange; 2 to 6 mm deep.
3	**Pores**	Cinnabar red; round or angular, spaced at 2 to 4 per mm.
4	**Spores**	Cylindrical or slightly allantoid (sausage-shaped), smooth, 5–6 × 2–2.5 μm; inamyloid.
5	**Spore print**	White.
6	**Odor /taste**	Not distinctive.
7	**Habitat and Ecological role**	Saprophytic on hardwoods, particularly Beech and birches.
8	**Season**	Late summer and autumn.
9	**Similar species**	*Fistulina hepatica* is bright red when young; its spores are pinkish yellow.

it becomes reddish-yellow to black. A detailed examination of numerous traits is required to define the Pycnoporus kind. The size of the pile is important to distinguish between *P. cinnabarinus* and *P. sanguineus, P. cinnabarinus* has a 5–15 mm thick body, whereas *P. sanguineus* has a 1–5 mm thick body. Furthermore, *P. sanguineus* has a scarlet tint that does not fade easily. The exposed portion of many specimens collected by *P. cinnabarinus* has a reddish-orange hue. Finally, the pores of *P. cinnabarinus* are 1 mm larger (2–4 pores) than those of *P. sanguineus*, with 4–6 pores per millimeter (Table 12.1).

Fruits: 2–11 cm wide, 2–7 cm wide, 0.4–2 cm thick, kidney or fan type; it shall be skin, dry; When they are still small, the upper surface gives them a soft, cool look that covers fine hair; then flattened and collected, cinnabar red, declining in color and age (Figure 12.1).

Spores: White, oblong-ellipsoid, non-amyloid, 5.5–6 × 2–4 μm.

Spore print: White

Tubes: 2–5 mm long, cinnabar red.

Pores: 3–4 per mm, round or angular, cinnabar-red.

Season: Summer and autumn. It is rare

Feast: Not a feast.

Habitat: Resides in evergreen shrubs, mainly cherry, beech, and birch.

Distribution: America, Asia, and Europe.

MEDICINAL USES

Fungi create bioactive chemicals that may or may not have health advantages. Extraction, identification, and evaluation of the effects of bioactive fungal substances in vivo and in vitro have all been attempted. It's critical to find novel fungal species that can create antimicrobial chemicals in insects, as well as conditions that will allow them to produce more of them (Kuang et al. 2021). Furthermore, there is a pressing need to design a more efficient and cost-effective storage solution. Antioxidant, anticancer, antiviral, antihypercholesterolemic, antihyperlipidemic, antihyperglycemic, antiallergic, and antifungal benefits have been discovered, as well as antiallergic and fungal effects. A 20-day water culture of *Pycnoporus cinnabarinus* demonstrated good antibiotic efficacy against Gram-positive

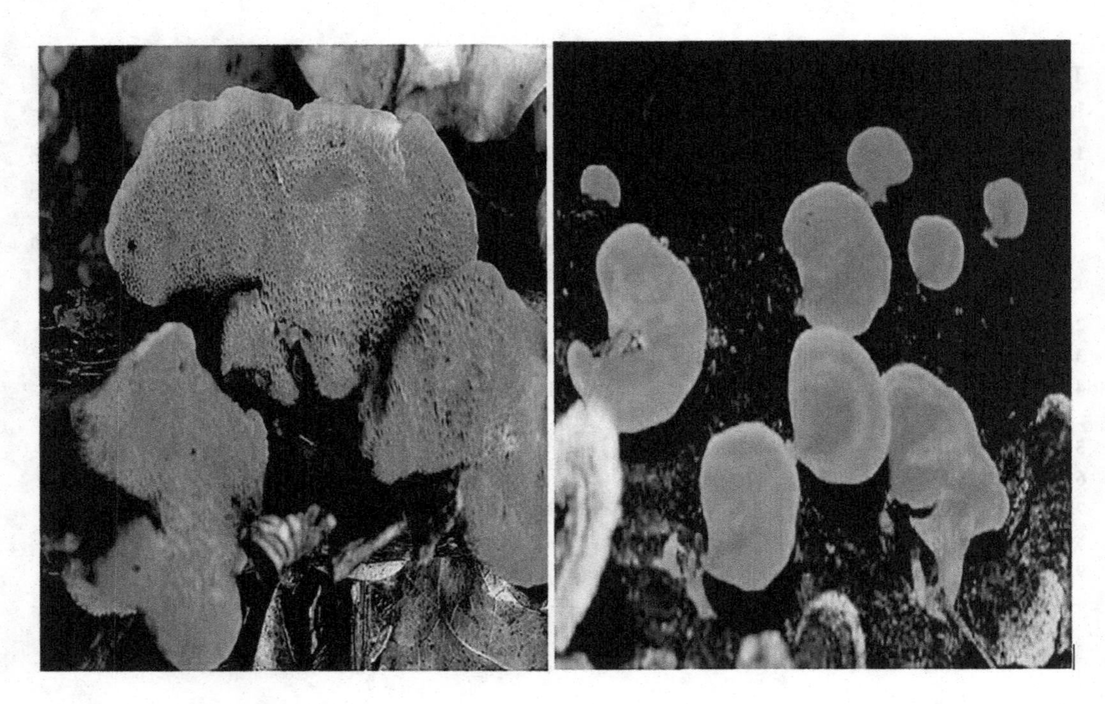

FIGURE 12.1 *Pycnoporus cinnabarinus.*

Staphylococcus aureus as well as Gram-negative *Escherichia coli* and *Pseudomonas aeruginosa* bacteria, in another investigation.

Cinnabarine, cinnabaric acid, and tramesanguine are the primary antioxidants, free radical scavengers, and antifungals found in *Pycnoporus* species pigments (Borderes et al. 2011). Anticarcinogens (Smânia et al. 2003) are immunomodulatory, larvicidal; have lechmanicidal activity (Correa et al. 2006); and antiviral, antibacterial (Smânia et al. 1995, 2003) and anti-inflammatory activity. Cinnabaric acid has also been demonstrated to have a powerful influence on apoptosis, resulting in increased oxygen production and the loss of mitochondrial membrane potential, as well as a reduction in caspase development. It is a modulator or immunomodulator of the immune response to autoimmune diseases, but its active activity is unknown. Cinnabar polypore, also known as *Pycnoporus cinnabarinus*, boosts immunity and aids in the treatment of blood problems. The washing industry uses cinnabarine, which is taken from the lining, to whiten particular colors. *Pycnoporus cinnabarinus* can aid with arthritis, gout, constipation, sore throats, ulcers, toothaches, fever, and bleeding.

PHYTOCHEMICAL PROFILE OF *PYCNOPORUS CINNABARINUS*

The study of phytochemical compounds, which are chemicals originating from plants, is known as phytochemistry. Phytochemicals are combined by plants for a variety of purposes, including pest protection and disease resistance. Alkaloids, phenylpropanoids, polichids, and terpenoids are all present in different plants, but they all fit into one of four biosynthetic classes: alkaloids, phenylpropanoids, polichids, and terpenoids. Weed-killing chemicals are produced by a variety of plants. The extraction, categorization, and characterization of natural products (MS, 1D, and 2D NMR), as well as various chromatographic methods, are the most often employed techniques in the subject of phytochemistry (MPLC, HPLC, and LC-MS). Animal and animal illness research demonstrates that eating entire fruits and vegetables is beneficial to health. Polyphenols, alkaloids, glucan compounds, flavonoids, phenols (p-phenylacetic acid, ferulic acid, p-benzoic acid, o-coumaric acid, and chrysin), cinnabarinic acid, citric acid, pigments, and enzymes are all bioactive molecules with antimicrobial action found in these foods (Eggert 1997; Adebayo et al. 2012). *Pycnoporus cinnabarinus*, a white basidiomycete, was recently discovered to conceal new clusters. Phenol oxidases are frequently formed in situations that promote ligninolysis. *P. cinnabarinus* genomic structure revealed a genetic complex found on fragile basidiomycetes. Thus, *P. cinnabarinus* belongs to an old family of enzymes that break down cellulose and hemicellulose, but its pectinolytic abilities appear to be limited. *P. cinnabarinus* also possesses a wide range of enzymatic capabilities that can harm lignin. Several genes that bind to three types of ligninolytic peroxidase have been discovered, including lignin peroxidase, peroxidase manganese, and different peroxidase.

PHARMACOLOGICAL EFFECTS OF *PYCNOPORUS CINNABARINUS*

Since 1945, when Bose revealed that the natural juice of orange or orange has a strong antibacterial activity against the antibacterial Gram of *P. cinnabarinus*, cinnabarine has been reported to have antibacterial activity (Lemberg 1952). Phenoxazone molecules appear to be part of a system that protects mammalian tissue against oxidative damage, according to a 1995 study (Eggert et al. 1995). Furthermore, the generation of cinnabarine (in culture) and its tests against a variety of viruses (Smânia EFA et al. 1997). Cinabarin (dosage 0.0625 mg / ml) has been demonstrated to be effective against *Leuconostoc plantarum* and *Bacillus cereus* (Smânia EFA et al. 1998). It was proposed by these workers that 1 is an antibiotic substance, which was consistent with previous findings (Smânia EFA et al. 1998). *P. sanguineus* ergosterol peroxide has been shown to have a leishmanicidal impact on Leishmania (Viannia) panamensis amastigotes (Correa et al. 2006). The Pharmacological Effects of *P. cinnabarinus* are presented in Table 12.2.

TABLE 12.2
Pharmacological Effects of *Pycnoporus cinnabarinus*

Sr.#	Effect	Description	References
1	Antibacterial effects	*Pycnoporus cinnabarinus* demonstrated good antibiotic efficacy against Gram-positive *S.aureus* as well as Gram-negative *E. coli* and *Pseudomonas aeruginosa* bacteria	(Khezerlou et al. 2018)
2	Anti-inflammatory effects	The anti-inflammatory and Gram-positive Streptococcus strain *P. cinnabarinus* has been demonstrated to examine the involvement of different strains of the Streptococcus strain.	Shittu et al. (2005)
3	Antitumor effects	Polysaccharides produced from *P. cinnabarinus* mycelial culture and directed intraperitoneally	(Hayashi et al. 2001)
4	Antimicrobial effects	Polyphenols, alkaloids, glucan compounds, flavonoids, phenols (p-phenylacetic acid, ferulic acid, p-benzoic acid, o-coumaric acid and chrysin), cinnabarinic acid, citric acid, pigments, and enzymes are all bioactive molecules with antimicrobial action found in these foods	(Rames and Pattar 2010, Adebayo et al. 2012)
5	Antiradical activity	*Pycnoporus cinnabarinus* plays a potential role in the DPPH receptor. Phenolic chemicals have a positive impact on life and help to prevent radical emptying (Razali et al. 2008). Phenolic synthesis inhibits the absorption of fats, oils, and fatty meals, lowering the risk of heart disease by reducing the oxidation of tiny lipoproteins.	(Sima-Obiang et al. 2017)

Antibacterial Effects

The fruit bodies of *P. cinnabarinus* have been tested and found to have antibacterial properties (Fajana et al. 1999). Shittu et al. (2005) investigated mycelial proliferation and synthesis of antibacterial metabolites. After four days of growth, antibacterial activity against *B. subtilis* (as determined by the agar cup distribution system) is at its peak. The anti-inflammatory and Gram-positive Streptococcus strain *P. cinnabarinus* has been demonstrated to examine the involvement of different strains of the Streptococcus strain. The fungus chooses a bag that transforms 3-hydroxyanthranilic acid to cinnabaric acid, which is essential for the formation of a bacterial complex. The research work of *P. cinnabarinus* focuses on the in vitro production of practically all cinnabaric acids as well as pure lacquer (Eggert 1997). A 20-day water culture of *P. cinnabarinus* demonstrated good antibiotic efficacy against Gram-positive *S.aureus* as well as Gram-negative *E. coli* and *Pseudomonas aeruginosa* bacteria, in another investigation. The weight of three plant pathogenic plants, Colletitrichum gloeosporioides (Glomerella cingulate) Botrytis cinerea, and Colletotrichum miyabeanus, has a beneficial anti-inflammatory impact when culture filtrate is utilized against mycelial growth (Tassoo ana Imtiaj 2007).

Pycnoporus cinnabarinus was found to have antioxidant and antibacterial properties by Adebayo et al. (2012). (JF736658). Free radical removal was successful with the fungal specimens discussed. The resistance area against *Staphylococcus aureus* (30 mm) was the highest, whereas the resistance area against *Escherichia coli* was the lowest (7 mm). *Pycnoporus cinnabarinus* was shown to have more potent inflammatory action than *Pleurotus pulmonarius* in this investigation. Cinnabarine and cinnabaric acid have antibacterial properties, but they are also responsible for the growth of these fungi. According to Eggert (1997), *P. cinnabarinus* selects a bag that converts 3-hydroxyanthranilic acid to cinnabaric acid, the latter having antibacterial effect, primarily in contrast to Gram-positive bacteria of the Streptococcus species.

Antitumor Effects

The disease is a skin condition that develops in an unexpected way. Tumors can be life-threatening (unhealthy) or non-fatal (healthy) (unhealthy). Boils are most commonly caused by the body's cells dividing and growing too quickly. Cell development and division are regulated by the body in general. To replace old or new cells, new ones are generated. Cells that have been damaged or are no longer needed are healed to allow for medical replacement. Tumors can form if the cell balance is disrupted by death.

Polysaccharides produced from *P. cinnabarinus* mycelial culture and directed intraperitoneally to mice at a rate of 250 mg/kg suppressed the growth of 180 severe sarcomas in Ehrlich by 90% (Ohtsuka et al. 1973). Cinnabaric acid has also been demonstrated to have a powerful influence on apoptosis, resulting in increased oxygen production and the loss of mitochondrial membrane potential, as well as a reduction in caspase development. It is a modulator or immunomodulator of the immune response to autoimmune diseases, but its active activity is unknown (Hayashi et al. 2001).

Antiradical Activity

The initial introduction looks into the factors that cause radical damage. All of the papers feature the work of a well-known artist. *Pycnoporus cinnabarinus* extract has excellent antioxidant activity (IAA = 3.13), as well as efficient (IAA = 0.90) and inefficient (IAA = 0.22) ethanol extraction. Borderes et al. (2011) discovered that *P. cinnabarinus* plays a potential role in the DPPH receptor. Phenolic chemicals have a positive impact on life and help to prevent radical emptying (Razali et al. 2008). Phenolic synthesis inhibits the absorption of fats, oils, and fatty meals, lowering the risk of heart disease by reducing the oxidation of tiny lipoproteins. Properties that are antiviral, antiangiogenic, and carcinogenic (Sima-Obiang et al. 2017).

Insecticidal Activity

Cane extract has been used to study the activities of EC insects generated from *P. cinnabarinus* and *Pleurotus pulmonarius* in Diatraea magnifactella larvae. The ruling party's assets had no bearing. After injection, 24 hours, 48 hours, 72 hours, and 96 hours *Pleurotus pulmonarius* EC demonstrates no pneumonia or adverse effects. Insect activity against Macrosiphum rosae (rose aphids) was studied using an ethyl acetate extract of fruit-bearing fruit, mycelium, and filtrate from Pleurotus ostreatus, according to Noshad et al. (2015). Diluted filtrate has an LC50 of 25.03 g/ml. EC from *P. cinnabarinus* killed larvae after 72 hours of culture containing 120, 144, 192, and 240 h, and other EC killed larvae after 72 hours of culture containing 120, 144, 192, and 240 h, and other EC killed larvae after 72 hours of culture containing 120, 144, 192, and 240 h, and other.

TOXICOLOGICAL BEHAVIOR OF *PYCNOPORUS CINNABARINUS*

Basidiomycetes are considered one of the sweetest fungus because of their capacity to multiply in harsh environments while continuously destroying natural lignocellulose (Lee et al. 2014). They can also be classified as white-on-white fungus or white-on-white fungi, depending on how much the plant has been injured. Plant polysaccharides can be broken down by the brown food, but the white fungus can break down anything in the plant, including lignin. Due to the presence of an enzyme supplement containing peroxidase manganese (MnP, EC 1.11.1.13), lignin peroxidase (LiP, EC 1.11.1.14) and blocking, pure corrosion fungi are known for actively destroying lignin (Lac, EC1.10.3 is for. 2). These enzymes are involved in the oxidation of lignin in plants as well as xenobiotic substances such as synthetic and polycyclic aromatic hydrocarbons (PAHs). Due to the depletion of soil, water, and other natural resources, pollution with polycyclic aromatic hydrocarbons (PAHs) is becoming a serious rising problem for humans and the environment.

Polycyclic Aromatic Hydrocarbons (PAHs) are a plant chemical with two or more benzene rings that have been found as long-term air pollutants that cause significant stress due to their poisonous, mutagenic, and carcinogenic properties. The most harmful PAHs, according to the US Environmental Protection Agency (US EPA), are 16 of them (Zhang et al. 2015). During this time, a number of techniques for overcoming these issues were developed. Air pollution's microbiological mechanism is a viable way for solving environmental issues. The use of microbial agents to eliminate PAH is a valuable "green solution." (Gogoi et al. 2003). The process through which microbes (bacteria, fungus, and algae) degrade or modify and break down contaminants such as carbon dioxide and water is known as bioremediation. When opposed to chemical systems that produce harmful compounds and have limited bioavailability for pollutants, bioremediation is a simple, cost-effective, ever-changing, environmentally friendly, and efficient method. White fungi's extracellular ligninolytic enzymes destroy PAHs by reactive reactions involving suitable quinone radicals (Bamforth and Singleton 2005). Laccase's catalase reaction is based on monoelectronic oxidation, which uses a mediator to transform the grain into active material (LI XZ et al. 2010). PAH depletion has been observed to improve when different mediators (ABTS, HBT, and phenol) are used (Covino et al. 2010). Surfactants can boost the bioavailability of PAH by increasing its surface air and increasing its toxicity by cutting micelles and solutions. In the critical micellar condition, this pattern is visible (CMC) (Wilson and Jones 1993).

EXAMPLES OF ENVIRONMENTAL POLLUTANTS DEGRADED BY *PYCNOPORUS CINNABARINUS*

Chlorinated Aromatic Compounds

- Pentachlorophenol
- 4-chloroaniline
- 2,4,5-trichlorophenoxyacetic acid
- Polychlorinated biphenyls
- Dioxin

Polycyclic Aromatic Compounds

- Benzo[a]pyren
- Pyrene
- Anthracene
- Chrysene

Dyes

- Crystal violet
- Azure blue

Pesticides

- DDT [1,1, l-trichloro- 2,2-bis(4-chlorophenyl)ethane]
- Lindane
- Chlordane
- Toxaphene

Munitions

- TNT (2,4,6-trinitrotoluene)
- RDX (cyclotrimethylenetrinitroamine)
- HMX (cyclotetramethylenetetranitramine

Others

- Cyanides
- Azide
- Aminotriazole
- Carbon tetrachloride

ECONOMIC IMPORTANCE OF *PYCNOPORUS CINNABARINUS*

Fungi are home to hundreds of animal species that are vital to human survival. Of course, fungi play an important role in our lives. There is never a period when these live things do not immediately benefit or influence us. Food is extremely important in the formation of the environment. Fungi play a vital function as spoilers and distributors in these habitats, supplying other kingdom members food and living food. Hundreds of animal species are found in fungi, all of which are necessary for human survival. The voice does, in fact, play a vital role in our lives. These living things will never directly benefit or affect humans. Food has a significant impact on the environment. Food serves as plunderers and distributors in these locations, as well as providing food and living food to the other kingdoms.

PYCNOPORUS AS A GENETIC COMPANY FOR THE PRODUCTION OF MANY USEFUL ENZYMES

Hydrolases

Pycnoporus, as well as Basidiomycete fungi, have strong glycosyl hydrolytic potency. (Esposito et al. 1993; Gomez-Alarcon et al. 1989;). Extracellular fluid from *P. cinnabarinus* produced chitinase and N-acetylglucosaminidase, which are essential for full chitin hydrolysis (Ohtakara 1988). *Pycnoporus cinnabarinus* hydrolyzed chitinase chitin, which acts on chitooligosaccharides, is endo-type, mainly hydrolyzing the second β-Nacetylglucosamine amide bond at its non-reducing end (Ohtakara 1988). β.-N-acetylhexosaminidase degraded chitoligosaccharides from *P. cinnabarinus* to an undifferentiated terminus. Exo and N-acetylglucosamine.

Laccase

Laccases (p-diphenol/oxygen-oxygen reductases, EC 1.10.3.2) are type 1 copper enzymes that use redox energy to oxidise substrates and three other copper enzymes that carry electrons in O2 and reduce their water content. Lacases are extracellular monomeric glycoproteins produced in marine culture (Lomascolo et al. 2003; Eggert et al. 1996) as well as solid agrotoxins in Pycnoporus genes (Vikineswary et al. 2006; Meza et al. 2006). Pycnoporus species from numerous geographical zones, primarily tropical areas, have been observed to yield up to 18,000 U/l salmon, or 75 mg/l (Uzan et al. 2010; Lomascolo et al. 2002). In the presence of ethanol as a catalyst, the monocaryotic form, *P. cinnabarinus* CIRM-BRFM 137, has been discovered as a strong inhibitor of lacase (266,600 U / l, i.e., 1 g / l) (Lomascolo et al. 2003). The fruit was 2,2'-azinobis acid (3-ethylbenzthiazoline-6-sulfonic acid (ABTS)) and the responses were monitored at pH 4. Pycnoporus varnishes offer biological features that make them excellent for biotechnological applications, such as strong heat performance (50–65 ° C), high stability, and salt and solvent resistance.

Cellobiose Dehydrogenase

Cellobiose dehydrogenase (CDH) is an extracellular oxidoreductase that also contains heme and a prosthetic flavin based on protoporphyrin-IX. It has a cellulose-binding domain and is a bifunctional enzyme. CDH may break down cellulose, hemicellulose, and lignin, as well as engage in plant destruction by a red fungus that produces hydroxyl radicals, which are important in the demethylation of lignin. *P. cinnabarinus* cells were used to make CDH cells (Moukha et al. 1999). It has been demonstrated that mRNA binds to *P. cinnabarinus* coding. CDR is cellulose, and it is inhibited in culture by cellobiose or glucose. Purification and identification of the matching protein (92 kDa) (Sigoillot et al. 2002).

Tyrosinase

Tyrosinases are three types of copper proteins involved in the early stages of melanin synthesis (o-diphenol, monophenol: oxygen oxidoreductase, EC 1.14.18.1). (Halaouli et al. 2005) were the first to show that Pycnoporus species can produce tyrosinase. Tyrosinase produces 45.4 protein at 153.6 U/g per day in *P. cinnabarinus* strain CBS 614.73, the best tyrosinase inhibitor for diphenols and monophenolase. This tyrosinase has been proven to be useful in the production of natural anti-oxidants in the protein synthesis process (Halaouli et al. 2005).

BIOTECHNOLOGICAL APPLICATIONS OF *PYCNOPORUS CINNABARINUS*

The usage and conversion of aromatic compounds in cell walls as a beneficial product is the nature of Pycnoporus as a functional tool of scent. Natural crops used to come from the earth. The active compounds, on the other hand, are frequently minuscule and only found in the wood, making it impossible to distinguish between pricey products. When grown on a normal medium or in front of antecedents, pure basidiomycetes, primarily Pycnoporus species, are biotechnological agents required for the production of tastes from scratch or bioconversion for industry (Lomascolo et al. 2002; Asther et al. 1998). Because of the increased demand for natural sweeteners, the biotransformation of vanillin could be a source of chemicals. Ferulic acid (4-hydroxy-3-methoxycinnamic acid), which is found in plant cell walls and has a chemical structure similar to vanillin, functions as a catalyst in the formation of vanillin. The dicharyotic animal feed *P. cinnabarinus*, type I-937, was chosen among the 300 basidiomycete diets because it could extract 64 mg/l vanillin from 300 mg/l ferulic acids after six days. as well as fruit seeds 27.5 percentage points.

In *P. cinnabarinus* I-937, the metabolism of ferulic acid in vanillin was described. Vanilla acid, a destructive product of ferulic acid, is created by oxidative decarboxylation or by reducing vanillin and alcohol and is then added to methoxyhydroquinone by oxidative decarboxylation or by reducing vanillin and alcohol (Falconnier et al. 1994). The risk of three negatively regulated pathways was reduced by reducing the biotransformation of vanillin from ferulic acid: (1) by selecting the p-type. cellobiose and fungal culture, (2) by directing the watering of vanillic acid by the reduction method (blockage 2) (Lesage-Meessen et al. 1997), and (3) by adding a selective agent (hydrophobic crosslinked polystyrene copolymer resin XAD-2) (Stentelaire et al. 1998).

PRODUCTION OF PURE VANILLIN

Two new ways for producing pure vanillin in large amounts from hazardous corn related to *Aspergillus niger* and *Pseudomonas cinnabarinus* have been devised. Hand-made corn has been described as having two strategies from the standpoint of the economic system. The ability of *A. niger* cultivated in beet pulp to produce high-quality polysaccharide enzymes, including feruloyl esterases, and the conversion of ferulic acid to vanillic acid were first well coupled in the autoclave to yield free ferulic acid (Bonnin et al. 2002). After 10 days of conversion, vanillic acid is extracted and converted to vanillin by a monocaryotic form of *P. cinnabarinus* MUCL 38534, yielding 767 mg/l biotechnological vanillin in the presence of cellobiose and XAD-2 resin. four, with a 71 percent increase after that (Figure 12.2).

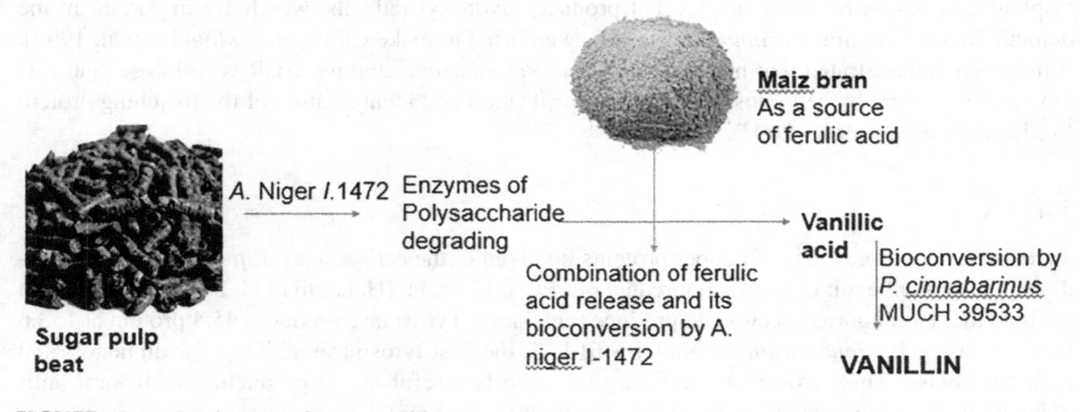

FIGURE 12.2 Production of pure vanillin.

BIOPOLYMER SYNTHESIS

Pycnoporus oxidative enzymes (*P. cinnabarinus lacase* and *P. sanguineus tyrosinase*) have also been used to successfully produce biopolymers suited for the food sector, such as agro-residues like beet pulp or maize sugar. Ferulic acid ester, like the rest of arabinofuranose and corn grains, binds to polysaccharides in sugar beet or corn pectins (Thibault and Saulnier 1999). As a result, these polysaccharide complexes can gel when ferulic acid and oxidizing systems like laccases are combined. *Pycnoporus cinnabarinus laccase* has been used for sugar beet pectins (Micard and Thibault 1999), maize bran arabinoxylans (de Wilde et al. 2008) and corn arabinoxylans soluble (Figueroa-Espinoza and Rouau 1998).

CONCLUSION

Pycnoporus cinnabarinus also possesses a wide range of enzymatic capabilities that can harm lignin. Antioxidant, anticancer, antiviral, antihypercholesterolemic, antihyperlipidemic, anti-hyperglycemic, antiallergic, and antifungal benefits have been discovered, as well as antiallergic and fungal effects. A 20-day water culture of *P. cinnabarinus* demonstrated good antibiotic efficacy against Gram-positive *S.aureus* as well as Gram-negative *E. coli* and *P.aeruginosa* bacteria. *Pycnoporus cinnabarinus*, a white basidiomycete, was recently discovered to conceal new clusters. Phenol oxidases are frequently formed in situations that promote ligninolysis. *Pycnoporus cinnabarinus* genomic structure revealed a genetic complex found on fragile basidiomycetes. Thus, *P. cinnabarinus* belongs to an old family of enzymes that break down cellulose and hemicellulose, but its pectinolytic abilities appear to be limited.

REFERENCES

E.A. Adebayo, J.K. Oloke, A.A. Ayandele, C.O. Adegunlola, 2012. Phytochemical, antioxidant and antimicrobial assay of mushroom metabolite from *Pleurotus pulmonarius*-LAU 09 (JF736658). *J. Microbio. Biotech. Res. 2*(2): 366–374.

P. Angelini, R. Venanzoni, G. Angeles Flores, B. Tirillini, G. Orlando, et al., 2020. Evaluation of antioxidant, antimicrobial and tyrosinase inhibitory activities of extracts from *Tricholosporum goniospermum*, an edible wild mushroom. *Antibiotics 9*(8): 513.

M. Asther, A. Lomascolo, M. Asther, S. Moukha, L. Lesage-Meessen, 1998. Metabolic pathways of biotransformation and biosynthesis of aromatic compounds for the flavour industry by the basidiomycete *Pycnoporus cinnabarinus*. *Micologia Neotropical Aplicada 11*: 69–76.

S.M. Bamforth, I. Singleton, 2005. Bioremediation of polycyclic aromatic hydrocarbons: current knowledge and future directions. *J. Chem. Technol. Biotech. 80*(7): 723–736.

E. Bonnin, L. Saulnier, M. Brunel, C. Marot, L. Lesage-Meessen, et al., 2002. Release of ferulic acid from agroindustrial by-products by the cell wall-degrading enzymes produced by *Aspergillus niger* I-1472. *Enzyme Microb. Technol. 31*(7): 1000–1005.

J. Borderes, A. Costa, A. Guedes, L.B.B. Tavares, 2011. Antioxidant activity of the extracts from *Pycnoporus sanguineus* mycelium. *Braz. Arc. Bio. Tech. 54*(6): 1167–1174.

E. Correa, D. Cardona, W. Quiñones, F. Torres, A.E. Franco, et al., 2006. Leishmanicidal activity of *Pycnoporus sanguineus*. *Phytother. Res., 20*(6): 497–499.

S. Covino, K. Svobodová, Z. Křesinová, M. Petruccioli, F. Federici, et al., 2010. In vivo and in vitro polycyclic aromatic hydrocarbons degradation by *Lentinus (Panus) tigrinus* CBS 577.79. *Bioresour. Technol. 101*(9): 3004–3012.

C. De Wilde, E. Uzan, Z. Zhou, K, Kruus, M. Andberg, et al., 2008. Transgenic rice as a novel production system for*Melanocarpus* and *Pycnoporus laccases*. *Transgenic Res. 17*(4): 515–527.

C. Eggert, U. Temp, J.F.D. Dean, K.E.L. Eriksson, 1995. Laccase-mediated formation of the phenoxazinone derivative, cinnabarinic acid. *FEBS Lett., 376*: 202–206.

C. Eggert, U. Temp, K.E.L. Eriksson, 1996. The lignolytic system of the white-rot fungus *Pycnoporus cinnabarinus*: purification and characterization of the laccase. *Appl. Environ. Microbiol. 62*:1151–1158.

C. Eggert, 1997. Laccase-catalyzed formation of cinnabarinic acid is responsible for antibacterial activity of *Pycnoporus cinnabarinus*. *Microbiol Res. 152*(3): 315–318. DOI: 10.1016/S0944-5013(97)80046-8

E. Esposito, L.H. Innocentini-Mei, A. Ferraz, V.P. Canhos, N. Duran, 1993. Phenoloxidases and hydrolases from *Pycnoporus sanguineus* (UEC-2050 strain): applications. *J Biotechnol. 29*:219–228.

O.B. Fajana, F.V. Alofe, G.O. Onawunmi, A.O. Ogundaini, A. O., T.A. Olugbade, 1999. Antimicrobial studies on Nigerian higher fungi. *Niger. J. Nat. Prod. Med. 3*: 64–65.

B. Falconnier, C. Lapierre, L. Lesage-Meessen, G. Yonnet, P. Brunerie, B. Colonna Ceccaldi, G. Corrieu, M. Asther, 1994. Vanillin as a product of ferulic acid biotransformation by the white-rot fungus *Pycnoporus cinnabarinus* I-937: identification of metabolic pathways. *J. Biotechn. 37*:123–132.

M.C. Figueroa-Espinoza, X. Rouau, 1998. Oxidative cross-linking of pentosans by a fungal laccase and horseradish peroxidase: mechanism of linkage between feruloylated arabinoxylans. *Cereal Chem. 75*:259–265.

B.K. Gogoi, N.N. Dutta, P. Goswami, T.K. Mohan, 2003. A case study of bioremediation of petroleum-hydrocarbon contaminated soil at a crude oil spill site. *Adv. Environ. Res. 7*(4): 767–782.

G. Gomez-Alarcon, C. Saiz-Jimenez, R. Lahoz, 1989. Influence of tween 80 on the secretion of some enzymes in stationary cultures of the white-rot fungus *Pycnoporus cinnabarinus. Microbios 60*:183–192.

P. Gong, S. Wang, M. Liu, F. Chen, W. Yang, et al., 2020. Extraction methods, chemical characterizations and biological activities of mushroom polysaccharides: a mini-review. *Carbohyd. Res. 494*: 108037.

G. Guzmán, 2004. Los hongos de El Edén Quintana Roo: introducción a la micobiota tropical de México. *Revista do Instituto de Medicina Tropical de São Paulo, 46*(5): 282- 282.

S. Halaouli, M. Asther, K. Kruus, L. Guo, M. Hamdi, et al., 2005. Characterization of a new tyrosinase from *Pycnoporus* species with high potential for food technological applications. *J. Appl. Microbiol., 98*(2): 332–343.

D.S. Hibbett, M. Binder, J.F. Bischoff, M. Blackwell, P.F. Cannon, et al., 2007. A higher-level phylogenetic classification of the Fungi. *Mycolog. Res. 111*(5): 509–547.

M.S. Hyder, S.D. Dutta, 2021. Mushroom-derived polysaccharides as antitumor and anticancer agent: a concise review. *Biocatal. Agric. Biotechnol. 35*: 102085.

A. Khezerlou, M. Alizadeh-Sani, M. Azizi-Lalabadi, A. Ehsani 2018. Nanoparticles and their antimicrobial properties against pathogens including bacteria, fungi, parasites and viruses. Microb. Pathog. *123*: 505–526.

T. Krupodorova, M. Sevindik, 2020. Antioxidant potential and some mineral contents of wild edible mushroom *Ramaria stricta. AgroLife Sci. J. 9*(1): 186–191.

Y. Kuang, B. Li, Z. Wang, X. Qiao, M. Ye. 2021. Terpenoids from the medicinal mushroom *Antrodia camphorata*: chemistry and medicinal potential. *Natural Product Reports, 38*(1): 83–102.

H. Lee, Y. Jang, Y.S. Choi, M.J. Kim, J. Lee, et al., 2014. Biotechnological procedures to select white rot fungi for the degradation of PAHs. *J. Microb. Method. 97*: 56–62.

R. Lemberg, 1952. Nitrogenous pigments from the fungus *Coriolus sanguineus* (*Polystictus cinnabarinus*). *Aust. J. Exper. Bio. Med. Sci. 30*: 271–278.

L. Lesage-Meessen, M. Haon, M. Delattre, J.F. Thibault, B. ColonnaCeccaldi, M. Asther, 1997. An attempt to channel the transformation of vanillic acid into vanillin by controlling methoxyhydroquinone formation in *Pycnoporus cinnabarinus. Appl. Microbiol. Biotechnol. 47*:393–397.

A. Levasseur, A. Lomascolo, O. Chabrol, F.J. Ruiz-Dueñas, E. Boukhris-Uzan, et al., 2014. The genome of the white-rot fungus *Pycnoporus cinnabarinus*: a basidiomycete model with a versatile arsenal for lignocellulosic biomass breakdown. *BMC Genomics, 15*(1): 1–24. 10.1186/1471-2164-15-486

H. Li, Y. Tian, N. Menolli Jr, L. Ye, S.C. Karunarathna, et al., 2021. Reviewing the world's edible mushroom species: a new evidence-based classification system. *Compr. Rev. Food Sci. Food Saf. 20*(2): 1982–2014.

X. Li, X. Lin, R. Yin, Y. Wu, H. Chu, et al., 2010. Optimization of laccase-mediated benzo [a] pyrene oxidation and the bioremedial application in aged polycyclic aromatic hydrocarbons-contaminated soil. *J. Health Sci. 56*(5): 534–540.

A. Lomascolo, J.L. Cayol, M. Roche, G.U.O. Lin, J.L. Robert, E. Record, et al., 2002. Molecular clustering of Pycnoporus strains from various geographic origins and isolation of monokaryotic strains for laccase hyperproduction. *Mycological Res. 106*(10): 1193–1203.

A. Lomascolo, E. Record, I. Herpoël-Gimbert, M. Delattre, J.L. Robert, et al., 2003. Overproduction of laccase by a monokaryotic strain of *Pycnoporus cinnabarinus* using ethanol as inducer. *J. App. Microbiol. 94*(4): 618–624.

A. Lomascolo, E. Uzan-Boukhris, I. Herpoël-Gimbert, J.C. Sigoillot, L. Lesage-Meessen, 2011. Peculiarities of Pycnoporus species for applications in biotechnology. *Appl. Microbio. Biotech. 92*(6): 1129–1149. 10.1007/s00253-011-3596-5.

J.C. Meza, J.C. Sigoillot, A. Lomascolo, D. Navarro, R. Auria, 2006. New process for fungal delignification of sugar-cane bagasse and simultaneous production of laccase in a vapor phase bioreactor. *J. Agri. Food Chem. 54*(11): 3852–3858.

V. Micard, J.F. Thibault, 1999. Oxidative gelation of sugar-beet pectins: use of laccases and hydration properties of the crosslinked pectins. *Carbohydr. Polym. 39*:265–273.

S.M. Moukha, T.J. Dumonceaux, E. Record, F.S. Archibald, 1999. Cloning and analysis of *Pycnoporus cinnabarinus* cellobiose dehydrogenase. *Gene 234*:23–33.

A. Noshad, M. Iqbal, Z. Iqbal, H. Bibi, S. Bibi, H.U. Shah 2015. Aphidicidal potential of ethyl acetate extract from Pleurotus ostreatus. *Sarhad Journal of Agriculture*, *31*(2): 101–105.

A. Ohtakara, 1988. Chitinase and β-*N*-acetylhexosaminidase from *Pycnoporus cinnabarinus*. *Methods Enzymol. 161*:462–470

S. Ohtsuka, S. Ueno, C. Yoshikumi, F. Hirose, Y. Ohmura, et al., 1973. Polysaccharides having an anti-carcinogenic effect and a method of producing them from species of Basidiomycetes. *UK Patent*, *1331513*(26): 02.

N. Razali, R. Razab, S.M. Junit, A.A. Aziz, 2008. Radical scavenging and reducing properties of extracts of cashew shoots (*Anacardium occidentale*). *Food Chem. 111*(1): 38–44.

L. Saulnier, J.F. Thibault, 1999. Ferulic acid and diferulic acids as components of sugar-beet pectins and maize bran heteroxylans. *J. Sci. Food Agric. 79*: 396–402

M. Sevindik, H. Akgul, Z. Selamoglu, N. Braidy, 2020. Antioxidant and antigenotoxic potential of infundibulicybe geotropa mushroom collected from Northwestern Turkey. *Oxid. Med. Cell. Longev. 2020.*

O.B. Shittu, F.V. Alofe, G.O. Onawunmi, A.O. Ogundaini, T.A. Tiwalade, 2005. Mycelial growth and antibacterial metabolite production by wild mushrooms. *African J. Biomed. Res. 8*(3): 157–162.

C. Sigoillot, A. Lomascolo, E. Record, J.L. Robert, M. Asther, J.C. Sigoillot, 2002. Lignocellulolytic and hemicellulolytic system of*Pycnoporus cinnabarinus*: isolation and characterization of a cellobiose dehydrogenase and a new xylanase. *Enzyme Microbial Technol. 31*(6), 876–883.

C. Sima-Obiang, R.L. Ngoua-Meye-Misso, G.R. Ndong-Atome, J.P. Ondo, L.C. Obame-Engonga, E. Nsi-Emvo, 2017. Chemical composition, antioxidant and antimicrobial activities of stem barks of *Englerina gabonensis* Engler and *Sterculia tragacantha* Lindl from Gabon. *Int. J. Phytomed. 9*(3): 501–510.

E.A. Smânia, A. Smânia Junior, C. Loguercio-Leite, M.L. Gil, 1997. Optimal parameters for cinnabarin synthesis by *Pycnoporus sanguineus*. *J. Chem. Tech. Biotech. 70*: 57–59

E.F.A. Smânia, A. Smânia Junior, C. Loguerico-Leite, 1998. Cinnabarin synthesis by *Pycnoporus sanguineus* strains and antimicrobial activity against bacteria from food products. *Revista de Microbiologia. 29*: 317–320

A. Smânia, F. Delle Monache, E.F.A. Smânia, M.L. Gil, L.C. Benchetrit, F.S. Cruz, 1995. Antibacterial activity of a substance produced by the fungus *Pycnoporus sanguineus* (Fr.) Murr. *J. Ethnopharm. 45*(3): 177–181.

A. Smânia, C.J.S. Marques, E.F.A. Smânia, C.R. Zanetti, S.G. Carobrez, et al., 2003. Toxicity and antiviral activity of cinnabarin obtained from *Pycnoporus sanguineus* (Fr.) Murr. *Phytother. Res. 17*(9): 1069–1072.

C. Stentelaire, L. Lesage-Meessen, M. Delattre, M. Haon, J.C. Sigoillot, et al., 1998. By-passing of unwanted vanillyl alcohol formation using selective adsorbents to improve vanillin production with *Phanerochaete chrysosporium. World J. Microbiol. Biotechnol. 14*:285–287.

E. Uzan, P. Nousiainen, V. Balland, J. Sipila, F. Piumi, et al., 2010. High redox potential laccases from the ligninolytic fungi *Pycnoporus coccineus* and *Pycnoporus sanguineus* suitable for white biotechnology: from gene cloning to enzyme characterization and applications. *J. App. Microbiol. 108*(6): 2199–2213.

S. Vikineswary, N. Abdullah, M. Renuvathani, M. Sekaran, A. Pandey, et al., 2006. Productivity of laccase in solid substrate fermentation of selected agro-residues by *Pycnoporus sanguineus*. *Bioresour. Technol. 97*(1): 171–177.

S.C. Wilson, K.C. Jones, 1993. Bioremediation of soil contaminated with polynuclear aromatic hydrocarbons (PAHs): a review. *Environ. poll. 81*(3): 229–249.

S. Zhang, Y. Ning, X. Zhang, Y. Zhao, X. Yang, et al., 2015. Contrasting characteristics of anthracene and pyrene degradation by wood rot fungus *Pycnoporus sanguineus* H1. *Inter. Biodeter. Biodegrad. 105*: 228–232.

J. Zhou, M. Chen, S. Wu, X. Liao, J. Wang, et al., 2020. A review on mushroom-derived bioactive peptides: preparation and biological activities. *Food Res. Int.*, *134*: 109230.

13 The Genus *Ramaria* (Basidiomycota, Agaricales) *Diversity, Edibility, and Bioactivity*

Kamal Ch. Semwal
Department of Biology, College of Sciences, Eritrea Institute of Technology, Mai Nafhi, Asmara, Eritrea (East Africa)

Vinod K Bhatt
Ingenious Research and Development Foundation, VIP Enclave, Chandrabani, Dehradun, Uttarakhand, India

Ajmal Hussen
Wolaita Sodo University, Wolaita, Ethiopia (East Africa)

Avnish Chauhan
Dept. of Environmental Science, Graphic Era Hill University, Dehradun, Uttarakhand, India

CONTENTS

DOI: 10.1201/9781003259763-13

INTRODUCTION

The term "Mushroom" is generally used to describe those fungi which form visible fruitbodies usually above ground (Chang and Miles, 2004; Kalac, 2009). These fruiting bodies have various shapes and colours and easily attract the attention of visitors to forests, meadows, and gardens. They have an umbrella-like cap at the top, which is knitted either with the plate-like structures (known as gills) or pores (the reproductive part of the mushroom), a stem to which the cap is attached, and sometimes a sac-like structure at the base of the stem. Some mushrooms form coral, jelly, ball, inverted bell, and star-like structures and are categorised in different taxonomic ranks based on these readily apparent morphological characteristic features. At present about 1,500,000 to 6,000,000 species of fungi species are estimated to exist by different researchers, and around 144,000 of these have been described and named so far at the rate of addition per year of almost 2,000 species (Kirk et al., 2008; Hawksworth, 2012; Blackwell, 2011; Taylor et al., 2014; Zied and Pardo–Gimenez, 2017; Hawksworth and Lücking, 2017, Niskanen et al., 2018). This number of species included here involves all forms of fungi (i.e., unicellular, multicellular, and mushrooms). Throughout the world, more than 3000 species are assumed as edible (Chang and Miles, 2004).

The major constituents of the mushroom fruitbody include ash, carbohydrates, chitin, fat, fibre, iron, minerals, oligosaccharide polysaccharides, proteins, selenium, sodium, vitamins, and zinc (Wong and Cheung, 2001; Dundar et al., 2008). The low-fat value and the low proportion of polyunsaturated fatty acids attribute to the maximum benefits of mushrooms (Malik et al., 2017). Moreover, the lack of starch in the fruit body of mushrooms makes it a good option for the diet of diabetic people (Olawale et al., 2013).

The use of wild fungi as food and medicine is valued all over the world due to these organisms having uncountable health benefits. The latter include preventing the formation of cancer cells from mutations in their DNA (Miles and Chang, 2004; O'Neil et al., 2013), making a person feel satiated, thus lowering their overall calorie intake, helping to lower blood pressure, and decreasing the potential risk of high blood pressure and cardiovascular diseases (Diekemann et al., 2005), modulating the effect on immune system function (Weigand–Heller et al., 2012), and promoting T-cell functions and production of cytokine (Kino et al., 1989; Tanaka et al., 1989). The bioactive metabolites present in different species of the two major divisions of fungi (Basidiomycota and Ascomycota) possess the properties of anti-aging (Wasser, 2005; Jing et al., 2018), anti-allergic (Bahl, 1983; Hetland et al., 2020), anti-asthma (Mizuno, 1995; Davis et al., 2020), anti-bacterial (Wang et al., 2012; Joshi et al., 2019), anti-cancer (Anusiya et al., 2021; Patel and Goyal, 2012), anti-dermatophytic (Ogidi and Oyetayo, 2016), anti-diabetic (Chaiyasut and Sivamaruthi, 2017; Nowacka–Jechalke et al., 2018; Yu et al., 1993), antifungal (Reis et al., 2011; Wang and Ng, 2006), anti-hypercholesterolemic (Wasser, 2005; Zhao et al., 2015) anti-hyperglycemic (Patel et al., 2012; Xu et al., 2019), anti-hypertensive (Hadi and Bremner, 2001; Carrasco–González et al., 2017), anti-inflammatory (Chen et al., 2018), anti-microbial (Dugler, 2004; Alves et al., 2012), anti-obesity (Lin et al., 2009; Gasecka et al., 2018), anti-oxidant (Barros et al., 2009; Kim and Lee, 2003; Shaffique et al., 2021), anti-parasitic (Wasser, 2002; Onyango et al., 2011), anti-proliferation (Xu et al., 2011; Deo et al., 2019), anti-radiation (Evgeny et al., 2005), anti-thrombotic (Rahman and Choudhury, 2012; Choi et al., 2020), anti-tumor (Moore et al., 1985; Meng et al., 2016), anti-viral (El–Mekkawy et al., 1998; Seo and Choi, 2021), cardiovascular protection (Gao et al., 2004; Rahman et al., 2016), hepatoprotective (Chen et al., 2018; Al–Dbass et al., 2012), hypolipidemic (Yeh et al., 2014; Mizutani et al., 2010) and immunostimulatory (Niu et al., 2009; Zhang et al., 2019). They have shown the ability to restore neurotransmitters associated with memory and thrombosis inhibition (Yang et al., 2015), used to produce soothing sleep, enhance memory, and improve sexual rejuvenators (Sharma, 2008). Even recent studies have shown the potential of several different bioactive compounds of mushrooms are proved effective against the SARS-CoV-2 which causes the current pandemic Coronavirus disease-19 (Rangsinth et al., 2021). According to Tsai–Teng et al., (2016), the lion's mane mushroom (*Hericium erinaceus*) is used for the treatment of Alzheimer's disease and as a possible treatment of Parkinson's disease (Kuo et al., 2016). A couple of species of mushrooms have been investigated for anti-HIV activities, such as the Shiitake mushroom (*Lentinus edodes*) (Tochikura et al., 1988), the jelly ear mushroom (*Auricularia polytricha*) (Sillapachaiyaporn et al., 2019), the Button mushroom *Agaricus blazei*, Chaga mushroom *Inonotus obliquus* and Willow bracket mushroom *Phellinus igniarius* which has shown anti-HIV activities (Choengpanya et al., 2021).

Many ethnic groups of different countries in the world regard mushrooms as a nutritional food source and medicine, and the species of *Ramaria* are among these (Hrudayanath and Singdevsachan, 2014; Semwal et al., 2014; Rammeloo and Walleyn, 1993; Kaul and Kachroo, 1974; Sarkar et al., 1988; Rai et al., 1993; Boa, 2004; Boruah and Singh, 2001; Pradhan et al., 2010).

TAXONOMY OF *RAMARIA*

Ramaria Fr. ex Bonordm, is one of the 35 genera of clavarioid fungi (Verma and Pandro, 2018) belonging to the *Phylum* Basidiomycota, *Class* Agaricomycetes, *Order* Gomphales, and *Family* Gomphaceae, commonly known as "Coral fungi" and members of the genus were originally assigned to the genus *Clavaria*. In general terminology, these *Coral* fungi are recognized as "clavarioid fungi" and comprise nine families (Clavariadelphaceae, Gomphaceae, Clavariaceae, Clavulinaceae, Aphelariaceae, Lentariaceae, Lachnocladiaceae, Typhulaceae, and Pterulaceae). In these Clavaroid fungi, a total of 16 genera are reported from India. These are, *Artomyces, Aphelaria, Clavariadelphus, Clavaria, Clavulinopsis, Clavulina, Gloeocantharellus, Deflexula, Lentaria, Multiclavula, Lachnocladium, Phaeoclavulina, Ramariopsis, Ramaria, Typhula* and *Scytinopogon* (Verma and Pandro, 2018). Among these most of the species of *Ramaria* have been reported from mid-altitudinal Himalayan ranges of Uttarakhand, West Bengal, and Himachal Pradesh states (Verma and Pandro, 2018).

The members of the order Gomphales consist of toothed, gilled, resupinate (crust-like) fungi, and the genus *Ramaria* shares the spore appearances of other members of this order (Villegas et al., 2005; Hosaka et al., 2006). The name *Ramaria* was proposed in 1790 by Holmskjöld and later raised to the genus level in 1933 (Humpert et al., 2001); it is considered to be one of the main genera of the clavaroid fungi. The genus contains 336 species worldwide and 39 species from India (Verma and Pandro, 2018; Ghosh et al., 2021; Dattaraj et al., 2020) (Table 13.1).

MORPHOLOGICAL AND MICROSCOPICAL CHARACTERISTICS

The species of the genus *Ramaria* are characterized by erect, repeatedly branched, coral-like fruit-bodies, with dichotomously branched or many branches or short stems (Figure 13.1). Branching begins right from the base of the stipe in a dichotomous or polychotomous manner, giving rise to multiple rows of branches. The fungi can be easily observed and identified in the forest based on the finger-like projections, coral-like structures, and variously coloured fruitbodies. Their colour ranges from yellow, creamish, pinkish, purplish, blue, brownish, greyish, orange, and reddish-brown. However, they can be confused with other coral fungi which have morphologically similar fruit-bodies. Members of the genus *Ramaria* can be distinguished from look-alike species of *Clavaria* and *Clavulina* (Agaricales, Clavariaceae) based on their repeatedly branched coral-like structure, yellow-brown to rusty-brown spore prints, the absence of cystidia (sterile cells situated between the basidia in the hymenium) and a spore wall that reacts with cotton blue to give a purplish-blue, known as cyanophilous reaction (Corner, 1950; Knudson, 2012; Kotlaba and Pouzar, 1964). The spores in *Ramaria* are generally ornamented and differently termed as *Echinulate* (spike-like or prickles-like small projections). Another type of spore ornamentation in which the spores are rounded and pronounced warts is known as tuberculate-nodulose. The spores have warts or lobes, which are rounded, and regularly linked to form ridges or waves known as *verrucose*. In *striate* spores, the ornamentation consists of shallow linear ridges over the surface of the spore (Petersen, 1975, 1981).

TABLE 13.1
Species of *Ramaria* Documented from India

S.No.	Species name	References
1.	*R. apiculata* (Fr.) Donk	Thite et al., 1976; Patil and Thite, 1977; Mohanan, 2011
2.	*R. aurea* (Schaeff.) Quél.	Sharma and Jandaik, 1978; Das, 2009
3.	*R. brevispora* Corner, K.S. Thind & Dev	Corner et al., 1958; Das, 2009
4.	*R. camelicolor* Corner, K.S. Thind & Anand	Corner et al., 1956; Sharda and Thind, 1986
5.	*R. clarobrunnea* Corner, K.S. Thind & Anand	Corner et al., 1956
6.	*R. concolor* (Corner) R.H. Petersen =*R. stricta* (Pers.) Quél. var. *concolor* Corner	Thind and Sukh Dev, 1957a, b; Sharma and Jandaik, 1978
7.	*R. echinovirens* Corner, K.S. Thind & Dev	Corner et al., 1957
8.	*R. eumorpha* (P. Karst.) Corner ≡*Clavariella eumorpha* P. Karst.	Mohanan, 2011; Thind et al., 1983
9.	*R. flava* (Schaeff.) Quél. =*R. flava var. sanguine*	Corner, 1956; Thind and Sukh Dev, 1957a; Mohanan, 2011
10.	*R. flaviceps* Corner, K.S. Thind & Anand	Corner et al., 1956
11.	*R. flavoalba* Corner	Sharma et al., 1977
12.	*R. flavobrunnescens* (Coker) Corner	Thind and Sukh Dev, 1957a; Sharda and Thind, 1986
13.	*R. flavoviridis* Corner & K.S. Thind	Corner and Thind, 1961
14.	*R. formosa* (Pers.) Quél. =*Clavaria formosa* Pers.	Berkeley, 1856; Mohanan, 2011; Thind and Anand, 1956b
15.	*R. gracilis* (Pers.) Quél. ≡ *Clavaria gracilis* Pers.	Sharma and Jandaik, 1978; Mohanan, 2011
16.	*R. holorubella* (G.F. Atk.) Corner ≡*Clavaria holorubella* G.F. Atk.	Thind and Rattan, 1967
17.	*R. kisantuensis* (Sacc.) Corner =*R. kisantuensis* var. *indica* Khurana & K.S. Thind	Khurana and Thind, 1979
18.	*R. laevispora* Corner & K.S. Thind	Corner, 1966
19.	*R. moelleriana* (Bres. & Roum.) Corner	Thind and Sukh Dev, 1957b
20.	*R. obtusissima* (Peck) Corner	Thind and Sukh Dev, 1957a, b; Sharda and Thind, 1986
21.	*R. ochrochlora* Furrer–Ziogas & Schild	Thind et al., 1983
22.	*R. pallida* (Schaeff.) Ricken ≡*Clavaria pallida* Schaeff.	Mohanan, 2011
23.	*R. perbrunnea* Corner & K.S. Thind	Corner, 1966
24.	*R. petersenii* K.S. Thind & Sharda	Thind and Sharda, 1984
25.	*R. pura* Corner & K.S. Thind	Corner, 1966
26.	*R. purpurissima* R.H. Petersen & Scates ≡*R. fumigata var. gigantea* K.S. Thind & Anand	Thind and Anand, 1956a
27.	*R. pusilla* Corner	Thind and Sukh Dev, 1957b
28.	*R. rasilispora* Marr & D.E. Stuntz	Sharda and Thind, 1986
29.	*R. rubrogelatinosa* Corner & K.S. Thind	Corner, 1966
30.	*R. sandaracina* Marr & D.E. Stuntz	Sharda and Thind, 1986
31.	*R. sanguinea* (Coker) Corner	Thind and Sukh Dev, 1957b
32.	*R. sikkimia* S.S. Rattan & Khurana	Rattan and Khurana, 1978
33.	*R. stricta* (Pers.) Quél. =*Clavaria stricta* Pers.	Berkeley, 1856
34.	*R. subalpina* K. Das & K. Acharya	Das et al., 2016
35.	*R. subaurantiaca* Corner	Thind and Sukh Dev, 1957b
36.	*R. subbotrytis* (Coker) Corner	Thind and Anand, 1956a

TABLE 13.1 *(Continued)*
Species of *Ramaria* Documented from India

S.No.	Species name	References
37.	*R. subgelatinosa* Corner	Thind and Sukh Dev, 1957b
38.	*R. suecica* (Fr.) Donk	Sharda and Thind, 1986
39.	*R. synaptopoda* Marr & D.E. Stuntz	Sharda and Thind, 1986

FIGURE 13.1 (A) *Ramaria stricta*; (B) *R. botrytis*; (C) *R. formosa*; (D) *R. aurea*; (E) Collection of different fungal species along with *Ramaria* species from the high altitude forest in Garhwal Himalaya (Uttarakhand, India). Photos by Kamal Ch. Semwal.

HOW TO DISTINGUISH DIFFERENT SPECIES AMONG GENUS *RAMARIA*?

Morphologically the apparent characteristic features are looking similar in species of *Ramaria* so the researchers use a couple of key micromorphological characteristics to differentiate those species. The type of hyphal system, clamp connection in such hyphae, and spore shape and size. The hyphal system in *Ramaria* is of two kinds—generative and skeletal hyphae. In generative hyphae, the hyphal wall can be thin or thick-walled, which may or may not have clamp connections and septa. While skeletal hyphae are generally profound thick-walled, lacking clamp connections and septa. Usually, the thickness in skeletal hyphae is much more in contrast to the thick-walled generative hyphae (Corner, 1950). The most readily apparent macroscopic feature is the colour of the fruitbody.

HABITAT AND DISTRIBUTION

Ramaria Fr. ex Bonordm fruits during the rainy seasons all over the world in coniferous and broadlleaf forests and are commonly distributed in Europe, America, and Asia over different elevational ranges (Semwal and Bhatt, 2019; Semwal et al., 2014; Lee, 1988; Park and Lee, 1997). A few species of *Ramaria* have been collected from the Garhwal Himalaya forests (Uttarakhand, India) during the fungal forays in the monsoon season of the year 2021 and earlier by two of the co-authors (Figure 13.1E - see middle; Semwal et al., 2014; Semwal and Bhatt, 2019). Ecologically, species of *Ramaria* are beneficial for the microclimate of the forest as is the case for other wild fungi. The genus *Ramaria* has been reported from different habitats and substrates all over the world. These fungi are commonly found as saprobes in humicolous soil, upon rotting logs, and linked with various kinds of tree roots, including the conifers *Pinus roxburghii* Sarg., *Cedrus deodara* (Roxb.) Don., and *Cryptomeria japonica* (Thunb. ex L.f.) Don, and broadleaf *Quercus leucotrichophora* A. Camus, *Myrica esculenta*, and *Rhododendron arboreum* Smith, where they form mycorrhizal associations (Thind and Sharda, 1985; Bilgrami et al., 1991; Semwal et al., 2014). *R. fumigata* was found ectomycorrhizically associated with the lower elevation trees *Eucalyptus globulus* Labill., and *Acacia auriculiformis* A. Cunn. ex Benth. (Pradhan et al., 2013). *R. formosa*, *R. flava*, *R. aurea*, *R. ochraceovirens*, *R. largentii*, *R. fumigata*, and *R. palmata* are other species reported as ectomycorrhizal (Baier et al., 2006; Trappe, 1962; Scattolin et al., 2008). This group of species is valued as important decomposers and plays a vital role in forest ecology. Many species of *Ramaria* form an ectomycorrhizal association with tree roots. About 46 species of *Ramaria* have been reported as forming mycorrhizal associations with broadleaf and coniferous trees in the Fennoscandian countries, which include Norway and Finland (Bendiksen et al., 2015).

MYCORRHIZAL ASSOCIATIONS

Many species of *Ramaria* are thought to be mycorrhizal and form ectomycorrhizal associations in which the fungal hyphae form the sheath around the roots of trees (Humpert et al., 2001). These mycorrhizal associations are generally formed with broadleaf and coniferous trees, mainly in temperate to subtemperate climatic regions (Agerer et al., 2012). At present, about 970 edible mycorrhizal fungi (EMF) have been reported thus far from all over the world (Pérez–Moreno et al., 2021) including the species of *Ramaria*. Ectomycorrhizal fungi are valued in the forest due to playing a significant role in the proper working of the forest microclimate since such mutual connections help in nutrient cycling, in which fungi receive carbohydrates and growth factors from the plant, and the plant is provided with an extended absorptive area since the hyphae of the fungus are distributed throughout a large area in the soil; the nutrients available there are thus able to reach the plant. Due to the extended absorptive area, the roots of the plant have an increased uptake of minerals or nutrients so that plant can be protected and grow well. Moreover, in such symbiotic association, fungi may not degrade wood.

ETHNOMYCOLOGY

Different species of *Ramaria* are among the most trusted wild species eaten by many ethnic groups in the world (Boa, 2004; Garibay–Orijel et al., 2006; Smith and Bonito, 2012; Ko et al., 2013; Liu et al., 2016; Rana, 2016; Kalita et al., 2016; Semwal et al., 2014; Semwal and Bhatt, 2019; Khadka et al., 2020). Examples include *R. flaccida, R. botrytis, R. subalpina, R. versatilis, R. stricta,* and *R. aurea*. In the Uttarakhand Himalaya region of India and Japan, this mushroom is locally known as

TABLE 13.2
Common or Traditional Names and Uses of Species of *Ramaria* in Different Regions of the World

Country	Species	Traditional name	Uses	References
India	*Ramaria* spp.	*Ungli- cheun* (Garhwali language – Uttarakhand)	Edible	Semwal et al., 2014
		Siun; *Shuntu, Chinmuh* and *Kyalmangmuh* (Pahadi language – Himachal Pradesh)	Edible	Semwal et al., 2014
		Shairee (Kashmiri language)	Edible	Semwal et al., 2014
		Rai Saad (Kashmiri language)	Generally used as blood purifier and enhances skin colour; to cure asthma, other respiratory problems, and eye diseases by locals.	Malik et al., 2017
		Tit lbonghati (Meghalaya)	Edible	Kabita et al., 2014
	R. formosa	Panz anguj or Hapat paanji	Local tribes used it to cure cardiac and diabetic ailments; For body pain among Gujjar and Bakerwalas ethnic group	Pala et al., 2013; Shah et al., 2015
Japan	*R. botrytis*	Houkitake	Edible	Yaoita et al., 2007
Guatemala	*R. araiospora*	Cabeza de gallo (Spanish) Rujolon äk, deer horns (local Kaqchikel dialect)	Edible	Mérida et al., 2019
Nepal	*R. aurea*	Thakre chyau (Nepali)	Edible	Khadka et al., 2020
Korea	*R. botrytis*	Ssaribeoseot (Korean Language)	Edible	Kim and Song, 2014
Mexico	*Ramaria* spp.	Xelhuasnandcatl (brush or broom, mushroom broom)	Edible	Montoya et al., 2003

"ungli-cheun" and *"Houkitake,"* respectively, as its fruitbody looks like the fingers of human and branches (*ungli* = finger and *cheun* = mushroom (Semwal et al., 2014; Yaoita et al., 2007) (Table 13.2). In Himachal Pradesh, it is known as "Siun," "*Chinmuh*" and "*Kyalmangmuh*" in the local languages. In Jammu & Kashmir, northern states of India, locally it is known as "*Shairee.*" Most of the members of this genus are edible and consumed all over the world. (Atri et al., 2019). *R. apiculata R. aurea*, *R. botrytoides*, *R. formosa*, *R. flavobrunnescens* var. *aurea, R. stricta, R. subalpina*, *R. flavo-brunnescens* var. *longisperma,* and *R. zippelii* are commonly consumed in India (Atri et al., 2019; Das et al., 2016; Rai et al., 2013). Traditionally, *R. madagascariensis* is consumed by local people in China (Dong–Ze Liu et al., 2015). *R. formosa* has been valued for the body pain in the Indian tribal groups Gujjar and Bakerwalas (Shah et al., 2015). The Bakerwalas are the herders involved in guiding livestock herds in grassland or forests.

PHYTOCHEMISTRY, BIOACTIVE COMPOUNDS AND THEIR HEALING EFFECTS

More than 100 medicinal functions have been reported by numerous mushrooms (Valverde et al., 2015), and the genus *Ramaria* is one such example. However, studies of the healing or therapeutical properties of *Ramaria* species are sparse or focused only on a few taxa that include *R. aurea, R. largentii, R. formosa, R. botrytis, R. flava, R. versatilis*, and *R. cystidiophora.* However, recently some other species of *Ramaria* have been investigated, including *R. versatilis* (Dattaraj et al., 2020) and a recently reported new species (*R. subalpina*) from India (Acharya et al., 2017).

Several therapeutical applications have been reported from the different kinds of extracts of species of *Ramaria.* It includes anti-cancer, anti-viral, anti-oxidant, anti-microbial, anti-inflammatory, anti-parasitic, immune system booster, and anti-hyperlipidemic (Wasser, 2017).

MAJOR BIOACTIVE COMPOUNDS OF SPECIES OF *RAMARIA*

Different components (e.g., polysaccharides, alkaloids, triterpenoids, sterols, proteins, and others) from species of *Ramaria* have been found to account for a wide array of bioactive properties.

Polysaccharides

Polysaccharides derived from different mushrooms have proven to be successful at shielding the human body from the damage to protein, DNA, and the cell itself during its growth and development, which consequently leads to aging (Xu et al., 2009, Chen et al., 2014). These polysaccharides also possess antitumor and immunomodulating properties (Patel and Goyal, 2012). These bioactive molecules can be separated from fruitbodies, mycelia, and fermented liquid broth of species of *Ramaria* or other medicinal mushrooms which play a vital role in cell metabolisms (Yang et al., 2013). Bhanja et al. (2014), isolated two glucans namely PS-I and PS-II from the alkali extract of *Ramaria botrytis.* The property of those polysaccharides was water-insoluble. Glucans found inside the cell wall of the fungal fruitbodies are the most copious forms of polysaccharides (Chan et al., 2009). This glucan showed immunostimulating activity (Bhanja et al., 2014) in which the glucans stimulated the immune system upon exposure to either a pathogen or xenobiotics. According to Hou et al. (2016), *R. flaccida* is composed of different kinds of amino acids, carbohydrates, and trace elements that are valuable to the human body. Polysaccharides RBP was isolated from the species of *R. botrytis.* These polysaccharides show a promising ability of hydroxyl radical scavenging bioactivity in which it protects cells from damage and probably it could be a novel natural anti-oxidant in the drug remedies or food industry (Li, 2017). The polysaccharide RF-1 was isolated from *R. flaccida* (Dong et al., 2020) and found to possess anti-tumour bioactivity *in vivo* mice models (Ghosh et al., 2021).

Terpenoids

Various kinds of terpenoids have been segregated from fungi and analysed for their anti-cancer, anti-tumour, and anti-microbial activities. These include 44 diterpenes, 70 sesquiterpenes, 5 monoterpenes, and 166 triterpenes (Duru and Cayan, 2015). Recently, sesquiterpene Ramarin A and Ramarin B were isolated from the fruitbodies of *R. formosa.* (Kim et al., 2015). Sesquiterpene is a class of terpenes composed of three isoprenes. Labdane diterpenes, a kind of terpene has been isolated from *R. formosa* and found to demonstrate an inhibiting activity against the human neutrophil elastase enzyme (Lee et al., 2015). HNE is a protease enzyme that when expressed abnormally can cause emphysema. This includes a breakdown of lung structure and increased airspaces which causes shortness of breath. Generally, diterpenes are known to be anti-microbial and anti-inflammatory (Breitmaier, 2006).

Sterol

Sterols are the major components of fungal cell walls instead of the cholesterol found in human or other mammalian cells. They are by-products of triterpenoids. From the fruitbody of *R. flava* six sterols comprising ergosterol peroxide were isolated. According to Liu et al. (2012), these sterols of this mushroom species can be considered for use as an anti-tumour agent since they have been proved as promising growth inhibitors of the human breast cancer cell line (MDA–MB–231).

Protein

Protein is always an important macromolecule of the fungal cell wall which serves many tasks including ion exchange, shape, cell rigidity, metabolism, interaction with the host cell, and many more. Zhou et al. (2017) and their team isolated "*Ramaria botrytis* ubiquitin-protein (RBUP)" from *R. botrytis* species and examined for the anti-tumour, hemagglutination, and DNase activities. They observed significant inhibition of the growth and tempted apoptosis in A549 cells. Apoptosis is a kind of cell death in the body used to get rid of abnormal cells, and A549 is one of the most commonly used cell lines used in research applications and is present in lung carcinoma epithelial cells. Hemagglutination could be understood as a reaction that causes clumping of red blood cells while DNase (deoxyribonuclease) is an enzyme that catalyses to cleave phosphodiester bonds in the DNA backbone, thus degrading DNA. In addition to this, a novel ribonuclease has been isolated from *R. formosa* and further investigated for its anti-viral activity against HIV-1 reverse transcriptase enzyme and found to have a 93% inhibition capacity, which suggests the potential use of this ribonuclease which could play an important role in preventing the vigorous spread of HIV disease (Zhang et al., 2015; Elkhateeb et al., 2021). *R. botrytis* was also examined for protein content by Lee and Han (2001).

Other Compounds

Total phenolic compounds, β-glucans, and phenolic acids such as cinnamic, gallic, and chlorogenic were isolated from *R. flava*, in Mexico (López et al., 2018). A new ceramide was isolated from the fruitbodies of *R. botrytis* (Yaoita et al., 2007). The ceramide is a kind of waxy lipid molecule that can influence the lateral organization in biological membranes (Nyholm et al., 2010). Different kinds of chemical compounds have been isolated from species of *Ramaria* including hemiacetals (Dong–Ze Liu et al., 2015) and two unknown alkaloids from *R. madagascariensis* (Dong–Ze Liu et al., 2014). Hemiacetal is a by-product of the acetal production process. In addition to this ascorbic acid, ß-carotene, flavonoids, phenolic compound, and lycopene are found in certain species of *Ramaria,* species namely *R. botrytis* and *R. subalpina* (Barros et al., 2008, Acharya et al., 2017, Gursoy et al., 2010). Though, the quantity of those bioactive molecules is very in different species.

KEY BIOACTIVITIES OF THESE COMPOUNDS

Anti-microbial

Various species of *Ramaria* exhibited anti-microbial activities in response to different extracts, including acetone, ethyl acetate, ethanol, methanol, and water extract. *R. aurea* (Rai et al., 2013), *R. botrytis* (Alves et al., 2012, Giri et al., 2012), *R. flava* (Gezer et al., 2006; Liu et al., 2013), *R. formosa* (Ramesh and Pattar, 2010; Pala et al., 2019), *R. zippelli* (Bala et al., 2011), and *R. madagascariensis* (Dong–Ze Liu et al., 2014) are known for their antimicrobial activities.

Some other edible species (e.g., *R. cystidiophora, R. rubripermanens, R. stricta*, and *R. flavescens*) are also regarded as anti-microbial against various micro-pathogens i.e., mycobacteria, bacteria, and fungi (Centko et al., 2012; Barros et al., 2008; Sharma and Gautam, 2017). A butenolide namely Ramariolides A was isolated from methanol extracts of *R. cystidiophora* by Centko et al., (2012) with descent anti-microbial potential against *Mycobacterium tuberculosis* and *M. smegmatis*.

Anti-oxidants

Antioxidants are compounds that can inhibit the activity of other chemicals and prevent oxidation. These other chemicals include free radicles, which are the kinds of unstable molecules that are made during cell metabolism and they can damage other molecules, which ultimately may increase the risk of cancer. In a hydroxyl radical scavenging activity, an anti-oxidant molecule protects cells from being damaged due to free radical activity. *R. aurea* (Khatua et al., 2015; Sharma and Gautam, 2017; Rai and Acharya, 2012), *R. botrytis* (Li, 2017; Barros et al., 2008; Han et al., 2017; Sharma and Gautam, 2017; Kim and Lee, 2003; An et al., 2020), *R. flava* (Gursoy et al., 2010; Gezer et al., 2006; Liu et al., 2013; Sharma and Gautam, 2017), *R. formosa* (Ramesh and Pattar, 2010; Kim et al., 2016), *R. flavescens* (Sharma and Gautam, 2017), *R. patagonica* (Toledo et al., 2016), *R. rubripermanens* (Sharma and Gautam, 2017), *R. versatilis* (Dattaraj et al., 2020), *R. subalpina* (Acharya et al., 2017), *R. stricta* (Sharma and Gautam, 2017), and *R. largentii* (Aprotosoaie et al., 2017) all show anti-oxidant properties in different extraction methods. The anti-oxidant compounds Tocopherols, phenolic compounds, ascorbic acid, β-carotene were isolated from *R. botrytis* (Bhanja et al., 2020; Han et al., 2017; Li, 2017).

OTHER COMPOUNDS AND BIOACTIVITIES REPORTED IN SPECIES OF *RAMARIA*

Antibiotic and anti-cancer activity was reported in *R. flava* and *R. formosa* (Anusiya et al., 2021; Ghosh et al., 2021). The compounds that are effective in the treatment of cancerous cells and tumor cells are known as anti-cancer and anti-tumour properties of those compounds, respectively. *R. flava* has exhibited anti-tumour activity (Popovic et al., 2013) and anti-fungal activities against three pathogenic fungi, namely *Cercosporella albomaculans*, *Fusarium auenaceum,* and *F. graminearum* (Liu et al., 2013). Anti-proliferative activity has been reported in *R. cystidiophora* (Deo et al., 2019), *R. botrytis* (Chung, 1979), and *R. flava* (Sadi et al., 2016). Anti-proliferative activity can be understood as what happens when a biomolecule prevents or retards the spread of malignant cells.

R. botrytis was examined for hepatoprotective activity in mice against liver toxicity by Kim and Lee (2003). The results suggested that the methanolic extract of *R. botrytis* is promising in high enzyme activities. In hepatoprotective activity, the biomolecules prevent damage to the liver.

The anti-cancer activity was exhibited by *R. flava* ethanol extract when examined in a couple of human cancer cell lines (BGC–803, NCI–H520, and MDA–MB–231). The inhibition percentages were varied in accordance to aforesaid human cancer cells as 33.83%, 54.63%, and 71.66%, respectively, and the mushroom extract concentration of 200 μg/mL was more promising (Elkhateeb et al., 2021). Furthermore, activity against the HIV–1 reverse transcriptase enzyme was reported for *R. formosa* (Zhang et al., 2015; Elkhateeb et al., 2021). ACE inhibitory activities have been tested in *R. botrytoides* water extract. However, the bioactive compound oligopeptide shows only 37.8% activity (Geng et al., 2015). The angiotensin-I converting enzyme (ACE) plays a vital role in blood pressure regulation. ACE promotes the conversion of angiotensin-I into the strong vasoconstrictor angiotensin-II as well as inactivating the bradykinin, a vasodilator (Kim et al., 2010). Vasodilators are the agents which promote the widening of the blood vessels, leading to increase blood flow. The twin roles of ACE cause a surge in blood pressure and finally lead to the risk of hypertension (Murray et al., 2007).

NUTRITIONAL AND DIETARY COMPOSITION

The different kinds of macromolecules present in different quantities in mushroom cells determine the nutritional quality and contribute to the delicacy of the mushroom. The quantities of carbohydrate, protein, fat, dietary fibres and moisture content have been variously observed by different researchers in *R. botrytis* (Sharma and Gautam, 2017; Barros et al., 2008), *R. brevispora*

(Murugkar and Subbulakshmi, 2005), *R. largentii* (Ouzouni et al., 2009), *R. patagonica* (Toledo et al., 2016), *R. aurea, R. rubripermanens, R. flava, R. flavescens,* and *R. stricta* (Sharma and Gautam, 2017). The variation in nutritional macromolecules may be due to the substrate or habitat and ecological regions where the mushrooms grow. The amount of lipids/fats ranges from 0.14% to 5.67% on a dry weight basis in species of *Ramaria* (Ghosh et al., 2021).

TOXICOLOGY

Although, most of the species recorded in the genus *Ramaria* are good edibles and many ethnic groups consume them without fear, toxicity and poisonous cases also have been reported from around the world. These include *R. flavo-brunnescens* (Atk.) Corner (Fidalgo and Kauffman–Fidalgo, 1970; Kommers and Santos, 1995; Barros et al., 2006; Scheid et al., 2022). *R. spinulosa* (Pers.: Fr.) Quel, *R. mairei* Donk (Jesús and Ferrera-Cerrato, 1995), *R. pallida* (Łuszczynski, 2009; Nasim et al., 2008) and *R. formosa* (Pers.) Quél. (O'Reilly, 2011; Jesús and Ferrera-Cerrato, 1995). However, *R. formosa* is said to be edible in Nepal (Adhikari and Durrieu, 1996). This may be a misidentification of the species of *Ramaria* involved. Stomach pains and diarrhoea were reported in a case of *R. formosa* poisoning. In *R. spinulosa* gastrointestinal poisoning has been recorded in Mexico. The victims developed stomach cramps, vomiting, and shortly thereafter violent diarrhoea. They recovered after two days of ingestion when given medical treatment. *R. flavo-brunnescens* poisoning was reported in cattle. *R. pallida* is mildly poisonous and cause nausea, vomiting, and diarrhea. Recently, an organoarsenic compound called homoarsenocholine has been found in *R. subbotrytis, R. largentii,* and *R. pallida* fruitbodies collected from the Czech Republic, Austria, and Slovakia. The occurrence of the arsenic compound in terrestrial mushroom fruitbody is a good example of the biotransformation mechanisms of arsenic compounds and this finding will provide an opportunity to understand more completely the geo-biochemical pathways of arsenic compounds in the environment (Braeuer et al., 2018).

CULTIVATION

There is no tangible information relating to the domestication of species of *Ramaria* However, a patent had been filed under patent no. KR1020060113034 and with the title "Artificial cultivation method of *R. botrytis* by employing wood and sawdust of needle-leaf tree and broadleaf tree, cottonseed meal, cottonseed hulls, and sugar cane" (Chul, 2006). The characteristics of the mycelial growth of *R. botrytis*, studied by Lee and Han (2005), demonstrated that the optimal temperature and recommended pH are $24^{0}C$ and 5.0, respectively for the same species.

CONCLUSIONS AND FUTURE PERSPECTIVES

There is a desperate need to increase the opportunities to cultivate edible species of the genus *Ramaria* and expand the research on nutraceutical, nutritional and phytochemical compounds in *Ramaria* species worldwide. However, there are constraints to growing *Ramaria* artificially due to its specific growth habits, but if scientists have succeeded in growing morels (*Morchella esculanta*), which also are ectomycorrhizal fungi and the peculiar phenomenon of a lightning effect is adhered to with fruitbody development, then it should be possible to culture *Ramaria* successfully *in vitro* since some of the species are saprophytic in the genus *Ramaria*. Furthermore, species of *Ramaria* can be investigated for new drug discoveries since they have a pharmacological potential as we have noted in the present article and reported by several other researchers (Elkhateeb, et al., 2021; Ghosh, et al., 2021). In recent years, macrofungi have become an imperative natural resource for the development of pharmaceutical mediators because of their varied and potentially significant pharmacological properties, and hundreds of *Ramaria* species exist worldwide, while only a few have been researched in detail for pharmacological purposes. Therefore, there is an inordinate need to carry out research on specific pharmacologically significant compounds of species of *Ramaria*. Other authors also commented on the need for research opportunities for more nutraceutical and nutritional compounds for species of *Ramaria* (Elkhateeb et al., 2021).

REFERENCES

K. Acharya, K. Das, S. Paloi, A.K. Dutta, M.E. Hembrom, S. Khatua, A. Parihar. 2017. Exploring a novel edible mushroom *Ramaria subalpina*: Chemical characterization and antioxidant activity. *Pharmacognosy J* 9(1): 30–34. doi:10.5530/pj.2017.1.6

M.K. Adhikari, G. Durrieu. 1996. Ethnomycologie Nepalaise. *Bulletin Societé Mycologique de France* 112: 31–41.

R. Agerer, J.M.C. Christan, E. Hobbie. 2012. Isotopic signatures and trophic status of *Ramaria*. *Mycological Progress* 11: 47–59.

A.M. Al–Dbass, S.K. Al–Daihan, R.S. Bhat. 2012. *Agaricus blazei* Murill as an efficient hepatoprotective and antioxidant agent against CCl4–induced liver injury in rats. *Saudi J Biol Sci* 19(3): 303–309.

M.J. Alves, I.C. Ferreira, A. Martins, M. Pintado. 2012. Antimicrobial activity of wild mushroom extracts against clinical isolates resistant to different antibiotics. *J Appl Microbiol* 113(2):466–475.

M.J. Alves, I.C. Ferreira, J. Dias, V. Teixeira, A. Martins, M. Pintado. 2012. A review on antimicrobial activity of mushroom (Basidiomycetes) extracts and isolated compounds. *Planta Med* 78: 1707–1718. doi: 10.1055/s–0032–1315370.

G.–H. An, J.–G. Han, J.H. Cho. 2020. Comparison of the antioxidant activity and nutritional contents of ectomycorrhizal mushroom extracts in Korea. *J Mushrooms* 18: 164–173. 10.14480/JM.2020.18.2.164.

G. Anusiya, P.U. Gowthama, N.V. Yamini, N. Sivarajasekar, K. Rambabu, G. Bharath, F. Banat. 2021. A review of the therapeutic and biological effects of edible and wild mushrooms. *Bioengineered*, doi: 10.1080/21655979.2021.2001183

A.C. Aprotosoaie, D.A. Zavastin, C.T. Mihai, G. Voichita, D. Gherghel, M. Silion, A. Trifan, A. Miron. 2017. Antioxidant and antigenotoxic potential of *Ramaria largentii* Marr and D.E. Stuntz, a wild edible mushroom collected from Northeast Romania. *Food Chem Toxicol* 108, 429–437, 10.1016/j.fct.2017.02.006

N.S. Atri, Y.P. Sharma, Kumar, Mridu. 2019. Wild edible mushrooms of north west Himalaya: Their nutritional, nutraceutical, and sociobiological aspects. T. Satyanarayana et al., (eds.), Microbial Diversity in Ecosystem Sustainability and Biotechnological Applications, 10.1007/978–981–13–8487–5_20

N. Bahl. 1983. Medicinal value of edible fungi. In: Proceeding of the International Conference on Science and Cultivation Technology of Edible Fungi. *Ind Mush Sci* II, pp. 203–209.

R. Baier, J. Ingenhaag, H. Blaschke, A. Göttlein, R. Agerer. 2006. Vertical distribution of an ectomycorrhizal community in uppersoil horizons of a young Norway spruce (*Picea abies* [L.] Karst.) stand of the Bavarian Limestone Alps. *Mycorrhiza* 16: 197–206.

N. Bala, A.B. Elizabeth Aitken, N. Fechner, A. Cusack, K.J. Steadman. 2011. Evaluation of antibacterial activity of Australian basidiomycetous macrofungi using a high–throughput 96–well plate assay. *Pharmaceut Biol* 11: 1–9. 10.3109/13880209.2010.526616

L. Barros, M. Duenas, I.C. Ferreira, P. Baptista, C. Santos–Buelga. 2009. Phenolic acids determination by HPLC–DAD–ESI/MS in sixteen different Portuguese wild mushrooms species. *Food Chem Toxicol* 47(6):1076–1079.

L. Barros, B.A. Venturini, P. Baptista, L.M. Estevinho, I.C. Ferreira. 2008. Chemical composition and biological properties of Portuguese wild mushrooms: A comprehensive study. *J Agricul Food Chem* 56: 3856–3862. 10.1021/jf8003114.

R.R. Barros, L.F. Irigoyen, G.D. Kommers, R.R. Rech, R.A. Fighera, C.S.L. Barros. 2006. Poisoning by *Ramaria flavo–brunnescens* (Clavariaceae) in cattle. *Pesquisa Veterinária Brasileira* 26(2):87–96.

K. Bendiksen, I. Kytövuori, M. Toivonen, E. Bendiksen, T.E. Brandrud. 2015. Ectomycorrhizal *Ramaria* species in nutrient poor Fennoscandian conifer forests including a note on the *Ramaria botrytis* complex. *Agarica* 36: 89–108.

M.J. Berkeley. 1856. Decades of fungi. Decades I–LXII. Indian fungi. Hooker's London. *J Bot Kew Garden Miscellany* 8: 174–280.

S.K. Bhanja, D. Rout, P. Patra, I.K. Sen, C.K. Nandanet, S.I. Islam. 2014. Water insoluble glucans from the edible fungus *Ramaria botrytis*. *Bioact Carbohyd Diet Fibre* 3: 52–58.

S.K. Bhanja, S.K. Samanta, B. Mondal, S. Jana, J. Ray, A. Pandey, T. Tripathy. 2020. Green synthesis of Ag–Au bimetallic composite nanoparticles using a polysaccharide extracted from *Ramaria botrytis* mushroom and performance in catalytic reduction of 4–nitrophenol and antioxidant, antibacterial activity. *Envi Nanot Monit Manag* 14, 100341.

K.S. Bilgrami, S. Jamaluddin, M.A. Rizwi. 1991. Fungi of India. Today and Tomorrow's Printers and Publ, New Delhi, India, pp. 433–434.

M. Blackwell. 2011. The fungi: 1, 2, 3 … 5.1 million species? *Am J Bot* 98: 426–438.

E. Boa. 2004. Wild edible fungi: a global overview of their use and importance to people. Non–wood Forest Products, No. 17, FAO, Forestry Department, Rome, Italy.

P. Boruah, R.S. Singh. 2001. Edible fungi of medicinal value from the eastern Himalaya region. *Int J Med Mushrooms* 3: 124.

E. Breitmaier. 2006. Terpenes: Flavors, Fragrances, Pharmaca, Pheromones. pp. 52–81. doi:10.1002/978352 7609949.ch4

S. Braeuer, J. Borovička, T. Glasnov, de la Cruz Guedes, K.B. Jensen, W. Goessler. 2018. *Talanta* 188(1): 107–110. doi:10.1016/j.talanta.2018.05.065.

J.A. Carrasco–González, S.O. Serna–Saldívar, J.A. Gutı errez–Uribe. 2017. Nutritional composition and nutraceutical properties of the *Pleurotus* fruiting bodies: potential use as food ingredient. *J Food Compost Anal* 58: 69–81. 10.1016/j.jfca.2017.01.016

R.M. Centko, S. Ramon–Garcia, T. Taylor, B.O. Patrick, C.J. Thompson, V.P. Miao, R.J. Andersen. 2012. Ramariolides A–D, Antimycobacterial butenolides isolated from the mushroom *Ramaria cystidiophora. J Nat Prod* 75: 2178–2182.

C. Chaiyasut, B.S. Sivamaruthi. 2017. Anti–hyperglycemic property of *Hericium erinaceus* – A mini review. Asian Pac. *J Tropical Biomed* 7 (11): 1036–1040.

G.C.F. Chan, W.K. Chan, D.M.Y. Sze. 2009. The effects of β–glucan on human immune and cancer cells. *J Hemat Oncol* 2: 25 10.1186/1756–8722–2–25.

S.T. Chang, P.G. Miles. 2004. Culture preservation. In: ST Chang, PG Miles (eds.), Mushrooms cultivation, nutritional value, medicinal effect and environmental impact. CRC Press, Boca Raton, pp 189–201.

H.P. Chen, Z.Z. Zhao, Z.H. Li, Y. Huang, S.B. Zhang, Y. Tang, J.N. Yao, L. Chen, M. Isaka, T. Feng, J.K. Liu. 2018. Anti–Proliferative and Anti–Inflammatory Lanostane Triterpenoids from the Polish Edible Mushroom *Macrolepiota procera. J Agric Food Chem* 66: 3146–3154. 10.1021/acs.jafc.8b00287.s002

T.Q. Chen, J.G. Wu, Y.J. Kan, C. Yang, Y.B. Wu, J.Z. Wu. 2018. Antioxidant and hepatoprotective activities of crude polysaccharide extracts from lingzhi or reishi medicinal mushroom, *Ganoderma lucidum* (Agaricomycetes), by ultrasonic–circulating extraction. *Int J Med Mush* 20(6) 581–593.

T.Q. Chen, Y.B. Wu, J.G. Wu, L. Ma, Z.H. Dong, J.Z. Wu. 2014. Efficient extraction technology of antioxidant crude polysaccharides from *Ganoderma lucidum* (Lingzhi), ultrasonic–circulating extraction integrating with superfne– pulverization. *J Taiwan Inst Chem* 45: 57–62

K. Choengpanya, S. Ratanabunyong, S. Seetaha, L. Tabtimmai, K. Choowongkomon. 2021. Anti–HIV–1 reverse transcriptase property of some edible mushrooms in Asia. *Saudi J Biol Sci* 28: 2807–2815. 10.1 016/j.sjbs.2021.02.012

E. Choi, J. Oh, G.–H. Sung. 2020. Antithrombotic and Antiplatelet Effects of *Cordyceps militaris. Mycobiology* 48: 228–232.

Y.S. Chul. 2006. Artificial cultivation method of *Ramaria botrytis* Ricken, by employing wood and sawdust of needle–leaf tree and broad leaf tree, cotton seed meal, cottonseed hulls and sugar cane (Korea, patent no. KR1020060113034), https://patentscope.wipo.int/search/en/detail.jsf;jsessionid=7CD159C40169DCC13E3 78FB9E56FC9E5.wapp1nB?docId=KR904580&_cid=P11–K5ZSBL–72162–2

K.S. Chung. 1979. The effects of mushroom components on the proliferation of HeLa cell line in Vitro. *Arch Pharmacal Res* 2: 25–33, 10.1007/BF02856430

E.J.H. Corner. 1950. A monograph of Clavaria and allied genera. *Ann Bot* 1: 1–740.

E.J.H. Corner. 1966. Species of Ramaria (clavariaceae) without clamps. *Trans Br Mycol Soc* 49(1): 101–113.

E.J.H. Corner, K.S. Thind. 1961. Dimitic species of *Ramaria* (Clavariaceae). *Trans Br Mycol Soc* 44: 233–238.

E.J.H. Corner, K.S. Thind, G.P.S. Anand. 1956. The Clavariaceae of the Mussoorie Hills (India) II. *Trans Br Mycol Soc* 39(4): 475–484.

E.J.H. Corner, K.S. Thind, S. Dev. 1957. The Clavariaceae of the Mussoorie Hills India–VII. *Trans Br Mycol Soc* 40: 472–476.

E.J.H. Corner, K.S. Thind, S. Dev. 1958. The Clavariaceae of the Mussoorie Hills India–IX. *Trans Br Mycol Soc* 41: 203–206.

K. Das. 2009. Mushroom of Sikkim I: Barsey *Rhododendraon* Sanctuary. Sikkim State Biodiversity Board, Gangtok and Botanical Survey of India, Kolkata, p. 160

K. Das, H.E. Manoj, A.K. Dutta, A. Parihar, S. Paloi, K. Acharya. 2016. *Ramaria subalpina* (Gomphaceae): a new edible fungus from India. *Phytotaxa* 246(2): 137–144 10.11646/phytotaxa.246.2.5

H.R. Dattaraj, K.R. Sridhar, B.R. Jagadish, M. Pavithra. 2020. Bioactive potential of the wild edible mushroom *Ramaria versatilis*. *Studies Fungi* 5: 73–83. 10.5943/sif/5/1/7.

R. Davis, A. Taylor, R. Nally, K.F. Benson, P. Stamets, G.S. Jensen. 2020. Differential Immune Activating,Anti–Inflammatory, and Regenerative Properties of the Aqueous, Ethanol, and Solid Fractions of a Medicinal Mushroom Blend. *J Inflamm Res* 13: 117–131.

G.S. Deo, J. Khatra, S. Buttar, W.M. Li, L.E. Tackaberry, H.B. Massicotte, K.N. Egger, K. Reimer, C.H. Lee. 2019. Antiproliferative, immuno–stimulatory, and anti–inflammatory activities of extracts derived from mushrooms collected in Haida Gwaii, British Columbia (Canada). *Int J Med Mush* 21: 629–643.

C.L. Diekemann, L.L. Bauer, E.A. Flickinger, G.C. Fahey. 2005. Effects of stage of maturity and cooking on the chemical composition of select mushroom varieties. *J Agric Food Chem* 53: 1130–1138.

L. Dong–Ze, J.G. Li, M.W. Zhang, G. Liu. 2015. New bicyclic hemiacetals from the edible mushroom *Ramaria madagascariensis*. *J Antibiotics* 68: 137–138. doi:10.1038/ja.2014.102

M. Dong, H. Yiling, D. Xiang. 2020. Structure identification, antitumor activity and mechanisms of a novel polysaccharide from *Ramaria flaccida (Fr.)* Quél. *Oncology Lett* 20: 2169–2182. doi: 10.3892/ol.2020.11761

L. Dong–Ze, J.G. Li, M.W. Zhang, L. Gang. 2014. Two new alkaloids from the edible macrofungus *Ramaria madagascariensis*. *J Basic Microbiol* 54: S70–S73. doi:10.1002/jobm.201301060

B. Dugler, A. Gonuz, F. Gucin. 2004. Antimicrobial activity of the macrofungus *Cantharellus cibarius*. *JBS* 7(9):1535–1539.

A. Dundar, A. Hilal, A. Yildiz. 2008. Yield performance and nutritional contents of three oyster mushroom species cultivated on wheat stalk. *Afr J Biotechnol* 7: 3497–3501.

E.M. Duru, T.G. Cayan. 2015. Biologically active terpenoids from mushroom origin: a review. *Rec Nat Prod* 9(4):456–483

W. Elkhateeb, M. Elnahas, L. Wenhua, M.C.A. Galappaththi, G.M. Daba. 2021. The coral mushrooms *Ramaria* and *Clavaria*. *Studies Fungi* 6(1): 495–506. doi:10.5943/sif/6/1/39

S. El–Mekkawy, M.R. Meselhy, N. Nakamura, Y. Tezuka, M. Hattori, N. Kakiuchi, K. Shimotohno, T. Kawahata, T. Otake. 1998. Anti–HIV–1 and Anti–HIV–1–Protease Substances from *Ganoderma lucidum*. *Phytochemistry* 49:1651–1657. doi:10.1016/S0031–9422(98)00254–4

V. Evgeny, O. Bent, J.Q.D. Petersen, W. Solomon. 2005. The Structure of the Polysaccharides Produced by Higher Basidiomyces *Tremella mesenterica* Ritz. Fr. and *Inonotus levis* P. Karst. *Int J Med Mush* 7(3):480–481.

O. Fidalgo, M.E.P. Kauffman–Fidalgo. 1970. A poisonous *Ramaria* from southern Brazil. *Rickia* 5: 71–91.

H. Gao, E. Chan, F. Zhou. 2004. Immunomodulating activities of *Ganoderma*, a mushroom with medicinal properties. *Food Rev Int* 20: 123–161.

R. Garibay–Orijel, J. Cifuentes, A. Estrada–Torres, J. Caballero. 2006. People using macro–fungal diversity in Oaxaca, Mexico. *Fung Divers* 21: 41–67.

M. Gasecka, Z. Magdziak, M. Siwulski, M. Mleczek. 2018. Profile of phenolic and organic acids, antioxidant properties and ergosterol content in cultivated and wild–growing species of *Agaricus*. *Eur Food Res Technol* 244: 259–268.

X. Geng, G. Tian, W. Zhang, Y. Zhao, L. Zhao, M. Ryu, H. Wang, T.B. Ng. 2015. Isolation of an Angiotensin I–converting enzyme inhibitory protein with antihypertensive effect in spontaneously hypertensive rats from the edible wild mushroom *Leucopaxillus tricolor*. *Molecules* 20: 10141–10153.

K. Gezer, M.E. Duru, I. Kivrak, A. Turkoglu, N. Mercan, H. Turkoglu, S. Gulcan. 2006. Free–radical scavenging capacity and antimicrobial activity of wild edible mushroom from Turkey. *Afr J Biotechnol* 5: 1924–1928.

S. Ghosh, C. Tribeni, A. Krishnendu. 2021. Revisiting *Ramaria* species: The Coral Fungi as Food and Pharmaceuticals. *Biointer Res Appl Chem* 11(3): 10790–10800.

S. Giri, G. Biswas, P. Pradhan, S.C. Mandal, K. Acharya. 2012. Antimicrobial activities of basidiocarps of wild edible mushrooms of West Bengal, India. *Int J Pharmtech Res* 4: 1554–1560.

N. Gursoy, C. Sarikurkeu, B. Tepe, M.H. Solak. 2010. Evaluation of antioxidant activities of 3 edible mushrooms: *Ramaria flava* (Schaef.: Fr.) Qu'el., *Rhizopogon roseolus* (Corda) T.M. Fries., and *Russula delica* Fr.. *Food Sci Biotechnol* 19(3):691–696. 10.1007/s10068–010–0097–8.

S. Hadi, J.B. Bremner. 2001. Initial studies on alkaloids from Lombok Medicinal Plants. *Molecules* 6: 117–129.

S.R. Han, K.H. Kim, H.J. Kim, S.H. Jeong, T.J. Oh. 2017. Comparison of biological activities using several solvent extracts from *Ramaria botrytis*. *Indian J Sci Technol* 9: 1–6. 10.17485/ijst/2016/v9i41/103921.

D.L. Hawksworth. 2012. Global species number of fungi: Are tropical studies and molecular approaches contributing to a more robust estimate? *Biodivers Conserv* 21: 2425–2433.

D.L. Hawksworth, R. Lücking. 2017. Fungal diversity revisited: 2.2 to 3.8 million species. *Microbiol Spectrum*. 10.1128/microbiolspec.FUNK–0052–2016

G. Hetland, J. Tangen, F. Mahmood, M.R. Mirlashari, L.S.H. Nissen–Meyer, I. Nentwich, S.P. Therkelsen, G.E. Tjønnfjord, E. Johnson. 2020. Antitumor, anti-inflammatory and antiallergic effects of *Agaricus blazei* mushroom extract and the related medicinal Basidiomycetes mushrooms, *Hericium erinaceus* and *Grifola frondosa*: A review of preclinical and clinical studies. *Nutrients* 12: 1339. 10.3390/nu12 051339.

K. Hosaka, S.T. Bates, R.E. Beever, M.A. Castellano, et al. 2006. Molecular phylogenetics of the gomphoidphalloid fungi with an establishment of the new subclass: Phallomycetidae and two new orders. *Mycologia* 98: 949–959.

Y.L. Hou, L. Liu, X. Ding, D.Q. Zhao, W.R. Hou. 2016. Structure elucidation, proliferation effect on macrophage and its mechanism of a new heteropolysaccharide from *Lactarius deliciosus* Gray. *Carbohydr Polym* 152: 648–657.

T. Hrudayanath, S.K. Singdevsachan. 2014. Diversity, nutritional composition and medicinal potential of Indian mushrooms: A review. *Afr J Biotech* 13(4): 523–545. 10.5897/AJB2013.13446

A.J. Humpert, E.L. Muench, A.J. Giachini, M.A. Castellano, J.W. Spatafora. 2001. Molecular phylogenetics of *Ramaria* and related genera: Evidence from nuclear large subunit and mitochondrial small subunit rDNA sequences. *Mycologia* 93: 465–477.

P.M. Jesús, R. Ferrera-Cerrato. 1995. A review of mushroom poisoning in Mexico. *Food Additives Contaminants* 12(3): 355–360. doi: 10.1080/02652039509374315

H. Jing, J. Li, J. Zhang, W. Wang, S. Li, Z. Ren, Z. Gao, X. Song, X. Wang, L. Jia. 2018. The antioxidative and anti–aging effects of acidic– and alkalic–extractable mycelium polysaccharides by *Agrocybe aegerita* (Brig.) Sing. *Int J Biol Macromol* 106: 1270–1278. 10.1016/j.ijbiomac.2017.08.138

M. Joshi, A. Sagar, S.S. Kanwar, S. Singh. 2019. Anticancer, antibacterial and antioxidant activities of *Cordyceps militaris*. *Indian Exp Biol* 57: 15–20.

D. Kabita, L. Albinus, P. Dibyendu, J. Lalit. 2014. Ethnomycological knowledge on wild edible mushroom of Khasi Tribes of Meghalaya, Northeastern India. *Eur Acad Res* 2(3): 3433–3442.

P. Kalac. 2009. Chemical composition and nutritional value of European species of wild growing mushrooms: a review. *Food Chem* 113: 9–16.

K. Kalita, R.N. Bezbaroa, R. Kumar, S. Pandey. 2016. Documentation of wild edible mushrooms from Meghalaya, Northeast India. *Curr Res Envi Appl Mycol* 6(3): 238–247. 10.5943/cream/6/4/1

T.N. Kaul, J.L. Kachroo. 1974. Common edible mushrooms of Jammu and Kashmir. *Ind Mush Sci* 71: 26–31.

B. Khadka, H.P. Aryal. 2020. Traditional knowledge and use of wild mushrooms in Simbhanjyang, Makwanpur district, Central Nepal. *Studies Fungi* 5(1): 406–419. 10.5943/sif/5/1/22

S. Khatua, P. Mitra, S. Chandra, K. Acharya. 2015. In vitro protective ability of *Ramaria aurea* against free radical and identification of main phenolic acids by HPLC. *J Herbs Spices Med Plants* 21: 380–391. 10.1080/10496475.2014.994085

I.P.S. Khurana, K.S. Thind. 1979. Species of ramaria aphyllophorales with dimitic fruitbody context from india and observations on their hyphal system. *Nippon Kingakukai Kaiho* 20(3): 279–298.

H. Kim, M.J. Song. 2014. Analysis of traditional knowledge for wild edible mushrooms consumed by residents living in Jirisan National Park (Korea). *J Ethnopharmacol* 153: 90–97. 10.1016/j.jep.2013 .12.041

H.J. Kim, K.R. Lee. 2003. Effect of *Ramaria botrytis* methanol extract on antioxidant enzyme activities in Benzo (α) Pyrene–treated mice. *Korean J Food Sci Technol* 35: 286–290. 10.4489/KJM.2003.31.1.034

K.C. Kim, Ik–Soo Lee, Ick–Dong Yoo, Byung–Jo Ha. 2015. Sesquiterpenes from the fruiting bodies of *Ramaria formosa* and their human neutrophil elastase inhibitory activity. *Chem Pharma Bull* 63(7): 554–557.

K.C. Kim, Y.B. Kwon, H.D. Jang, J.W. Kim, J.C. Jeong, I.S. Lee, B.J. Ha, I.D. Yoo. 2016. Study on the antioxidant and human neutrophil elastase inhibitory activities of mushroom *Ramaria formosa* extracts. *J Soc Cosmet Sci Korea* 42, 269–278, 10.15230/SCSK.2016.42.3.269.

S.K. Kim, I. Wijesekara. 2010. Development and biological activities of marine-derived bioactive peptides: A review. *J Funct Foods* 2: 1–9.

K.Y. Kino, K. Yamaoka, J. Watanabe, K. Shimizu, H. Tsunoo. 1989. Isolation and characterization of a newly immunomodulatory protein Ling Zhi–8 (LZ–8) from *Ganoderma lucidum*. *J Biol Chem* 264: 472–478.

P.M. Kirk, P.F. Cannon, D.W. Minter, J.A. Stalpers. 2008. Dictionary of the fungi, 10th edn. CAB International, Wallingford

A.G. Knudson. 2012. The Genus *Ramaria* in Minnesota. Thesis. https://conservancy.umn.edu/bitstream/11299/122161/1/Knudson_Alicia_February2012.pdf

P.Y. Ko, S.H. Kang, G.P. Song, Y.C. Jeun. 2013. Traditional knowledge on wild mushrooms in the surROUNDING VILLAges Hallyeo–Haesang National Park. *Kor J Mycol* 1(2): 127–131. 10.4489/KJM.2013.41.2.127

G.D. Kommers, M.N. Santos. 1995. Experimental poisoning of cattle by the mushroom *Ramaria flavo–brunnescens* (Clavariaceae): A study of the morphology and pathogenesis of lesions in hooves, tail, horns and tongue. *Vet Hum Toxicol* 37: 297–302.

F. Kotlaba, Z. Pouzar. 1964. Preliminary results of staining spores. *Trans Br Mycol Soc* 64: 822–829.

H.C. Kuo, C.C. Lu, C.H. Shen, S.Y. Tung et al. 2016. *Hericium erinaceus* mycelium and its isolated erinacine A protection from MPTP–induced neurotoxicity through the ER stress, triggering an apoptosis cascade. *J Transl Med* 14: 78. 10.1186/s12967–016–0831–y.

I.S. Lee, K.C. Kim, I.D. Yoo, B.J. Ha. 2015. Inhibition of human neutrophil elastase by labdane diterpenes from the fruiting bodies of *Ramaria formosa. Biosci Biotechnol Biochem* 79(12): 1921–1925.

J. Y. Lee. 1988.*Coloured Korean Mushrooms*. Academic Publishing Co. Ltd., Seoul, pp. 204–213.

T.H. Lee, Y.H. Han. 2001. Enzyme activities of the fruit body of *Ramaria botrytis*. *Mycobiology* 29(3): 173–175.

T.H. Lee, Y.H. Han. 2005. Cultural characteristics for the enhanced mycelial growth of *Ramaria botrytis*. *Mycobiology* 33(1): 12–14

H. Li. 2017. Extraction, purification, characterization and antioxidant activities of polysaccharides from *Ramaria botrytis* (Pers.) Ricken. *L. Chem Cent J* 11, 24 10.1186/s13065–017–0252–x

X. Lin, L. Ma, S.B. Racette, C.L.A. Spearie, R.E. Ostlund. 2009. Phytosterol glycosides reduce cholesterol absorption in humans. *Am J Physiol Gastrointest Liv Physiol* 296, 931–935.

K. Liu, Wang J, Zhao L, Xu D et al. 2012. Chemical constituents of *Ramaria flava*. *Chinese Pharm J* 1285–1288.

K. Liu, J. Wang, L. Zhao, O. Wang. 2013. Anticancer, antioxidant and antibiotic activities of mushroom *Ramaria flava*. *Food Chem Toxicol* 58: 375–380.

Y. Liu, D. Chen, Y. You, S. Zeng, Y. Li, Q. Tang, et al. 2016. Nutritional composition of boletus mushrooms from Southwest China and their antihyperglycemic and antioxidant activities. *Food Chem* 211: 83–91, 10.1016/j.foodchem.2016.05.032.

V.E. López, F.P. García, M.G. Canales, E.M. Otazo Sánchez, J.R. Villagómez Ibarra. 2018. Chemical characterization, nutraceutical, and evaluation of the geomicological activity of *Ramaria flava* wild and edible in the state of Hidalgo, Mexico. *Int J Curr Res* 10(1): 64214–64227.

J. Łuszczynski. 2009. *Ramaria fagicola* (Fungi, Basidiomycota): The first record for Poland, and from a new substratum. *Acta Soc Bot Pol* 78: 287–289.

A.R. Malik, A.H. Wani, M.Y. Bhat, S. Parveen. 2017. Ethnomycological knowledge of some wild mushrooms of northern districts of Jammu and Kashmir, India. *Asian J Pharm Clin Res* 10(9):399–405.

X. Meng, H. Liang, L. Luo. 2016. Antitumor polysaccharides from mushrooms: A review on the structural characteristic, antitumor mechanisms and immunomodulating activities. *Carbohydr Res* 424: 30–41.

P. Mérida, H. Calderón, M.A. Comandini, et al. 2019. Ethnomycological knowledge among Kaqchikel, indigenous Maya people of Guatemalan Highlands. *J Ethnob Ethnomed* 15: 36. 10.1186/s13002–019–0310–7

P.G. Miles, S.T. Chang. 2004. Mushrooms: Cultivation, nutritional value, medicinal effect, and environmental impact, CRC Press, Boca Raton, Florida. ISBN 0- 8493-1043-1.

T. Mizuno. 1995. *G. lucidum* and *G. tsugae*: Bioactive substances and medicinal effects. *Food Rev Int* 11(1):151–166.

T. Mizutani, S. Inatomi, A. Inazu, E. Kawahara. 2010. Hypolipidemic effect of *Pleurotus eryngii* extract in fat–loaded mice. *J Nutr Sci Vitaminol* 56: 48–53.

C. Mohanan. 2011. Macrofungi of Kerala. Kerala Forest Research Institute, Hand Book # 27, Kerala, India, pp. 597.

A. Montoya, O. Hernández–Totomoch, A. Estrada–Torres, A. Kong, J. Caballero. 2003. Traditional knowledge about mushrooms in a Nahua Community in the State of Tlaxcala, México. *Mycologia* 95(5): 793–806.

D. Moore, L.A. Casselton, D.A. Wood, J.C. Frankland. 1985. Development biological of higher fungi. British Mycological Society Symposium. Cambridge University Press, Cambridge

B.A. Murray, R.J. FitzGerald. 2007. Angiotensin converting enzyme inhibitory peptides derived from food proteins: Biochemistry, bioactivity and production. *Curr Pharm Des* 13: 773–791.

A.D. Murugkar, G. Subbulakshmi. 2005. Nutritional value of edible wild mushrooms collected from the Khasi Hills of Meghalaya. *Food Chem* 89: 599–603. 10.1016/j.foodchem.2004.03.042

G. Nasim, M Ali, A. Shabbir. 2008. A study of genus *Ramaria* from Ayubia national park, Pakistan. *Mycopath* 6: 43–46.

T. Niskanen, D. Brian, P. Kirk, P. Crous, R. Lücking, P. Brandon Matheny, L. Cai, K.D. Hyde, M. Cheek. 2018. State of the World's Fungi. https://stateoftheworldsfungi.org (accessed 1 Feb 2022).

Y.C. Niu, J.C. Liu, X.M. Zhao, F.Q. Su, H.X. Cui. 2009. Immunostimulatory activities of a low molecular weight antitumoral polysaccharide isolated from *Agaricus blazei* Murill (LMPAB) in Sarcoma 180 ascitic tumor–bearing mice. *Pharmazie* 64: 472–476.

N. Nowacka–Jechalke, R. Nowak, M. Juda, A. Malm, M. Lemieszek, W. Rzeski et al. 2018. New biological activity of the polysaccharide fraction from *Cantharellus cibarius* and its structural characterization. *Food Chem* 268: 355–361.

T.K.M. Nyholm, Pia–Maria Grandell, Bodil Westerlund, J. Peter Slotte. 2010. Sterol affinity for bilayer membranes is affected by their ceramide content and the ceramide chain length. *Biochim Biophys Acta* 1798(5):1008–1013. 10.1016/j.bbamem.2009.12.025

C.E. O'Neil, T.A. Nicklas, V.L. Fulgoni. 2013. Mushroom intake is associated with better nutrient intake and diet quality: 2001–2010 National Health and Nutrition Examination Survey. *J Nutr Food Sci.* 3: 5.

O.C. Ogidi, V.O. Oyetayo. 2016. Phytochemical property and assessment of antidermatophytic activity of some selected wild macrofungi against pathogenic dermatophytes. *Mycology* 7(1): 9–14. 10.1080/21501203.2016.1145608

A.F. Olawale, N.M. Muhammed. 2013. Fungi: A review on mushrooms. In: Subha Gangully (ed.), Current trends in advancement of scientific research and opinion in applied microbiology and biotechnology. First Edition, Chapter 5. Science and Education Development Institute, Nigeria.

B.O. Onyango, V.A. Palapala, P.F. Arama, S.O. Wagai, B.M. Gichimu. 2011. Morphological characterization of Kenyan native wood ear mushroom [*Auricularia auricula* (L. ex Hook.)], the effect of supplemented millet and sorghum grains in spawn production. *Agric Biol J North America* 2(3): 407–414.

P. O'Reilly. 2011. Fascinated by fungi: Exploring the history, mystery, facts, and fiction of the underworld kingdom of mushrooms. *First Nature*. 450 p.

P.K. Ouzouni, D. Petridis, W.D. Koller, K.A. Riganakos. 2009. Nutritional value and metal content of wild edible mushrooms collected from West Macedonia and Epirus, Greece. *Food Chem* 115: 1575–1580. 10.1016/j.foodchem.2009.02.014.

S.A. Pala, A.H. Wani, M.Y. Bhat. 2013. Ethnomycological studies of some wild medicinal and edible mushrooms in the Kashmir Himalayas (India). *Int J Med Mushrooms* 15(2):211–220.

S.A. Pala, A.H. Wani, B.A. Ganai. 2019. Antimicrobial potential of some wild macromycetes collected from Kashmir Himalayas. *Plant Science Today* 6: 137–146. 10.14719/pst.2019.6.2.503.

W. H. Park, H. D. Lee. 1997. *Illustrated Book of Korean Medicinal Mushrooms*. Kyo–Hak Publishing Co., Ltd., Seoul, pp. 576.

S. Patel, A. Goyal. 2012. Recent developments in mushrooms as anti-cancer therapeutics: A review. *Biotech* 2: 1–15. 10.1007/s13205–011–0036–2

Y. Patel, R. Naraian, V.K. Singh. 2012. Medicinal properties of *Pleurotus* species (Oyster mushroom): A review. *World J Fungal Plant Biol* 3(1):1–12.

M.S. Patil, A.N. Thite. 1977. Fungal flora of Radhanagri, Kolhapur. *J Shivaji Univ* 17: 149–162.

J. Pérez–Moreno, A. Guerin–Laguette, A.C. Rinaldi, F. Yu, A. Verbeken, F. Hernández–Santiago, M. Martínez–Reyes. 2021. Edible mycorrhizal fungi of the world: What is their role in forest sustainability, food security, biocultural conservation and climate change? *Plants People Planet* 3(5): 471–490. 10.1002/ppp3.10199

R.H. Petersen. 1975. *Ramaria* subgenus *Lentoramaria* with Emphasis on North American Taxa. *J Cramer*. Vaduz, Liechtenstein. 161p.

R.H. Petersen. 1981. *Ramaria* subgenus *Echinoramaria*. *J Cramer*. Vaduz, Liechtenstein. 261 p.

V. Popovic, Jelena Zivkovic, Slobodan Davidovic, Milena Stevanovic, Dejan Stojkovic. 2013. Mycotherapy of cancer: An update on cytotoxic and antitumor activities of mushrooms, bioactive principles and molecular mechanisms of their action. *Curr Top Medi Chem* 13 (21): 2791–2806. 10.2174/15680266113136660198

P. Pradhan, S. Banerjee, A. Roy, K. Acharya. 2010. Role of wild edible mushrooms in the Santal livelihood in lateritic region of West Bengal. *J Bot Soc Ben* 64(1): 61–65.

P. Pradhan, A.K. Dutta, A. Roy, K. Acharya. 2013. Notes on *Ramaria fumigata* regarding its occurrence and plant association in West Bengal. *India Envi Ecol* 31(1A): 243—246.

M.A. Rahman, N. Abdullah, N. Aminudin. 2016. Interpretation of mushroom as a common therapeutic agent for Alzheimer's disease and cardiovascular diseases. *Crit Rev Biotechnol* 36: 1131–1142.

T. Rahman, M.B.K. Choudhury. 2012. Shiitake mushroom: A tool of medicine. *Bangladesh J Med Biochem* 5(1): 24–32.

B.K. Rai, S.S. Ayachi, A. Rai. 1993. A note on ethno-myco-medicines from Central India. *Mycologist* 7: 192–193.

M. Rai, K. Acharya. 2012. Proximate composition, free radical scavenging and NOS activation properties of *Ramaria aurea*. *Res J Pharm Tech* 5: 1421–1427.

M. Rai, S. Sen, K. Acharya. 2013. Antimicrobial activity of four wild edible mushrooms from Darjeeling hills, West Bengal, India. *Int J PharmTech Res* 5(3): 949–956.

C.H. Ramesh, M.G. Pattar. 2010. Antimicrobial properties, antioxidant activity and bioactive compounds from six wild edible mushrooms of Western Ghats of Karnataka, India. *Pharmacog Res* 2: 107–112.

J. Rammeloo, R. Walleyn. 1993. The edible fungi of Africa South of the Sahara: A literature survey. *Scripta Botanica Belgica* 5: 1–62.

R. Rana. 2016. Systematic studies on wild edible mushroom *Ramaria botrytis* (Fr.) Ricken, collected from Shimla Hills situated in Himachal Pradesh, India. *Int J Innovative Res Sci* 3: 4569–4573.

P. Rangsinth, C. Sillapachaiyaporn, S. Nilkhet, T. Tencomnao, T.A. Ung, S. Chuchawankul. 2021. Mushroom–derived bioactive compounds potentially serve as the inhibitors of SARS–CoV–2 main protease: An *in–silico* approach. *J Trad Compl Med* 11: 158–172. 10.1016/j.jtcme.2020.12.002

S.S. Rattan, I.P.S. Khurana. 1978. The Clavarias of the Sikkim Himalayas. *Bibliotheca Mycologica* 66: 1–68.

F.S. Reis, E. Pereira, L. Barros, M.J. Sousa, A. Martins, I.C.F.R. Ferreira. 2011. Biomolecule profiles in inedible wild mushrooms with antioxidant value. *Molecules* 16(6):4328–4338.

G. Sadi, A. Kaya, H.A. Yalcin, B. Emsen, D.İ. Kartal, A. Altay. 2016. Wild edible mushrooms from Turkey as possible anticancer agents on HepG2 cells together with their antioxidant and antimicrobial properties. *Int J Med Mush* 18: 83–95. 10.1615/IntJMedMushrooms.v18.i1.100.

B.B. Sarkar, D.K. Chakraborty, A. Bhattacharjee. 1988. Wild edible mushroom flora of Tripura. *Indian Agriculturist* 32: 139–143.

L. Scattolin, L. Montecchio, E. Mosca, R. Agerer. 2008. Vertical distribution of the ectomycorrhizal community in the top soil of Norway spruce stands. *Eur J Forest Reso* 127: 347–357.

H.V. Scheid, E.S.V. Sallis, F. Riet-Correa, A.L. Schild. 2022. *Ramaria flavo-brunnescens* mushroom poisoning in South America: A comprehensive review. *Toxicon* 205, 91–98, 10.1016/j.toxicon.2021.12.001.

K.C. Semwal, V.K. Bhatt. 2019. A report on diversity and distribution of macrofungi in the Garhwal Himalaya, Uttarakhand, India. *Biodiv Res Conserv* 53: 7–32. Doi 10.2478/biorc–2019–0002

K.C. Semwal, S.L. Stephenson, V.K. Bhatt, R.P. Bhatt. 2014. Edible mushrooms of the Northwestern Himalaya, India: A study of indigenous knowledge, distribution and diversity. *Mycosphere* 5(3): 440–461. 10.5943/mycosphere/5/3/7.

D.J. Seo, C. Choi. 2021. Antiviral bioactive compounds of mushrooms and their antiviral mechanisms: A review. *Viruses* 13: 350. 10.3390/v13020350

S. Shaffique, S.M. Kang, A.Y. Kim, M. Imran, M. Aaqil Khan, I.J. Lee. 2021. Current knowledge of medicinal mushrooms related to anti-oxidant properties. *Sustainability* 13: 7948. 10.3390/su13147948

A. Shah, A.B. Kumar, J. Ahmad, M.P. Sharma. 2015. New ethnomedicinal claims from Gujjar and Bakerwals tribes of Rajouri and Poonch districts of Jammu and Kashmir, India. *J Ethnopharmac* 166: 119–128. 10.1016/j.jep.2015.01.056i

R.M. Sharda, K.S. Thind. 1986. Genus *Ramaria* in the Eastern Himalaya: subgenus *Laeticolora* – II. *Proc Ind Acad Sci Plant Sci* 96(6): 519–529.

A.D. Sharma, C.L. Jandaik. 1978. Genus *Ramaria* Holmsk. in Himachal Pradesh. *Ind J Mush* 4: 5–7.

A.D. Sharma, C.L. Jandaik, R.L. Munjal. 1977. Some fleshy fungi from Himachal Pradesh. *Ind J Mush* 3: 12–15.

S.K. Sharma, N. Gautam. 2017. Chemical and bioactive profiling and biological activities of coral fungi from Northwestern Himalayas. *Sci Rep* 7: 46570. 10.1038/srep46570

T.K. Sharma. 2008. Vegetable caterpillar, Science Reporter. 5th May ISBN 0036–8512. National institute of science communication and information resources (NISCAIR), CSIR, pp. 33–35.

C. Sillapachaiyaporn, S. Nilkhet, A.T. Ung, S. Chuchawankul. 2019. Anti-HIV-1 protease activity of the crude extracts and isolated compounds from Auricularia polytricha. *BMC Complement Altern Med* 19(351). 10.1186/s12906-019-2766-3

E. Smith Matthew, G.M. Bonito. 2012. Systematics and ecology of edible ectomycorrhizal mushrooms. In: A. Zambonelli and G.M. Bonito (eds.), Edible Ectomycorrhizal Mushrooms. *Soil Biology* 34, 10.1007/978–3–642–33823–6_2, Springer–Verlag Berlin Heidelberg 2012

S.K. Tanaka, K. Ko, K.K. Kino, K.K. Tsuchiya, A. Yamashite, A. Marasugi, S.S. Sakuma, H. Tsunoo. 1989. Complete amino acid sequence of an immunomodulatory protein Ling Zhi (LZ– 8), an immunomodulator from a fungus *Ganoderma lucidum*, having similarity to immunoglobulin variable regions. *J Biological Chem* 264: 16372–16377.

D.L. Taylor, T.N. Hollingsworth, J.W. McFarland, N.J. Lennon, C. Nusbaum, R.W. Ruess. 2014. A first comprehensive census of fungi in soil reveals both hyperdiversity and fine-scale niche partitioning. *Ecol Monog* 84: 3–20.

K.S. Thind, Sukh Dev. 1957a. The Clavariaceae of the Mussoorie Hills–VI. *J Ind Botanical Soc* 36: 92–103.

K.S. Thind, Sukh Dev. 1957b. The Clavariaceae of the Mussoorie Hills–VIII. *J Ind Botanical Soc* 36: 475–485.

K.S. Thind, G.P.S. Anand. 1956a. The Clavariaceae of the Mussoorie Hills–I. *J Ind Botanical Soc* 35: 92–102.

K.S. Thind, G.P.S. Anand. 1956b. The Clavariaceae of the Mussoorie Hills–IV. *J Ind Botanical Soc* 35: 323–332.

K.S. Thind, S.S. Rattan. 1967. The Clavariaceae of India–XI. *Proc Indian Acad of Sci* 66B: 143–156.

K.S. Thind, R.M. Sharda. 1984. Three new species of clavarioid fungi from the Himalayas. *Indian Phytopathol* 37: 234–240.

K.S. Thind, R.M. Sharda. 1985. Genus *Ramaria* in the Eastern Himalaya: Subgenus Laeticolora–I. *Proc Ind Acad Sci* (*Plant Sci*) 95: 271–281.

K.S. Thind, I.P.S. Khurana, S.C. Kaushal. 1983. The Clavariaceae of India–XIII. Three species of Ramaria subgenus *Echinoramaria* new to India. *Kavaka* 11: 32–35.

A.N. Thite, M.S. Patil, T.N. More. 1976. Some fleshy fungi from Maharashtra. *Botanique* 7 (2–3): 77–88.

T.S. Tochikura, H. Nakashima, Y. Ohashi, N. Yamamoto. 1988. Inhibition (in vitro) of replication and of the cytopathic effect of human immunodeficiency virus by an extract of the culture medium of *Lentinus edodes* mycelia. *Med Microbiol Immunol* 177: 235–244.

C.V. Toledo, C. Barroetaveña, Â. Fernandes, L. Barros, I.C.F.R. Ferreira. 2016. Chemical and antioxidant properties of wild edible mushrooms from native *Nothofagus* spp. Forest, Argentina. *Molecules* 21, 10.3390/molecules21091201.

J.M. Trappe. 1962. Fungus associates of ectotrophic mycorrhizae. *Botanical Review* 28: 538–606.

T. Tsai–Teng, C. Chin–Chu, L. Li–Ya, C. Wan–Ping et al. 2016. Erinacine A – Enriched *Hericium erinaceus* mycelium ameliorates Alzheimer's disease-related pathologies in APPswe/PS1dE9 transgenic mice. *J Biomed Sci* 23 (1): 1–12.

M.E. Valverde, T. Hernández–Pérez, O. Paredes–López. 2015. Edible mushrooms: Improving human health and promoting quality life. *Int J Microbiol* 376387 pp. 1–14. 10.1155/2015/376387

R.K. Verma, V. Pandro. 2018. Diversity and distribution of clavarioid fungi in India, three fungi from Central India. *Int J Curr Microbiol App Sci* 7(12): 2129–2147. 10.20546/ijcmas.2018.712.242

M. Villegas, J. Cifuentes, A.E. Torres. 2005. Sporal characters in Gomphales and their significance for phylogenetics. *Fung Divers* 18: 157–175.

H. Wang, T. Ng. 2006. Ganodermin, an antifungal protein from fruiting bodies of the medicinal mushroom *Ganoderma lucidum. Peptides* 27(1): 2730. 10.1016/j.peptides.2005.06.009.

Y. Wang, L. Bao, D. Liu et al. 2012. Two new sesquiterpenes and six nor sesquiterpenes from the solid culture of the edible mushroom *Flammulina velutipes*. *Tetrahedron* 68: 3012–3018.

S. Wasser. 2002. Medicinal mushrooms as a source of antitumor and immunomodulating polysaccharides. *Appl Microb Biotech* 60(3): 258–274.

S.P. Wasser. 2005. Reishi or Lingzhi (*G. lucidum*). In: P. Coates, M.R. Blackman, G. Cragg, M. Levine, J. Moss, J. White (eds.), Encyclopaedia of dietary supplements. Marcel Dekker, New York, pp. 603–622.

S.P. Wasser. 2017. Medicinal mushrooms in human clinical studies. Part I. Anticancer, oncoimmunological, and immunomodulatory activities: A review. *Int J Med Mush* 19(4): 1–10.

A.J. Weigand–Heller, P.M. Kris–Etherton, R.B. Beelman. 2012. The bioavailability of ergothioneine from mushrooms (*Agaricus bisporus*) and the acute effects on antioxidant capacity and biomarkers of inflammation. *Prev Med* 54: S75–S78

W.C. Wong, P.C.K. Cheung. 2001. Food and nutritional sciences programme. The Chinese University of Hong Kong, Shatin, New Territories, Hong Kong, China.

H.S. Xu, Y.W. Wu, S.F. Xu, H.X. Sun, F.Y. Chen, L. Yao. 2009. Antitumor and immunomodulatory activity of polysaccharides from the roots of *Actinidia Eriantha*. *J Ethnopharmacol* 125: 310–317

X. Xu, H. Yan, J. Chen, X. Zhang. 2011. Bioactive proteins from mushrooms. *Biotechnol Adv* 29(6):667–674.

Z. Xu, L. Fu, S. Feng et al. 2019. Chemical composition, antioxidant and antihyperglycemic activities of the wild *Lactarius deliciosus* from China. *Molecules* 24: 1357. 10.3390/molecules24071357

H. Yang, S. Guo, A. Sagitov. 2013. The collection and separation of the special edible and medicinal mushrooms in China and Kazakhstan and efficacy evaluation. *Sci Tech Infor Dev Econom* 23: 132–136.

W. Yang, J. Yu, L. Zhao, N. Ma, Y. Fang, F. Pei, et al. 2015. Polysaccharides from *Flammulina velutipes* improve scopolamine-induced impairment of learning and memory of rats. *J Funct Foods* 18: 411–422. 10.1016/j.jff.2015.08.003

Y. Yaoita, Y. Satoh, M. Kikuchi. 2007. A new ceramide from *Ramaria botrytis* (Pers.) Ricken. *J Nat Med* 61: 205–207.

M.–Y. Yeh, W.–C. Ko, L.–Y. Lin. 2014. Hypolipidemic and antioxidant activity of enoki mushrooms (*Flammulina velutipes*). *BioMed Res Int* (11): 352385, 6 p. 10.1155/2014/352385

L. Yu, D.G. Fernig, J.A. Smith, J.D. Milton, J.M. Rhodes. 1993. Reversible inhibition of proliferation of epithelial cell lines by *Agaricus bisporus* (edible mushroom) lectin. *Cancer Res* 53(19): 4627–4632.

J. Zhang, C. Wen, Y. Duan, H. Zhang, H. Ma. 2019. Advance in *Cordyceps militaris* (Linn) Link Polysaccharides: Isolation, structure, and bioactivities: A review. *Int J Biol Macromol* 132, 906–914.

R. Zhang, G. Tian, Y. Zhao, et al. 2015. A novel ribonuclease with HIV–1 reverse transcriptase inhibitory activity purified from the fungus *Ramaria formosa*. *J Basic Microbiol* 55: 269–275.

S. Zhao, C. Rong, Y. Liu, F. Xu, S. Wang, C. Duan, J. Chen, X. Wu. 2015. Extraction of a soluble polysaccharide from *Auricularia polytricha* and evaluation of its anti–hypercholesterolemic effect in rats. *Carbohydr Polym* 122: 39–45.

R. Zhou, Ya–Jie Han, Min–Hui Zhang, Ke–Ren Zhang, Tzi Bun Ng, Fang Liu. 2017. Purification and characterization of a novel ubiquitin-like antitumour protein with hemagglutinating and deoxyribonuclease activities from the edible mushroom *Ramaria botrytis*. *AMB Expr* 7: 47. 10.1186/s13568–017–0346–9

D.C. Zied, A. Pardo–Gimenez. 2017. Edible and Medicinal Mushrooms: Technology and Applications. Wiley, Hoboken, p. 592.

LIST OF ABBREVIATIONS

ACE	Angiotensin-I converting enzyme
EMF	Edible mycorrhizal fungi
HNE	Human Neutrophil Elastase Enzyme
RBP	*Ramaria botrytis* Polysaccharides
RBUP	*Ramaria botrytis* ubiquitin-protein
RF-1	*Ramaria flaccida*

14 *Schizophyllum commune*

Shweta Singh and Chandramani Raj
ICAR-Indian Institute of Sugarcane Research, Lucknow, Uttar Pradesh, India

Harvinder Kumar Singh
Indira Gandhi Krishi Viswavidyalaya, Raipur, Chhattisgarh, India

R. Gopi
ICAR-Sugarcane Breeding Institute Research Centre, Kannur, Kerala, India

Susheel Kumar Sharma and Chandan Kapoor
ICAR-Indian Agricultural Research Institute, New Delhi, India

Matber Singh
ICAR-Indian Institute of Soil & Water Conservation, Dehradun, Uttarakhand, India

Swati Allen
Central Procurement Cell, Indian Council of Medical Research, V. Ramalingaswami Bhawan, Ansari Nagar East, New Delhi, India

CONTENTS

INTRODUCTION

Schizophyllum commune Fr. (Family: Schizophylaceae) is one of the most commonly and widely grown edible mushroom that grows naturally on dead and decayed woods, particularly during the monsoon. The genus *Schizophyllum* basically means "split gill," hence it is also called as split gill mushroom. The mushroom causes white rot on dead and decaying woods making it distinctive

DOI: 10.1201/9781003259763-14

from other gill-based fungus (Imtiaj et al., 2008). *S. commune* has the potential to colonize at least 150 genera of woody plants, but it prefers soft wood and grass silage (de Jong, 2006).

TAXONOMY

Domain: Eukarya
Kingdom: Fungi
Division: Basidiomycota
Class: Agaricomycetes
Order: Agaricales
Family: Schizophyllaceae
Genus: *Schizophyllum*
Species: *S. commune*
Botanical name: *Schizophyllum commune*
Local names: The mushroom is also acknowledged as the split gill mushroom and is widely consumed in North East India, where it is known as *Kanglayen* in Manipur, *pasi* in Mizoram, and *chamrae cheoae* in Sikkim.

MUSHROOM CHARACTERISTICS

In nature, the fruiting bodies of split gill mushrooms grow as sessile brackets and can been seen in different tiers of dead logs (hardwood) las well as branches (Figure 14.1). This wood-decaying fungus can breakdown even complex plant biomass such as lignin, into simple sugars and also cause white rot on logs. The mushroom fruiting body is white in colour and appears to be a small bracket fungus from the dorsal surface. However, radial gill like shape are found in underside of the cap. Each of them is centrally split. The main fruiting body is about 1–4 cm wide, spongy as well as dense in texture, flabelliform shaped, with small stipes which become stipeless at maturity. The fruiting bodies on the upper surface are covered with small white to greyish hairs and laterally attached to the substratum (Figure 14.2). The pileus at young stage is hairy and white which upon maturity turns pale yellow to brownish or even when water is splashed over the fruiting body of the mushroom. The mature basidiocarp can be characterized into two major parts namely pilear region and the gills. However, one of the most characteristic features of this mushroom is that the gills are split through mid-region on the underside. The so-called split gills are usually pinkish grey in colour and are found radiating from the nearby attachment point. During the dry weather, in order to protect the fertile surfaces, also known as hymenium, the gills curl back to the body. The fruiting body of this mushroom also show typical gymnocarpus characteristic as sporiferous part as well as the hymenophore are only found on the underside part of the cap. It also has distinct folded hymenophore and unique longitudinally split gills down the middle. The mushroom's stems are crude, stunted, and rarely noticeable when seen above the substrate. The initial pin head in domesticated fruiting bodies are about 2 mm in size, typically stipeless or else they are only up to 1 mm in length and growing scattered (Singh et al., 2021).

DISTRIBUTION

With exception to Antarctica, since there is no timber or log to use as substrate, *S. commune* is a common white rot fungus with a global distribution (Khatua et al., 2013). However in several South Asian countries such as Taiwan, North eastern parts of India, Vietnam, Thailand, Southern part of China and Malaysia this mushroom is also used in medicine as well as food. In a study carried out by Raper *et al.,* in 1958, this mushroom has been isolated from the dead or decayed wood of as many as 150 distinctive plant species, primarily as white rot fungus but

FIGURE 14.1 Fruiting bodies of *Schizophyllum commune* growing in tier-like arrangement on hardwood logs.

also found as a human and plant pathogen. (Schmidt and Liese, 1980). The annual mushroom production is approximately 2.5 million tonnes (Kothe, 2001). It is a known food source to several indigenous people worldwide particularly in Africa and Asia. However as per a study carried out by Cooke 1961, in few countries, it is also consumed as chewing gum. Lumbsch et al. (2008) proposed two alternative theories for the worldwide distribution of fungus species: first is the vicariance which has resulted in more heterogeneous globally distributed population which can be attributed to geographically different populations or frequent long-distance scattering actions. In a study on phylogenetic investigation of the intergenic spacer (IGS) region of 195 *S. commune* isolates isolated from across the globe revealed three distinct geographical clades: the first comprised the isolates mostly from North and Central America, the second set of isolates was the South American, and the last group of isolates were from European and Asia/Australia. Therefore, due to frequent and long-distance dispersal of *S. commune*, the worldwide distribution of *S. commune* has brought up heterogeneity in many parts of world (James et al., 2001).

FIGURE 14.2 Upper or dorsal surface of flabelliform basidiocarp.

MEDICINAL USES

Split gill mushrooms have exceptional immunostimulatory, antimicrobial, antitumor, antifungal, antineoplastic, immunomodulatory, and chemoprotective effects for a medicinal mushroom (Patel and Goyal, 2012). *S. commune* extracts have antibacterial and antifungal properties and can thus be used as a potential antimicrobial agent. The polysaccharide from the mushroom Schizophyllan, has significant anticancer activity and is the most important compound that makes *S. commune* a medicinal mushroom. Schizophyllan is also known as sizofiran or sonifilan in the pharmaceutical industry. Extensive research on this polysaccharide has contributed to its widespread use as a biologically active polysaccharide in pharmaceuticals. Schizophyllan acts like a biological response modifier (BRM) as well as non-specific immune system stimulator (Rau, 2002). Other schizophyllan bioactivities include hepatoprotective effects (Kukan et al., 2004), antineoplastic, antibacterial, and antiparasitic properties, and anti-inflammatory effects (Wasser and Weis, 1999).

The most promising of these reported bioactivities are their antitumor and immunobiological properties. It also has chemotherapeutic effects on leukaemia, melanoma, and fibrosarcoma (Jacques, 1982; Chihara, 1984). Schizophyllan has been shown to inhibit the growth of 180 sarcoma tumours. It is endowed with the capacity to treat recurrent and inoperable gastric cancer and escalates the survival capacity of head as well as neck cancer patients without any side effects. The anticancer effect is host-mediated and the effect is due to increased interferon synthesis which leads to the activation of macrophages as well as T-lymphocyte (Deng et al., 2000). Scientific advancements such as ultrasonic treatment of schizophyllan resulted in better anticancer as well as immune-boosting properties (Zhong et al., 2015). The compound schizophyllan is useful in treating variety of diseases, including AIDS. It is also known to enhance the efficacy of vaccines as well as antitumor treatments (Leathers, Nunnally and Price, 2006).

Schizophyllan nanofibers are well known as an excellent biocompatible biomaterial for wound dressing because they promote the development of the dermis and epidermis layers. The electrospinning technique assisted in creation of nano-fibrous scaffold of schizophyllan/polyvinyl alcohol and are utilized for evaluating the indirect cytotoxicity with respect to mouse fibroblasts (L929). The cultivation of L929 cells resulted in no cytotoxicity. The schizophyllan/polyvinyl alcohol nanofiber mats accelerate cell adhesion and proliferation in addition to sustaining flexibility as well as provide tensile strength during lab cell culture. As a result, this filament can be efficiently used in wound dressing or skin recovery (Safaee-Ardakani et al., 2019).

The protective effect of dietary schizophyllan was observed on damaged livers of mice through induction of SIRT-3 which is a NAD+-dependent deacetylase causing deacetylation of SDHA and SOD2 (Lee et al., 2020). A similar study was conducted to mark the impact of schizophyllan-containing polysaccharide blend on horses having active ulceration and in this case the horses showed recovery in ulcerative areas, better appetite, increased weight, and positive behaviour (Slovis, 2017). The polysaccharide is also nucleic acid carrier with capacity to form complexes with different polynucleotides with the help of hydrogen bonding therefore minimizing the cytotoxicity due to positive charge density (Zhang et al., 2013). Currently, there are several Japanese pharmaceutical companies which are commercially producing schizophyllan and the same is extensively being used in clinical trials against antitumor therapy (Daba and Ezeronye, 2003).

S. commune antioxidant peptide fractions and protein hydrolysate have been reported to be useful bioactive compounds in preventing oxidative damage and impairment to the human intestinal cell line (HT-29) (Wongaem et al., 2021). It also has free radical scavenging activities, which prevents free radicals from playing a negative role in various diseases, including cancer. Peptides from the transport/signalling protein family, such as NADH dehydrogenase subunit 3, polysaccharide lyase family 8 protein, a glycosyltransferase family 69 protein, and an HD2 mating type protein isolated from *S. commune*, are known to stabilise free radicals owing to their activity in donating or absorbing electrons from those free radicals. However, in each case, the reactivity of free radicals is significantly reduced (Ren et al., 2008).

The lectin isolated from *S. commune* has mitogenic activity in a variety of body cells. In a study conducted on mouse, the lectin demonstrated mitogenic activity on splenocytes, antiproliferative activity against tumour cell lines, and inhibitory activity against HIV-1 reverse transcriptase. As the lectin has sequence homology to cell division protein, it shows mitogenic activity toward spleen cells. The polysaccharides and phenolic compounds of *S. commune* contain a variety of antioxidant properties, making it suitable for consumption in daily diet and preventing or reducing oxidative damage to human health.

PHYTOCHEMISTRY

S. commune is known to produce biopolymers in aqueous culture medium, primarily as three compounds—24kDa hydrophobin, 17kDa protein, and schizophyllan (Martin et al., 1999), a gel-forming (13)—d-glucan with a mean molecular weight (MW) of 1,00,000 to 1,50,000. (Tabata et al., 1988; Sakurai and Shinkai, 2000). Hydrophobins are small proteins made up of eight conserved cysteine residues that self-assemble at hydrophilic/hydrophobic interfaces to affect the hydrophobicity of surfaces.

In *S. commune*, schizophyllan is the most studied water-soluble triple-stranded helix glucan. Although schizophyllan is unusually heat stable, the three strands will separate above 135 °C (Brant et al., 1997). It is also referred to as SPG, Sizofiran, Sonilifilan, and Sizofilan. Schizophyllan is a non-ionic, water-soluble homoglucan made up of different sizes and degrees of branching of glucose polymers. It is composed of (13)—D-glucan main chains and 1,6--d-glucosyl side groups (-d-glucoside links at position 6 and frequently repeating) that form triple helix structures for cell structural rigidity and melting points ranging from 5 to 20 °C (Mueller et al., 2000).

The triple helix is rigid up to $M = 5 \times 10^5$ g/mol (Yanaki et al., 1980), tougher in structure than other triple helix structures such as collagen, and similar in diameter and pitch to -glucan lentinan from the shiitake mushroom Lentinus edodes. These glucans' immunomodulatory properties are determined by their branching pattern, molecular pattern, molecular weight, and structural confrontation, such as triple helix. In cancer patients, glucans with more branching have higher activity and strong BRM activity (Tateishi et al., 1997). SPG (triple helix) and SPG-OH (single strand) have the same antitumor effects and hematopoietic response to cyclophophamide in mice. SPG has higher zymosan-mediated hydrogen peroxide synthesis on peritoneal macrophages than SPG-OH, while SPG-OH has stronger lipopolysaccaride priming of tumour necrosis factor synthesis, nitric oxide synthesis, and hydrogen peroxide synthesis in mice (Ohno et al., 1995).

Schizophyllan is a compound produced by the solid and submerged cultures of *S. commune* Fries. It is then precipitated from the clear 4- to 8-day-old culture filtrate. The filtrate is then concentrated to 35–45% concentration by adding water-miscible organic solvents such as methanol. The typical culture solution contains the following ingredients: 3.0% glucose; 0.3% yeast extract; 0.15% $(NH_4)_2HPO_4$; 0.1% KH_2PO_4; 0.05% $MgSO_4.7H2O$; pH 4.8. (Kikumoto et al., 1970). In addition to schizophyllan, another compound of interest in *S. commune* is Schizocommunin, which is also an indole derivative. Indigotin (indigo), indirubin, isatin, and tryptanthrin can all be isolated from *S. commune* aqueous culture medium. Schizocommunin has also been shown to be cytotoxic to murine lymphoma cells (Hosoe et al., 1999).

PHARMACOLOGY

The utilization of mushrooms and their metabolites has emerged as a prolific area of research, potentially curing various human health conditions. Mushrooms also have a low fat content, a high protein content, minerals, vitamins, and a diverse range of bioactive compounds such as phenols, nucleotides, polysaccharides, steroids, terpenoids and glycoproteins. The split gill mushroom is the most powerful and well-studied medicinal mushroom (Mizuno, 1999; Chang and Miles, 2004).

Immunomodulatory, cell-mediated immunity, anticancer and antitumor properties, hepatoprotective, antifungal, antiviral properties, protection against infections, chemo- and radiation therapy are all benefits of split-gilled mushroom. When the SPG binds to immune cells like PMNs and peripheral blood mononuclear cells, it triggers the production of cytokines like IL-8 (Suzuki et al., 2002; Kubala et al., 2003). In vitro and in vivo, SPG activates macrophages, increasing T-cell activity and the sensitivity of cytotoxic LAK and NK cells to Il-2 (Kano et al., 1996). Peritoneal macrophage levels and IFN- production have been reported to increase following i.m injection of SPG in a group of 62 patients undergoing laparotomy for intraperitoneal recurrence of ovarian cancer (Chen et al., 1990). SPG and carboxymethylglucan-activated phagocytes and lymphocytes, increasing levels of IL-6, IL-8, TNF-, lymphocytes, CD62L, PMN leukocyte/monocyte CD11, and CD62L surface receptor expression in cultured human blood leukocytes. SPG's increased phagocytic stimulation is most likely due to its higher branching frequencies, polymer charge, or varied solution confirmation (Kubala et al., 2003). The studies on cytokine modulation due to SPC were carried out on mice by pretreating them with SPC orally as part of their diet, which revealed increased production of IL-12 and IFN- (Kawaguchi et al., 2004). In mice, schizophyllan administration increased IL-1 and Il-1 mRNA expression (Nemoto et al., 1993). In studies in which single (SPG-OH) or triple helix conformations of SPG were given to mice along with cyclophosphamide, it induced gene expression of stem cell factor and colony stimulating factor, as well as increased IL-6 production (Tsuzuki et al., 1999).

The SPG protected mouse bone marrow cells by inhibiting chromosomal damage caused by chemotherapeutic agents such as cyclophosphamide, adriamycin, and mitomycin C, as well as radiations. The SPG also restored bone marrow cell mitosis, which had previously been suppressed by anticancer drugs (Yang et al., 1993). When squamous cell inoculated mice were given SPG, they lived longer, had less tumour growth, and had fewer metastatic foci (Arika et al., 1992). It also inhibits tumour growth when combined with cisplatin in chemically induced rat ovarian adenocarcinoma (Sugiyama et al., 1995) and when administered prior to photodynamic therapy in mice with squamous cell carcinoma (Sugiyama et al., 1995; Krosl and Korbelik, 1994).

Schizophyllan has antitumor activity in both solid (Tabata et al., 1981) and ascites forms of Sarcoma180 (Komatsu et al., 1969), Lewis lung carcinoma (Yamamoto et al., 1981). It also improved cellular immunity in mice with tumour implants by resuming suppressed killer-cell activity (Oka, 1985). It is also important to note that SPG's cytokine production and antitumor effects are dose dependent and become inactive when administered in high doses (Miura et al., 2000). In mice, the compound protects against *Staphylococcus aureus*, *Klebsiella pneumoniae Pseudomonas aeruginosa* and *Escherichia coli* infections (Komatsu et al., 1973). SPG also improves immune response to viruses by producing IFN- and increasing the proliferation of peripheral blood mononuclear cells, as in the case of chronic hepatitis B infected patients (Kakumu et al., 1991).

Sizofiran, another bioactive molecule from *S. commune*, protects against chemo and radiation therapy by restoring cell mitosis and decreasing sister chromatid exchanges (SCEs) in mouse bone marrow after radio- and chemotherapy treatment (Yang et al., 1993).

MYCOLOGY

Split gill mushroom, also known as *S. commune*, is a wood-rotting basidiomycetes fungus found in the environment. The genus Schizophyllum is derived from the terms "schiza," which means "split," due to the appearance of centrally split, gill-like folds, radial, and "commune," which means "common or shared ownership" or "ubiquitous". The scientific name of the mushroom was assigned by Elias Magnus Fries, a Swedish mycologist in 1815.

The fruiting bodies of the *S. commune* can be easily grown in the laboratory on Malt extract agar (MEA) as well as on Potato dextrose agar (PDA) media (Kumar et al., 2018; Singh et al., 2021). Mycelia grow easily on PDA, forming dense white woolly mycelial growth with a characteristic bump formation in the centre and yellow to brown reverse on the opposite side (Singh et al., 2021) (Figure 14.3).

The mushroom can be easily multiplied on PDB media, where it grows in form of white cottony mycelium until 20 days of inoculation and later forms vaguely curled up like finger shaped basidiocarps which are measuring 4 cm in length and 1 cm width (Figure 14.4). When grown on malt agar plates, addition of tannic acid yields positive reaction for oxidase as well as laccase while produces negative reaction for tyrosinase. The colony's outline was white and there was no odour. However, the reverse of the plate was darkened and clamps were present. The hyphae were hyaline, thin walled and septate with 2 × 3.125 mm thickness. The mycelial mat was purely white in colour, beginning near the inoculum and spreading across the medium's surface, with a smooth, dull texture and curled surface.

The microscopic examination of spore prints reveal that the pilear cuticle is made up of more than one layer (multistratous), where the fungal hyphae are longitudinal and are closely packed. The hyphal penetration from the surface makes the pilear region look hairy (Figure 14.5). The pilear medulla has loosely arranged hyphae and therefore appears to be multistratous.

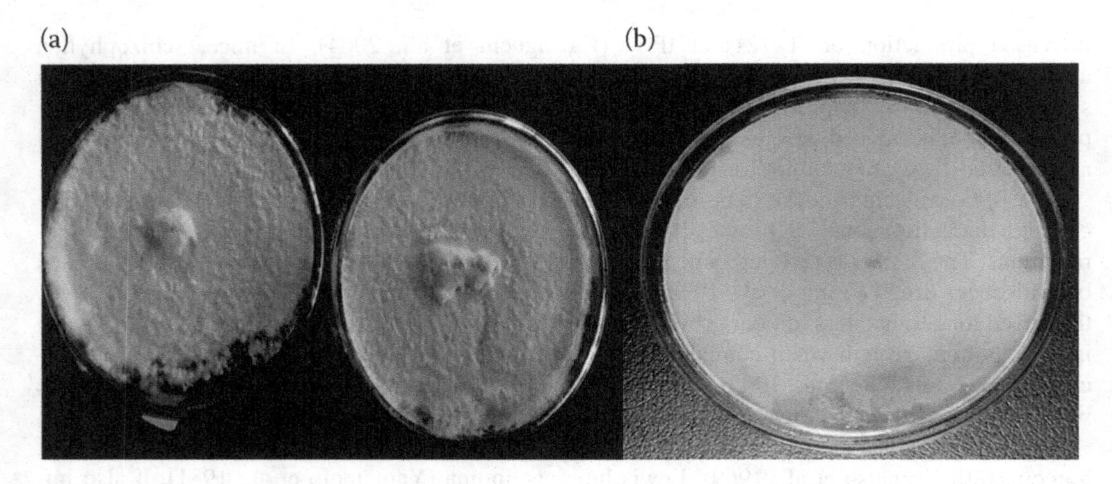

FIGURE 14.3 (a) Pure culture of *S. commune* growing as white and densely woolly mycelia with bump at the centre. (b) The reverse side of culture of *S. commune* showing yellowish-brown colour.

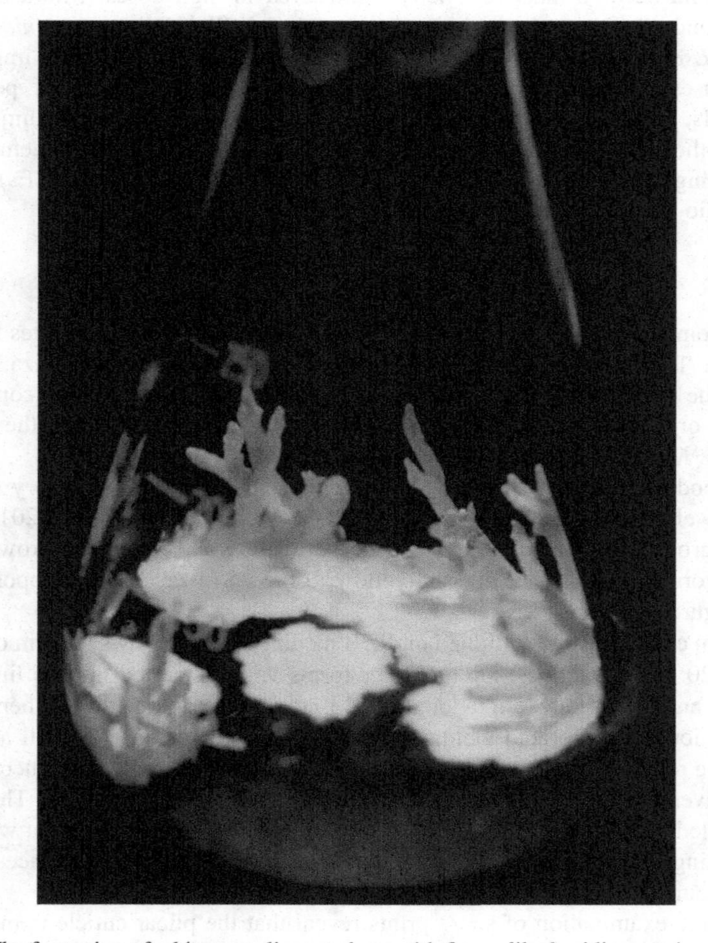

FIGURE 14.4 The formation of white mycelia mat along with finger-like basidiocarps in submerged culture.

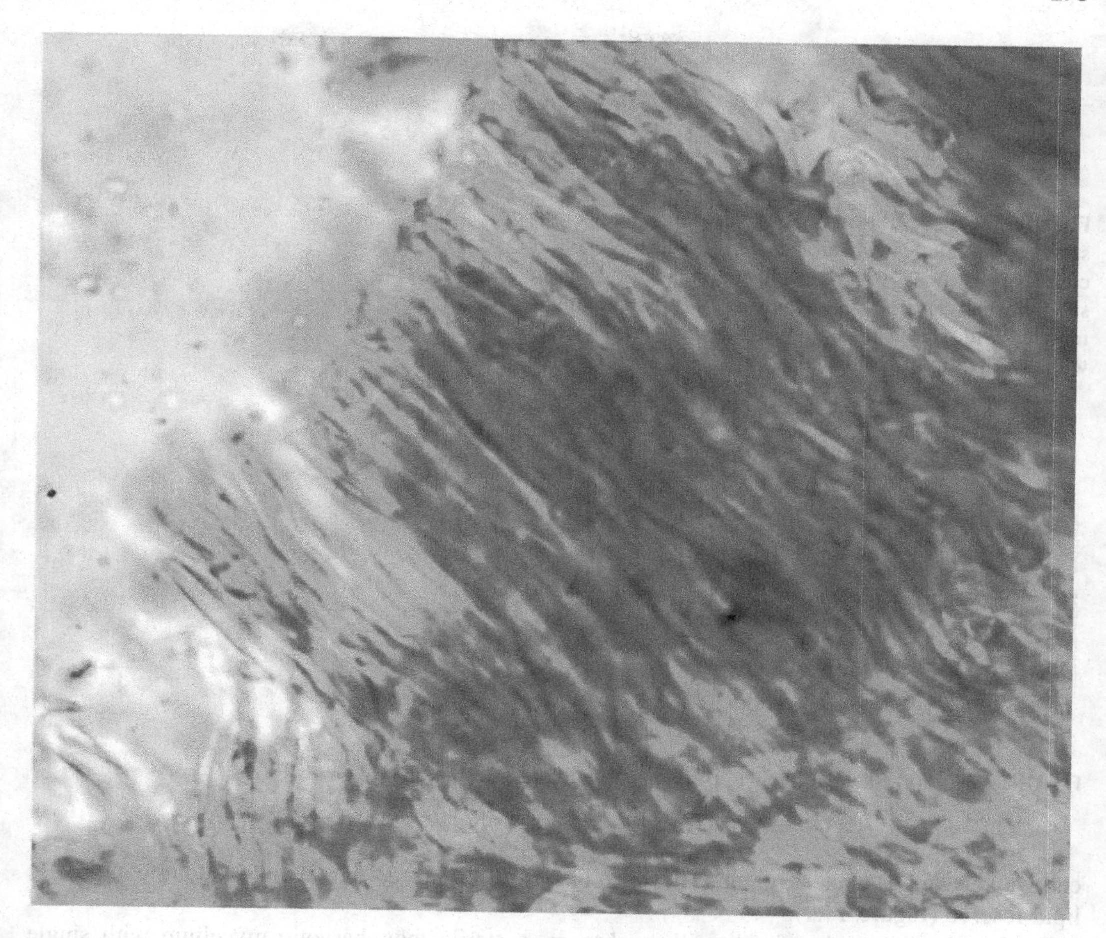

FIGURE 14.5 The presence of tramal tissues with interwoven longitudinal hyphae of *S. commune.*

This is very similar to hymenophoreal trama, which has huge oval to globose cells known as "sphaerocysts" dispersed among hyphae. In the hymenophoral trama region, the sphaerocysts are tightly packed, disposing the pseudo-parenchymatous tissue which have air spaces or lacunae in their pilear region. Although, the hymenial layer which is formed from cystidia is parallel and compactly arranged. The thin sterile paraphyses arises from it which appears as extended cum elongated sac-like structures with swollen hyphal tip (Figure 14.6). Polypore contains a trimitic hyphal system with three hyphae such as generative, binding, as well as skeletal. The septate, thin-walled generative hyphae are always clamped and segregated into thin basidia (Figure 14.6). The basidia are thin-walled, hyaline, clavate, club-shaped, and have four sterigmata (Figure 14.6).

The skeletal hyphae are broad and bulged in the centre, with few septa, whereas the binding hyphae are relatively thick walled as well as branched (Figure 14.7). The generative hyphae are hyaline, with a septate clamp connection and dikaryon formation in the hyphae (Figure 14.8). The basidiospores are hyaline, smooth as well as cylindrical with a short protuberance present at the proximal end which signifies the previous unification (hilum) with the fruiting body's sterigma (Figure 14.9). The transparency of cells reveals distinction between germinated and ungerminated basidiospores. Ungerminated basidiospores have a transparent cytoplasmic area along with dense and transparent nuclear area. They also have more carbohydrate reserve along with disorganised mitochondria. In contract, germinating basidiospores have much darker colour owing to their dense

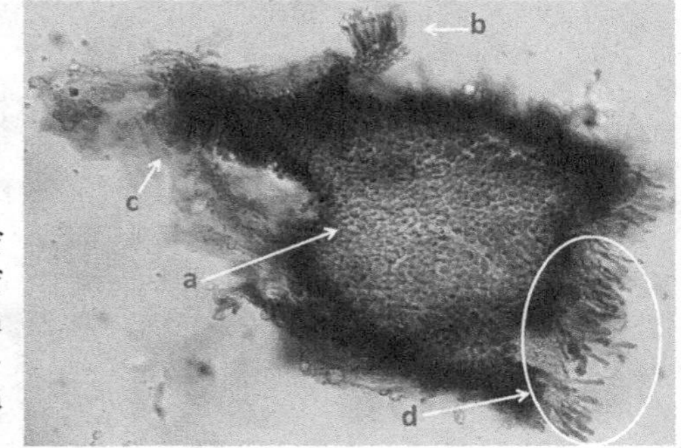

FIGURE 14.6 a. Pilear region with large number of sphaerocysts along with air lacuna. b. The presence of club-shaped young basidia with sterigma arising from sub-hymenial layer of *S. commune*. c. Cystidia intermingled with sterile hyphae of *S. commune* d. Cystidia with swollen tip of *S. commune*.

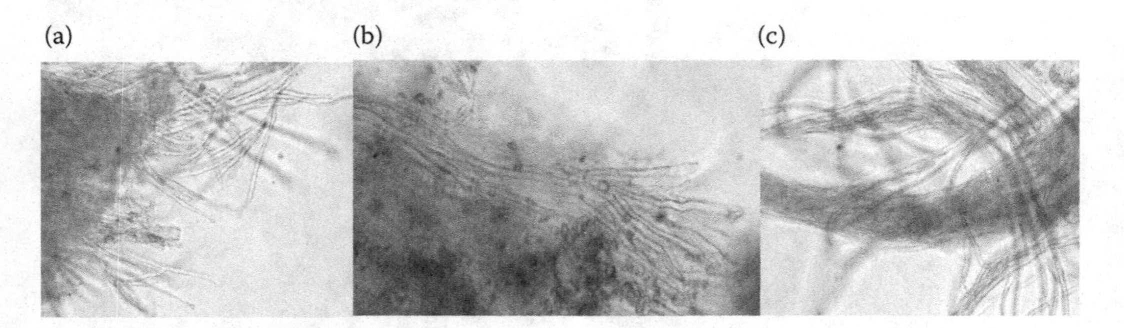

FIGURE 14.7 (a) Skeletal hyphae, (b) generative hyphae, (c) binding hyphae.

cell wall as well as nuclear region. They also have well-organized mitochondria along with excess lipid reserve (Figure 14.9).

During the life cycle of split gill mushroom, a sterile monokaryotic mycelium with single nucleus in each compartment is formed from the germinating meiospores. Initially, the mycelium grows beneath the substrate, leading to production of aerial hyphae upon few days of germination. However, a fertile dikaryon is produced from the fusion of encountering monokaryons of different partners of alleles of the mating-type loci known as matA and matB. The fruiting is favoured by brief exposure to light and inhibited by elevated concentration of carbon dioxide (CO_2) and high temperatures (30–37 °C). The mature fruiting bodies are formed from fruiting-body primordia which is comprised of aggregated aerial dikaryotic hyphae leading to development of mushrooms. The mature fruiting body's basidia become the site of karyogamy and meiosis, resulting in the formation of basidiospores, which produce new monokaryotic mycelia.

GENOME SEQUENCING

The basidiomycete *S. commune* can be easily grown on defined media, and an array of molecular tools have been used to examine its growth and development. *S. commune* is the only mushroom-forming fungus where homologous recombination has been reported to inactivate genes. It is also considered as a model system for studies owing to the ability of its recombinant DNA constructs which can easily express themselves in other mushroom-forming fungi. Besides, it is also studied for gene expression, differentiation, and tetrapolar sexuality in higher basidiomycetes (Ullrich et al., 1991). The process of dikaryotisation in *S. commune* is a complex process that necessitates four multiallelic master regulatory loci (A, A, B, and B) to regulate sexual development in this mushroom (Ullrich et al., 1991; Stankis et al., 1992).

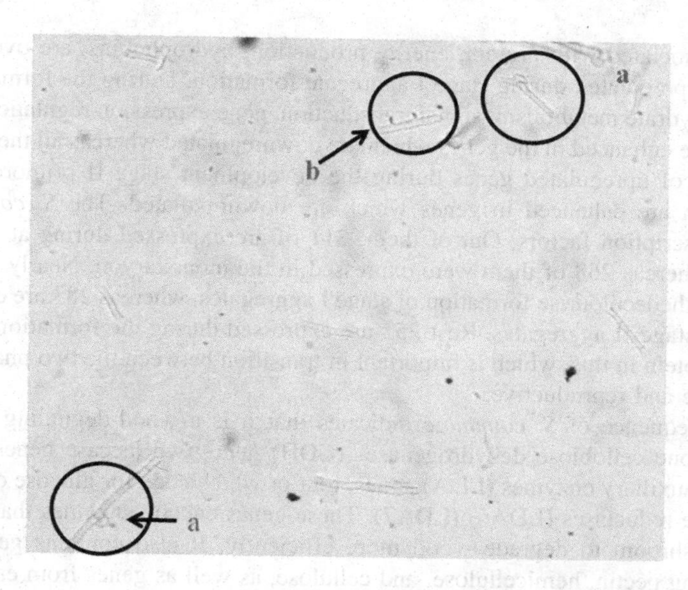

FIGURE 14.8 a. Formation of clamp connection in *S. commune*, b. Formation of dikaryon in the hyphae of *S. commune*.

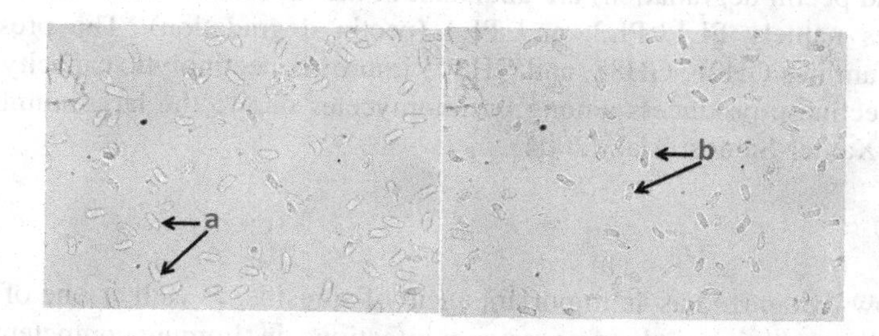

FIGURE 14.9 Cylindrical, hyaline basidiospores of *S. commune*. a. ungerminated basidiospores, b. germinating basidiospores.

S. commune's genome is 38.5 megabases long and contains 11.2% repeats. The assembly is made up of 36 scaffolds that represent 14 chromosomes (Asgeirsdottir et al., 1994). The genome contains 13,210 predicted genes, 42% of which are assisted by expressed sequence tags (ESTs) and almost 69% of which are similar to proteins from other organisms. *S. commune* protein clustering with other fungi revealed 7055 groups with at least one *S. commune* protein, with approximately 39% of these having orthologs in dikarya and thus conserved in Ascomycetes and basidiomycetes (Ohm et al., 2010).

The expression of the entire genome can be studied at its four developmental stages namely monokaryons, followed by stage I aggregates, followed by stage II primordia, and finally mature fruiting bodies. Approximately, 59.8% of the 13,210 predicted genes were present in at least one developmental stage. The differentially expressed genes which are expressed during the four developmental stages and are overrepresented in the pool of genes appear to be unique to *S. commune*. Anti-sense transcription is a common occurrence in *S. commune*. Among the tags associated with gene models, 18.7% are derived from an antisense transcript, along with 42.3% of predicted genes exhibit antisense expression at one or more developmental stages. Among the four developmental stages, the second has the most uniquely expressed genes in the anti-sense direction.

Gene terms associated with protein, energy production, hydrophobins, are overrepresented in genes which are upregulated during stage I aggregate formation. During the formation of stage II primordia, carbohydrate metabolism, signal transduction, gene expression regulation as well as cell wall biogenesis are enhanced in the genes which are downregulated whereas all these processes are enhanced in case of upregulated genes during the development stage II primordia. Protein and energy production are enhanced in genes which are downregulated. The *S. commune* genome encodes 471 transcription factors. Out of them, 311 of are expressed during at least one developmental stage whereas 268 of them were expressed in the monokaryon. Nearly 200 of them are expressed during thedecolourise formation of stage I aggregates, whereas 283 are expressed during the formation of stage II aggregates. Rest 253 are expressed during the formation of mushrooms. There is a fst4 protein in this, which is important in transition between the two phases of life cycle namely vegetative and reproductive.

The genome sequence of *S. commune* indicates that it is a wood-degrading fungus with 16 FOLyme genes, one cellobiose dehydrogenases (CDH) gene, two laccase genes (LO1), and 13 lignin degrading auxiliary enzymes (LDA) genes, four of which code for glucose oxidases (LDA6) and benzoquinone reductases (LDA6) (LDA7). These genes encode enzymes that degrade lignin, allowing the mushroom to degrade wood more efficiently. It also contains genes that encode enzymes degrading pectin, hemicellulose, and cellulose, as well as genes from each family. They also play key role in breakdown of polysaccharides which are present in plant cell wall. The members of the glycosyl hydrolase families GH93 (hemicellulose degradation) and GH43 (hemicellulose and pectin degradation) are abundant in the mushroom genome, as are members of the lyase families namely PL1, PL3, and PL4 (pectin degradation). The presence of pectin hydrolases from families GH28, GH88, and GH105 improves pectinolytic capacity. *S. commune* is one of the best pectinase producers among basidiomycetes due to the large number of predicted pectinase genes (Xavier-Santos et al., 2004).

TOXICOLOGY

S. commune is now recognized as an important agent of sinusitis, as well as one of the filamentous basidiomycetes that colonize and cause severe infections in immunocompetent humans. Few studies suspect *S. commune* is an emerging pathogen, owing to an increase in reports of illnesses such as chronic lung disease, sinusitis with chronic and acute inflammation, allergic broncho-pulmonary mycosis, bronchial asthma, atypical meningitis, and onychomycosis of toenails (Kamei et al., 1994; Clark et al., 1996; Rihs et al., 1996; Shaw et al., 2000; Kawano et al., 2003). Intraconazol, an antifungal agent, can be used to treat sinusitis (Sigler et al., 1997).

ECONOMIC IMPORTANCE

S. commune and its polysaccharide schizophyllan have been extensively studied for their numerous applications. Besides, it has a high potential as a petroleum recovery agent, emulsifier to improve oil recovery, cosmetic lotion thickener, oxygen impermeable films used for food preservation, and application in high-value pharmaceuticals (Sutivisedsak et al., 2013; Zhang et al., 2013). It is a biopolymer of great industrial application owing to its properties like high viscosity, film formation, and thermal stability. Because of its triple helical structure and intramolecular hydrogen bonds amongst its helical pattern, SPG is highly stable at high temperatures and is thus used for enhanced oil recovery at high temperatures.

S. commune also produces enzymes such as xylanases, lignin peroxidase, lipases, cellulases, laccases, manganese peroxide and pectinases (Paice et al., 1978; Sornlake et al., 2017; Gautam et al., 2018; Zhu et al., 2016; Kam et al., 2016; Kumar et al., 2015; Asgher et al., 2016). These enzymes have enormous utilization in industrial applications, including chemicals, brewing and wine, food, pulp and paper, textile and laundry, feed, and biofuel (Kumar et al., 2019; Yadav et al.,

2018; Bharti et al., 2018). When applied to fruit pulp (apple), pectin lyase, an enzyme produced by *S. commune*, increased juice yield by up to 40%. The enzyme petin lyase is also used in detergents to increase their capacity (Mehmood et al., 2018). *S. commune* has tremendous potential as a dye-degrading agent and in wastewater treatment. The mushroom (carbon and nitrogen sources) along with maltose and ammonium sulphate can completely decolourise Solar Brilliant red 80 (Asgher et al., 2013). It is also capable of decolonizing xenobiotic dyes such as malachite green, crystal violet, rose Bengal, orange G, and Remazol brilliant blue R (van Brenk and Wösten, 2021).

CONCLUSION AND FUTURE PERSPECTIVES

Schizophyllum commune is a fungus with a wide range of applications for humans due to its fruiting bodies, bioactive compounds, and complete set of enzymes, making it important in many areas of life. The mushroom genome is an effective tool for deciphering the mechanism of natural substrate degradation and fruiting body formation. It can degrade all lignocellulosic biomass components. The various enzymes found in *S. commune* which are involved in plant cell wall degradation and their modification make it as an excellent candidate for studying and exploiting the mechanism of biomass degradation. It also presents opportunities to enhance the efficiency of traditional industrial processes such as bioconversion of agricultural byproducts, lignocellulosic ethanol production, and biodegradation of xenobiotics and pollutants. It is a fungal cell factory which can synthesize a diverse range of valuable metabolites from low-cost ligno-cellulosic substrate.

REFERENCES

Arika T, Amemiya K, Nomoto K. Combination therapy of radiation and Sizofiran (SPG) on the tumor growth and metastasis on squamous-cell carcinoma NR-S1 in syngeneic C3H/He mice. *Biotherapy*. 1992. 4(2):165–170.

Asgeirsdottir SA, Schuren FH, Wessels JG. Assignment of genes to pulse-field separated chromosomes of *Schizophyllum commune*. *Mycological Research*. 1994. 98(6):689–693.

Asgher M, Yasmeen Q, Iqbal HM. Enhanced decolorization of Solar brilliant red 80 textile dye by an indigenous white rot fungus *Schizophyllum commune* IBL-06. *Saudi Journal of Biological Sciences*. 2013 Oct 1;20(4):347–352.

Asgher M, Wahab A, Bilal M, Iqbal HM. Lignocellulose degradation and production of lignin modifying enzymes by *Schizophyllum commune* IBL-06 in solid-state fermentation. *Biocatalysis and Agricultural Biotechnology*. 2016. 6:195–201.

Bharti AK, Kumar A, Kumar A, Dutt D. Exploitation of *Parthenium hysterophorous* biomass as low-cost substrate for cellulase and xylanase production under solid-state fermentation using *Talaromyces stipitatus* MTCC 12687. *Journal of Radiation Research and Applied Sciences*. 2018. 11(4):271–280.

Brant D, McIntire T, Gascoigne L. Understanding why stiff triple-stranded polysaccharides form cyclic structures. Book of Abstracts, 213th ACS National Meeting, San Francisco, April 1997. Comp-072.

Chang ST, Miles PG. *Mushrooms: cultivation, nutritional value, medicinal effect, and environmental impact*. CRC Press; 2004 March. 451 pp.

Chen JT, Teshima H, Shimizu Y, Hasumi K, Masubuchi K. Effect of sizofiran or recombinant interferon gamma on the activation of human peritoneal macrophage function; an approach for the prophylaxis of intraperitoneal recurrence of ovarian cancer. *Nihon Sanka Fujinka Gakkai Zasshi*. 1990. 42(2):179–184.

Chihara, G. In P Rohlich, & E Bacsy (Eds.), *Tissue culture and RES*. 1984. Budapest: Akademia Kiado.

Clark S, Campbell CK, Sandison A, Choa DI. *Schizophyllum commune*: an unusual isolate from a patient with allergic fungal sinusitis. *Journal of Infection*. 1996. 32(2):147–150.

Cooke D, Flegg PB. The relation between yield of the cultivated mushroom and stage of maturity at picking. *Journal of Horticultural Science*. 1962. 37(3):167–174.

Daba AS, Ezeronye OU. Anti-cancer effect of polysaccharides isolated from higher basidiomycetes mushrooms. *African Journal of Biotechnology*. 2003. 2(12):672–678.

de Jong JF. Aerial hyphae of *Schizophyllum commune*: their function and formation. PhD thesis, Univ. Utrecht. 2006.

Deng C, Yang X, Gu X, Wang Y, Zhou J, Xu H. A β-D glucan from the sclerotia of *Pleurotus tuber-regium* (Fr.) Sing. *Carbohydrate Research*. 2000. 328, 629–633.

Gautam A, Kumar A, Bharti AK, Dutt D. Rice straw fermentation by *Schizophyllum commune* ARC-11 to produce high level of xylanase for its application in pre-bleaching. *Journal of Genetic Engineering and Biotechnology*. 2018. 16(2):693–701.

Hobbs C. The chemistry, nutritional value, immunopharmacology, and safety of the traditional food of medicinal split-gill fugus *Schizophyllum commune* Fr.: Fr. (Schizophyllaceae). A literature review. *International Journal of Medicinal Mushrooms*. 2005. 7(1&2).

Hosoe T, Nozawa K, Kawahara N, Fukushima K, Nishimura K, Miyaji M, Kawai KI. Isolation of a new potent cytotoxic pigment along with indigotin from the pathogenic basidiomycetous fungus *Schizophyllum commune*. *Mycopathologia*. 1999. 146(1):9–12.

Imtiaj A, Jayasinghe C, Lee GW, Kim HY, Shim MJ, Rho HS, Lee HS, Hur H, Lee MW, Lee UY, Lee TS. Physicochemical requirement for the vegetative growth of *Schizophyllum commune* collected from different ecological origins. *Mycobiology*. 2008. 36(1):34–39.

Jacques, P. In B Serrou (Ed.), *Current concepts in human immunology and cancer immunomodulation*. 1982. Amsterdam: Elsevier Biomedical Press B.V.

James TY, Moncalvo JM, Li S, Vilgalys R. Polymorphism at the ribosomal DNA spacers and its relation to breeding structure of the widespread mushroom *Schizophyllum commune*. *Genetics*. 2001. 157(1):149–161.

Kakumu S, Ishikawa T, Wakita T, Yoshioka K, Ito Y, Shinagawa T. Effect of sizofiran, a polysaccharide, on interferon gamma, antibody production and lymphocyte proliferation specific for hepatitis B virus antigen in patients with chronic hepatitis B. *International Journal of Immunopharmacology*. 1991. 13(7):969–975.

Kam YC, Hii SL, Sim CY, Ong LG. *Schizophyllum commune* lipase production on pretreated sugarcane bagasse and its effectiveness. *International Journal of Polymer Science*. 2016. 2016

Kamei K, Unno H, Nagao K, Kuriyama T, Nishimura K, Miyaji M. Allergic bronchopulmonary mycosis caused by the basidiomycetous fungus *Schizophyllum commune*. *Clinical Infectious Diseases*. 1994. 18(3):305–309.

Kano Y, Kakuta H, Hashimoto J. Augmentation of antitumor effect by combined administration with interleukin-2 and sizofiran, a single glucan, on murine EL-4 lymphoma. *Biotherapy*. 1996. 9(4):241–247.

Kawaguchi MA, Shiomi RY, Tabata KE, Kosuge YO, Mizumachi KO, Kurisaki JI. Enhancement of Th1 immune response in mice by the ingestion of Suehirotake (*Schizophyllum commune* Fries) mycelium. *Journal of the Japanese Society for Food Science and Technology (Japan)*. 2004.

Kawano T, Matsuse H, Iida K, Kondo Y, Machida I, Saeki S, Tomari S, Miyazaki Y, Kohno S. Two cases of allergic bronchopulmonary mycosis caused by *Schizophyllum commune* in young asthmatic patients. *Nihon Kokyuki Gakkai Zasshi the Journal of the Japanese Respiratory Society*. 2003. 41(3):233–236.

Khatua S, Paul S, Acharya K. Mushroom as the potential source of new generation of antioxidant: a review. *Research Journal of Pharmacy and Technology*. 2013.6(5):496–505.

Kikumoto S. Polysaccharide produced by *Schizophyllum commune*, I. Formation and some properties of an extracellular polysaccharide. *Nippon Nogeikagaku Kaishi*. 1970. 44:337–342.

Komatsu N, Nagumo N, Okubo S, Koike K. Protective effect of the mushroom polysaccharide schizophyllan against experimental bacterial infections. *The Japanese Journal of Antibiotics*. 1973. 26(3):277–283.

Komatsu N, Okubo S, Kikumoto S, Kimura K, Saito G, Sakai S. Host-mediated antitumor action of schizophyllan, a glucan produced by *Schizophyllum commune*. *GANN Japanese Journal of Cancer Research*. 1969. 60(2):137–144.

Kothe E. Mating-type genes for basidiomycete strain improvement in mushroom farming. *Applied Microbiology and Biotechnology*. 2001. 56(5):602–612.

Krosl G, Korbelik M. Potentiation of photodynamic therapy by immunotherapy: the effect of schizophyllan (SPG). *Cancer Letters*. 1994. 84(1):43–49.

Kubala L, Ruzickova J, Nickova K, Sandula J, Ciz M, Lojek A. The effect of (1→ 3)-β-D-glucans, carboxymethylglucan and schizophyllan on human leukocytes in vitro. *Carbohydrate Research*. 2003. 338(24):2835–2840.

Kukan M, Szatmáry Z, Lutterova M, Kuba D, Vajdova K, Horecky J. Effects of sizofiran on endotoxin-enhanced cold ischemia-reperfusion injury of the rat liver. *Physiological Research*. 2004. 53(4):431–437.

Kumar A, Gautam A, Khan H, Dutt D. Organic acids: microbial sources, production and application. In *Microbial enzyme and additives for the food industry*. 2019 (pp. 261–276). Nova Science Publishers Inc.

Kumar VP, Naik C, Sridhar M. Morphological and phylogenetic identification of a hyper laccase producing strain of *Schizophyllum commune* NI-07 exhibiting delignification potential. *Indian Journal of Biotechnology*. 2018. 17: 302–315.
Kumar VP, Naik C, Sridhar M. Production, purification and characterization of novel laccase produced by *Schizophyllum commune* NI-07 with potential for delignification of crop residues. *Applied Biochemistry and Microbiology*. 2015. 51(4):432–441.
Leathers TD, Nunnally MS, Price NP. Co-production of schizophyllan and arabinoxylan from corn fiber. *Biotechnology Letters*. 2006. 28(9):623–626.
Lee D, Kim YR, Kim JS, Kim D, Kim S, Kim SY, Jang K, Lee JD, Yang CS. Dietary schizophyllan reduces mitochondrial damage by activating SIRT3 in mice. *Archives of Pharmacal Research*. 2020. 43(4):449–461.
Lumbsch HT, Buchanan PK, May TW, Mueller GM. Phylogeography and biogeography of fungi. *Mycological Research*. 2008. 112(4):423–424.
Martin GG, Cannon GC, McCormick CL. Adsorption of a fungal hydrophobin onto surfaces as mediated by the associated polysaccharide schizophyllan. Biopolymers: Original Research on *Biomolecules*. 1999. 49(7):621–633.
Mehmood T, Saman T, Irfan M, Anwar F, Salman M. Pectinase production from *Schizophyllum commune* through central composite design using citrus waste and its immobilization for industrial exploitation. *Waste and Biomass Valorization*. 2018. 10:2527–2536.
Miura T, Miura NN, Ohno N, Adachi Y, Shimada S, Yadomae T. Failure in antitumor activity by overdose of an immunomodulating β-glucan preparation, sonifilan. *Biological and Pharmaceutical Bulletin*. 2000. 23(2):249–253.
Mizuno T. The extraction and development of antitumor-active polysaccharides from medicinal mushrooms in Japan. *International Journal of Medicinal Mushrooms*. 1999.1(1).
Mueller A, Raptis J, Rice PJ, Kalbfleisch JH, Stout RD, Ensley HE, Browder W, Williams DL. The influence of glucan polymer structure and solution conformation on binding to (1→3)- β-d-glucan receptors in a human monocyte-like cell line. *Glycobiology*. 2000. 10:339–346.
Nemoto J, Ohno N, Saito K, Adachi Y, Yadomae T. Expression of interleukin 1 family mRNAs by a highly branched (1→ 3)-β-D-glucan, OL-2. *Biological and Pharmaceutical Bulletin*. 1993. 16(10):1046–1050.
Ohm RA, De Jong JF, Lugones LG, Aerts A, Kothe E, Stajich JE, De Vries RP, Record E, Levasseur A, Baker SE, Bartholomew KA. Genome sequence of the model mushroom *Schizophyllum commune*. *Nature Biotechnology*. 2010. 28(9):957–963.
Ohno N, Miura NN, Chiba N, Adachi Y, Yadomae T. Comparison of the immunopharmacological activities of triple and single-helical schizophyllan in mice. *Biological and Pharmaceutical Bulletin*. 1995.18(9):1242–1247.
Oka T. Antitumor effects and augmentation of cellular immunity by schizophyllan and bestatin. *Journal of Okayama Medical Association*. 1985. 97(5-6):527–541.
Paice MG, Jurasek L, Carpenter MR, Smillie LB. Production, characterization, and partial amino acid sequence of xylanase A from *Schizophyllum commune*. *Applied and Environmental Microbiology*. 1978. 36(6):802–808.
Patel S, Goyal A. Recent developments in mushrooms as anti-cancer therapeutics: A Review. *3 Biotech*. 2012. 2(1):1–5.
Raper JR, Miles PG. The genetics of *Schizophyllum commune*. *Genetics*. 1958. 43(3):530.
Rau, U. Schizophyllan. In E J Vandamme, S D Baets, & A Steinbuchel (Eds.), *Biopolymers. Vol. 6. Polysaccharides II: Polysaccharide from Eukaryotes, Weinheim* (pp. 61–91). 2002. Weinheim: Wiley-VCH.
Ren J, Zhao M, Shi J, Wang J, Jiang Y, Cui C, Kakuda Y, Xue SJ. Optimization of antioxidant peptide production from grass carp sarcoplasmic protein using response surface methodology. *LWT-Food Science and Technology*. 2008. 41(9):1624–1632.
Rihs JD, Padhye AA, Good CB. Brain abscess caused by *Schizophyllum commune*: an emerging basidiomycete pathogen. *Journal of Clinical Microbiology*. 1996. 34(7):1628–1632.
Safaee-Ardakani MR, Hatamian-Zarmi A, Sadat SM, Mokhtari-Hosseini ZB, Ebrahimi-Hosseinzadeh B, Rashidiani J, Kooshki H. Electrospun schizophyllan/polyvinyl alcohol blend nanofibrous scaffold as potential wound healing. *International Journal of Biological Macromolecules*. 2019. 127:27–38.
Sakurai K, Shinkai S. Molecular recognition of adenine, cytosine, and uracil in a single-stranded RNA by a natural polysaccharide: schizophyllan. *Journal of American Chemical Society*. 2000. 122: 4520–4521.
Schmidt, O, Liese, W. Variability of wood degrading enzymes of *Schizophyllum commune*. *Holzforschung*. 1980. 34, 67–72.

Shaw CK, McCleave M, Wormald PJ. Unusual presentations of isolated sphenoid fungal sinusitis. *The Journal of Laryngology & Otology*. 2000. 114(5):385–388.

Sigler L, Estrada S, Montealegre NA, Jaramillo E, Arango M, De Bedout C, Restrepo A. Maxillary sinusitis caused by *Schizophyllum commune* and experience with treatment. *Journal of Medical and Veterinary Mycology*. 1997. 35(5):365–370.

Singh S, Raj C, Singh HK, Avasthe RK, Said P, Balusamy A, Sharma SK, Lepcha SC, Kerketta V. Characterization and development of cultivation technology of wild split gill *Schizophyllum commune* mushroom in India. *Scientia Horticulturae*. 2021. 289:110399.

Slovis N. Polysaccharide treatment reduces gastric ulceration in active horses. *Journal of Equine Veterinary Science*. 2017. 50:116–120.

Sornlake W, Rattanaphanjak P, Champreda V, Eurwilaichitr L, Kittisenachai S, Roytrakul S, Fujii T, Inoue H. Characterization of cellulolytic enzyme system of *Schizophyllum commune* mutant and evaluation of its efficiency on biomass hydrolysis. *Bioscience, Biotechnology, and Biochemistry*. 2017. 81(7):1289–1299.

Stankis MM, Specht CA, Yang H, Giasson L, Ullrich RC, Novotny CP. The A alpha mating locus of *Schizophyllum commune* encodes two dissimilar multiallelic homeodomain proteins. *Proceedings of the National Academy of Sciences*. 1992. 89(15):7169–7173.

Sugiyama T, Nishida T, Kumagai S, Imaishi K, Ushijima K, Kataoka A, Yakushiji M. Combination treatment with Cisplatin and schizophyllan for 7, 12-dimethylbenz (a) anthracene-induced rat ovarian adenocarcinoma. *Journal of Obstetrics and Gynaecology*. 1995. 21(5):521–527.

Sutivisedsak N, Leathers TD, Price NP. Production of schizophyllan from distiller's dried grains with solubles by diverse strains of *Schizophyllum commune*. *SpringerPlus*. 2013. 2(1):1–6.

Suzuki T, Tsuzuki A, Ohno N, Ohshima Y, Adachi Y, Yadomae T. Synergistic action of β-glucan and platelets on interleukin-8 production by human peripheral blood leukocytes. *Biological and Pharmaceutical Bulletin*. 2002. 25(1):140–144.

Tabata K, Ito W, Kojima T, Kawabata S, Misaki A. Ultrasonic degradation of schizophyllan, an antitumor polysaccharide produced by *Schizophyllum commune* Fries. *Carbohydrate Research*. 1981. 89(1):121–135.

Tabata K, Yanaki T, Ito W, Kojima T. Conformation of schizophyllan and its antitumor activity. *Seito Gijutsu Kenkyu Kaishi*. 1988. 36: 69–75.

Tateishi T, Ohno N, Adachi Y, Yadomae T. Increases in hematopoietic responses caused by β-glucans in mice. *Bioscience, Biotechnology, and Biochemistry*. 1997. 61(9):1548–1553.

Tsuzuki A, Tateishi T, Ohno N, Adachi Y, Yadomae T. Increase of hematopoietic responses by triple or single helical conformer of an antitumor (1→ 3)-β-d-glucan preparation, Sonifilan, in cyclophosphamide-induced leukopenic mice. *Bioscience, Biotechnology and Biochemistry*. 1999. 63(1):104–110.

Ullrich RC, Specht CA, Stankis MM, Yang H, Giasson L, Novotny CP. Molecular biology of mating-type determination in *Schizophyllum commune*. *Genetic engineering*. 1991:279–306.

van Brenk B, Wösten HA. A screening method for decoloration of xenobiotic dyes by fungi. *Journal of Microbiological Methods*. 2021. 188:106301.

Wasser SP, Weis AL. Medicinal properties of substances occurring in higher basidiomycetes mushrooms: current perspectives. *International Journal of Medicinal Mushrooms*. 1999. 1(1).

Wongaem A, Reamtong O, Srimongkol P, Sangtanoo P, Saisavoey T, Karnchanatat A. Antioxidant properties of peptides obtained from the split gill mushroom (*Schizophyllum commune*). *Journal of Food Science and Technology*. 2021. 58(2):680–691.

Xavier-Santos S, Carvalho CC, Bonfá M, Silva R, Capelari M, Gomes E. Screening for pectinolytic activity of wood-rotting basidiomycetes and characterization of the enzymes. *Folia Microbiologica*. 2004. 49(1):46–52.

Yadav M, Sehrawat N, Kumar A. Microbial laccases in food processing industry: current status and future perspectives. *Research Journal of Biotechnology*. 2018. 13:108–113.

Yamamoto T, Yamashita T, Tsubura E. Inhibition of pulmonary metastasis of Lewis lung carcinoma by a glucan, schizophyllan. *Invasion & Metastasis*. 1981. 1(1):71–84.

Yanaki T, Norisuye T, Fujita H. Triple helix of *Schizophyllum commune* polysaccharide in dilute solution. 3. Hydrodynamic properties in water. *Macromolecules*. 1980. 13(6):1462–1466.

Yang ZB, Tsuchiya Y, Arika T, Hosokawa M. Inhibitory effects of sizofiran on anticancer agent-or X-ray-induced sister chromatid exchanges and mitotic block in murine bone marrow cells. *Japanese Journal of Cancer Research*. 1993. 84(5):538–543.

Zhang Y, Kong H, Fang Y, Nishinari K, Phillips Go. Schizophyllan: a review on its structure, properties, bioactivities and recent developments. *Bioactive Carbohydrates and Dietary Fibre*. 2013. 1(1):53–71.

Zhong K, Tong L, Liu L, Zhou X, Liu X, Zhang Q, Zhou S. Immunoregulatory and antitumor activity of schizophyllan under ultrasonic treatment. *International Journal of Biological Macromolecules*. 2015. 80:302–308.

Zhu N, Liu J, Yang J, Lin Y, Yang Y, Ji L, Li M, Yuan H. Comparative analysis of the secretomes of *Schizophyllum commune* and other wood-decay basidiomycetes during solid-state fermentation reveals its unique lignocellulose-degrading enzyme system. *Biotechnology for Biofuels*. 2016. 9(1):1–22.

15 *Trametes versicolor*

Thiribhuvanamala Gurudevan
Department of Plant Pathology, Centre for Plant Protection Studies, Tamil Nadu Agricultural University, Coimbatore, India

Parthasarathy Seethapathy
Department of Plant Pathology, Amrita School of Agricultural Sciences, Amrita Vishwa Vidyapeetham, Coimbatore, India

CONTENTS

INTRODUCTION

Recent efforts to develop self-sustaining technological solutions and materials reuse, as defined by the paradigm of a circular bio-economy, have emphasized the crucial roles that microorganisms and enzymes play in facilitating efficient and environmentally friendly lignocellulosic biomass processing. In recent years, several initiatives have been undertaken. Numerous studies have been conducted on wood-rotting polypore fungi due to their capacity to degrade cellulose, hemicellulose, and lignin, which are all components of plant cell walls. Consequently, they play a vital role in the majority of biological systems' nutrient recycling processes. Moreover, certain polypores are commercially valuable forest pathogens and medicinal mushrooms (Bains et al. 2021) and considerably contribute to biotechnological processes (Téllez-Téllez et al. 2016). These results have been reported in Biotechnology and Bioengineering. These medicinal mushrooms are edible representatives of the kingdom of fungi and have long been associated with health-promoting properties. They have been used for many years to treat a range of diseases, particularly in the traditional medicine of Asia and the folklore of Eastern Europe. Turkey tail mushrooms, which are members of the *Trametes* genus, are considered medicinal mushrooms. The principal medicinal components of turkey tail mushrooms are their mycelium, fruiting bodies, and brewing substrate.

DOI: 10.1201/9781003259763-15

In terms of safety, turkey tail mushroom-based products have an impeccable track record. *T. versicolor* includes a vast array of complex polysaccharides, many of which enhance the defences and immunity of the host. Some of these polysaccharides, such as glucans, proteoglycans, and heteroglucans that comprise the chitin-based cell wall of the fungus, have been the topic of substantial research. They are renowned for prolonging human life, treating infectious diseases and cancer, and enhancing overall health and wellness. Recent research has focused chiefly on its extensive immunological effects.

On decaying wood, polypores such as turkey tail fungus play a crucial part in the biological cycle of nutrients. They can convert lignified cells into coarse woody substances by both enzymatic and non-enzymatic mechanisms (Cui and Chisti 2003). They harm root systems and standing tree stumps by continuously weakening the host's defences. This permits wood-rotting organisms to enter via the incisions, resulting in a substantial loss of the wood's structural integrity. Although they are primarily employed in the pulp and paper industries, they are detrimental to root systems and standing tree stumps. White rot fungi can convert and mineralize a wide range of xenobiotic compounds, including synthetic colours, chlorinated solutes, polycyclic aromatic hydrocarbons (PAHs), brominated flame retardants, and medical drugs, as well as ultraviolet filters. They utilize a non-specific enzymatic system consisting of extracellular lignin-modifying enzymes (primarily laccase, peroxidase, manganese peroxidase, and versatile peroxidase) and intracellular lignin-modifying enzymes such as cytochrome P450 (Mir-Tutusaus et al. 2014). Although it can be found throughout the year, the turkey tail fungus is most evident during winter, when hardwoods are dormant. This fungus is quite varied and prefers to grow on dead hardwoods, including tree trunks, standing dead trees, and falling branches. Several of the examples feature dazzlingly vivid colouring. Numerous scientific studies have reached the same result on *T. versicolor* ability to destroy wood: the fungus is capable of simultaneously digesting lignin, cellulose, and hemicelluloses (Chen et al. 2017). In addition, this fungus has a long history of use in the Asian therapeutic area (Córdoba and Róos 2012). This makes it more attractive as a bioagent for the treatment of industrial pollution and as a therapeutic technique (Rameshaiah and Jagadish Reddy 2015).

HISTORY AND ETHNOPHARMACOLOGY

In recent years, the medicinal mushroom *Trametes versicolor* (L.) (Syn. *Polyporus versicolor*), formerly known as *Coriolus versicolor* (L.) Quél., has gained the most significant interest. The following is the Latin etymology of *T. versicolor*: The prefix tram- relates to the woof (or weft in weaving), or trametes, which means "thin" Although the mushroom is sometimes referred to as *C. versicolor* or, less frequently, these days, *P. versicolor*, taxonomists believe that *Trametes*, not *Coriolus* or *Polyporus*, is the correct name. In China, the fungus is known as yunzhi, which translates as "cloud mushroom." In Japan, it is known as "Kawaratake," which means "next to the river mushroom." It is known as "rainbow bracket" or "rainbow fungus" in Australia (Hlerema et al. 2017). The genus is abundant throughout the world's biota and has been employed in traditional Chinese medicine (Atilano-Camino et al. 2020, Córdoba and Ros 2012). Even though *T. versicolor* is the most researched species in the genus, more species have been included in recent studies. *T. versicolor* exploits low-molecular-mass compounds, reactive oxygen species, and enzymes like oxidoreductases and hydrolases to attack plant cell walls (Aguiar et al. 2014). Its medical efficacy as a component of traditional Chinese medicine dates back at least 2,000 years and comprises a variety of health-promoting characteristics, such as stamina and longevity. In addition, the mushroom possesses various advantageous properties, including the potential to cure, enhance organ function, and increase vitality (Habtemariam 2020). In addition to proteins and amino acids, additional proteoglycans such as polysaccharide peptide (PSP) and polysaccharide-K (Krestin, PSK). These include anti-cholinesterases, anti-oxidant, anti-bacterial, anti-cancer, anti-diabetic, anti-inflammatory, anti-viral, immune-enhancing activities, increased superoxide dismutase (SOD), activities of lymphocytes and the thymus, induction of analgesia, induced obesity, protective effects of alcoholic liver injury, and prebiotic properties (Córdoba and Ros 2012, Knežević et al. 2018).

DISTRIBUTION AND ECOLOGY

T. versicolor, a white-rot fungus, grows in clusters or rosettes and can colonize and proliferate by degrading the cellulose, hemicellulose, and lignin in wood lots tree trunks, and branches. The fungus is found on all continents and infects numerous plant species, including nearly all genera of hardwood trees and a significant proportion of conifers. Geographically, the fungus has spread to all temperate regions of Asia, Latin America, and Europe, including the United Kingdom. More than 120 *T. versicolor* isolates are included in the Compendium of Chinese Medicine. (Rau et al. 2009). *T. versicolor* is abundant on hilly paths in India, especially in the Himalayan trails of Uttarakhand and the Western Ghats and in southern Nepal (Bhatt et al. 2018, Veena and Pandey 2012). The species' growing requirements are particularly appropriate for Indian conditions. *T. versicolor* can thrive in any environment that provides sufficient moisture, heat, and organic substrates. The availability of agricultural wastes, environmental circumstances, and human resources make the circumpolar parts of the world's temperate, subtropical, and tropical regions ideal for raising medicinal mushrooms. In India, the fungus is typically saprophytic on dead or fallen hardwood trees of a wide variety of deciduous species, including Abies, Acacia, Birch, Cherries, Maple, Oaks, Prunus, Pine, Quercus, and Walnut (Alanazi 2018, Bari et al. 2019, Gautam 2013). As with other species, *T. versicolor* is an abundant wood decomposer that causes substantial white rot in colonized wood. It has a dynamic relationship with nature since it contributes to the breakdown of dead plant matter. The fungus possesses both a vegetative and a reproductive stage for spore dispersal. Mycelium is a strand or thread of hyphae with many nuclei. *T. versicolor* is an axylotrophic fungus that produces and secretes extracellular lignin-modifying enzymes that can alter the chemical structure of the xylem. These enzymes include cellobiose dehydrogenase, laccase, manganese peroxidases, and lignin peroxidases (Atilano-Camino et al. 2020, Tišma et al. 2021).

BOTANY

Conventionally, basidiomycetes are classed according to the morphology of their basidiocarps. Identification can be complicated by the absence of basidiocarps, structural flexibility, and cryptic species (Bhatt et al. 2018). Uncommonly known are taxonomic investigations on the relationships and categorization of *Tremetes* species. The colourful bracket fungus *T. versicolor* is sometimes known as Turkey Tail. This species was formerly known as *Boletus versicolor* and was renamed *T. versicolor* (L.) Lloyd 1920 (Cincinnati) Mycological Writings, 6: 1029–1101. Basionym: *Boletus versicolor* L. 1753; the name was given in 1939 by Czech mycologist Albert Pilát (1903–1974). It belongs to the famous order Polyporales and the family Polyporaceae, with over 50 species. Turkey Tail is the most common name in the Western hemisphere, and it is recognizable by its concentric multicoloured with tones of greys and browns, often with bluish, greenish, reddish, or white zones on the outer portion of the cap (conk), and spore-bearing polypores on the lower surface. It is among the most effective white-rot fungus in the division Basidiomycota, class Agaricomycetes. It is a fungus with a complex morphology that assumes many structural forms throughout its existence. Hyphae are trimitic, generative, thin-walled with clamps, and 1.5 to 2.5 μm in diameter; skeletal hyphae are thick-walled, non-septate, and 3.0 to 4.0 μm in diameter; and binding hyphae are also thick-walled, non-septate, and extensively branched, with a diameter of 1.5 to 2.0 μm. The conk is a hemispherical, imbricate, fan-shaped, or kidney-shaped bracket with a tomentose, fuzzy, velvety texture, measuring 10 cm in diameter and between 1.6 and 6.8 cm in breadth. The striped upper side of the conk is coated with delicate, silky hairs (Figure 15.1). Caps are sessile, typically 1 to 3 mm thick, grow singly, sometimes overlapping in a row or rosette, are flat to wavy, and exhibit multicoloured concentric zoning on the upper surface (red, yellow,

FIGURE 15.1 Tremetes versicolor.

green, blue, brown, black, and white). The side with tiny shallow pores has 3 to 6 pores per mm, rounded to angular pores, 35 pores per mm, is hard, fibrous, up to 4 mm thick, is unaffected by bruising, and is a creamy white colour. The pores do not discolour appreciably when injured. Although these annual bracket fungi are present throughout the year, they are most numerous in fall and winter, when they release their spores. The spore print was white, and the spores were cylindrical/elliptical/sausage-shaped, inamyloid, smooth, and hyaline, ranging from 4.69 μm (3.46–6.05 μm) to 2.54 μm (1.9–2.7 μm). In addition, basidiospores and fusoid cystidioles are present (Gautam 2013, Veena and Pandey 2012).

ENVIRONMENTAL SIGNIFICANCE

Instead of the forest, agriculture, or energy grasses, it is preferable to employ lignocellulose waste products to minimize adverse environmental and biological repercussions. *T. versicolor* is utilized in numerous industrial processes to produce substantial quantities of biomass-derived from lignocellulose products such as barley straw, corn stalks, corn straw, oat straws, sawdust, and wheat straw, or as by-products such as brewer's spent grain, sugar beet pulp, wheat bran, rice bran, and oil cakes (Durak 2018, Rameshaiah and Jagadish Reddy 2015, Wang et al. 2012, Xu et al. 2020). These materials can be used as feedstock in lignocellulosic biorefineries, which have gained popularity in recent years as a viable alternative to the unsustainable chemical and fuel yield from fossil fuels by incorporating multiple conversion and separation steps and utilizing all process substituents and by-products. To fully comprehend the economic potential of lignocellulosic biomass, all of its structural elements, including cellulose, hemicellulose, and lignin, must be exploited (Suryadi et al. 2022). It is crucial to select the proper pretreatment technique for lignocellulosic biomass (Adekunle et al. 2017, Rameshaiah et al. 2020). Aside from the fact that T. versicolor pretreatment is environmentally friendly and energy-efficient, one of the most significant advantages for biofuel and chemical production is the absence of side products, especially phenolic compounds that are known inhibitors of subsequent fermentation processes, which is frequently a problem when employing other techniques (Tima et al. 2021). It is crucial to utilize white-rot fungi, such as T. versicolor, since their enzymes may degrade lignocellulosic biomass and convert it into different phenolic chemicals. This is essential for producing bio-based goods and a variety of environmental applications, such as wastewater treatment (Hu et al. 2020, 2021).

BIODEGRADATION PROSPECTS

T. versicolor is a white-rot fungus that concurrently degrades all wood constituents (Dwivedi 2006, Pop et al. 2018, Xu et al. 2020). However, when grown on non-woody substrates, this fungus is selective for de-lignification (Adekunle et al. 2017). Extensive research has been undertaken on the relationship between wood characteristics and chemical compositions by chemically eliminating chemical components from wood cell walls. In addition, it has been found that the chemical makeup of wood has a substantial impact on its mechanical capabilities. A reduction in hemicellulose compromises the integrity of the cell wall polymers and decreases their resistance to mechanical stresses. However, only a few biological natural degradation mechanisms can remove the chemical makeup of wood cell walls (Suryadi et al. 2022). The parasitic and saprotrophic life strategies of *T. versicolor* produce simultaneous and selective deterioration in naturally colonized wood. During wood degradation, they can employ various tactics, most likely dependent on the substrate environment (Bari et al. 2019). *T. versicolor* extensively degrades the lignin component of the substrate, hence lowering its strength. Enzyme activity and lignin degradation are affected by several variables, such as the nutritional makeup of the fungal strain (nitrogen, manganese, and copper), the amount of moisture in the culture medium, and the pH of the medium, and the temperature. *T. versicolor* degrades lignocellulosic materials and has the potential to enhance the conversion of polysaccharides into sugars that can be used to produce ethanol and bio-pulp (Rameshaiah et al. 2020, Rameshaiah and Jagadish Reddy 2015, Singh et al. 2013).

CULTIVATION

In nature, *T. versicolor* lives on many genera of hardwood trees (oak, prunus) and some conifers (fir and pine), with the basidium occurring primarily on stubs and trunks year-round (Janjušević et al. 2018). With the potential for biodegradation solid-state fermentations (SSFs) with this fungus have enhanced the enzymatic hydrolysis products of bamboo culms and residues by reducing their lignin concentration (Yu et al. 2010, Zhang et al. 2007). In a separate investigation, biodegradation by *T. versicolor* followed by an alkaline/oxidative treatment of maize straw increased the production of reducing sugar after enzymatic hydrolysis (Yu et al. 2010). These authors found that *T. versicolor* had the most impressive ability to break down lignin and attained the lowest sugar yield due to a very high cellulose loss caused by biotreatment. *T. hirsuta* and *T. ochracea*, which belong to the same genus, have similarly improved the enzymatic hydrolysis efficiency of lignocellulosic wastes (Mir-Tutusaus et al. 2014). Bio-pulping, which employs enzymes to decompose wood chips and other non-woody materials prior to mechanical pulping to reduce the energy requirements of refining and improve the characteristics of the pulp, is yet another promising bioprocess utilizing *T. versicolor*. Additionally, the white-rot fungus can be utilized before chemical pulping to reduce chemical usage (Aguiar et al. 2014). *T. versicolor* infiltrated the lumen of various wood cell types, entering cell walls via pits, generating erosion troughs and boreholes, and destroying all cell layers, consequently causing the Chinese fir to deteriorate. The white-rot fungus can modify the chemical characteristics of a mass proportion of Chinese fir: hemicellulose > lignin > cellulose. The most resilient chemical compounds were lignin, cellulose, and hemicellulose (Chen et al. 2017).

Growth Conditions

Nutrients and the conditions under which the mushroom is grown are two of the most critical factors determining vegetative development and the creation of basidiocarps. On the other hand, very few studies have been carried out to optimize the culture conditions for the mycelial growth and fructification of *T. versicolor*. It was discovered that dextrin and yeast extract was the most effective carbon and nitrogen sources. Other minor components employed for its growth included vitamins thiamine-HCl and biotin, organic acids succinic acid, lactic acid, citric acid, and mineral salts $MgSO_4.7H_2O$. (Jo et al. 2010). Another study has shown that carbon sources, such as fructose and xylose, and nitrogen sources, such as peptone and yeast extract, all contribute to the growth of *T. versicolor* in a vegetative state that is beneficial. In order for mycelium to grow, it is necessary for there to be a favourable balance between the sources of carbon and nitrogen, specifically in the proportion of 3:1. (Nguyen et al. 2021). *T. versicolor* can grow in an extensive pH range, extending from 4 to 9. (Jo et al. 2010). According to Veena and Pandey (2012), the ideal conditions for fructification are a temperature of 122°C and relative humidity of 80–85%. For the cultivation of *T. versicolor*, the only acceptable basal substrate identified thus far is sawdust. The substrate selection for mushroom cultivation is typically made to make use of the agricultural and industrial waste already present in the region.

Spawn Preparation and Wood-log Cultivation

This study utilized pure culture of *T. versicolor*, potato dextrose agar, and malt extract agar quite frequently (Veena and Pandey 2012). At 24°C, the *T. versicolor* strain was incubated after being inoculated into sterile culture media (20 ml/90 mm diameter Petri plate). The spawn was produced on sorghum grain, and an attempt at fructification was made using a substrate mixture that included sawdust with 90% rice bran and 10% rice bran. The spawn run required 18–20 days at a temperature of 25.2°C, and fructification was better when the temperature was 25.2°C with 80–85% relative humidity. In a separate line of research, sorghum grain spawn was created in glass jars by following conventional procedures. *T. versicolor* was seeded into sterile jars that held 250 grams of sorghum kernels, and the jars were then heated to 24°C for the duration of the mycelial colonization process. In this study, the following substrates were tested (in terms of their dry weight): oak sawdust with (homogeneous mixture: oak sawdust 78%, sorghum kernels 10%, wheat bran 10%, and gypsum 2%). Each substrate was then placed in polypropylene plastic bags measuring 15 by 39 cm and given a moisture level of between 60 and 70%. After that, these bags were heated to 121°C for two hours to disinfect them. Each bag was sterilely injected with 70–80 grams of spawn, and then it was placed in an incubator at 22°C and kept in the dark (Guerrero et al. 2011). To cultivate *T. versicolor*, sawdust and other agro-industrial wastes such as rice husk, cotton waste, and maize cob can be employed as the primary substrates. The substrate mixture consisting of 62% sawdust, 30% rice husk, 3% wheat bran, and 1% $CaCo_3$ produced the highest yield (Nguyen et al. 2021).

Fermentation Process

Submerged fermentation (SmF) and SSF are viable cultivation methods for *T. versicolor*. During its normal development on woody materials, *T. versicolor* can degrade lignocellulose. In addition, agricultural by-products have been used successfully in the cultivation of *T. versicolor* in order to produce mycelial composites and mycelium-derived chitin, both of which can be used to dispose of residues following SSF or wastewater treatment (mycelia with or without contaminants) (Tima et al. 2021). Composites manufactured from mycelial biomass are attracting interest from businesses and academic institutions due to their eco-friendly and sustainable production technique. However, their production is impeded by the slow growth of biomass, which hurts the economics of large-scale production and must compete with the quick production of synthetic materials. The growth rate of filamentous fungi is very variable between species. It is mainly determined by the characteristics of the hyphae in conjunction with the conditions of their environment and the chemical nutrition they receive. It has been demonstrated that trimitic hyphal systems have a higher rate of hyphal extension than monomitic hyphal systems; therefore, this fungus may also be a good producer of biomass for composites. *T. versicolor* has a trimitic hyphal system consisting of generative, binding, and skeletal hyphae (Jones et al. 2019). *T. versicolor* can be produced by either an SmF or an SSF. The efficiency of these systems is contingent on several factors, such as the genetic traits of the strain that is producing the organism, the composition of the substrate, and the operational settings (Montoya et al. 2021).

Submerged Fermentation (SmF)

When *T. versicolor* is used for SmF, the medium must be a liquid and include either dissolved or suspended nutrients readily available to the microbe. The nutrient media used to create inoculum typically differ from the media used for the manufacture of bioproducts. The most popular bioreactor used for submerged cultivation is the stirred tank bioreactor, which can operate in batch, fed-batch, or continuous mode. Other types of bioreactors, including bubble columns and air-lift bioreactors, are occasionally utilized. According to Znidarsic and Pavko's (2001) research, the morphological type and the corresponding physiology are the critical variables required to explain and forecast the performance of a bioreactor and design a bioprocess. The synthesis of *T. versicolor* enzymes is the most common use of SmF, and the liquid broth is frequently used as an inducer during this process. In addition to that, mycelial generation for subsequent use as an inoculum in SSF can be accomplished using SmF. Significant work has been put into the research and development of technologically advanced bioreactors for the treatment of wastewater using *T. versicolor* (Hu et al. 2020).

Solid-state Fermentation (SSF)

Most biological pretreatments involving white-rot fungi have been done by SSF. Under SSF conditions, *T. versicolor* was selected to de-lignify oil palm trunk chips, yielding enhanced properties for the pulps (Singh et al. 2013). The *T. versicolor* treatment of Norway spruce wood chips followed by a thermomechanical pulping procedure did not impact the mechanical qualities of the resulting hand sheets. Still, it created a less hazardous effluent, boosting biodegradability (van Beek et al. 2007). *T. versicolor* is cultured in SSF on inert or non-inert solid substrates. Among the latter, lignocellulosic biomass is the most prevalent substrate used for inoculum and bio-product production. In many circumstances, the output of SSF is bio-transformed lignocellulosic biomass, which can then be utilized to manufacture biofuels and bio-based goods, or as animal feed. Considering the variability of supplements, it is tough to regulate elements impacting fungal growth, such as temperature, pH, initial inoculum concentration, oxygen availability, and water activity, in the design of SSF bioreactors. It is tough to assure the presence of a gas phase between substrate particles that may be constricted or coagulated, causing hot spots to be generated. It is necessary to know the chemical composition to give an appropriate C/N ratio for chemically complex and heterogeneous substrates such as lignocellulosic biomass. A high C concentration can increase hyphal branching and decrease hyphal extension rate, boosting oxygen uptake rate. Abundant carbon supplies avoid catabolic repression (inhibition of enzyme synthesis), a fundamental aspect of SSF and an advantage over SmF (Lizardi-Jiménez and Hernández-Martínez 2017). SSF of corn stalk, tea residues, olive leaves, corn-steep liquor, and steam-exploded corn stalk substantially increased the laccase yield (Suryadi et al. 2022, Wang et al. 2014, Xu et al. 2020), since the steam explosion purification disrupted the compressed lignocellulosic architecture, increased the surface area, and made soluble molecules more accessible to *T. versicolor* for biological utilization. Additionally, the preprocessing treatment freed many phenolic compounds, including putative laccase inducers (Adekunle et al. 2017).

PHYTOCHEMISTRY

A combination of macromolecules that surround the cell and are located outside the plasma membrane makes up the fungal cell wall. It is well recognized that the fungus contains potentially active secondary metabolites of low molecular weight in addition to the primary macromolecules (proteins, carbohydrates, and lipids) and minerals. Glucans, chitin, glycoproteins, chitosan, inorganic salts, polyuronids, tyrosol, friedelin, triterpenoids, alnusenone, and pigments are the primary components. Around 80% of the dry matter is composed of skeletal polysaccharides such as glucans and chitin, while the protein composition ranges from 3 to 20% on average. These cell wall elements can vary significantly between fungal species and even between different cell types within the same fungal species. In terms of the bioactive compounds that have been linked to biological processes in T. versicolor, the majority of studies that have been conducted to date have concentrated on the polysaccharide fraction. This fraction contains β-(1,4) glucan main chain as well as β-(1,3) and β-(1,6)-linked side chains, polymers of D-glucose in a mixture or not with units of glucuronic acids, arabinose, D-mannose, L-abequo (Li et al. 2011). In addition to polysaccharides, 18 different amino acids were found, some of which are listed below: aspartic acid, threonine, serine, glutamic acid, glycine, alanine, valine, and leucine (Cruz et al. 2016, Cui and Chisti 2003). (Janjušević et al. 2018) looked at the phenolic make-up of T. versicolor basidiocarp collected from various locations across Europe. They were able to identify a large number of phenolic compounds and hydroxy-cinnamic acids by the use of liquid chromatography-mass spectrometry. Although solvent extracts contain the highest concentrations of these phenolics, water extracts contain significant amounts of baicalein, baicalin, quercetin, isorhamnetin, catechin, amentoflavone, p-hydroxybenzoic acid, and cyclohexane carboxylic acid. Baicalein and baicalin are phenolic compounds that inhibit the growth of cancer cells. However, it has not been demonstrated that these chemicals are significant components of fungi; hence, additional research is required to determine the possible significance of these compounds for recognized biological function. However, very little research has investigated the composition of total phenolic compounds and total flavonoid content (flavones, flavonols, flavanones, flavanols, coumarins, biflavonoid, and isoflavonoid), both of which are essential for the antioxidant properties of the plant (Pop et al. 2018). In cultures of *T. versicolor*, the following sesquiterpenes were discovered: tramspiroins A–D, one new rosenonolactone 15,16-acetonide, isodrimenediol, and funatrol D. Additionally, one new rosenonolactone 15,16-acetonide was discovered (Wang et al. 2015).

PHARMACOLOGICAL VALUE

Traditional Chinese medicine has utilized *T. versicolor* for at least 2,000 years due to its medicinal significance, which includes effects typically good to one's health, such as enhanced endurance and lifespan. It has been mentioned for millennia in the Shen Non-Compendium of Medicine (Wasser 2010, Wasser 2014), and a more recent study has proved its therapeutic value (Wasser 2010, Wasser 2014, Rameshaiah et al. 2020). Initially, a large percentage of the research's focus was centred on its bioactivity to improve immunological function (Li et al. 2011). Extracts derived from the fruiting bodies and mycelium of the *T. versicolor* fungus have an unusually high anticancer potential (Habtemariam, 2020). Several investigations discovered that the extracts prevent the growth of leukaemia, lymphoma, and melanoma cell lines and reduce prostate, stomach, cervix, colon, and lung cancers. PSK, commonly known as "Krestin," and PSP are two essential compounds identified from mycelial cultures of the fungus and the topic of contemporary research. In China, the "COV-1" strain is the principal source of polysaccharide S (PSS), but in Japan, the "CM-101" strain is the key source of polysaccharide K (PSK). In order to make the covalent bond, they are composed of polysaccharides that have been O- or N-glycosidically bonded to peptides. In Japan in 1977 and China in 1987, PSK was approved for use as an adjuvant in treating patients with stomach cancer following considerable clinical study. PSK has been shown to boost cellular activity in the host organism directly and indirectly by acting cytostatically and cytotoxically on cancer cells. In addition, PSK has been demonstrated to possess antiviral effects by boosting interferon production. PSP is both a potent immunostimulant and a cancer-fighting drug. The China State Food and Drug Administration (SFDA) has approved thirteen different medications derived from *T. versicolor* for clinical and commercial use and one product produced from *T. versicolor*. The Health and Welfare Ministry of Japan, Japan's equivalent to the Food and Drug Administration in the United States, licensed Krestin as the first anticancer drug derived from mushrooms. In 1987, "Krestin" was responsible for 25% of Japan's total expenditures on anti-cancer agents. In Japan, the purchase of Krestin is covered by every single available health insurance plan. Overall, it has been proven that the polysaccharides generated by *T. versicolor* can induce death in cancer cells and directly inhibit their proliferation. In recent years, mycelial extracts of *T. versicolor*, also known as Yunzhi, have attracted much attention due to their extensive array of health-improving properties. Currently, this is being explored and investigated as a possible treatment for Alzheimer's disease and other neurodegenerative conditions (Trovato-Salinaro et al. 2016). *T. versicolor* exhibits enhanced LXA4, a molecule with well-known anti-inflammatory properties similar to lion's mane mushroom (Trovato-Salinaro et al. 2016). LXA4 could likely be a therapy for any neurodegenerative disorder linked with brain inflammation. By stimulating the production of superoxide dismutase (SOD), glutathione peroxidase, and interleukin-6, PSP and PSK can treat tumours and limit the growth of many malignancies. They can also stimulate the immune system by raising the production of interleukin-6, interferons, Immunoglobulin-G, macrophages, and T lymphocytes; neutralizing the immunosuppressive effects of chemotherapy, radiation, and blood transfusion; and stimulate the production of natural killer cells (Cui and Chisti 2003). As a prebiotic, *T. versicolor* has improved the human gut microbiota's equilibrium, resulting in more productive interactions with host cells (Pallav et al. 2014, Yu et al. 2013). In addition, it can induce vitagenes, which are proteins involved in regulating cellular stress responses and redox homeostasis, as well as the improvement of glucose tolerance, the mitigation of diabetes-related pathologies, the relief of colitis symptoms, and the production of analgesic effects. Extracts of *T. versicolor* have demonstrated antibacterial and antifungal action against a wide range of common pathogens, including *Escherichia coli*, *Pseudomonas aeruginosa*, *Staphylococcus aureus*, *Candida albicans*, *Klebsiella pneumoniae*, *Listeria monocytogenes*, and *Streptococcus pneumoniae*. Additionally, *T. versicolor* extracts have been demonstrated to limit the growth of Additionally, intestinal health may be enhanced. The presence of helpful prebiotics limits the growth of harmful bacteria, such as *E. coli*. Additionally, it increases the number of beneficial bacteria in the digestive tract. Similar to other forms of mushrooms, turkey tails include a certain quantity of dietary fibre. Additionally, this can help promote good digestion. The pharmacological preparations or polysaccharides derived from them were well tolerated in clinical trials. Any adverse effects that occurred may have been attributed to other chemotherapeutic drugs provided concurrently. In certain cases, this can be accomplished by using a coarse powder, while in others, it can be accomplished through the extraction of bioactive compounds. Therefore, the medicinal properties of a certain product derived from *T. versicolor* depend on the sections of the fungus that were utilized, the type of fermentation substrate used, and the extraction or preparation technique employed. In addition to their nutritional value, medicinal mushrooms have recently gained popularity as a source of pharmaceuticals and as adjuvants to conventional forms of chemotherapy and radiation therapy. They are believed to either increase the efficacy of these treatments or reduce their adverse side effects.

ENZYMATIC POTENTIAL

T. versicolor is an exogenous, non-specific, and influential producer of ligninolytic enzymes. These enzymes include laccase, lignin peroxidases (LiPs), manganese peroxidases (MnPs), endoxylanase, β-glucosidase, and laccase. When grown under optimal conditions, all enzymes have significant activity as secondary metabolites. These lignin-modifying enzymes can degrade compounds that include aromatic rings. Some examples of these molecules include insecticides, phenolic chemicals, chlorophenols, PAHs, medicinal products, and industrial colours (Necochea et al. 2005). Additionally, the digestibility of plant wastes for ruminants can be improved by delignifying them with fungal ligninases. Ruminants are animals that have four stomach chambers. These enzymes have a wide range of possible applications, including degrading cyanide, stabilizing wine, baking, bio-pulping and bio-bleaching, dye-decolorization, denim finishing, and the manufacture of bioethanol (Rameshaiah et al. 2020). Laccases are N-glycosylated blue multicopper oxidases in the immediate sense. These enzymes mineralize lignin and a variety of other refractory aromatic compounds, including synthetic colours. It is possible to regard it as the most critical enzyme in the process of oxidation, modification, or breakdown of lignin (Aguiar et al. 2014). On microspheres made of poly(glycidyl methacrylate), laccase enzymes derived from the fungi *T. versicolor* were successfully co-immobilized. In order to break down diazinon, a biocatalyst that had laccases connected to it was utilized (Vera et al. 2020). Other oxidases and peroxidases, such as MnP, LiP, and multifunctional peroxidase, are produced *by T. versicolor* in addition to laccase. These other oxidases include aryl-alcohol oxidase, pyranose oxidase, and polyphenol oxidase (Wyman et al. 2018). Many other enzymes are involved in the process of lignocellulose degradation. These enzymes include various dehydrogenases (aryl-alcohol dehydrogenase, cellobiose dehydrogenase), reductases (NADPH-hemoprotein reductase, NADH: quinone reductase), hydrolases (cellobiohydrolase, glucanase, glucoamylase, and manganese). The handling of waste has been the focus of most contemporary studies with *T. versicolor*. The utilization of *T. versicolor* and purified or crude laccase extracts in free or immobilized form can be separated into two groups. The first category is the free form. Fungi are the primary agents in the bioprocess known as mycoremediation, which removes soil and water pollutants, contaminants, and poisons. Fungi are more resistant to poisonous substances than bacteria due to their chitin-rich cell walls and extracellular enzymes (Tima et al. 2021). Therefore, it is preferable to utilize fungi rather than bacteria in this situation. In addition, the nature of their hyphal growth makes it possible for *T. versicolor* to extend a great distance from its initial starting place, which is very helpful for the bioremediation of soils (Vera et al. 2020). In order to evaluate the extracellular activities and consequent degradation of wood components, such as enzymes and oxalic acid, as well as the development of water-soluble compounds with $Fe3+$ reducing activity, the purpose of this study is to determine the extracellular activities (Aguiar et al. 2014). Because of its ability to utilize several biochemical and physical reactions for intermolecular bond breaking, demethylation, hydroxylation, dechlorination, and aromatic ring-opening, *T. versicolor* is helpful in the removal of micro-pollutants in wastewater treatment. This is because *T. versicolor* can utilize these reactions (Dalecka et al. 2021). Because of their capacity to break down a wide variety of organic pollutants through the production of non-specific intracellular and extracellular enzymes, white-rot fungi in biochemical approaches have garnered much attention in recent years.

OTHER APPLICATIONS

The *T. versicolor* fungus has the potential to be used in a wide variety of contexts and applications. Researchers looked into how the fungus *T. versicolor* degraded the insecticides imiprothrin (IP) and cypermethrin (CP), as well as the insecticide/nematicide carbofuran (CBF), and the agricultural antibiotic oxytetracycline (OTC) (Mir-Tutusaus et al. 2014). The health benefits of the bioactive chemicals identified from *T. versicolor* came from their ability to suppress the effects of aflatoxins. Tramesan is a branching fungal glycan isolated from the culture filtrate of the asexual mycelia of *T. versicolor*. It can influence the activation of anti-oxidant defences in the proliferation and secondary metabolism (such as aflatoxins) of plant pathogenic fungus (Scarpari et al. 2017). The consumption of Tramesan improves both the resistance of durum wheat and the production of melanin in melanoma. Notably, this polysaccharide causes an increase in the production of genes involved in the body's response to oxidative stress, such as peroxidases and Nrf-2. Tramesan can trigger an antioxidant response even in these species and mycotoxin-producing fungus, most likely by modulating gene expression. Wheat reportedly benefits from endophytic *T. versicolor* in terms of growth stimulation and grain yield. In P-rich environments, it was also able to demonstrate an increased capacity for phosphorus uptake. These findings provided further evidence that phosphorus acquisition and transportation are essential components of the interaction between plants and endophytic fungus and significant contributors to the enhancement of plant production (Taghinasab et al. 2018). According to Janjušević et al. (2018), even though *T. versicolor* is responsible for producing ferulic acid, protocatechuic acid, and syringic acid, this organism is also a good option for the development of health-promoting food supplements. In tests conducted using bioreactors, it was shown that *T. versicolor* pellets could break down pharmaceutically active chemicals found in both sterile and non-sterile waste.

The most prevalent types of reactor configurations are the fluidized-bed, fixed-bed, membrane, and flask types (Tormo-Budowski et al. 2021). To produce ethanol, any of these organisms can be inoculated *with T. versicolor* pellets or mycelial suspensions, and any of these organisms can be employed. Mycelial pellets produced by *T. versicolor* effectively remove a wide variety of pollutants from sludge's. At a semi-industrial level, research was conducted on the possibility of co-composting used coffee grounds, the wastewater sludge from olive mills, and chicken manure. Composts were inoculated with *T. versicolor* at the beginning of the maturation period to lessen the harmful effects of the phenolic component and boost the degree of composting humification (Hachicha et al. 2012). There has been some investigation into the viability of *T. versicolor* as a raw material for bioethanol production. During times of aggressive growth and in the presence of hypoxia, *T. versicolor* is capable of converting xylose and hexose into ethanol. Still, the fermentation rate needs to be increased in order for it to be commercially exploited for the manufacture of bioethanol. This can be done by increasing the amount of inoculum used, as well as by employing metabolic engineering to make improvements in the xylose utilizing pathway. The fact that *T. versicolor* can grow in many different inhibitors and eliminate them demonstrates that it has the potential to be utilized as a detoxifying agent in the manufacturing of ethanol (Kudahettige et al. 2012). In the past, *T. versicolor* has been utilized for the pretreatment of lignin-biomass in batch trials in preparation for the production of biogas (Akyol et al. 2019; Wyman et al. 2018). To produce biogas, a pilot-scale semi-continuous method was developed by Tima et al. 2021. This process involved treating corn silage with *T. versicolor* and combining it with cow dung. While additional study is required to treat ligno-biomass with *T. versicolor* to produce bioethanol, biogas production using an SSF plant that uses this fungus is considerably closer to becoming a commercially viable option. The generation of biogas, feed, and biofertilizer is an auspicious use for *T. versicolor*.

CONCLUSION

T. versicolor is a fungus with much potential for use in various applications. It can use lignocellulosic waste biomass, which is readily available, to produce biofuels, biofertilizers, feeds, and crude laccase extracts for biotransformation processes and wastewater treatment. It has been demonstrated that *T. versicolor* may increase the nutritional value of various lignocellulosic biomasses. It has been demonstrated that the mycelial mass that contains pollutants may be conveniently utilized in anaerobic digestion, which is another step toward reaching the aims of the circular bioeconomy. In addition to this, the mycelial mass can be used as an adsorbent for pollutants. Separate research needs to be carried out to complete the cycle and integrate this fungus into either existing or newly constructed lignocellulosic bio refineries to contribute to the expansion of circular economies around the world.

REFERENCES

Adekunle, A. E., Zhang, C., Guo, C., & Liu, C. Z. (2017). Laccase production from *Trametes versicolor* in solid-state fermentation of steam-exploded pretreated cornstalk. *Waste and Biomass Valorization*, *8*(1), 153–159. 10.1007/s12649-016-9562-9

Aguiar, A., Gavioli, D., & Ferraz, A. (2014). Metabolite secretion, Fe3+-reducing activity and wood degradation by the white-rot fungus *Trametes versicolor* ATCC 20869. *Fungal Biology*, *118*(11), 935–942. 10.1016/j.funbio.2014.08.004

Akyol, Ç., Ince, O., Bozan, M., Ozbayram, E. G., & Ince, B. (2019). Biological pretreatment with *Trametes versicolor* to enhance methane production from lignocellulosic biomass: A metagenomic approach. *Industrial Crops and Products*, *140*(July). 10.1016/j.indcrop.2019.111659

Alanazi, M. (2018). The Transcriptional Response of *Trametes versicolor* to Growth on Maple Chips and Miscanthus Straw. Eastern Illinois University

Atilano-Camino, M. M., Álvarez-Valencia, L. H., García-González, A., & García-Reyes, R. B. (2020). Improving laccase production from *Trametes versicolor* using lignocellulosic residues as cosubstrates and evaluation of enzymes for blue wastewater biodegradation. *Journal of Environmental Management*, *275*. 10.1016/j.jenvman.2020.111231

Bains, A., Chawla, P., Kaur, S., Najda, A., Fogarasi, M., & Fogarasi, S. (2021). Bioactives from mushroom: Health attributes and food industry applications. *Materials*, *14*(24). 10.3390/ma14247640

Bari, E., Daryaei, M. G., Karim, M., Bahmani, M., Schmidt, O., Woodward, S., Tajick Ghanbary, M. A., & Sistani, A. (2019). Decay of Carpinus betulus wood by *Trametes versicolor* - An anatomical and chemical study. *International Biodeterioration and Biodegradation*, *137*, 68–77. 10.1016/j.ibiod.2018.11.011

Bhatt, M., Mistri, P., Joshi, I., Ram, H., Raval, R., Thoota, S., Patel, A., Raval, D., Bhargava, P., Soni, S., Bagatharia, S., & Joshi, M. (2018). Molecular survey of basidiomycetes and divergence time estimation: An Indian perspective. *PLoS ONE*, *13*(5), 1–17. 10.1371/journal.pone.0197306

Chen, M., Wang, C., Fei, B., Ma, X., Zhang, B., Zhang, S., & Huang, A. (2017). Biological degradation of chinese fir with *Trametes versicolor* (l.) lloyd. *Materials*, *10*(7), 1–13. 10.3390/ma10070834

Córdoba M, K. A., & Ríos H, A. (2012). Biotechnological applications and potential uses of the mushroom Tramestes versicolor. *Vitae*, *19*(1), 70–76.

Cruz, A., Pimentel, L., Rodríguez-Alcalá, L. M., Fernandes, T., & Pintado, M. (2016). Health benefits of edible mushrooms focused on Coriolus versicolor: A review. *Journal of Food and Nutrition Research*, *4*(12), 773–781. 10.12691/jfnr-4-12-2

Cui, J., & Chisti, Y. (2003). Polysaccharopeptides of Coriolus versicolor: Physiological activity, uses, and production. *Biotechnology Advances*, *21*(2), 109–122. 10.1016/S0734-9750(03)00002-8

Dalecka, B., Strods, M., Cacivkins, P., Ziverte, E., Rajarao, G. K., & Juhna, T. (2021). Removal of pharmaceutical compounds from municipal wastewater by bioaugmentation with fungi: An emerging strategy using fluidized bed pelleted bioreactor. *Environmental Advances*, *5*, 100086. 10.1016/j.envadv.2021.100086

Durak, H. (2018). *Trametes versicolor* (L.) mushrooms liquefaction in supercritical solvents: Effects of operating conditions on product yields and chromatographic characterization. *Journal of Supercritical Fluids*, *131*, 140–149. 10.1016/j.supflu.2017.09.013

Dwivedi, R. C. (2006). Extracellular proteins from lignocellulose degrading Basidiomycetes: Redox enzymes from *Trametes versicolor* and Coprinopsis cinerea. *Georg-August-University Göttingen*, 162. https://ediss.uni-goettingen.de/bitstream/handle/11858/00-1735-0000-0006-B0F9-7/dwivedi.pdf?sequence=1

Gautam, A. K. (2013). Notes on wood rotting fungi from India (1): *Trametes versicolor*-The Turkey Tail. *Journal on New Biological Reports*, *2*(2), 67–70.

Guerrero, D. G., Martínez, V. E., & de la Torre Almaráz, R. (2011). Cultivation of Trametes versicolor in Mexico. *Micologia Aplicada International*, *23*(2), 55–58.

Habtemariam, S. (2020). Trametes versicolor (Synn. Coriolus versicolor) polysaccharides in cancer therapy: Targets and efficacy. *Biomedicines*, *8*(5), 135. 10.3390/biomedicines8050135

Hachicha, R., Rekik, O., Hachicha, S., Ferchichi, M., Woodward, S., Moncef, N., Cegarra, J., & Mechichi, T. (2012). Co-composting of spent coffee ground with olive mill wastewater sludge and poultry manure and effect of *Trametes versicolor* inoculation on the compost maturity. *Chemosphere*, *88*(6), 677–682. 10.1016/j.chemosphere.2012.03.053

Hlerema, I. N., Eiasu, B. K., & Koch, S. H. (2017). Pineapple (Ananas comusus) plant material as supplement for maize residue-based oyster mushroom substrate and reduction of cadmium soil contamination. *HortScience*, *52*(4), 667–671. 10.21273/HORTSCI10880-16

Hu, K., Peris, A., Torán, J., Eljarrat, E., Sarrà, M., Blánquez, P., & Caminal, G. (2020). Exploring the degradation capability of *Trametes versicolor* on selected hydrophobic pesticides through setting sights simultaneously on culture broth and biological matrix. *Chemosphere, 250*, 126293. 10.1016/j.chemosphere.2020.126293

Hu, K., Sarrà, M., & Caminal, G. (2021). Comparison between two reactors using *Trametes versicolor* for agricultural wastewater treatment under non-sterile condition in sequencing batch mode. *Journal of Environmental Management, 293*. 10.1016/j.jenvman.2021.112859

Janjušević, L., Pejin, B., Kaišarević, S., Gorjanović, S., Pastor, F., Tešanović, K., & Karaman, M. (2018). *Trametes versicolor* ethanol extract, a promising candidate for health–promoting food supplement. *Natural Product Research, 32*(8), 963–967. 10.1080/14786419.2017.1366484

Jo, W.-S., Kang, M.-J., Choi, S.-Y., Yoo, Y.-B., Seok, S.-J., & Jung, H.-Y. (2010). Culture Conditions for Mycelial Growth of *Coriolus versicolor. Mycobiology, 38*(3), 195. 10.4489/myco.2010.38.3.195

Jones, M. P., Lawrie, A. C., Huynh, T. T., Morrison, P. D., Mautner, A., Bismarck, A., & John, S. (2019). Agricultural by-product suitability for the production of chitinous composites and nanofibers utilising *Trametes versicolor* and *Polyporus brumalis* mycelial growth. *Process Biochemistry, 80*, 95–102. 10.1016/j.procbio.2019.01.018

Knežević, A., Stajić, M., Sofrenić, I., Stanojković, T., Milovanović, I., Tešević, V., & Vukojević, J. (2018). Antioxidative, antifungal, cytotoxic and antineurodegenerative activity of selected Trametes species from Serbia. *PLoS ONE, 13*(8), 1–18. 10.1371/journal.pone.0203064

Kudahettige, R. L., Holmgren, M., Imerzeel, P., & Sellstedt, A. (2012). Characterization of bioethanol production from hexoses and xylose by the white rot fungus *Trametes versicolor. Bioenergy Research*, *5*(2), 277–285. 10.1007/s12155-011-9119-5

Li, F., Wen, H. A., Zhang, Y. J., An, M., & Liu, X. Z. (2011). Purification and characterization of a novel immunomodulatory protein from the medicinal mushroom *Trametes versicolor. Science China Life Sciences*, 54(4), 379–385. 10.1007/s11427-011-4153-2

Lizardi-Jiménez, M. A., & Hernández-Martínez, R. (2017). Solid state fermentation (SSF): diversity of applications to valorize waste and biomass. *3 Biotech, 7*(1). 10.1007/s13205-017-0692-y

Mir-Tutusaus, J. A., Masís-Mora, M., Corcellas, C., Eljarrat, E., Barceló, D., Sarrà, M., Caminal, G., Vicent, T., & Rodríguez-Rodríguez, C. E. (2014). Degradation of selected agrochemicals by the white rot fungus *Trametes versicolor. Science of the Total Environment, 500–501*, 235–242. 10.1016/j.scitotenv.2014.08.116

Montoya, S., Patiño, A., & Sánchez, Ó. J. (2021). Production of lignocellulolytic enzymes and biomass of *Trametes versicolor* from agro-industrial residues in a novel fixed-bed bioreactor with natural convection and forced aeration at pilot scale. *Processes, 9*(2), 1–19. 10.3390/pr9020397

Necochea, R., Valderrama, B., Díaz-Sandoval, S., Folch-Mallol, J. L., Vázquez-Duhalt, R., & Iturriaga, G. (2005). Phylogenetic and biochemical characterisation of a recombinant laccase from *Trametes versicolor. FEMS Microbiology Letters, 244*(2), 235–241. 10.1016/j.femsle.2005.01.054

Nguyen, B. T. T., Van Le, V., Nguyen, H. T. T., Nguyen, L. T., Tran, T. T. T., & Ngo, N. X. (2021). Nutritional requirements for the enhanced mycelial growth and yield performance of *Trametes versicolor. Journal of Applied Biology and Biotechnology*, *9*(1), 1–7. 10.7324/JABB.2021.9101

Pallav, K., Dowd, S. E., Villafuerte, J., Yang, X., Kabbani, T., Hansen, J., Dennis, M., Leffler, D. A., & Kelly, C. P. (2014). Effects of polysaccharopeptide from *Trametes versicolor* and amoxicillin on the gut microbiome of healthy volunteers: A randomized clinical trial. *Gut Microbes, 5*(4), 458–467. 10.4161/gmic.29558

Pop, R. M., Puia, I. C., Puia, A., Chedea, V. S., Leopold, N., Bocsan, I. C., & Buzoianu, A. D. (2018). Characterization of *Trametes versicolor*: Medicinal mushroom with important health benefits. *Notulae Botanicae Horti Agrobotanici Cluj-Napoca, 46*(2), 343–349. 10.15835/nbha46211132

Rameshaiah, G. N., & Jagadish Reddy, M. L. (2015). Applications of ligninolytic enzymes from a White-Rot fungus *Trametes versicolor. Universal Journal of Environmental Research and Technology, 5*(1), 1–7. http://www.environmentaljournal.org/5-1/ujert-5-1-1.pdf

Rameshaiah, G. N., Jagadish Reddy, M. L., Suresh, G., Cabezudo, I., Pulicharla, R., Cuprys, A., Rouissi, T., Brar, S. K., Suryadi, H., Judono, J. J., Putri, M. R., Eclessia, A. D., Ulhaq, J. M., Agustina, D. N., Sumiati, T., Chen, M., Wang, C., Fei, B., Ma, X., …Ríos H, A. (2020). Evaluation of biological pretreatment with white rot fungi for the enzymatic hydrolysis of bamboo culms. *Journal of Environmental Management, 8*(1), 1–7. 10.4489/myco.2010.38.3.195

Rau, U., Kuenz, A., Wray, V., Nimtz, M., Wrenger, J., & Cicek, H. (2009). Production and structural analysis of the polysaccharide secreted by Trametes (Coriolus) versicolor ATCC 200801. *Applied Microbiology and Biotechnology, 81*(5), 827–837. 10.1007/s00253-008-1700-2

Scarpari, M., Reverberi, M., Parroni, A., Scala, V., Fanelli, C., Pietricola, C., Zjalic, S., Maresca, V., Tafuri, A., Ricciardi, M. R., Licchetta, R., Mirabilii, S., Sveronis, A., Cescutti, P., & Rizzo, R. (2017). Tramesan, a novel polysaccharide from *Trametes versicolor*. Structural characterization and biological effects. *PLoS ONE, 12*(8), 1–22. 10.1371/journal.pone.0171412

Singh, P., Sulaiman, O., Hashim, R., Peng, L. C., & Singh, R. P. (2013). Evaluating biopulping as an alternative application on oil palm trunk using the white-rot fungus *Trametes versicolor*. *International Biodeterioration and Biodegradation, 82*, 96–103. 10.1016/j.ibiod.2012.12.016

Suryadi, H., Judono, J. J., Putri, M. R., Eclessia, A. D., Ulhaq, J. M., Agustina, D. N., & Sumiati, T. (2022). Biodelignification of lignocellulose using ligninolytic enzymes from white-rot fungi. *Heliyon, 8*, e08865. 10.1016/j.heliyon.2022.e08865

Taghinasab, M., Imani, J., Steffens, D., Glaeser, S. P., & Kogel, K. H. (2018). The root endophytes *Trametes versicolor* and Piriformospora indica increase grain yield and P content in wheat. *Plant and Soil, 426*(1–2), 339–348. 10.1007/s11104-018-3624-7

Téllez-Téllez, M., Villegas, E., Rodriguez, A., Acosta-Urdapilleta, M. L., O'Donovan, A., & Diaz-Godinez, G. (2016). Mycosphere essay 11: Fungi of Pycnoporus: morphological and molecular identification, worldwide distribution and biotechnological potential. *Mycosphere, 7*(10), 1500.10.5943/mycosphere/si/3b/3

Tišma, M., Žnidaršič-Plazl, P., Šelo, G., Tolj, I., Šperanda, M., Bucić-Kojić, A., & Planinić, M. (2021). *Trametes versicolor* in lignocellulose-based bioeconomy: State of the art, challenges and opportunities. *Bioresource Technology, 330*. 10.1016/j.biortech.2021.124997

Tormo-Budowski, R., Cambronero-Heinrichs, J. C., Durán, J. E., Masís-Mora, M., Ramírez-Morales, D., Quirós-Fournier, J. P., & Rodríguez-Rodríguez, C. E. (2021). Removal of pharmaceuticals and eco-toxicological changes in wastewater using *Trametes versicolor*: A comparison of fungal stirred tank and trickle-bed bioreactors. *Chemical Engineering Journal, 410*. 10.1016/j.cej.2020.128210

Trovato-Salinaro, A., Siracusa, R., Di Paola, R., Scuto, M., Fronte, V., Koverech, G., Luca, M., Serra, A., Toscano, M. A., Petralia, A., Cuzzocrea, S., & Calabrese, V. (2016). Redox modulation of cellular stress response and lipoxin A4 expression by Coriolus versicolor in rat brain: Relevance to Alzheimer's disease pathogenesis. In *NeuroToxicology* (Vol. 53). Elsevier B.V. 10.1016/j.neuro.2015.09.012

van Beek, T. A., Kuster, B., Claassen, F. W., Tienvieri, T., Bertaud, F., Lenon, G., Petit-Conil, M., & Sierra-Alvarez, R. (2007). Fungal bio-treatment of spruce wood with *Trametes versicolor* for pitch control: Influence on extractive contents, pulping process parameters, paper quality and effluent toxicity. *Bioresource Technology, 98*(2), 302–311. 10.1016/j.biortech.2006.01.008

Veena, S. S., & Pandey, M. (2012). Physiological and cultivation requirements of *Trametes versicolor*, a medicinal mushroom to diversify Indian mushroom industry. *Indian Journal of Agricultural Sciences, 82*(8), 672–675.

Vera, M., Nyanhongo, G. S., Guebitz, G. M., & Rivas, B. L. (2020). Polymeric microspheres as support to co-immobilized Agaricus bisporus and *Trametes versicolor* laccases and their application in diazinon degradation. *Arabian Journal of Chemistry, 13*(2), 4218–4227. 10.1016/j.arabjc.2019.07.003

Wang, F., Hu, J. H., Guo, C., & Liu, C. Z. (2014). Enhanced laccase production by *Trametes versicolor* using corn steep liquor as both nitrogen source and inducer. *Bioresource Technology, 166*, 602–605. 10.1016/j.biortech.2014.05.068

Wang, S. R., Zhang, L., Chen, H. P., Li, Z. H., Dong, Z. J., Wei, K., & Liu, J. K. (2015). Four new spiroaxane sesquiterpenes and one new rosenonolactone derivative from cultures of Basidiomycete *Trametes versicolor*. *Fitoterapia, 105*, 127–131. 10.1016/j.fitote.2015.06.017

Wang, X. C., Xi, R. J., Li, Y., Wang, D. M., & Yao, Y. J. (2012). The species identity of the widely cultivated ganoderma, "*G. lucidum*" (ling-zhi), in China. *PLoS ONE, 7*(7). 10.1371/journal.pone.0040857

Wasser, S. P. (2010). Medicinal mushroom science: History, current status, future trends, and unsolved problems. *International Journal of Medicinal Mushrooms, 12*(1). 10.1615/intjmedmushr.v12.i1.10

Wasser, S. P. (2014). Medicinal mushroom science: Current perspectives, advances, evidences, and challenges. *Biomedical Journal, 37*(6), 345–356. 10.4103/2319-4170.138318

Wyman, V., Henríquez, J., Palma, C., & Carvajal, A. (2018). Lignocellulosic waste valorisation strategy through enzyme and biogas production. In *Bioresource Technology* (Vol. 247). 10.1016/j.biortech.2017.09.055

Xu, L., Sun, K., Wang, F., Zhao, L., Hu, J., Ma, H., & Ding, Z. (2020). Laccase production by *Trametes versicolor* in solid-state fermentation using tea residues as substrate and its application in dye decolorization. *Journal of Environmental Management, 270*, 110904. 10.1016/j.jenvman.2020.110904

Yu, H., Zhang, X., Song, L., Ke, J., Xu, C., Du, W., & Zhang, J. (2010). Evaluation of white-rot fungi-assisted alkaline/oxidative pretreatment of corn straw undergoing enzymatic hydrolysis by cellulase. *Journal of Bioscience and Bioengineering, 110*(6), 660–664. 10.1016/j.jbiosc.2010.08.002

Yu, Z. T., Liu, B., Mukherjee, P., & Newburg, D. S. (2013). *Trametes versicolor* extract modifies human fecal microbiota composition in vitro. *Plant Foods for Human Nutrition*, *68*(2), 107–112. 10.1007/s11130-013-0342-4

Zhang, X., Yu, H., Huang, H., & Liu, Y. (2007). Evaluation of biological pretreatment with white rot fungi for the enzymatic hydrolysis of bamboo culms. *International Biodeterioration and Biodegradation*, *60*(3), 159–164. 10.1016/j.ibiod.2007.02.003

Žnidaršič, P., & Pavko, A. (2001). The morphology of filamentous Fungi in submerged cultivations as a bioprocess parameter. *Food Technology and Biotechnology*, *39*(3), 237–252.

16 *Tremella fuciformis*

Arun Kumar Gupta
Department of Food Science and Technology, Graphic Era (Deemed to be University), Bell Road, Clement Town, Dehradun, Uttarakhand, India

Muzamil A. Rather and Shuvam Bhuyan
Department of Molecular Biology and Biotechnology, Tezpur University, Tezpur, Assam, India

Bindu Naik
Department of Food Science and Technology, Graphic Era (Deemed to be University), Bell Road, Clement Town, Dehradun, Uttarakhand, India

Mukesh S. Sikarwar
College of Pharmacy, Teerthanker Mahaveer University, Moradabad, Uttar Pradesh, India

Poonam Mishra
Department of Food Engineering and Technology, Tezpur University, Tezpur, Assam, India

CONTENTS

DOI: 10.1201/9781003259763-16

INTRODUCTION

Tremella fuciformis is a tremellaceae-family basidiomycotan fungus (Table 16.1). It has a parasitic mode of nutrition due to its poor ability to decompose cellulose and lignin. Since ancient times, it has been employed in traditional Chinese medicine and food. It is a classic dimorphic fungus. Its primary life cycle is that of a slow-growing yeast (anamorph) which is gelatinized in appearance, and buds as slime-like chains of individual cells which can metamorphosize into hyphal stage spontaneously on the outer margins of the culture upon encountering a suitable host where white filamentous mycelia emerge. In its sexual stage (teleomorph), basidia are formed upon its fruiting body.

T. fuciformis, like *T. mesenterica*, the type species of the genus *Tremella*, often known as witch's butter, has an odd ability to dry and rehydrate. Upon a few showers, the young fruitbodies swell to resemble blobs of translucent whitish jelly. After a few dry spells, these fruitbodies collapse but upon rehydration, their reform and regrows to their original shape, eventually forming ear-shaped leaflets, and as they mature, these leaflets resemble the petals of a chrysanthemum flower. It darkens with age, and upon over maturing, the leaflets soften and wilt. In China, *T. fuciformis* was considered a luxury dish only for the tables of the rich. It has a crunchy but bland taste (The Mala Market, 2021). Owing to its neutral taste, it takes up the flavor of the recipe it is added. They are also made into candies by boiling down in sugar water (Stamets, 2000; Chen, 1978).

Botanical name: *Tremella fuciformis* Berk.

Local name: Yin Er (Chinese), Shiro Kikurage (Japanese)

Common name: Snow ear, snow fungus, silver ear fungus, white jelly mushroom, white jelly leaf, chrysanthemum mushroom (Stamets, 2000).

CHARACTERISTICS

It has a gelatinous, lobbed fruiting body measuring >10 cm across and 4 cm high. The fruiting body (basidiome) is frond-like and clustered by flat flaky leaflets. The margins of the leaflets are lobed and curved. Each leaflet may reach 5–15 × 4–12 cm and 0.5–0.6 mm in thickness. The surface is smooth, shiny, translucent, and white. The basidiocarp is made up of thin septate hyphae with binucleate cells. The hymenium spreads over the whole surface of the fruit body and is very rarely papillate. The probasidium is globose to subglobose. With maturity, it becomes tetra-longitudinally septate (cruciate) and measures around 6–11.4 × 7.2 μ. The basidium varies from 24.5 to 28 μm and consists of hypobasidium and epibasidium. Each hypobasidium gives rise to one epibasidium. The epibasidium measures 12–16 × 2.5–3 μ. Basidiospore is found on each epibasidium. Paraphyses and cystidia are present. Clamp connections are present in some hyphae. The spore print is white. The basidiospores measure 4–8.5 × 6.8–14 μm and are subglobose to obovate and smooth (Kuo, 2008; Chen, 1978; Ghosh et al., 2016).

RESOURCE AVAILABILITY (WILD/CULTIVATED)

T. fuciformis is a parasite that forms its edible basidiome only by parasitizing another fungus. Traditionally it is believed that in wild, *T. fuciformis* feeds on the mycelium of *Annulohypoxylon archeri* (previously *Hypoxylon archeri*) which is a saprobic fungal species and grows on hardwood logs after heavy showers. However, the recent morphological and molecular investigations identify *A. stygium* as the preferred host of *T. fuciformis* (Deng et al., 2016). *A. stygium* is a white-rot ascomycetes fungus from the Xylariaceae family that is highly efficient in degrading lignin and carbohydrates of deadwood.

The traditional method for the cultivation of *T. fuciformis* began in China and Taiwan on logs. Initially, cultivation began by inserting broken fruitbodies directly into holes drilled into logs. The use of fruit bodies was soon replaced by pure culture sawdust spawn which increased yields. Commercial cultivation of *T. fuciformis* employs the mixed culture method for producing spawn

TABLE 16.1
Systematic Classification of *Tremella fuciformis*

Kingdom	Fungi
Division	Basidiomycota
Class	Tremellomycetes
Order	Tremellales
Family	Tremellaceae
Genus	*Tremella*
Species	*fuciformis*

where both *T. fuciformis* and its preferred fungal host are inoculated at a ratio of <1000:1 under optimal growth conditions (Stamets, 2000).

SUSTAINABILITY

Cultivating mushroom, in general, require very less amounts of growing materials, land, water, and energy. Upcycled agricultural wastes are the preferred choice of substrates for mushrooms and thus their cultivation leaves a very smaller footprint on the environment. Also, mushrooms emit very low levels of CO_2.

T. fuciformis can adapt to a wide variety of substrate compositions. Generally, broadleaf hardwood sawdust is supplemented with 20% bran, 1% gypsum, or 1% sugar for its cultivation. As an alternative to sawdust, sugarcane bagasse is also used along with 20% bran, 1% $CaCO_3$, and 1% soybean powder.

When cottonseed hulls are utilized for cultivation, *T. fuciformis* shows a 1:1 ratio of yield: to substrate which is biologically very efficient. Also, its cultivation demands a substrate with 65–70% moisture only (Stamets, 2000). Typical produce of 1 kg rice demands an estimated 2500 L of water (Bouman, 2009).

DISTRIBUTION

It is geographically distributed along tropical and temperate climatic zones. It is found in South-Eastern Asia, North America, the Caribbean and Central America, South America, Sub-Saharan, and South Africa, Australia, and the Pacific Islands (Chen, 1978; Hemmes and Desjardin, 2002; Roberts, 2001; Lowy, 1971).

FOLKLORE

According to Chinese folklore, Yang Guifei, the beloved consort of Emperor Xuanzong from the Tang dynasty, attributed her pristine, radiant, and youthful complexion to snow mushroom's enriching properties (Shahrajabian, 2020; New World Encyclopedia, 2020).

MEDICINAL USE

The polysaccharides of *T. fuciformis* have been clinically used as a naturally occurring nontoxic active product. The primary chain is mannose, whereas the branch chain comprises of glucuronic acid, fucose, xylose, and minor quantities of glucose. Its molecular weight is 1.86×10^6 Da (Wang et al., 2019). In traditional Chinese medicine, *T. fuciformis* is utilized as an immune tonic. It is believed to enhance beauty and support the stomach, the brain, the heart, the kidneys, and the lung (Shahrajabian et al., 2020).

Modern-day findings have shown that sulfation of low molecular weight *T. fuciformis* polysaccharide (LTP) augments the antioxidant activity of natural LTP and *T.fuciformis* polysaccharide (TP). As such, reports suggest that sulfated LTP *in vitro* has better-scavenging activity than LTP and TP (Wu et al., 2007). Further, TP upregulated SIRT1 expression. SIRT1 alleviates oxidative stress and skin fibroblast apoptosis induced by H_2O_2 and through eIF2α deacetylation protects the heart (Shen et al., 2017; Prola et al., 2017). Regeneration of endogenous collagen is also promoted by TP. To protect the structure of the skin from UV damage, TP maintains the I/III collagen ratio (Wen et al., 2016). Besides skin protection, TP extracts also have skin whitening, anti-wrinkling (Lee et al., 2016), and skin wound healing promoting properties (Khamlue et al., 2012). High concentration ethanol precipitates of TPs can prevent obesity. It inhibits the differentiation of 3T3-L1 adipocytes by inhibiting triglyceride accumulation and mRNA expression of PPARγ, C/EBPα, and leptin (Jeong et al., 2008). Mice fed with hypercholesterolemic diets supplemented with 5% dried *T. fuciformis* powder showed a reduction in the levels of serum LDL cholesterol and triglyceride (Cheung, 1996). Cho et al. (2007) reported *T. fuciformis* exopolysaccharide to exhibit hypoglycemic effect and improved insulin sensitivity. Administration of a high dosage of TP in mice with cyclophosphamide-induced immunodeficiency significantly improved the levels of IL-2, IL-12, INF-γ, and IgG in mice blood serum while reducing TGF- β levels. TP also upregulated the expression of immune-related genes such as *IL-1β, IL-4,* and *IL-12* while downregulating *TGF- β* in the liver and the spleen (Zhou et al., 2018). The proliferation of $CD4^+CD25^{(high)}$ Treg $CD4^+T$ lymphocytes is also inhibited by TP in post-burn sepsis mice with *Pseudomonas aeruginosa* infection by reducing IL-10 (Shi et al., 2014). TP reduced inflammatory cell infiltration, inhibit expressions of pro-inflammatory cytokines, and restored functions of intestinal epithelial barrier and mucus barrier in mice with ulcerative colitis, thereby, asserting its role in anti-inflammation too (Xiao et al., 2021). Besides enhancing cognition in subjective cognitive impaired individuals (Ban et al., 2018), *T. fuciformis* also help in nerve regeneration (Hsu et al., 2013) and neuroprotection against glutamate-induced cytotoxicity (Jin et al., 2016).

PHYTOCHEMISTRY

Important Bioactive Secondary Metabolites/Natural Products and/or the Known Pharmacological Active Components

The acidic polysaccharide obtained by hydrothermal method from *T. fuciformis* includes mannose, glucuronic acid, glucose, xylose, and fucose. *T. fuciformis* polysaccharide has a major chain of mannose, with branch chains of glucuronic acid, fucose, xylose, and a fraction of glucose. Polysaccharides from *T. fuciformis* may be built into nanostructures using chitosan because glucuronic acid confers a negative charge on polysaccharides in an aqueous solution (Deqiang Wang et al., 2019). Physicochemical characteristics of chains and conformation, *T. fuciformis* fruit bodies were studied in depth. It existed as flexible chains with a persistence length of 9.20 nm and a chain diameter of 0.97 nm in 0.15 M NaCl (pH 7.4) solution. These properties suggested that it may be utilized to construct food, pharmaceutical, and cosmetic microstructure systems and moisture-holding components (Xiaoqi et al., 2020).

Total sugar, protein, and nucleic acid were discovered in polysaccharide fractions. Furthermore, SAE-1 and SAE-2 fractions had higher iron chelating and superoxide radical scavenging capabilities. This shows that sonicated polysaccharides have strong antioxidant properties (Zou and Hou, 2017).

T3a-T3d are four acidic heteroglycans identified from *T. fuciformis* body. T3a was 550 kDa, T3b was 420 kDa, T3c was 55 kDa, and T3d was 48 kDa. The mannan backbone is made up of 3-linked Man p and side chains with glucosyl, mannosyl, fucosyl, xylosyl, and glucuronic acid residues connected through O-2, O-4, or O-6 of nearly half of the backbone mannosyl residues, according to glycosidic linkage analysis (Gao et al., 1996).

A simple in-situ synthesis technique was designed to synthesize a complex of *T. fuciformis* (TF) and gold nanoparticles (Au NPs). The TF and Au NPs complex was utilized as a surface-enhanced Raman scattering (SERS) substrate, and the shrinking of the TFs produced by drying concentrated dyes on their fruiting bodies resulted in an increase in dye Raman signals. The employment of fungal materials in optical detection of targets might be aided by in-situ production of Au NPs on TF (Bin et al., 2018).

Various Class of Secondary Metabolites

Volatile Oil

The researchers used hydrodistillation and solid-phase microextraction to extract volatile oils from Tremella fuciformis (SPME). Nineteen components were discovered in hydrodistillation and sixty-eight components in SPME. The extract volatile oil contained aromatic chemicals (93.5%), terpenes (5.7%), alkanes (0.4%), and alcohols (0.3%). The greatest concentration was found in butylated hydroxytoluene (92.5%). Acetic acid, borneol, and (-)-α-terpineol were found (Liu et al., 2019).

Sterol

Despite the fact that a GC/MS analysis of *T. fuciformis* oil indicated the presence of five components, lanosterol and lupeol were identified to be the two primary therapeutic chemicals responsible for the mushroom's dermatological and neutraceutical potentials (Ohiri, 2017).

Disaccharide Sugar

A disaccharide $C_{12}H_{22}O_{11}.2H_2O$ {systematic name: 6,60-oxybis[2-(hydroxymethyl)-3,4,5,6-tetrahydro-2H-pyran-3,4,5- triol] dihydrate}, has been extracted from *T. fuciformis.* Both of the six-membered rings in the molecule assume a chair conformation (Liu et al., 2012).

DIETARY COMPOSITION

In myofibrillar protein-TPS gels, the polysaccharide fraction (TPS) of *T. fuciformis* improved hydrogen bonding, resulting in higher myofibrillar protein hardness and WHC. As a result, TPS exhibited a clear ability to improve the qualities of low-fat pig muscle gels, and it had promising application potential in the production of low-fat meat products (Ma et al., 2021).

The emulsifying characteristics of *T. fuciformis* (TFS) in oil-in-water (O/W) emulsions were compared to those of lotus seed, purple sweet potato, and gum arabic. When compared to lotus seed, purple sweet potato, and gum arabic, the results revealed that TFS had outstanding emulsifying stability and reduced droplet size. TFS was shown to be a better emulsifier than the others in the zeta-potential test. The droplet size of TFS emulsions was the smallest. When it came to freeze-thaw cycles, the cream index of 4.0% TFS was the highest. As a result, the TFS may be utilized in the food sector as an emulsifier and thickening (Zhang et al., 2019).

A study was conducted to produce tremella polysaccharide soft sweets. Single-factor and orthogonal experiments were utilized to confirm the optimal formula and process. The study showed that the soft sweets produced under this condition have a smooth surface, plumping sweet body, chewy taste, and pure flavor (Yang et al., 2017).

Because of their health advantages and protein-interaction capabilities, *T. fuciformis* polysaccharides might be used as a new natural improver of low-fat yoghurt. Polyphenols from Flos Sophorae extract supplemented with polysaccharides from the fungus *T. fuciformis* might be used to improve the health-promoting characteristics of low-fat yoghurts while also addressing their flaws (Renyong, 2020).

ISOLATION TECHNIQUES

Isolation of trehalose: Ethanol was used to extract *T. fuciformis* (5 kg) air-dry powder. After extraction, the solvent was ejected, leaving just the crude extract. Following that, the crude extract was purified according to standard techniques, yielding trehalose. The crystalline form of the extract was obtained by slowly evaporating a solution of hydrous ethanol at room temperature (Liu et al., 2012).

Polysaccharide identified in *T. fuciformis* includes polysaccharide from the fruit body, the cell wall, extracellular polysaccharide, and intracellular polysaccharide. The most typical approach for extracting *Tremella* polysaccharides is ethanol precipitation, however it is time-consuming and resulting in a high rate of polysaccharide loss. In addition, the extract includes a number of contaminants, including pigments, nucleic acids, and proteins, which further have to be isolated and refined (Ma et al., 2021).

The *Tremella* polysaccharide was extracted using ultrasonic-assisted hot-water extraction from *T. fuciformis*. Ultrasonic frequency, extraction material to liquid ratio, extraction temperature, and extraction duration were all factors that influenced the extraction rate of the *Tremella* polysaccharide. The extraction approach was shown to be more successful after orthogonal testing (Xie et al., 2021).

Polysaccharides from *T. fuciformis* were extracted by using an aqueous polyethylene glycol solution followed by ultrafiltration. When using optimum polyethylene glycol-based ultrasound-assisted extraction instead of hot water extraction, the extraction yield was higher. In comparison to ethanol precipitation, polysaccharide recovery and purity were higher (Zhang, 2016).

PHARMACOLOGICAL POTENTIAL

Figure 16.1 and Table 16.2 discuss the various pharmacological activities of *Tremella fuciformis*.

MYCOLOGY

The Tremellales order and family Tremellaceae includes *T. fuciformis*. Some of the popular names for this fungus are snow fungus, snow ear, silver ear fungus, white jelly mushroom, and white

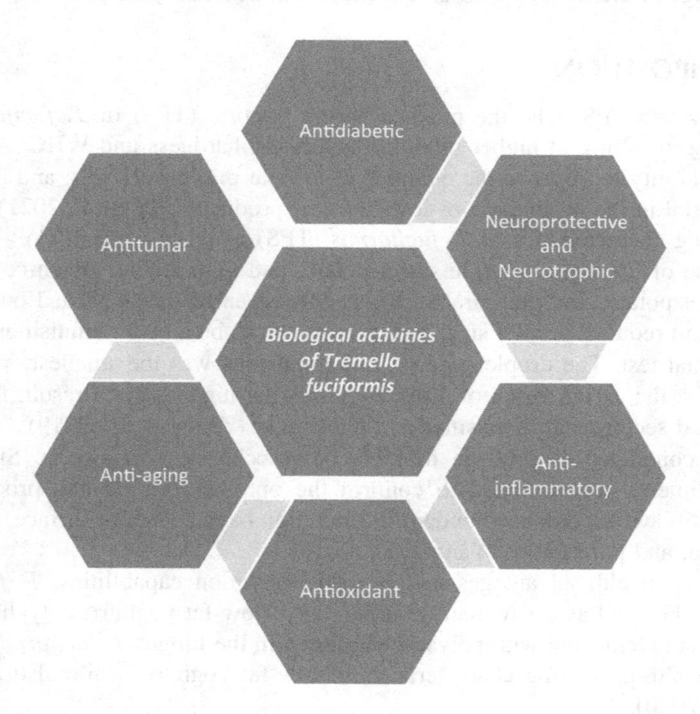

FIGURE 16.1 Biological activities of *Tremella fuciformis*.

TABLE 16.2
Biological Activities of *Tremella fuciformis*

Sr. No.	*Tremella fuciformis*	Biological activities related to medicinal uses	References
1.	Antioxidant activity of protein-*T. fuciformis* polysaccharide complexes	Electrostatic and covalent methods were used to prepare Low-density lipoprotein-*T. fuciformis* polysaccharide (LDL-TFP) complexes and its antioxidant activity was investigated. Results showed that the structure of LDL-TFP complexes could be related to the antioxidant activity of the LDL-TFP complexes on the HepG2.	Chen et al. (2022)
2.	*T. fuciformis* berk polysaccharide to control diabetes	Exopolysaccharides extracted from *T. fuciformis* are beneficial in the control of D type 1 diabetes mellitus until blood glucose 130 mg/dL is achieved by reducing cholesterol, triglyceride, GPT, urea, and increasing HDL cholesterol however Exopolysaccharides in reducing is not satisfactory blood glucose level is above 200 mg/dL.	Bach et al. (2015)
3.	Anti-inflammatory effect of the polysaccharide derived from the mycelium of *T. fuciformis* in mice	In mice, a polysaccharide produced from *T. fuciformis* mycelium had an anti-inflammatory impact on DSS-induced colitis. DSS substantially raised the protein expression of IL-6, TNF-, and COX-2 in colon tissue compared to the control group.	Yoo et al. (2021)
4.	Antioxidant and anti-inflammatory activities of *T. fuciformis* methanol extracts	The edible white jelly mushroom (*T. fuciformis*) subfractions with the highest antioxidant activity, total phenolic content, and flavonoids content were found in methanol extract chloroform subfractions. The chloroform subfraction has the greatest ABTS (+) radical scavenging activity of all the subfractions tested. This subfraction also had the greatest levels of DPPH radical scavenging activity and inhibition of LDL oxidation. Inhibition of nitric oxide generation and inducible nitric oxide synthase expression were also shown to have anti-inflammatory properties in this subfraction.	Li et al. (2014)
5.	*T. fuciformis* polysaccharides and its antioxidant and antitumor activities *in vitro*.	A three-level three-factor Box-Behnken design was used to extract *T. fuciformis* polysaccharides and study them. A temperature of 100 °C produced the best yield of polysaccharides extract. Polysaccharides from T. fuciformis have been shown to scavenge superoxide anion and hydroxyl radicals, according to research. With increasing polysaccharide concentrations, antitumor activity of *T. fuciformis* polysaccharides rose from 73.4% to 92.1%.	Chen (2010)

(*Continued*)

TABLE 16.2 *(Continued)*
Biological Activities of *Tremella fuciformis*

Sr. No.	*Tremella fuciformis*	Biological activities related to medicinal uses	References
6.	Antioxidation Evaluation of Polysaccharides from *T. fuciformis*	The study suggested that polysaccharides of *Tremella fuciformis* extracted by sonication possess high antioxidant activities.	Zou and Hou (2017)
7.	Polysaccharides separated from *T. fuciformis* and their neuroprotective effect	Polysaccharides from *Tremella fuciformis* were purified and characterized and studies confirmed the protective effect of TL04 against glutamate-induced neurotoxicity in DPC12 cells. Caspase-dependent mitochondrial pathway was found to be the mechanism to be associated in this activity.	Jin et al. (2016)
8.	Protective effect of *T. fuciformis* Berk extract on LPS-induced acute inflammation via inhibition of the NF-κB and MAPK pathways	Study showed that oral administration of *Tremella fuciformis* Berk extract significantly inhibited lipopolysaccharide (LPS)-induced IL-1β, IL-6, and TNF-α production and iNOS and COX-2 expression. Gentisic acid, protocatechuic acid, 4-hydroxybenzoic acid, and coumaric acid were found to be the major bioactive compounds from *Tremella fuciformis* Berk extract. In summary, the study suggests that *Tremella fuciformis* Berk extract is a promising anti-inflammatory agent.	Lee et al. (2016)
9.	*T. fuciformis* on UV-induced photoaging	The primary components of *T. fuciformis* water extract were mannose and 10.77 percent (w/w) uronic acid. Polysaccharides from *T. fuciformis* effectively reduce water and collagen losses in the skin while also inhibiting the growth of glycosaminoglycans, making them a promising functional dietary supplement for skin function protection.	Lingrong et al. (2016)
10.	*T. fuciformis* polysaccharide on hydrogen peroxide-induced injury of human skin fibroblasts	Polysaccharide from *T. fuciformis* protects cells against oxidative stress and apoptosis caused by hydrogen peroxide, mostly via upregulating SIRT1 and boosting downstream signaling. This shows that *T. fuciformis* polysaccharide might be used to treat oxidative stress-related skin disorders and ageing.	Shen et al. (2017)
11.	Antioxidant activity of polysaccharides from *Tremella fuciformis*	A study showed that degraded samples of polysaccharides from *T. fuciformis* with lower molecular weight have higher antioxidant activities. Combination of Fe^{2+}, ascorbic acid and H_2O_2 was used as degradation regents to obtain the lower molecular weight product.	Zhang et al. (2014)
12.	Neuroprotective and Neurotrophic Effects of *T. fuciformis*	Neurite outgrowth was increased in PC12h cells by a hot water extract of *T. fuciformis*. *T. fuciformis* might be employed as a preventative in neurodegenerative illness, according to the findings.	Park et al. (2007)

TABLE 16.2 *(Continued)*
Biological Activities of *Tremella fuciformis*

Sr. No.	*Tremella fuciformis*	Biological activities related to medicinal uses	References
13.	*T. fuciformis* on trimethyltin-induced impairment of memory	The purpose of this research was to see how *T. fuciformis* affected learning and memory function, as well as brain activity, in rats with memory deficiencies caused by trimethyltin (TMT). *T. fuciformis* may improve cognitive performance via regulating the CREB signaling pathway and the cholinergic system in the hippocampus, according to the findings.	Park et al. (2012)
14.	*Tremella fuciformis* Polysaccharides in Mitigating Oxidative Stress and Inflammation	*T. fuciformis* polysaccharides prevent LPS-induced oxidative stress and inflammation in macrophages by reducing miR-155 expression and NFκ-B activation, indicating that it might be used as an anti-inflammatory reagent.	Ruan et al. (2018)
15.	Hypoglycemic activity of an acidic *Tremella fuciformis]*polysaccharide	Normal mice showed a substantial dose-dependent hypoglycemic response to glucuronoxylomannan from *T. fuciformis*. The activities of hepatic hexokinase and glucose-6-phosphatase dehydrogenase were dramatically raised when Glucuronoxylomannan was given to normal mice, whereas hepatic glucose-6-phosphatase activity was lowered.	Kiho et al. (1994)
16.	polysaccharides from *T. fuciformis* for antioxidant and moisture-preserving activities	High antioxidant and moisture-preserving activities of CATPs was observed from carboxymethylated polysaccharide of *T. fuciformis*.	Wang et al. (2015)
17.	Protective effect of polysaccharides isolated from *T. fuciformis* against radiation	The radioprotective properties of water-soluble homogenous polysaccharides from *T. fuciformis* were examined in mice. The number of micronuclei generated by 2-Gy irradiation in a group treated with 72 mg/kg b.w. was determined. Chromosome aberration caused by 3-Gy irradiation dropped from 56.01% to 28.13%, and water-soluble homogenous polysaccharides fell from 30.30 to 11.32%.	Xu et al. (2012)
18.	*T. fuciformis* Polysaccharides Inhibited Colonic Inflammation	*T. fuciformis* polysaccharides (TPs) revealed to have protective effect from dextran sulfate sodium (DSS)-induced colitis, and a high dose of TPs could restrain the colon from shortening, scaled down the activity of colonic myeloperoxidase and serum diamine oxidase (DAO), lowered D-lactate concentration, and mitigated damage of colonic tissue.	Xu et al. (2021)
19.	*T. fuciformis* Exopolysaccharide Anti-aging Effects on Skin Cells	To enhance extracellular-polysaccharide (EPS) production, *T. fuciformis* TFCUV5 was generated from TFC6 mycelia using ultraviolet mutagenesis. Because EPS extract improves skin cells' hydration and antiwrinkle factor, it might be utilized in cosmetics.	Jo et al. (2021)

(Continued)

TABLE 16.2 *(Continued)*
Biological Activities of *Tremella fuciformis*

Sr. No.	*Tremella fuciformis*	Biological activities related to medicinal uses	References
20.	*T. fuciformis* Inhibits Melanogenesis in B16F10 Cells and Promotes Migration of Human Fibroblasts and Keratinocytes	In B16F10 cells, *T. fuciformis* decreased melanin content and tyrosinase expression in a concentration-dependent manner. In human HaCaT keratinocytes and Detroit 551 fibroblasts, it also aided wound healing. This study demonstrated its promise as a new skin-whitening agent by suppressing melanogenesis and enhancing wound healing *in vitro*.	Chiang et al. (2022)

auricularia. It's a common macrofungi found in tropical areas on decaying branches of broadleaf trees. Despite its origins in Brazil, it has become one of China's and other Asian countries' most frequently cultivated fungal species. Fruiting bodies of *T. fuciformis* are viscous and pale yellow in color. They are joined together by limbs or brain-like structures, or by multiple flaky, corrugated leaflets (Chen, 1998; Chen et al., 2012; Bach et al., 2015).

The fruit body is transparent to opaque white, strongly gelatinous, agar-like at first, becoming soft and mucilaginous with age, and frequently gaining a sordid tint. The hymenium covers the whole surface of the fruiting body and is seldom papillate. The probasidia had cross septa and were spherical or hemispherical. The epibasidia ranged in size from ovoid to clavate. The nutgall type and the cock's comb type of fruiting bodies are the most common. The first has a bigger and thicker blade than the cock's comb kind and is more valuable economically. The basidia are obovate, measuring 7.2 µm in length, and tightly packed to create the hymenium, which is split longitudinally into four cells. Epibasidia is white, seldom yellow, subglobose to obovate 8–10×1.5 µm in diameter, and thin at one end. Basidiospores are obovate or hemispherical in shape and range in size from 5.3–6.8 m small conidia emerge from the basidiospores. Conidia can sometimes generate mycelium (Chang, 1998; Deng et al., 2016).

The Yellow Brain Fungus, *T. mesenterica*, is mostly seen in the winter when it develops on fallen branches of deciduous trees. This fungus turns a hard orange bracket in dry conditions, making it considerably more difficult to identify. Reproduction takes place through both Sexual and asexual means. It reproduces not only through basidiospores but also through the production of conidiospores. Anders Jahan Retzius (1742–1821), a Swedish botanist, first characterized this jelly fungus in 1769 and named it *T. mesenterica*, which mycologists still use today. *Exidia candida, T. albida, T. candida, T. lutescens Pers*., and *Hormomyces aurantiacus Bonord* are all synonyms for *T. mesenterica*. *T. mesenterica* is the genus Tremella's type species (Wong et al., 1985). When wet, the fruitbody is usually golden yellow and gelatinous, but when dry, it becomes orange and shrinks to a quarter of its original size; originally disc-like, the fruitbody eventually develops random contortions that only loosely resemble the structure of a brain. Fruitbodies range in size from 2 to 8 cm in diameter. Basidia are ellipsoidal, smooth, and cruciately septate (walls split them into four compartments, giving them the appearance of 'hot cross buns' when viewed from above). Basidiospores are inamyloid and measure 7–16 × 6–10 µm. Conidia (asexual spores) are 2–3 × 2–2.5 µm and are spherical, ovoid, or broadly ellipsoidal. Hyphae are found with clamp connections (Wong et al., 1985).

METABOLOMICS AND DNA SEQUENCING

Technological improvements have made high-throughput metabolomics studies of higher fungal metabolites conceivable. Metabolomic analysis is routinely integrated with other omics technologies, such as transcriptomic and proteomic research, to get more detailed data.

High-performance liquid chromatography (HPLC), ultraviolet (UV) spectroscopy, infrared (IR) spectroscopy, gas chromatography mass spectrometry (GC-MS), and nuclear magnetic resonance (NMR) were used to evaluate *T. fuciformis* polysaccharides. Yui et al. (1995) determined that the principal (or backbone) chain of extracted and purified TPS was mannose, with the C-2 of the main chain mannose attached to xylose, uluconic acid, and xylobiose. In 2016, Jin and coworkers purified A TPS, also known as TL04.

Less research has been done on the polysaccharide's main repeating unit structure, although the majority of TPS articles have only shown monosaccharide content and molecular weight. The polysaccharides such as T3a-1, T3a-2, T3a-3, T3a-4, T2a, T2c, T2d, SAE-1, SAE-2, TWE-1, TWE-2, A-BTF, PTF-M38, PTF-Y3, PTF-Y8, T-19, T-7, TAPA1, and TP have been reported from *T. fuciformis* previously by several authors (Wu et al., 2019). The two principal medicinal components identified by GC/MS are lanosterol and lupeol which may be responsible for dermatological and nutraceutical properties. This might be due to their unique properties or a synergistic impact of these triterpenoids and other medicinal non-volatile chemicals found in this fungus (Ohiri et al., 2017). The polysaccharides of *T. mesenterica* CBS 123296 include β-glucans, preferably a linear 3.4 β-glucan, and glucuronoxylomannan (Wasser, 2012). As per the literature survey, fewer reports are available on the metabolomics of both species.

T. fuciformis has a genomic size of 24.5 Mb and 1040 predicted genes. It exhibited a strong penchant for C and then G (C+G; 66.4 percent) at the third base pair position of utilized codons, despite the fact that the mean GC content of predicted genes was somewhat higher (58.5%) in comparison to the overall genome sequence average (55.9%). Despite the fact that *T. fuciformis* and *T. mesenterica* showed significantly higher overall sequence homology, only 7 in 21 optimum codons were similar, whereas *T. fuciformis* had 20 out of 21 codons in common with other species. Tremella uses a variety of codons, which need be determined for each species.

In the establishment of *T. fuciformis* as a bioreactor system, finding suitable and high expression related codons is thus crucial. 201 and 134 CAZymes were discovered in the genomes of *T. fuciformis* and *T. mesenterica*, respectively. Glycoside hydrolase, glycosyltransferase, polysaccharide lyases, and carbohydrate esterases are the most prevalent CAZymes in *T. fuciformis* and *T. mesenterica* (Fang et al., 2020).

ECONOMIC IMPORTANCE

T. fuciformis has a low calorie and fat content but is high in protein, polysaccharides, and dietary fiber. It is rich in nutrients, trace elements, and vitamins and most common culinary and herbal medicine component in Asian countries (Zhao et al., 2011). It is also used in used as a tonic. *T. fuciformis* is also rich in polyphenols having a strong antioxidant activity (Li et al., 2014). The polysaccharides of *T. fuciformis* are thought to be good elements for healthy meals and pharmaceutics (Wang et al., 2018) and are identified as the most active component concerning human nutrition and health (Zhang et al., 2017). Anti-aging, anti-ulcer, anti-thrombosis, and anti-mutagenicity are only a few of the bioactivities of polysaccharides. Numerous research have revealed that *T. fuciformis* polysaccharides (TPS) have anti-tumor, immunomodulatory, anti-oxidative, anti-aging, memory restoration, anti-inflammatory, hypoglycemic, and hypocholesterolemic activities (Wu et al., 2019).

T. mesenterica (Heterobasidiomycetes) is a popular, edible, and medicinal fungus in Orient, containing up to 20% of the polysaccharide glucuronoxylomannan (GXM) in its fruiting bodies (Reshetnikov et al., 2000). *T. mesenterica's* anti-hyperglycemic activities are aided by its fiber, polysaccharide, and other constituents. *T. mesenterica* may also have anti-hypertriglyceridemia capabilities, which might aid in the reduction of diabetes-related macrovascular complications (Lo et al., 2006). *T. mesenterica* is a yellow brain mushroom with a wide range of therapeutic qualities, including immune-stimulating, radiation protection, antidiabetic, anti-inflammatory, hypocholesterolemic, hepatoprotective, and antiallergic capabilities. Most of the above-mentioned therapeutic activities of *T. mesenterica* are dependent on glucuronoxylomannan (GXM) found in fruiting bodies or synthesized in pure culture conditions (Vinogradov et al., 2004).

TOXICOLOGY

Nutraceuticals from mushrooms are known as mushroom nutraceuticals. They are enhanced or modified functional foods that are taken as part of a regular diet to give health advantages. They are refined or partially defined extractives that are taken as a dietary supplement (not food) in the form of capsules or tablets and have potential therapeutic applications, such as improving disease resistance and triggering disease regression. Pharmaceuticals, on the other hand, are specified chemical compositions whose specifications are documented in a pharmacopeia and are mostly employed therapeutically; pharmaceuticals have great potency; therefore, the risk of overdose and toxicity necessitates government supervision. Mushrooms with a low risk of toxicity and overdose don't need to be regulated as strictly and can be sold as 'over-the-counter' medications To date no toxicity data is available for *T. mesenterica* and *T. fuciformis* (Rajarathnam and Shashirekha, 2003).

CONCLUSION AND FUTURE PROSPECTIVE

Many wild-grown mushrooms are an excellent diet for the consumer, with high protein, carbohydrate, necessary minerals, and low energy levels that may be compared to meat, eggs, and milk. The potential medical value is also considerable, with benefits such as immune system stimulation, blood lipid management, anticancer activity, and so on. More comprehensive and effective preservation procedures and culinary treatments are required to further maintain the nutrients. In light of the current circumstances, these components' study is lacking. To safeguard the safety of consumers, we should detect more hazardous wild-grown mushrooms and test for harmful compounds. As a result, additional research is needed to reduce waste, improve resource efficiency, and avoid overuse.

Diverse studies have identified a number of physiologically active chemicals from *T. mesenterica* that are responsible for numerous biological activities; however, the use of many of these components is debatable and has yet to be established in clinical trials. Research is needed to modify the structural properties of existing mushroom varieties to improve their biological properties through small chemical modifications. More knowledge on the mechanisms of action of several metabolites isolated from mushrooms for the treatment or prevention of illnesses, as well as preclinical and clinical proof based upon existing results, is needed to improve mushrooms' potential as a medicine.

REFERENCES

Bach, Erna E., Costa, Silvia G., Oliveira, Helenita A., Silva Junior, Jorge A., da Silva, Keisy M., de Marco, Rogerio M., Bach Hi, Edgar M., & Wad, Nilsa S. Y. (2015). Use of polysaccharide extracted from *Tremella fuciformis* berk for control diabetes induced in rats. Emirates Journal of Food and Agriculture, 27(7), 585–591. doi: 10.9755/ejfa.2015.05.307

Ban, S., Lee, S. L., Jeong, H. S., Lim, S. M., Park, S., Hong, Y. S., & Kim, J. E. (2018). Efficacy and safety of Tremella fuciformis in individuals with subjective cognitive impairment: A randomized controlled trial. Journal of Medicinal Food, 21(4), 400–407.

Bouman, B. (2009). How much water does rice use? Management, 69, 115–133.

Chang, H. Y. (1998). Mycelial properties of Tremella fuciformis and Hypoxylon sp. The Korean Journal of Mycology, 26(3), 321–326.

Chen, B. (2010). Optimization of extraction of *Tremella fuciformis* polysaccharides and its antioxidant and antitumour activities in vitro. Carbohydrate Polymers, 81(2), 420–424.
Chen, C. (1998). Morphological and Molecular Studies in the Genus Tremella. AGRIS.
Chen, P. C. (1978). The Biology and Cultivation of Edible Mushrooms || Tremella fuciformis, 629–643. doi: 10.1016/b978-0-12-168050-3.50035-9
Chen Xue, Yiling Tian, Jie Zhang, Yueqiu Li, Wenhui Zhang, Jing Zhang, Yuwei Dou, & Haiyang Dou (2022). Study on effects of preparation method on the structure and antioxidant activity of protein-*Tremella fuciformis* polysaccharide complexes by asymmetrical flow field-flow fractionation. Food Chemistry, 384, 132619. 10.1016/j.foodchem.2022.132619.
Chen, Y., Zhao, L., Liu, B., & Zuo, S. (2012). Application of response surface methodology to optimize microwave-assisted extraction of polysaccharide from Tremella. Physics Procedia, 24, 429–433.
Cheung, P. C. (1996). The hypocholesterolemic effect of two edible mushrooms: Auricularia auricula (tree-ear) and Tremella fuciformis (white jelly-leaf) in hypercholesterolemic rats1. Nutrition Research, 16(10), 1721–1725.
Chiang, J. H., Tsai, F. J., Lin, T. H., Yang, J. S., & Chiu, Y. J. (2022 Mar-Apr). *Tremella fuciformis* Inhibits Melanogenesis in B16F10 Cells and Promotes Migration of Human Fibroblasts and Keratinocytes. In Vivo, 36(2), 713–722. doi: 10.21873/invivo.12757. PMID: 35241526.
Cho, E. J., Hwang, H. J., Kim, S. W., Oh, J. Y., Baek, Y. M., Choi, J. W.,… & Yun, J. W. (2007). Hypoglycemic effects of exopolysaccharides produced by mycelial cultures of two different mushrooms Tremella fuciformis and Phellinus baumii in ob/ob mice. Applied Microbiology and Biotechnology, 75(6), 1257–1265.
Deng, Y., van Peer, A. F., Lan, F. S., Wang, Q. F., Jiang, Y., Lian, L. D., Lu, D. M., & Xie, B. (2016). Morphological and Molecular Analysis Identifies the Associated Fungus ("Xianghui") of the Medicinal White Jelly Mushroom, Tremella fuciformis, as Annulohypoxylon stygium. International Journal of Medicinal Mushrooms, 18(3), 253–260.
Fang, M., Wang, X., Chen, Y., Wang, P., Lu, L., Lu, J., … & Zhang, Y. (2020). Genome sequence analysis of Auricularia heimuer combined with genetic linkage map. Journal of Fungi, 6(1), 37.
Gao, Q. P., Jiang, R. Z., Chen, H. Q., Jensen, E., & Seljelid, R. (1996 Aug). Characterization and cytokine stimulating activities of heteroglycans from Tremella fuciformis. Planta Medica, 62(4), 297–302. doi: 10.1055/s-2006-957888. PMID: 8792658.
Ghosh, S. K., Mitra, S., & Mukherjee, S. (2016). Study of Jelly Mushroom-Tremella fuciformis In 24-Parganas (N), West Bengal, India. Australian Journal of Basic and Applied Sciences, 10(12), 457–461.
Hemmes, D. E., & Desjardin, D. E. (2002). Mushrooms of Hawai'i: An Identification Guide. Ten Speed Press. ISBN 1-58008-339-0.
Hsu, S. H., Chan, S. H., Weng, C. T., Yang, S. H., & Jiang, C. F. (2013). Long-term regeneration and functional recovery of a 15 mm critical nerve gap bridged by tremella fuciformis polysaccharide-immobilized polylactide conduits. Evidence-Based Complementary and Alternative Medicine, 2013.
Jeong, H. J., Yoon, S. J., & Pyun, Y. R. (2008). Polysaccharides from Edible Mushroom Hinmogi (Tremella fuciformis) Inhibit Differentiation of 3T3-L1 Adipocytes by Reducing mRNA Expression of PPARγ, C/EBPα, and Leptin. Food Science and Biotechnology, 17(2), 267–273.
Jin, Y., Hu, X., Zhang, Y., & Liu, T. (2016). Studies on the purification of polysaccharides separated from Tremella fuciformis and their neuroprotective effect. Molecular Medicine Reports, 13(5), 3985–3992. 10.3892/mmr.2016.5026
Jo, M. H., Kim, B., Ju, J. H. et al. (2021). *Tremella fuciformis* TFCUV5 Mycelial Culture-derived Exopolysaccharide Production and Its Anti-aging Effects on Skin Cells. Biotechnology and Bioprocess Engineering, 26, 738–748. 10.1007/s12257-020-0361-6
Khamlue, R., Naksupan, N., Ounaroon, A., & Saelim, N. (2012, September). Skin wound healing promoting effect of polysaccharides extracts from Tremella fuciformis and Auricularia auricula on the ex-vivo porcine skin wound healing model. In Proceedings of the 4th International Conference on Chemical, Biological and Environmental Engineering (Vol. 43, pp. 93–98). IACSIT Press: Singapore.
Kiho T., Tsujimura Y., Sakushima M., Usui S., & Ukai S. (1994 May) [Polysaccharides in fungi. XXXIII. Hypoglycemic activity of an acidic polysaccharide (AC) from Tremella fuciformis]. Yakugaku Zasshi, 114(5), 308–315. Japanese. doi: 10.1248/yakushi1947.114.5_308. PMID: 8014840.
Kuo, M. (2008, November). Tremella fuciformis. Retrieved from the MushroomExpert.Com Web site: http://www.mushroomexpert.com/tremella_fuciformis.html
Lee, J., Ha, S. J., Lee, H. J., Kim, M. J., Kim, J. H., Kim, Y. T., … & Jung, S. K. (2016). Protective effect of Tremella fuciformis Berk extract on LPS-induced acute inflammation via inhibition of the NF-κB and MAPK pathways. Food & Function, 7(7), 3263–3272.

Lee, K. H., Park, H. S., Yoon, I. J., Shin, Y. B., Baik, Y. C., Kooh, D. H., … & Kim, M. S. (2016). Whitening and anti-wrinkle effects of Tremella fuciformis extracts. Korean Journal of Medicinal Crop Science, 24(1), 38–46.

Li, H., Lee, H. S., Kim, S. H., Moon, B., & Lee, C. (2014 Apr). Antioxidant and anti-inflammatory activities of methanol extracts of *Tremella fuciformis* and its major phenolic acids. Journal Food Science, 79(4), C460–C468. doi: 10.1111/1750-3841.12393. Epub 2014 Feb 18. PMID: 24547933.

Lingrong Wen, Qing Gao, Chung-wah Ma, Yazhong Ge, Lijun You, Rui Hai Liu, Xiong Fu, & Dong Liu (2016). Effect of polysaccharides from Tremella fuciformis on UV-induced photoaging. Journal of Functional Foods, 20, 400–410. 10.1016/j.jff.2015.11.014.

Liu, W., Yan, J., Song, Q., Gou, X. J., & Chen, F. Z.. (2012 Aug 1). Trehalose dihydrate from Tremella fuciformis. Acta Crystallogr Sect E Struct Rep Online, 68(Pt 8), o2511. doi: 10.1107/S1600536812031947. Epub 2012 Jul 21. PMID: 22904949; PMCID: PMC3414962.

Liu, W., Tang, Q. T., Wei, Y. T., Han, L., Han, W., Feng, N., & Zhang, J. S. (2019). Chemical Compounds and Antioxidant Activity of Volatile Oil from the White Jelly Mushroom, Tremella fuciformis (Tremellomycetes). Int J Med Mushrooms, 21(3), 207–214. doi: 10.1615/IntJMedMushrooms.2019030099. PMID: 31002605.

Lo, H. C., Tsai, F. A., Wasser, S. P., Yang, J. G., & Huang, B. M. (2006). Effects of ingested fruiting bodies, submerged culture biomass, and acidic polysaccharide glucuronoxylomannan of Tremella mesenterica Retz.: Fr. on glycemic responses in normal and diabetic rats. Life Sciences, 78(17), 1957–1966.

Lowy, B. (1971). Flora Neotropica 6: Tremellales. New York: Hafner. ISBN 0-89327-220-5.

Ma, X., Yang, M., He, Y., Zhai, C., & Li, C. (2021). A review on the production, structure, bioactivities and applications of Tremella polysaccharides. International journal of immunopathology and pharmacology, 2021; 35, 20587384211000541. 10.1177/20587384211000541

Ohiri, R. C. (2017). GC/MS analysis of *Tremella fuciformis* (White jelly mushrooms) oil. Ukr. Biochem. J., 89(3), 46–51. doi: 10.15407/ubj89.03.046

Park, H. J., Shim, H. S., Ahn, Y. H., Kim, K. S., Park, K. J., Choi, W. K., Ha, H. C., Kang, J. I., Kim, T. S., Yeo, I. H., Kim, J. S., & Shim, I. (2012 Apr 1). *Tremella fuciformis* enhances the neurite outgrowth of PC12 cells and restores trimethyltin-induced impairment of memory in rats via activation of CREB transcription and cholinergic systems. Behav Brain Res., 229(1), 82–90. doi: 10.1016/j.bbr.2011.11.017. Epub 2011 Dec 14. PMID: 22185695.

Park, K. J., Lee, S. Y., Kim, H. S., Yamazaki, M., Chiba, K., & Ha, H. C. (2007). The Neuroprotective and Neurotrophic Effects of *Tremella fuciformis* in PC12h Cells. Mycobiology, 35(1), 11–15. 10.4489/MYCO.2007.35.1.011

Prola, A., Pires Da Silva, J., Guilbert, A., Lecru, L., Piquereau, J., Ribeiro, M., … & Lemaire, C. (2017). SIRT1 protects the heart from ER stress-induced cell death through eIF2α deacetylation. Cell Death & Differentiation, 24(2), 343–356.

Rajarathnam, S., & Shashirekha, M. N. (2003). Mushrooms and TrufflesClassification and Morphology In: Encyclopedia of Food Sciences and Nutrition, 2nd ed, Academic Press, United States, pp. 4040–4048.

Reshetnikov, S.V., Wasser, S.P., Nevo, E., Duckman, I., & Tsukor, K. (2000). Medicinal value of the genus Tremella Pers. (Heterobasidiomycetes). International Journal of Medicinal Mushroom, 2, 169–193.

Roberts, P. (2001). Heterobasidiomycetes from Korup National Park, Cameroon. Kew Bulletin, 56(1), 163–187. doi:10.2307/4119434

Ruan, Y., Li, H., Pu, L., Shen, T., & Jin, Z. (2018). *Tremella fuciformis* Polysaccharides Attenuate Oxidative Stress and Inflammation in Macrophages through miR-155. Analytical cellular pathology (Amsterdam), 2018, 5762371. 10.1155/2018/5762371

Shahrajabian, M. H., Sun, W., Shen, H., & Cheng, Qi. (2020). Chemical compounds and health benefits of Tremella, a valued mushroom as both cuisine and medicine in ancient China and modern era. Amaz. Jour. of Plant Resear, 4(3), 692–697.

Shen, T., Duan, C., Chen, B., Li, M., Ruan, Y., Xu, D., Shi, D., Yu, D., Li, J., & Wang, C. (2017). *Tremella fuciformis* polysaccharide suppresses hydrogen peroxide-triggered injury of human skin fibroblasts via upregulation of SIRT1. Molecular medicine reports, 16(2), 1340–1346. 10.3892/mmr.2017.6754

Shi, Z. W., Liu, Y., Xu, Y., Hong, Y. R., Liu, Q., Li, X. L., & Wang, Z. G. (2014). Tremella Polysaccharides attenuated sepsis through inhibiting abnormal CD4+ CD25high regulatory T cells in mice. Cellular Immunology, 288(1-2), 60–65.

Snow Fungus Jujube Dessert Soup (Yin'er Tang, 银耳汤) (2021, September 26). The Mala Market Blog. Retrieved from the blog.themalamarket.com website: https://blog.themalamarket.com/dessert-soup-yiner-tang/

Stamets, P. (2000). "Chapter 21: Growth Parameters for Gourmet and Medicinal Mushroom Species". Growing gourmet and medicinal mushrooms = [Shokuyo oyobi yakuyo kinoko no sabai] (3rd ed.). Berkeley, California, USA: Ten Speed Press. pp. 402–405. ISBN 978-1-58008-175-7.

Tang Renyong, Yurong Lu, Caiyun Hou, Jiaxuan Peng, Wei Wang, & Xiulan Guo. (2020). Co-Supplementation of Flos Sophorae Extract with *Tremella fuciformis* Polysaccharides Improves Physicochemical, Textural, Rheological, and Antioxidant Properties of Low-Fat Yogurts. Journal of Food Quality, 2020, Article ID 2048756, 9. 10.1155/2020/2048756

Tang Bin, Jun Liu, Linpeng Fan, Daili Li, Xinzhu Chen, Ji Zhou, & Jingliang Li. (2018). Green preparation of gold nanoparticles with Tremella fuciformis for surface enhanced Raman scattering sensing. Applied Surface Science, 427, Part A: 210–218. 10.1016/j.apsusc.2017.08.008.

Vinogradov, E., Petersen, B. O., Duus, J. Ø., & Wasser, S. (2004). The structure of the glucuronoxylomannan produced by culinary-medicinal yellow brain mushroom (Tremella mesenterica Ritz.: Fr., Heterobasidiomycetes) grown as one cell biomass in submerged culture. Carbohydrate research, 339(8), 1483–1489.

Wang, D., Wang, D., Yan, T., Jiang, W., Han, X., Yan, J., & Guo, Y. (2019). Nanostructures assembly and the property of polysaccharide extracted from Tremella Fuciformis fruiting body. International Journal of Biological macromolecules, 137, 751–760.

Wang Deqiang, Deguo Wang, Tingxuan Yan, Weifeng Jiang, Xinya Han, Jvfen Yan, & Yanrong Guo. (2019). Nanostructures assembly and the property of polysaccharide extracted from *Tremella Fuciformis* fruiting body. International Journal of Biological Macromolecules, 137, 751–760. 10.1016/j.ijbiomac.2019.06.198. doi: 10.3389/fimmu.2021.648162

Wang X., Zhang Z., & Zhao M. (2015 Jan). Carboxymethylation of polysaccharides from *Tremella fuciformis* for antioxidant and moisture-preserving activities. Int J Biol Macromol. 72, 526–530. doi: 10.1016/j.ijbiomac.2014.08.045. Epub 2014 Sep 4. PMID: 25194971.

Wang, Z., Xie, J., Shen, M., Nie, S., & Xie, M. (2018). Sulfated modification of polysaccharides: synthesis, characterization and bioactivities. Trends Food Sci. Technol., 74, 147–157.

Wasser, S. P. (2012). *U.S. Patent Application No. 13/256,043*.

Wen, L., Gao, Q., Ma, C. W., Ge, Y., You, L., Liu, R. H., … & Liu, D. (2016). Effect of polysaccharides from Tremella fuciformis on UV-induced photoaging. Journal of Functional Foods, 20, 400–410.

Wong, G. J., Wells, K., & Bandoni, R. J. (1985). Interfertility and comparative morphological studies of Tremella mesenterica. Mycologia, 77(1), 36–49.

Wu, Q., Zheng, C., Ning, Z. X., & Yang, B. (2007). Modification of low molecular weight polysaccharides from Tremella fuciformis and their antioxidant activity in vitro. International Journal of Molecular Sciences, 8(7), 670–679.

Wu, Y. J., Wei, Z. X., Zhang, F. M., Linhardt, R. J., Sun, P. L., & Zhang, A. Q. (2019). Structure, bioactivities and applications of the polysaccharides from Tremella fuciformis mushroom: a review. International journal of biological macromolecules, 121, 1005–1010.

Xiao, H., Li, H., Wen, Y., Jiang, D., Zhu, S., He, X., … & Liang, J. (2021). Tremella fuciformis polysaccharides ameliorated ulcerative colitis via inhibiting inflammation and enhancing intestinal epithelial barrier function. International Journal of Biological Macromolecules, 180, 633–642.

Xiaoqi Xu, Aijun Chen, Xinyan Ge, Sha Li, Tao Zhang, & Hong Xu (2020). Chain conformation and physicochemical properties of polysaccharide (glucuronoxylomannan) from Fruit Bodies of *Tremella fuciformis*. Carbohydrate Polymers, 245, 116354. 10.1016/j.carbpol.2020.116354.

Xie Ling-na, Hang Ping, & Du Zhi-yun. (2021). An Optimization of Ultrasonic Extraction Process of Tremella Polysaccharides and its Anti-inflammation Effect on BV2 cell[J]. Journal of Guangdong University of Technology, 38(02), 94–98.

Xu, W., Shen, X., Yang, F., Han, Y., Li, R., Xue, D., & Jiang, C. (2012). Protective effect of polysaccharides isolated from *Tremella fuciformis* against radiation-induced damage in mice. J Radiat Res., 53(3), 353–360. doi: 10.1269/jrr.11073. PMID: 22739004.

Xu Yingyin, Liyuan Xie, Zhiyuan Zhang, Weiwei Zhang, Jie Tang, Xiaolan He, Zhou Jie, & Weihong Peng (2021). *Tremella fuciformis* Polysaccharides Inhibited Colonic Inflammation in Dextran Sulfate Sodium-Treated Mice via Foxp3+ T Cells, Gut Microbiota, and Bacterial Metabolites, Frontiers in Immunology, 12, 1–15.

Yang Guifei. (2020, October 15). New World Encyclopedia. Retrieved from the newworldencyclopedia.org website: https://www.newworldencyclopedia.org/p/index.php?title=Yang_Guifei&oldid=104389

Yang, X. Z., Duan, X. Y., Hong, D., et al. (2017). Processing technology of tremella polysaccharide soft sweets. Value Engineering, 36(11), 135–137.

Yoo, Sun Hee, & Kang, Soon Ah. (2021). Anti-Inflammatory Effects of Polysaccharides Isolated from *Tremella fuciformis* Mycelium on Dextran Sulfate Sodium-Induced Colitis, Model. Korean J. Food Nutr. 34(2) 146–155. 10.9799/ksfan.2021.34.2.146

Yui, T., Ogawa, K., Kakuta, M., & Misaki, A. (1995). Chain Conformation of a Glucurono-xylo-mannan Isolated from Fruit Body of *Tremella fuciformis* Berk. Journal of Carbohydrate Chemistry, 14, 255–263.

Zhang, Jian, Zhang, Ya-Kun, Liu, Yong, & Wang, Jun-Hui (2019). Emulsifying Properties of *Tremella Fuciformis:* A Novel Promising Food Emulsifier. International Journal of Food Engineering, 15(3-4), 20180217. https://doi.org/10.1515/ijfe-2018-0217

Zhang, L., & Wang, M. (2016). Polyethylene glycol-based ultrasound-assisted extraction and ultrafiltration separation of polysaccharides from *Tremella fuciformis* (snow fungus). Food and Bioproducts Processing, 100, 464–468.

Zhang, Y.-K., Zhang, Q., Lu, J., Xu, J.-L., Zhang, H., & Wang, J.-H. (2017). Physicochemical properties of Tremella fuciformis polysaccharide and its interactions with myofibrillar protein. Bioact. Carbohydr. Diet. Fibre, 11, 18–25.

Zhang, Z., Wang, X., Zhao, M.,& Qi, H. (2014 Nov 4). Free-radical degradation by Fe2+/Vc/H2O2 and antioxidant activity of polysaccharide from *Tremella fuciformis*. Carbohydr Polym, 112, 578–582. doi: 10.1016/j.carbpol.2014.06.030. Epub 2014 Jun 18. PMID: 25129784.

Zhao, X., Hu, Y., Wang, D., Guo, L., Yang, S., Fan, Y., Zhao, B. et al. (2011). Abula, optimization of sulfated modification conditions of Tremella polysaccharide and effects of modifiers on cellular infectivity of NDV. Int. J. Biol. Macromol., 49(1), 44–49.

Zhou, Y., Chen, X., Yi, R., Li, G., Sun, P., Qian, Y., & Zhao, X. (2018). Immunomodulatory effect of tremella polysaccharides against cyclophosphamide-induced immunosuppression in mice. Molecules, 23(2), 239.

Zou, Y., & Hou, X. (2017). Extraction Optimization, Composition Analysis, and Antioxidation Evaluation of Polysaccharides from White Jelly Mushroom, *Tremella fuciformis* (Tremellomycetes). Int J Med Mushrooms, 19(12), 1113–1121. doi: 10.1615/IntJMedMushrooms.2017024590. PMID: 29431072.

17 Lesser Known Mushrooms

Hellvella crispa (Scop.) Fr., Lactarius volemus (Fr.), Clavulinopsis fusiformis (Sowerby), Hypholoma capnoides (Fr.), Psathyrella spadicea, and Suillus granulatus

Manju Nehra, Vikash Nain, and Amanjyoti
Chaudhary Devi Lal University, Sirsa, Haryana, India

CONTENTS

DOI: 10.1201/9781003259763-17

INTRODUCTION

Worldwide, gourmet cuisine has used mushrooms as an ingredient because of their distinctive flavour and the value they have historically held in the eyes of mankind as a gastronomic marvel. There really are about 2,000 different kinds of mushrooms in the world, and only approximately 25 of them are regularly eaten as foodstuffs, and only a small number are raised for commercial purposes. Due to their medicinal effectiveness, sensory appeal, and economic importance, mushrooms are recognised as a treat with high nutritional and functional quality and are also classified as nutraceutical foods. (S. T. Chnag and P. G. Miles, 2008; P. G. Ergon et al., 2013). However, it can be difficult to distinguish between edible and medicinal mushrooms because numerous of the more widely available culinary species also contain medicinal qualities (E. Guillamon et al., 2010). Mushrooms may include secondary metabolites such as quinolones, benzoic acid derivatives, anthraquinones, steroids, and terpenes in addition to some main metabolites such as proteins, peptides, and oxalic acid. Lentinus edodes, the species that has been the subject of the majority of studies, appears to have antibacterial action on both gram-positive and gram-negative bacteria (M. Alves et al., 2012). Being relatively high in protein, rich in fibre and essential amino acids, low in fat but high in essential fatty acids, and comparatively high in protein, they have outstanding nutritional value. It also contains some fat-soluble and water-soluble vitamins (B_1, B_2, B_{12}, C, D, and E) (S.A. Heleno et al., 2010; P. Mattila et al., 2001). Mushrooms are a good source of nutrients and bioactive components. (L.Barros et al., 2007, 2008; I.C.F.R. Ferreira et al., 2009; E. Pereira et al., 2012; J.A. Vaz et al., 2010).

Numerous types of mushrooms have been traditionally employed in many different cultures for health maintenance and the prevention and treatment of serious diseases due to their immunomodulatory and anti-cancer properties. The pharmacological potential of mushrooms has attracted a lot of attention in the past ten years, and it has been hypothesised that many of them serve as miniature pharmaceutical factories, creating chemicals with amazing biological capabilities (S. Patel and A. Goyal, 2012; I.C.F.R. Ferreira et al., 2010). Additionally, new medications against aberrant molecular and biochemical signals that cause cancer can be discovered thanks to the increased knowledge of the molecular causes of cancer development or metastasis (B.Z. Zaidman et al., 2005).

Mushrooms and fungi have more than 100 medicinal properties, with the most important ones being their antioxidant, anticancer, anticholesterolemic, antiallergic, immunomodulating, cardiovascular protector, antifungal, antiviral, antibacterial, antiparasitic, antidiabetic, detoxification, and hepatoprotective effects (S.T. Chang and S.P. Wasser; T.C. Finimundy et al., 2013; L.Zhang et al., 2011). The growth of tumours and inflammatory processes are inhibited by mushrooms. It is known that many macrofungi produce bioactive molecules, including polysaccharides, proteins, fats, minerals, glycosides, alkaloids, volatile oils, terpenoids, tocopherols, phenolics, flavonoids, carotenoids, folates, lectins, enzymes, ascorbic acid, and organic acids in general. Fruit tissues, cultured mycelium, and fermented broth all contain these beneficial compounds. The importance of carbohydrates in modern medicine is largest, and -glucan or versatile molecule with a wide spectrum of biologically active compounds. [S. Patel and A. Goyal, 2012, S.T. Chang and S.P. Wasser; T.C. Finimundy et al., 2013; J.Chen and R. Seviour, 2007).

A healthy, balanced diet is the best defence against disease, particularly oxidative stress. Oriental medicine has traditionally used mushrooms as both a disease preventative and a disease therapy. These days, it is used as dietary supplements due to its benefits, particularly for boosting the immune system and having anticancer activity (E. Guillamon et al., 2010, L.Barros et al., 2007, I.C.F.R. Ferreira et al., 2009; T.C. Finimundy et al., 2013; A.C. Brown and C.I. Waslien, 2003; H.G. Kim et al., 2007; C. Sarikurkcu et al., 2008; A. Synytsya et al., 2009; Z.Wang et al., 2004).

The nutritional content, chemical and nutraceutical composition, and market potentials of the most commonly cultivated consumable mushrooms globally were examined in this study.

Helvella crispa [Scop.] Fr

In both the terrestrial biomes of the Northern and Southern Hemispheres, *Helvella* is a common, speciose ascomycete (Pezizomycetes: Pezizales). Some bigger or more charismatic species of the order Pezizales can be found in this genus, which also includes species with folded and lobed caps resting on plain, ribbed or wrinkled stipes. The elaborate apothecia are cam- panulate, cupulate, saddle-shaped, convex, and more. Although obvious polymorphic apothecia easily distinguish *Helvella* species from other macrofungi, *Helvella* species are surprisingly difficult to distinguish from one another. While the microanatomy of the sterile and fertile structures has added a few traits that can be used to distinguish between species, historically, the shape, colour, and outer surface features have been prioritised in the process. N.S. Weber reached the following conclusion after conducting extensive research on Helvella samples from Michigan: "In general, most anatomical and morphological traits imply essentially continuous change in the organism as a whole. A particular region of the character variation spectrum for a given character may be shared by numerous species, and each organisms only represents one region of the spectrum. Consequently, a set of traits is required to identify between species.

The type specimen of the fungus genus *Helvella (Helvellaceae, Pezizales)* is *Helvella mitra (5 Helvella crispa [Scop.] Fr.),* which is distinguished macroscopically by saddle-shaped, cupulate or irregularly-lobed apothecia. These apothecia are generally separated by colours ranging from white to black and by presence of rib cage on the stipe, S.P. Abbott and R.S. Currah, 1997)). Approximately 60 *Helvella* have been described thus far with these traits. It has ellipsoidal or subfusoidal ascospores, a medullary excipulum and stipe inner layer made of cells of textura intricata, and an ectal excipulum and stipitipellis made of textura angularis cells, according to numerous sources (N.S. Weber, 1952; H.Dissing, 1972, S.P. Abbott and R.S. Currah, 1997; R.P. Korf, 1952) According to ecological theory, only a certain species of *Helvella* are important ectomycorrhizal (ECM) fungi in the environment and they can form ECM relationships with Pinaceae and Fagaceae plants (L.C. Maia et al., 1996, L. Tedersoo et al., 2006; J. Hwang et al 2015). Recent multigenic DNA studies have enabled the description of a significant number of novel *Helvella* entities from around the globe (J. Hwang et al 2015; N.H. Nguyen et al., 2013; Q.Zhao et al., 2015; S. Tibpromma et al., 2017). In the shadow regions of deciduous and coniferous forests, many *Helvella* taxa are widely spread (Q.Zhao et al., 2015, Y.Z. Wang and C.M. Chen, 2002).

The excipulum, according to him, has two layers, one of which is made up of intricately intertwined hyphae and the other of nearly isodiametric cells. Textura intricata, which consists the medullary excipulum, & textura angularis to prismatica, which consists the ectal excipulum, are the names later given to these tissues by R.P. Korf (1952). Additionally, *Helvella* ascospores are tetra-nucleate containing two nuclei at each end), have a big central guttula, and may have little apical guttulae.

Description

Helvella crispa has a cap that is 2–5 centimetres in diameter, is a creamy white colour, and measures 6–13 cm (212–5 in) in length. Due to the unevenly formed lobes on the crown and the substantial, creamy-white base (2–8×1–0.025 m in size), it is conspicuous. Its skin is as thin as paper and brittle. The stem measures 3–10 cm (114–4 in) in length and has elaborate ribs. It has a pinkish or white colour. Even though it smells beautiful, it cannot be eaten raw. The spore print is white, and the round spores are typically 19 x 11.5 mm in size. (R. Phillips, 2006) Forms with white caps can occasionally be encountered. It can be identified by its fuzzy cap undersurface and inrolled margins when young, unlike *Helvella lacunosa's* sporadic white forms. (J.F. Ammirati et al., 1985).

Helvella crispa is a fungus that grows in grass, hedges, and the talus of meadows, as well as in humid hardwoods like beech trees. They can be observed between the conclusion of the summer and the start of the fall. (Haas and Hans, 1969). Eastern North America, China, Japan, Europe, and eastern Africa are all places where you can find it. The comparable Helvella lacunosa takes its place in the west of the country (A. David, 1986). Despite the fact that this species is listed as edible in some guidebooks, (R. Phillips, 2006; W. Y. Zhuang, 2004). It is possible that it contains the drug monomethylhydrazine, which can be extremely intoxicating and possibly cancerous. When eaten raw, it has been noted to cause digestive problems (J.F. Ammirati et al., 1985).

Lactifluus volemus

A type of fungus belonging to the Russulaceae family is called *Lactifluus volemus*, formerly known as *L. volemus*. It is extensively distributed in the northern hemisphere and can be found in temperate areas of Europe, North America, and Asia as well as some subtropical and tropical areas of Central America and Asia. It is a mycorrhizal fungus that produces fruit bodies that can grow alone or in clusters from summer to autumn at the foot of many species of trees. It is valued as a mushroom that can be eaten and is available in Asian markets. While *L. volemus* resembles other *Lactifluus* mushrooms, particularly the highly associated edible species *L. corrugis*, they can be distinguished from one another by differences in habitat, external morphology, and microscopic characteristics. The spores of *L. volemus* are generally spherical, 7–8 micrometre in diameter, and leave a white spore print. The *L. volemus* mushroom's cap can grow up to 11 cm (4+12 in) wide and can range in colour from apricot to brown. On the underside of the cap, there are several closely spaced, occasionally forked pale golden yellow gills. Weeping milk cap and voluminous-latex milky are two common names for the mushroom because of the large amount of latex ("milk") that it exudes when the gills are damaged.

It also has a strong fishy smell, but it has no impact on the flavour. Chemical analysis of the fruit bodies revealed various sterols related to ergosterol. The fungus also includes a chemically defined natural rubber. According to phylogenetic analysis, *Lactifluus volemus* is more likely to represent a number of species or subspecies than just one taxon.

Taxonomy

Lactifluus volemus was first reported in the published research in Carl Linnaeus' 1753 book Organisms Plantarum under the name Genus lactifluus. (E.M. Fries, 1821) Elias Magnus Fries, a Swedish mycologist, named it Agaricus volemus in his Systema Mycologicum in 1821(E.M. Fries, 1838). He suggested a tribus, or tribe, of related species within the genus Agaricus, which he named Galorrheus, in this paper. In his 1838 work Epicrisis Systematis Mycologici, Fries later recognised Lactarius as a separate genus, using Galorrheus as a synonym (P. Kummer, 1871) Despite the fact that Linnaeus published the species before Fries, Fries' name has been approved and thus has primacy in terms of nomenclature. Paul Kummer designated the species Galorrheus volemus in 1871 after elevating the majority of Fries' tribes to generic classification (C.H. Peck, 1885). Charles Horton Peck recognised the variant *L. volemus* var. subrugosus in 1879 (W.C. Roody, 2003) but it is now considered to be a different species, *L. corrugis* (O. Kuntze, 1891) In 1891, Otto Kuntze transferred the species to *Lactifluus* (O. Kuntze, 1891). It was then confused for a considerable period of time with Lactarius before being recognised as a separate genus in 2008 and undergoing subsequent taxonomic reorganisations within the Russulaceae family. [B.Buyck, 2008, 2010; A. Verbeken).

Another etymological equivalent is Lactarius lactifluus, which Lucien Quélet adopted in 1886, renaming Agaricus lactifluus after Linnaeus. Lactarius wangii, which Hua-An Wen and Jian-Zhe Ying claimed to be a brand-new species from China in 2005 (H.A. Wen and J.Z. Ying, 2005), was paired with L. volemus two years later. The particular epithet "volemus" is derived from the Latin vola (H.S. Frieze, 1882), which means "the hollow of the hand," suggesting Fries's statement that the significant amount of latex "flowed enough to fill the hand." (S. Metzler and V. Metzler, 1992) Weeping milk cap, tawny milk cap, orange-brown milky, voluminous-latex milky, lactarius orange, fishy milkcap, and apricot milk cap are some of the common names for L. volemus. The term "leatherback" or "bradley" is used to describe the mushroom in the West Virginian highlands of the United States. Its German name, Brätling, may be where the latter name came from (S. Metzler and V. Metzler, 1992).

The current classification of *Lactifluus volemus*, which was the type species of the section Dulces in subgenus Lactarius, is Lactifluus section Lactifluus (A. Verbeken, 2012). This *L. volemus* is surrounded by species that have a dry cap, a lot of latex, and a white or light cream spore print (R. Singer, 1986). It has proven challenging to properly identify between the two species since the closely related *L. corrugis* shares overlapping physical characteristics, such as colouration in the cap and stem that is comparable. The fact that both species have a variety of colour forms makes it more challenging to distinguish between them: *L. volemus* specimens from Japan can have a red cap, a yellow cap with a long stem, or a surface texture like velvet; *L. corrugis* specimens can have either a red or, more frequently, rust-coloured cap. Japanese researchers used molecular phylogenetics to clarify the relationships between these two species and others in the section Dulces in 2005. They also compared variations in shape, taste, and fatty acid composition. The phylogenetically determined subclades in which the colour variations cluster lead to the conclusion that these variations should actually be classified as "different species, subspecies, or varieties." (Y. Shimono et al., 2007). Six of the 18 phylogenetic species identified in a 2010 molecular study of the L. volemus were described as new species: *Lactifluus acicularis, L. crocatus, L. distantifolius, L. longipilus, L. pinguis,* and *L. vitellinus*. The study examined 79 tested specimens of the *L. volemus* from northern Thailand (K. Van de Putt et al., 2010).

Macromorphology and Micromorphology

The Lactifluus volemus fruit body develops a different shape as it ages; initially convex with inwardly curling edges, it subsequently becomes centre dip in a smooth surface. The fruit cap has a smooth or velvety surface and is meaty and solid. Its diameter ranges from 5 to 11 cm, and its shows apricot to brown colour (R. Phillips, 1981). However, there is some variation in the cap colouration, as shown in specimens from Asia, Europe (J. Hellman Clausen, 1998), and North America. The stem is significantly paler in colour than the cap, ranging in height from 4 to 12 cm and 1 to1.5 cm (0.4 and 0.6 in) thick. It is solid and has a surface that is velvety or smooth with occasional longitudinal depressions going up and down its length. The gills are fragile, small, relatively closely spaced, adnate to slightly decurrent, and occasionally forked. The gills, which are typically a light golden yellow colour, turn brown when injured. Lamellulae, which are brief gills that don't reach the stem, are dotted among the gills. White and firm skin is present. The mushroom has a faintly fishy fragrance; one source claims it smells "like a dead shad," the most repulsive-smelling freshwater fish, as fishermen would attest (M. Kuo, 2010). When the fruit bodies are dried, the aroma is intensified. The mushroom's profuse latex is one of its most distinguishing characteristics; it is so abundant that even a little puncture on the gills will cause it to "weep" the milky material. Whatever the latex comes in contact with will usually end up stained brown (M. Kuo, 2010).

The spore print is light-coloured. The spores are typically 7.5–10.0 by 7.5–9.0 m in size, nearly spherical, and transparent (hyaline) (Bessette et al., 2009). Reticulate refers to ridges that cover the spore surface and create a seamless network. The ridges have noticeable projections up to 1.2 metres high and can reach heights of 0.8 metres. The club-shaped, four-spored, hyaline, and club-shaped basidia of the hymenium have dimensions of 40–62 by 7.2–10.4 m (L. Montoya et al., 1996). Cystidia, sterile cells, are scattered amid the basidia. The spindle- to club-shaped pleurocystidia (cystidia on the side of a gill) measure 48–145 by 5–13 m. The spindle-, club-, awl-, or intermediate between these shapes (subulate) cheilocystidia (cystidia on the edge of a gill) measure

27–60 by 5–7 m. Furthermore, cystidia can be found on the surface of the stem and cap (R. Phillips, 1981). When a drop of ferric sulphate is given to the mushroom tissue, the stain turns dark bluish-green very immediately (used to identify mushrooms) (Bessette et al., 2009).

VARIETIES

In their 1979 monograph on North American Lactarius species, Alexander H. Smith and Lexemuel Ray Hesler described the variant *Lactifluus volemus* var. flavus. This rare variety has a cap that stays yellow throughout growth and is only located from South Carolina through Florida and west to Texas. Additionally, the spores are 6.5–9.0 by 6–8 m, little smaller than the usual kind (Bessette et al., 2009). It also makes a tasty food (R. Phillips, 2010).

SIMILAR SPECIES

The appearance of *Lactifluus volemus* and *L. corrugis* is often similar. Although there are intermediate colour variations, More often than not, *L. corrugis* have little orange colour, deeper gills, a weaker or absent scent, and more surface wrinkles. Microscopically, the two can be identified more clearly: *L. corrugis* contains larger spores, which are generally 10.4–12.8 by 9.6–11.8 m in size and have a finer surface reticulum, as well as larger pleurocystidia (R. Phillips, 2010).The reticulations on the spores of the closely related species In comparison to *L. lamprocystidiatus, Lactifluus austrovolemus* is higher, sharper, and produces smaller meshes at the intersection of the reticulations. Papua New Guinea is home to the sole known populations of *L. austrovolemus* and *L. lamprocystidiatus*. Contrary to *L. volemus, Lactifluus hygrophoroides* possesses gills that are widely spaced apart and spores without surface reticulations (D.N. Pegler and J.P. Fiard, 1979).

In addition, some Lactarius species resemble one another: Although the tropical African *Lactarius chromospermus* and *L. volemus* appear to be similar at first glance, the former species may be distinguished from the latter due to its peculiar cinnamon-brown Russulaceae spore print and African distribution (D.N. Pegler, 1982). *Lactarius subvelutinus* resembles *Lactarius volemus* in that it has narrow gills, a white latex that does not change hue, and hues ranging from drab yellow-orange to vivid golden orange. It doesn't have the fishy smell, though (Bessette et al., 2009).

EDIBILITY AND OTHER USES

Despite the unpleasant fishy smell that is released when the mushroom is harvested, *L. volemus* is edible and recommended for use in cooking. However, it has a somewhat gritty texture like most milk caps, which some people could find unpleasant (D. Arora, 1986). During cooking, the smell goes away (M. Kuo, 2007). The flavour of the latex is hardly detectable (R. Phillips, 1981) The species is thought to be safe for casual mushroom collectors to eat (S. Metzler and V. Metzler, 1992) and is best served by slow cooking to avoid it being too hard (Haas, Hans, 1969); after drying, specimens that have been rehydrated and may need longer cooking periods to get rid of the gritty texture (M. Kuo, 2007). It has also been advised that the mushroom be used in casseroles and thick sauces (A. Bessette and D.H. Fischer, 1992). Due to the significant volumes of latex it emits, pan frying is not a suggested cooking method (AH. Smith and N.S. Weber, 1980). In Nepal, one of the many varieties of milk caps produced in Yunnan Province, China, *L. volemus* is popular mushroom species used for eating purpose and sale (M. Christensen et al., 2008). Bessette and colleagues refer to the fungus as "the best-known and most popular edible milk mushroom" in the eastern United States in their 2009 book on milk caps of North America. [61] According to a Turkish, the nutrient makeup of the fruit bodies, *L. volemus* is a rich source of both protein and carbs. In Turkey's central Anatolia, two elderly adults who consumed *L. volemus* experienced a brief pancreatitis. Both had previously consumed the Tirmit mushroom, which they were both familiar with. The problem resolved itself on its own (S. Karahan et al., 2016).

Bioactive Components

Volemolide, a special sterol molecule found in fruit bodies and fungal (sterol ergosterol), which is useful in the study of the chemotaxonomy of fungi (K. Kobata et al., 1994). Nine more sterols were found in a 2001 study, three of which were brand-new to science. The scientists claim that fungi rarely contain these kinds of highly oxygenated chemicals, which are comparable to the sterols seen in marine soft coral and sponges (J.M. Yue et al., 2001). Additionally, the mushroom contains volemitol (D-glycero-D-mannoheptitol), a seven-carbon sugar alcohol that was first isolated from the organism through the French scientist Émile Bourquelot in 1889. Numerous types of plants and brown algal organisms have volemitol as a free sugar (M. Sivakumar et al., 2005). Fruit bodies from *L. volemus* have a natural polyisoprene content of 1.1–7.7% by dry weight, which makes them useful for making rubber (Y. Tanaka et al., 1994; M. Litvinov, 2002). Rubber from mushrooms is chemically composed of a high molecular mass homologue of polyprenol, two trans isoprene units, a string of cis isoprenes (between 260 and 300 units), and a hydroxyl or fatty acid ester (Y. Tanaka et al., 1995). Trans, trans-farnesyl pyrophosphate is the starting material used in the biosynthesis of polyisoprene, and esterification of polyisoprenyl pyrophosphate is assumed to be the last step (Y. Tanaka et al., 1994). In L. volemus and numerous other milk cap species, the enzyme isopentenyl-diphosphate delta isomerase has been found to be necessary for the beginning of rubber synthesis (N. Ohya et al., 1997).

Ecology, Distribution, and Habitat

L. volemus develops ectomycorrhizae, advantageous symbiotic relationships with different tree species, such all milk caps (Bessette et al., 2009). The fungus grows around the plant's root and in the spaces between its cortical cells in this relationship, but does not really enter the cortical cells. The hyphae spread into the soil, helping the plant absorb nutrients from the surrounding area. This increases the surface area for absorption. Although it grows more frequently in deciduous forests, it can also be seen at the bases of broad-leaved and coniferous trees. It can occasionally be discovered in peat moss beds as well. The fruit bodies are widespread and develop between summer and autumn. In warm, humid weather, they can be seen growing singly or in bunches, and they are more prevalent (S. Metzler and V. Metzler, 1992).

Several kinds of fungi-dwelling mites and limoniid fly species, including *Discobola marginata* and *Limonia yakushimensis*, can live on fruit bodies. In a symbiotic relationship known as phoresis, in which the mites are mechanically conveyed by their host, the flies serve as the hosts for the mites. The insect hosts, on the other hand, are huge and can move the mites between their preferred eating sites. On the other hand, mites are tiny and unable to travel the considerable distances between mushrooms on their own (M.Sueyoshi et al., 2007).

In the Northern Hemisphere, *Lactifluus volemus* can be present in warm areas as well as some subtropical and tropical areas. It is extensively dispersed over Europe (Y. Shimono et al., 2007) but it is declining in several nations and has become extinct locally in the Netherlands (and Flanders) due to its rarity (Ghent university, 2010). The species' range in the Americas extends south to the East Coast of the United States, Mexico, and beyond into Central America. The northern limit of its range in the Americas is southern Canada east of the Great Plains (Guatemala) (Y. Shimono et al., 2007). It is also well-known in Asia, particularly in Japan, India, China, Korea, Nepal, and Vietnam (M. Christensen et al., 2008; X.H. Wang, 2000; S.S. Saini and N.S. Atri, 1993; H. Dörfeld et al., 2004). Additionally, collections from the Middle East, notably Iranand Turkey, have been produced (M. Saber, 1989).

Clavulinopsis fusiformis

Clavulinopsis fusiformis is a species of coral fungus in the *Clavariaceae* family. It is often referred to as golden spindles, spindle-shaped yellow coral, or spindle-shaped fairy club.

Taxonomy

James Sowerby, an English botanist, named the species *Clavaria fusiformis* in 1799 after collecting specimens in London's Hampstead Heath (J. Sowerby, 1799). In 1828, Elias Fries referred to it as a subspecies of *Clavaria inaequalis*. In 1950, E.J.H. Corner moved it to *Clavulinopsis*. In 1978, Ronald H. Petersen moved it to Ramariopsis.

The Latin-derived particular epithet fusiformis translates as "spindle-shaped." It goes by several names, including "golden spindles," "spindle-shaped yellow coral," "spindle-shaped fairy club," and most recently "French Fries Mushroom" (G. Konstantinidis, 2005; Mushroom identification Facebook, 2020).

Description

The fruit bodies are shaped like thin, 5–15 cm height, bright yellow clubs with sharp, pointy points. As the tissue matures, it becomes hard, brittle, and yellow before becoming hollow. The spores are 5–9 by 4.5–8.5 m and are smooth, broadly ellipsoid, or nearly spherical in shape. They have an apiculus that is between one and two metres long, and one or more sizable oil droplets may be present. The club-shaped basidia (spore-bearing cell) have a long, cylindrical base that is 1.5–2.5 m broad and measure 40–65 by 6–9 m. At the base, there is a clamp connection. Although there are rarely two- and three-spored variations, four-spored basidia predominate (E.J.H. Corner, 1950). Both expanded hyphae up to 12 m and thin hyphae up to 4 m make up the flesh (E.E. Tylukti, 1987). It is nonpoisonous and has been classified in field guides as both edible (A. Bessette et al., 1997; Rogar, Phillipe, 2010) and in edible (P. Roberts and S. Evans, 2011). In Nepal (M. Christensen et al., 2008), where it is regularly collected and consumed, fruit bodies are referred to as Kesari chyau (M.K. Adhikari et al., 2005).

Similar Species

Similar in size to *Clavulinopsis fusiformis* and also growing in thick clusters, *Clavaria amoenoides* is significantly more uncommon. By microscopic examination, it can be easily recognised from *C. fusiformis* because it possesses inflated hyphae without clamp connections (P. Roberts, 2008). While *Clavulinopsis laeticolor* is smaller, up to 5 cm high, lacks pointed tips, and grows singly, randomly, or in loose groups, they are similar in colour and form (R.M. Davis et al., 2012). Similar in colouration but smaller and less likely to grow in clusters are *C. helvola* and *C. luteoalba* (P. Roberts, 2011).

Habitat and Distribution

Clavulinopsis fusiformis is a saprophytic species (P. Roberts, 2011). Fruit cells grow on the ground as loose to dense clusters and distributed troops in grassy areas and among moss (P. Roberts, 2011). It has been reported in Asia from many locations, including Iran, (M.Saber, 1989), China (Y. Zhang et al., 2010), Nepal (M.Christensen et al., 2008) and Japan. It can be found in both Europe and North America. It is one of the most prevalent macrofungal species at 2,600–3,500 m in Fargesia spathacea–dominated community forests in China (8,500–11,500 ft) (Y. Zhang et al., 2010).

Hypholoma capnoides

Hypholoma capnoides is a species of edible fungus in the Strophariaceae family (P. Kummer, 1871). *H. Capnoides* grows in groups on rotting wood, such as tufts on old tree stumps, throughout the United States, Europe, and Asia, just like its toxic or dubious relatives H. *H. fasciculare*, or "sulphur tuft," lateritium ("brick caps") (S.Trudell and J. Ammirati, 2009).

Although it is edible, the poisonous sulphur tuft is more common in several locations. Due to the dark colour of its spores, H. capnoides exhibits grey gills, but sulphur tuft has green gills. It might also be mistaken for the poisonous Kuehneromyces mutabilis or the delicious Galerina marginata (E. Gerhardt, 2006). A really well pileus, a veil on the stipe that is variably developed but never forms a cell membrane annulus, spore prints that vary from violaceous to purple, basidiospores with a relatively thick wall and a prominent germ pore, and the existence of chrysocystidia are some of its distinctive characteristics. It differs from the closest genus, Stropharia (Fr.) Quél, since the rhizomorphs lack acanthocytes (D.F. Farr, 1980; L.L. Norvell and S.A. Redhead, 2000). Chrysocystidia are a characteristic that sets Chrysocystidia apart from Psilocybe (Fr.) P. Kumm., another closely related genus in the family (G.Guzmán, 1980, 1983, 1999). Hawksworth et al. (1995) reported that 30 species of Hypholoma in the world, which are found in temperate to tropical regions and grow on soil, living trees, mosses, or decaying wood (R.Singer, 1986).

Hypholoma species are venomous, and *H. fasciculare* are not edible (S.M. Badalyan et al., 1995). Additionally, this species' antioxidative (S.M. Badalyan, 2003) and hypoglycemic (S.M. Badalyan et al. 2002) properties have been studied. They are currently employed with in biological control of phytopathogenic fungi as well as the bioconversion of cellulose, fabric, and dye industrial residues (M.Hofrichter and W.Fritsche, 1997, K.T. Steffen et al., 2000). They are important players in forest ecosystems because they are active decomposers of wood and litter (S.M. Badalyan et al., 2002; B.Chapman et al., 2004)

Description

Cap: diameter (6 cm), matt yellow, yellow to orange-brownish, and occasionally viscid in colour.
Gills: At first pale orangish-yellow, then when grown, pale grey, and finally a darker purple/brown.
Spore powder: Dim brown or maroon spore powder.
Stipe: Rust-brown underneath with a yellowish tint.
Flavour: Mild in taste (other Hypholomas mostly have a bitter taste).

Psathyrella spadicea

The Coprinaceae family originally included the genera *Psathyrella*. The understanding of the connections within coprinoid fungus has significantly changed as a result of molecular phylogenetic analyses based on nu-rDNA sequences. The chestnut brittlestem, also known as sathyrella spadicea or Homophron spadiceum, is a species of agaric fungus in the Psathyrellaceae family. German mycologist Jacob Christian Schäffer first named the fungus Agaricus spadiceus in 1783. It was moved to the genus *Psathyrella* in 1951 by Rolf Singer, who placed it in the section Spadiceae (Species Fungorum, 2021). *P. spadicea* was transferred to the newly established genus Homophron in 2015 when rstadius & Larsson resurrected the term Homophron (which has been in use at the subgenus level since 1883) (L. Örstadius et al., 2015; H. Laessoe et al., 2019). *Psathyrella spadicea* can be found in North America and Europe. In North America, it occurs more frequently in the north than in the south, particularly in Alaska and the Yukon Territories. Reddish to red-brown spore prints are produced by fruit bodies (O.K.Miller and H. Miller, 2006; E.R. Boa, 2004).

There are 400 dark-spored taxa in Psathyrella (Fr.) Quel. (Psathyrellaceae, Agaricales), which has a distribution that is global (138). Only seven of these taxa—*P. atroumbonata, P. candolleana, P. coprinoceps, P. hymenocephala, P. piluliformis, P. rugocephala, and P. spadicea*—are known to be edible and are regularly consumed by people in Malaysia and the Democratic Republic of the Congo. However, no information about their nutritional and therapeutic benefits could be uncovered. *P. spadicea*, an edible species of Psathyrella, was discovered during a foray in the Leh area of the Ladakh region of Jammu & Kashmir (India). Here, an effort has been made to comprehend *P. spadicea's* morphological, anatomical, molecular, biochemical, physical, and antioxidative characteristics.

Numerous criteria were examined, including fluorescence analysis, HPLC, antioxidant activity, and molecular characterisation. According to the findings, the tested mushroom extract demonstrated a significant quantity of antioxidant activity and had a high capacity to scavenge DPPH radicals due to its significant concentrations of flavonoids, ascorbic acid, -carotene, and total phenolics.

This mushroom can therefore be used as a readily available sources of antioxidants and to treat a range of diseases associated with oxidative stress.

Suillus granulatus

Suillus granulatus (L.) Roussel, sometimes known as the "weeping bolete," is a palatable fungus with a white, quickly yellowish, and non-staining body. It tastes gentle and has a little fragrance to it. Despite the fact that this species (like with all of the Suillus species) is not one of the more popular delicacies like truffles or morels, it is widely gathered and consumed by the general public, especially by those who traditionally go mushroom hunting. It is frequently blended with other species to enhance taste and flavour because of its mild flavour. A porous mushroom belonging to the genus Suillus and family *Suillaceae* is called *Suillus granulatus*. It resembles the related *S. luteus* in appearance, but the ringless stem helps to recognise it. It is an edible mushroom that, like *S. luteus*, frequently develops in symbiosis (mycorrhiza) with pine. It's also been called the weeping bolete or the granulated bolete. That species, which was previously believed to reside in North America, has now been identified as the recently discovered *Suillus weaverae* (V.B. Mcknight and K.H. McKnight, 1987; iucn.ekoo.se. 2020).

Taxonomy

Carl Linnaeus first identified *Suillus granulatus* as a species of Boletus in 1753 (C. Linnaeus, 1753) Henri François Anne de Roussel, a French naturalist, gave it its present name in 1796 when he moved it to Suillus. An old name for fungi is suillus, which comes from the word "swine." The word "granulatus," which translates to "grainy," describes the glandular spots on the top of the stem (O'Reilly, pat, 2018; M. Kuo, 2018). The letter *S. granulatus* this kind was previously known as lactifluus, which translates to "oozing milk," due to the fact that glandular dots do not greatly distinguish it. The glandular spots, however, may not be seen in some specimens and may not darken with ageing (M. Kuo, 2018).

Description

Usually 4 to 12 cm in diameter, orange-brown to brown-yellow, viscid (sticky) when wet, and shiny when dry, the cap is round and has these characteristics. The stem is around 4–8 tall, 1–2 cm wide, and pale yellow in colour. It is of uniform thickness and has small brownish granules at the tip. It doesn't have a ring. When young, the tiny, light-yellow tubes and pores release pale milky drops. The flesh is also a light golden colour (R. Davis et al., 2012).

Suillus luteus, another abundant and extensively dispersed species found in the same habitat, is sometimes confused with *Suillus granulatus*. A noticeable partly veil and ring, as well as the lack of milky drops on the pores, help to identify *S. luteus*. (Y.J. Min et al., 2014). *Suillus brevipes* is also comparable; it does not drip droplets from the surface of the pores and has a short stipe in comparison to the cap. *Suillus pungens* is a species that is related (R. Davisv et al., 2012).

Bioleaching

Using living things to remove metals from ores is known as bioleaching, and it is often used when only a small quantity of the metal requires to be extracted. Suillus granulatus has been shown to be able to remove trace amounts of lead, titanium, calcium, potassium, and magnesium from apatite and wood ash (G.M. Gadd, 2010).

Distribution and Habitat

Occasionally grows in large numbers and co-occurs with Pinus (pine trees) in both calcareous and acid soils. The Suillus species most commonly associated with pines in warm areas is Suillus granulatus (D.M. Richardson, 2000). In Europe and Britain, it is typical. In South Korea, it is connected to Japanese red pine (Pinus densiflora) (Y.J. Min et al., 2014) The fungus, which is indigenous to the Northern Hemisphere, was brought into Australia, Chile, Africa, New Zealand, Hawaii, Argentina, and Africa (D.Simberloff and M. Rejmanek, 2010).

Edibility

According to your source, *Suillus granulatus* is edible and is either of good or poor quality (R. Phillipe, 2010). As with all Suillus species, the gelatinous pileipellis should be taken out first, and it is better to remove the tubes before boiling. In some situations, gastrointestinal discomfort has reportedly been caused by it (Jr. Miller and H. Hope, 2006). It occasionally appears in commercially available mushroom preserves. The fruit bodies can be regarded as a functional food because they are low in fat, high in fibre and carbs, and a source of nutraceutical chemicals (F.S.Reis et al., 2014).

CONCLUSION

Fungi like *Helvella crisp*, *Lactifluus volemus, Clavulinopsis fusiformis,* etc., are of great importance and found in various subcontinents and have their presence globally. These lesser known fungi have various bioactive compounds and possess a lot of therapeutic effects. Out of these fungi discussed here in this chapter have some edible fungi also and can be utilised globally as healthy food. They are widely used for the prevention and cure of various diseases by the local population of that area. Detailed research on theses fungi can explore a lot of opportunity on their diversified applications.

REFERENCES

S.P. Abbott, and R.S. Currah. 1997. "The Hevellaceae: systematic revision and occurrence in northern and northwestern North America," Mycotaxon **62**:1–125

M.K. Adhikari, S. Devokta, and R.D. Tiwari. 2005. "Ethnomycological knowledge on uses of wild mushrooms in western and central Nepal" (PDF). Our Nature. **3**: 13–19. doi:10.3126/on.v3i1.329.

M. Alves, I.F.R. Ferreira, J. Dias, V. Teixeira, A. Martins, and M. Pintado. 2012. "A review on antimicrobial activity of mushroom (Basidiomycetes) extracts and isolated compounds," Planta Medica. **78** (16): 1707–1718.

J. F. Ammirati, J. A. Traquair, and P. A. Horgen. 1985. Poisonous mushrooms of the northern United States and Canada. Minneapolis: University of Minnesota Press. p. 259. ISBN 0-8166-1407-5.

D. Arora. 1986. Mushrooms Demystified. Berkeley, California: Ten Speed Press. p. 78. ISBN 978-0-89815-169-5.

S.M. Badalyan. 2003. "Edible and medicinal higher basidiomycete mushrooms as a source of natural antioxidants," International Journal of Medicinal Mushrooms **5**: 153–162.

S.M. Badalyan, S. Rapior, J. Le Quang, L. Doko, M. Jacob, C. Andary, and J.-J. Serrano. 1995. "Investigation of fungal metabolites and acute toxicity studies from fruitbodies of Hypholoma species (Strophariaceae)," Cryptogamie, Mycologie **16**: 79–84.

S.M. Badalyan, G. Innocenti, and N.G. Garibyan. 2002. "Antagonistic activity of xylotrophic mushrooms against pathogenic fungi of cereals in dual culture," Phytopathologia Mediterranea **41**: 200–225.

L. Barros, P. Baptista, D.M. Correia, S. Casal, B. Oliveira, and I.C.F.R. Ferreira. 2007. "Fatty acid and sugar compositions, and nutritional value of five wild edible mushrooms from Northeast Portugal," Food Chemistry. **105** (1): 140–145.

L. Barros, D.M. Correia, I.C.F.R. Ferreira, P. Baptista, and C. Santos-Buelga. 2008. "Optimization of the determination of tocopherols in Agaricus sp. edible mushrooms by a normal phase liquid chromatographic method," Food Chemistry. **110** (4): 1046–1050.

Bessette et al. 2009, p. 5.

A. Bessette, and D.H. Fischer. 1992. Edible Wild Mushrooms of North America: a Field-to-Kitchen Guide. Austin, Texas: University of Texas Press. p. 68. ISBN 978-0-292-72080-0. Retrieved 2010-03-24.

A. Bessette, A.R. Bessette, and D.W. Fischer. 1997. Mushrooms of Northeastern North America. Syracuse, New York: Syracuse University Press. p. 421. ISBN 978-0815603887.

E.R. Boa. 2004. Wild Edible Fungi: A Global Overview of Their Use and Importance to People. Food & Agriculture Organization. p. 138. ISBN 978-92-5-105157-3.

A.C. Brown, and C.I. Waslien. 2003. "Stress and nutrition," in Encyclopedia of Food Sciences and Nutrition, L. Trugo and P.M. Finglas, Eds., London, UK: Academic Press.

B. Buyck, V. Hofstetter, U. Eberhardt, A. Verbeken, and F. Kauff. 2008. "Walking the thin line between Russula and Lactarius: the dilemma of Russula sect. Ochricompactae" (PDF). Fungal Diversity. **28**: 15–40.

B. Buyck, V. Hofstetter, A. Verbeken, and R. Walleyn. 2010. "Proposal to conserve Lactarius nom. cons. (Basidiomycota) with conserved type," Taxon. **59**: 447–453. doi:10.1002/tax.591031.

S.T. Chang, and P.G. Miles. 2008. Mushrooms: Cultivation, Nutritional Value, Medicinal Effect, and Environmental Impact. Boca Raton, Fla, USA: CRC Press, 2nd edition.

B. Chapman, G. Xiao, and S. Myers. 2004. Early results from field trails using Hypholoma fasciculare to reduce Armillaria ostoyae root disease. Canadian Journal of Botany **82**: 962–969.

J. Chen, and R. Seviour. 2007. "Medicinal importance of fungal - (1 → 3), (1 → 6)-glucans," Mycological Research. **111** (6): 635–652.

M. Christensen, S. Bhattarai, S. Devkota, and H.O. Larsen. 2008. "Collection and use of wild edible fungi in Nepal," Economic Botany. **62** (1): 12–23. doi:10.1007/s12231-007-9000-9. S2CID 6985365

E.J.H. Corner. 1950. A monograph of Clavaria and allied genera. Annals of Botany Memoirs. Oxford University Press. p. 367.

Arora David. 1986. Mushrooms demystified: a comprehensive guide to the fleshy fungi (2nd ed.). Berkeley: Ten Speed Press. p. 816. ISBN 0-89815-169-4.

R. M. Davis, R. Sommer, and A. J. Menge. 2012. Field Guide to Mushrooms of Western North America. Berkeley: University of California Press. pp. 331–332. ISBN 978-0-520-95360-4. OCLC 797915861.

H. Dörfeld, T.T. Kiet, and A. Berg. 2004. "Neue Makromyceten-Kollektionen von Vietnam und deren systematische und ökogeographische Bedeutung" [New collections of macromycetes from Vietnam and their systematic and ecogeographical significance]. Feddes Repertorium (in German). **115** (1–2): 164–177. doi:10.1002/fedr.200311034.

P.G. Ergon¨ ul, I. Akata, F. Kalyoncu, and B. Erg ¨ on¨ ul. 2013. "Fatty ¨ acid compositions of six wild edible mushroom species," The Scientific World Journal. **2013**, Article ID 163964, 4 pages, 2013.

D.F. Farr. 1980. The acanthocyte, a unique cell type in Stropharia (Agaricales). Mycotaxon **11**: 241–249.

I.C.F.R. Ferreira, L. Barros, and R.M.V. Abreu. 2009. "Antioxidants in wild mushrooms," Current Medicinal Chemistry. **16** (12): 1543–1560.

I.C.F.R. Ferreira, J.A. Vaz, M.H. Vasconcelos, and A. Martins. 2010."Compounds from wild mushrooms with antitumor potential," Anti-Cancer Agents in Medicinal Chemistry. **10** (5): 424–436, 2010.

T.C. Finimundy, G. Gambato, R. Fontana et al. 2013. "Aqueous extracts of Lentinula edodes and Pleurotus sajor-caju exhibit high antioxidant capability and promising in vitro antitumor activity," Nutrition Research. **33** (1): 76–84.

E.M. Fries. 1821. Systema Mycologicum (in Latin). Vol. 1. Lund, Sweden: Ex Officina Berlingiana. p. 69.

E.M. Fries. 1828. Elenchus Fungorum. Vol. 1. Greifswald, Germany: Ernestus Mauritius. p. 231.

E.M. Fries. 1838. Epicrisis Systematis Mycologici, seu Synopsis Hymenomycetum (in Latin). Uppsala, Sweden: Typographia Academica. p. 344.

H.S. Frieze. 1882. A Vergilian dictionary embracing all the words found in the Eclogues, Georgics, and Aeneid of Vergil: with numerous references to the text verifying and illustrating the definitions. New York, New York: D. Appleton and company. p. 227.

G. M. Gadd. 2010. "Metals, minerals and microbes: geomicrobiology and bioremediation," Microbiology. **156** (3): 609–643. doi:10.1099/mic.0.037143-0. PMID 20019082.

E. Gerhardt. 2006. "BLV Handbuch Pilze. BLV, München," Seite **244**. ISBN 3-8354-0053-3.

E. Guillamon, A. Garc ´ ´ıa-Lafuente, M. Lozano et al. 2010. "Edible mushrooms: role in the prevention of cardiovascular diseases," Fitoterapia. **81** (7): 715–723.

G. Guzmán. 1980. "Three new sections in the genus Naematoloma and a description of a new tropical species," Mycotaxon **12**: 235–240.

G. Guzmán. 1983. "The genus Psilocybe," Beiheft zur Nova Hedwigia **74**: 1–439.
G. Guzmán. 1999. "New combinations in Hypholoma and information on the distribution and properties of the species," Documents Mycologiques **114**: 65–66.
Haas, and Hans. 1969. The Young Specialist looks at Fungi. Burke. p. 184. ISBN 0-222-79409-7.
D.L. Hawksworth, P.M. Kirk, B.C. Sutton, and D.N. Pegler. 1995. Ainsworth & Bisby's Dictionary of the Fungi. 8th ed. Surrey: International Mycological Institute/CABI Publishing.
S.A. Heleno, L. Barros, M.J. Sousa, A. Martins, and I.C.F.R. Ferreira. 2010. "Tocopherols composition of Portuguese wild mushrooms with antioxidant capacity," Food Chemistry. **119** (4): 1443–1450, 2010.
J. Hellman-Clausen. 1998. The genus Lactarius. Vol. 2. Espergaerde, Denmark: Svampetryk, for the Danish Mycological Society. ISBN 978-87-983581-4-5.
M. Hofrichter, and W. Fritsche. 1997. "Depolymerization of lowrank coal by extracellular fungal enzyme fungi. II. The ligninolytic enzymes of coal-humic-acid-depolimerizing fungus Naematoloma frowardii," Applied Microbiology and Biotechnology **47**: 419–425.
J. Hwang, Q. Zhao, Z.L..Yang, Z. Wang, and J.P. Townsend. 2015. "Solving the ecological puzzle of mycorrhizal asso- ciations using data from annotated collections and environmental samples—an example of saddle fungi," Environ Microbiol Rep **7**:658–667, doi:10.1111/1758-2229.12303
S. Karahan, A. Erden, A. Cetinkaya, D. Avci, A. Irfan, H. Karagoz, K. Bulut, and M. Basak. 2016. "Acute Pancreatitis Caused By Mushroom Poisoning," J Investig Med High Impact Case Rep. **4** (1): 232470961562747. doi:10.1177/2324709615627474. PMC 4724762. PMID 26835473.
H.G. Kim, D.H. Yoon, W.H. Lee et al. 2007. "Phellinus linteus inhibits inflammatory mediators by suppressing redox-based NF-B and MAPKs activation in lipopolysaccharide-induced RAW 264.7 macrophage," Journal of Ethnopharmacology. **114** (3): 307–315.
P.M. Kirk, P.F. Cannon, D.W. Minter, and J.A. Stalpers. 2008. Ainsworth and Bisby's Dictionary of the Fungi. 10th ed. Wallingford, UK: CAB International. p. 771
K. Kobata, T. Wada, Y. Hayashi, and H. Shibata. 1994. "Studies on chemical components of mushrooms.3. Volemolide, a novel norsterol from the fungus Lactarius volemus," Bioscience, Biotechnology, and Biochemistry. **58** (8): 1542–1544. doi:10.1271/bbb.58.1542.
G. Konstantinidis. 2005. Elsevier's Dictionary of Medicine and Biology: In English, Greek, German, Italian and Latin. Elsevier. p. 607. ISBN 978-0-08-046012-3.
R.P. Korf. 1952. "A monograph of the Arachnopezizeae," Lloydia **14**:129–180.
P. Kummer. 1871. Der Führer in die Pilzkunde [The Mycological Guide] (in German) (1 ed.). p. 127.
O. Kuntze. 1891. Revisio Generum Plantarum (in Latin). Vol. 2. Leipzig, Germany: A. Felix. pp. 856–857.
M. Kuo. 2007. 100 Edible Mushrooms. Ann Arbor, Michigan: The University of Michigan Press. p. 181. ISBN 978-0-472-03126-9.
M. Kuo. 2010. "Lactarius volemus," MushroomExpert.com. Retrieved 2010-11-16.
M. Kuo. 2018. "Suillus granulatus," Mushroom Expert. Retrieved 26 October 2018.
C. Linnaeus. 1753. "Tomus II," Species Plantarum (in Latin). Vol. 12. Stockholm: Laurentii Salvii. p. 1177.
V.M. Litvinov. 2002. Spectroscopy of Rubber and Rubbery Materials. Shawbury, Shrewsbury, Shropshire, UK: iSmithers Rapra Technology. p. 431. ISBN 978-1-85957-280-1.
H. Læssøe, and Jens Petersen. 2019. Fungi of Temperate Europe. Princeton University Press. p. 578. ISBN 9780691180373.
L.C. Maia, A.M. Yano, and J.W. Kimbrough. 1996. "Species of Ascomycota-forming ectomycorrhiza," Mycotaxon **57**:371–390.
P. Mattila, K. Könkö, M. Eurola et al. 2001. "Contents of vitamins, mineral elements, and some phenolic compounds in cultivated mushrooms," Journal of Agricultural and Food Chemistry. **49** (5): 2343–2348, 2001.
V.B. McKnight, and K.H. McKnight. 1987. A Field Guide to Mushrooms: North America. Peterson Field Guides. Boston, Massachusetts: Houghton Mifflin. p. 208. ISBN 978-0-395-91090-0.
S. Metzler, and V. Metzler. 1992. Texas Mushrooms. Austin, Texas: University of Texas Press. p. 118. ISBN 978-0-292-75125-5.
K. O. Miller Jr., and H. H. Miller. 2006. North American Mushrooms: A Field Guide to Edible and Inedible Fungi. Guilford, CN: FalconGuide. p. 358. ISBN 978-0-7627-3109-1.
O.K. Miller, and H. Miller. 2006. North American Mushrooms: A Field Guide to Edible and Inedible Fungi. Falcon Guide. p. 239. ISBN 978-0-7627-3109-1.
Y.J. Min, M.S. Park, J.J. Fong, S.J. Seok, S.K. Han, Y.W. Lim. 2014. "Molecular Taxonomical Reclassification of the Genus Suillus Micheli ex S. F. Gray in South Korea," Mycobiology. **42** (3): 221–228. doi:10.5941/MYCO.2014.42.3.221. PMC 4206787. PMID 25346598.

L. Montoya, V.M. Bandala, and G. Guzmán. 1996. "New and interesting species of Lactarius from Mexico including scanning electron microscope observations," Mycotaxon. **57**: 411–424. Archived from the original on 2015-09-23. Retrieved 2010-03-23.

"Mushroom Identification Facebook," Retrieved 2020-10-18.

N.H. Nguyen, F. Landeros, R. Garibay-Orijel, K. Hansen, E.C. Vellinga. 2013. "The *Helvella* lacunosa species complex in western North America: cryptic species, misapplied names and parasites," Mycologia **105**, 1275–1286.

L.L. Norvell, and S.A. Redhead. 2000. "Stropharia albivelata and its basionym Pholiota albivelata," Mycotaxon **76**: 315–320

Pat O'Reilly. 2018. "Fascinated by Fungi," First Nature. Retrieved 26 October 2018.

N. Ohya, Y. Tanaka, K. Ogura, and T. Koyama. 1997. "Isopentenyl diphosphate isomerase activity in Lactarius mushrooms," Phytochemistry. **46** (6): 1115–1118. doi:10.1016/S0031-9422(97)00410-X.

L. Örstadius, M. Ryberg, and E. Larsson. 2015. "Molecular phylogenetics and taxonomy in Psathyrellaceae (Agaricales) with focus on psathyrelloid species: introduction of three new genera and 18 new species," Mycological Progress. **14** (5). doi:10.1007/s11557-015-1047-x. S2CID 16637489. Retrieved 2021-07-13.

S. Patel, and A. Goyal. 2012. "Recent developments in mushrooms as anticancer therapeutics: a review," 3 Biotech. **2** (1): 1–15, 2012.

C.H. Peck. 1885. "New York species of Lactarius. Report of the State Botanist (for 1884)," Annual Report of the New York State Museum. **38**: 111–133.

D.N. Pegler. 1982. "Agaricoid and boletoid fungi (Basidiomycota) from Malaŵi and Zambia," Kew Bulletin. **37** (2): 255–271. doi:10.2307/4109968. JSTOR 4109968.

D.N. Pegler, and J.P. Fiard. 1979. "Taxonomy and ecology of Lactarius (Agaricales) in the lesser Antilles," Kew Bulletin. **33** (4): 601–628. doi:10.2307/4109804. JSTOR 4109804.

E. Pereira, L. Barros, A. Martins, and I.C.F.R. Ferreira. 2012. "Towards chemical and nutritional inventory of Portuguese wild edible mushrooms in different habitats," Food Chemistry. **130** (2): 394–403.

R.H. Petersen. 1978. "Notes on clavarioid fungi. XV. Reorganization of Clavaria, Clavulinopsis and Ramariopsis," Mycologia. **70** (3): 660–671. doi:10.2307/3759402. JSTOR 3759402.

R. Phillips. 1981. Mushrooms and other Fungi of Britain and Europe. London, England: Pan Books. p. 88. ISBN 978-0-330-26441-9.

R. Phillips. 2006. Mushrooms. London: Pan Macmillan Ltd. p. 360. ISBN 0-330-44237-6.

R. Phillips. 2010. Mushrooms and Other Fungi of North America. Buffalo, NY: Firefly Books. p. 216. ISBN 978-1-55407-651-2.

L. Quélet. 1886. Enchiridion Fungorum in Europa media et praesertim in Gallia Vigentium (in Latin). Paris, France: O. Doin. p. 131.

F.S. Reis, D. Stojković, L. Barros, J. Glamočlija, A. Ćirić, M. Soković, A. Martins, M.H. Vasconcelos, P. Morales, and I.C. Ferreira. 2014. "Can Suillus granulatus (L.) Roussel be classified as a functional food?" (PDF). Food & Function. **5** (11): 2861–2869. doi:10.1039/C4FO00619D. hdl:10198/12054. PMID 25231126.

D.M. Richardson. 2000. Ecology and Biogeography of Pinus. Cambridge University Press. p. 334. ISBN 978-0-521-78910-3.

P. Roberts 2008. "Yellow Clavara species in the British Isles," Field Mycology. **9** (4): 142–145. doi:10.1016/S1468-1641(10)60593-2.

P. Roberts. and S. Evans. 2011. The Book of Fungi. Chicago, Illinois: University of Chicago Press. p. 494. ISBN 978-0226721170.

W.C. Roody. 2003. Mushrooms of West Virginia and the Central Appalachians. Lexington, Kentucky: University Press of Kentucky. p. 86. ISBN 978-0-8131-9039-6.

M. Saber. 1989. "The species of Lactarius in Iran," Iranian Journal of Plant Pathology (in Arabic). **25** (1–4): 13–16. ISSN 0006-2774.

S.S. Saini, and N.S. Atri. 1993. "Studies on genus Lactarius from India," Indian Phytopathology. **46** (4): 360–364. ISSN 0367-973X.

C. Sarikurkcu, B. Tepe, and M. Yamac. 2008. "Evaluation of the antioxidant activity of four edible mushrooms from the Central Anatolia, Eskisehir—Turkey: Lactarius deterrimus, Suillus collitinus, Boletus edulis, Xerocomus chrysenteron," Bioresource Technology. **99** (14): 6651–6655.

Y. Shimono, M. Hiroi, K. Iwase, and S. Takamats. 2007. "Molecular phylogeny of Lactarius volemus and its allies inferred from the nucleotide sequences of nuclear large subunit rDNA," Mycoscience. **48** (3): 152–157. doi:10.1007/s10267-006-0346-0. S2CID 85066524.

D. Simberloff, and M. Rejmanek. 2010. Encyclopedia of Biological Invasions. University of California Press. p. 470. ISBN 978-0-520-94843-3.

R. Singer. 1986. The Agaricales in Modern Taxonomy (4th ed.). Königstein im Taunus, Germany: Koeltz Scientific Books. p. 832. ISBN 978-3-87429-254-2.

M. Sivakumar, S.V. Bhat, and B.A. Nagasampagi. 2005. Chemistry of Natural Products. Berlin, Germany: Springer. p. 495. ISBN 978-3-540-40669-3.

A.H. Smith, and N.S. Weber. 1980. The Mushroom Hunter's Field Guide. University of Michigan Press. p. 257. ISBN 978-0-472-85610-7.

J. Sowerby. 1799. Coloured Figures of English Fungi. Vol. 2. London, UK: J. Davis. p. 98; plate 234.

K.T. Steffen, M. Hofrichter, and A. Hatakka. 2000. "Mineralisation of 14C-labelled synthetic lignin and ligninolytic enzyme activities of litter-decomposing basidiomycetous fungi," Applied Microbiology and Biotechnology **54**: 819–825.

M. Sueyoshi, K. Okabe, and T. Nakamura. 2007. "Host abundance of crane flies (Diptera: Limoniidae) and their role as phoronts of Acari (Arachnida) inhabiting fungal sporophores," Canadian Entomologist. **139** (2): 247–257. doi:10.4039/N06-016. S2CID 85947038

"Suillus weaverae," iucn.ekoo.se. Retrieved 2020-10-01.

A. Synytsya, K. M´ıckov, I. Jablonsk et al. 2009. "Glucans from ´ fruit bodies of cultivated mushrooms Pleurotus ostreatus and Pleurotus eryngii: structure and potential prebiotic activity," Carbohydrate Polymers. **76** (4): 548–556.

Y. Tanaka, S. Kawahara, A.H. Eng, A. Takei, and N. Ohya. 1994. "Structure of cis-polyisoprene from Lactarius mushrooms," Acta Biochimica Polonica. **41** (3): 303–309. doi:10.18388/abp.1994_4719. ISSN 0001-527X. PMID 7856401.

Y. Tanaka, S. Kawahara, E. Aikhwee, K. Shiba, and N. Ohya. 1995. "Initiation of biosynthesis in cis polyisoprenes," Phytochemistry. **39** (4): 779–784. doi:10.1016/0031-9422(95)00981-C.

L. Tedersoo, K. Hansen, B.A. Perry, and R. Kjøller. 2006. "Molecular and morphological diversity of pezizalean ectomycorrhiza," New Phytologist **170**:581–596, doi:10.1111/j.1469-8137.2006.01678.x

"The fishy milkcaps (Lactarius volemus sensu lato), cryptic species with a long and pandemic history," Department of Biology, Ghent University. 2010. Archived from the original on 2010-11-09. Retrieved 2010-11-18.

"The Homophron spadiceum page," Species Fungorum. Royal Botanic Gardens Kew. Retrieved 2021-07-13.

S. Tibpromma, K.D. Hyde, R. Jeewon, S.S.N. Maharachchikumbura et al. 2017. "Fungal diversity notes 491–602: taxonomic and phylogenetic contributions to fungal taxa," Fungal Diversity **83**, 1261.

S. Trudell, and J. Ammirati. 2009. Mushrooms of the Pacific Northwest. Timber Press Field Guides. Portland, OR: Timber Press. p. 206. ISBN 978-0-88192-935-5.

E.E. Tylukti. 1987. Mushrooms of Idaho and the Pacific Northwest. Vol 2. Non-gilled Hymenomycetes. Moscow, Idaho: The University of Idaho Press. pp. 87–88. ISBN 0-89301-097-9.

K. Van de Putte, J. Nuytinck, D. Stubbe, H. Thanh Le, and A. Verbeken. 2010. "Lactarius volemus sensu lato (Russulales) from northern Thailand: morphological and phylogenetic species concepts explored," Fungal Diversity. **45** (1): 99–130. doi:10.1007/s13225-010-0070-0. S2CID 25615396.

J.A. Vaz, S.A. Heleno, A. Martins, G.M. Almeida, M.H. Vasconcelos, and I.C.F.R. Ferreira. 2010. "Wild mushrooms Clitocybe alexandri and Lepista inversa: in vitro antioxidant activity and growth inhibition of human tumour cell lines," Food and Chemical Toxicology. **48** (10): 2881–2884, 2010.

A. Verbeken, and J. Nuytinck. 2013. "Not every milkcap is a Lactarius" (PDF). Scripta Botanica Belgica. **51**: 162–168.

A. Verbeken, K. Van de Putte, and E. De Crop. 2012. "New combinations in Lactifluus. 3. L. subgenera Lactifluus and Piperati," Mycotaxon. **120**: 443–450. doi:10.5248/120.443. hdl:1854/LU-3150382.

X.H. Wang. 2000. "A taxonomic study on some commercial species in the genus Lactarius (Agaricales) from Yunnan Province, China," Acta Botanica Yunnanica (in Chinese). **22** (4): 419–427. ISSN 0253-2700.

Y.Z. Wang, and C.M. Chen. 2002. "The Genus *Helvella* in Taiwan," Fungal Science **17**, 11–17.

Z. Wang, D. Luo, and Z. Liang. 2004. "Structure of polysaccharides from the fruiting body of Hericium erinaceus Pers," Carbohydrate Polymers. **57** (3): 241–247.

H.A. Wen, and J.Z. Ying. 2005. "Studies on the genus Lactarius from China II. Two new taxa from Guizhou," Mycosystema. **24** (2): 155–158.

J.M. Yue, S.N. Chen, Z.W. Lin, and H.D. Sun. 2001. "Sterols from the fungus Lactarius volemus," Phytochemistry. **56** (8): 801–806. doi:10.1016/S0031-9422(00)00490-8. PMID 11324907.

B.-Z. Zaidman, M. Yassin, J. Mahajna, and S.P. Wasser. 2005."Medicinal mushroom modulators of molecular targets as cancer therapeutics," Applied Microbiology and Biotechnology. **67** (4): 453–468.

L. Zhang, C. Fan, S. Liu, Z. Zang, and L. Jiao. 2011. "Chemical composition and antitumor activity of polysaccharide from Inonotus obliquus," Journal of Medicinal Plants Research. **5** (7): 1251–1260.

Y. Zhang, D.Q. Zhou, I. Zhao, T.X. Zhou, and K.D. Hyde. 2010. "Diversity and ecological distribution of macrofungi in the Laojun Mountain region, southwestern China," Biodiversity and Conservation. **19** (12): 3545–3563. doi:10.1007/s10531-010-9915-9. S2CID 24882278.

Q. Zhao, B. Tolgor, Y. Zhao, Z.L. Yang, and K.D. Hyde. 2015. "Species diversity within the *Helvella* crispa group (Ascomycota: *Helvella*ceae) in China," Phytotaxa **239**, 130–142.

W.Y. Zhuang. 2004. "Preliminary survey of the *Helvella*ceae from Xinjiang, China," Mycotaxon. **90** (1): 35–42.

Index

Printed in the United States
by Baker & Taylor Publisher Services